DEVELOPMENTS IN
PALAEONTOLOGY & STRATIGRAPHY 20

# UNDERSTANDING LATE DEVONIAN AND PERMIAN-TRIASSIC BIOTIC AND CLIMATIC EVENTS

## Towards an Integrated Approach

## FURTHER TITLES IN THIS SERIES

A.J. Boucot
Evolution and Extinction Rate Controls

W.A. Berggren and J.A. van Couvering
The Late Neogene-Biostratigraphy, Geochronology and Paleoclimatology of the Last 15 Million Years in Marine and Continental Sequences

L.J. Salop
Precambrian of the Northern Hemisphere

J.L.Wray
Calcareous Algae

A. Hallam (Editor)
Patterns of Evolution, as Illustrated by the Fossil Record

F.M. Swain (Editor)
Stratigraphic Micropaleontology of Atlantic Basin and Borderlands

W.C. Mahaney (Editor)
Quaternary Dating Methods

D. Janóssy
Pleistocene Vertebrates Faunas of Hungary

Ch. Pomerol and I. Premoli-Silva (Editors)
Terminal Eocene Events

J.C. Briggs
Biogeography and Plate Tectonics

T. Hanai, N. Ikeya and K. Ishizaki (Editors)
Evolutionary Biology of Ostracoda. Its Fundamentals and Applications

V.A. Zubakov and I.I. Borzenkova
Global Palaeoclimate of the Late Cenozoic

F.P. Agterberg
Automated Stratigraphic Correlation

J.C. Briggs
Global Biogeography

A. Montanari, G.S. Odin and R. Coccioni (Editors)
Miocene Stratigraphy: An Integrated Approach

F.M. Swain
Fossil Nonmarine Ostracoda of the United States

H. Okada and N.J. Mateer (Editors)
Cretaceous Environments of Asia

H. Yin, J.M. Dickins, G.R. Shi and J. Tong (Editors)
Permian-Triassic Evolution of Tethys and Western Circum-Pacific

G.S. Odin (Editor)
The Campanian-Maastrichtian Stage Boundary

*Developments in Palaeontology & Stratigraphy, 20*

# UNDERSTANDING LATE DEVONIAN AND PERMIAN-TRIASSIC BIOTIC AND CLIMATIC EVENTS

## Towards an Integrated Approach

Edited by

**D.J. OVER**
*Geological Sciences, SUNY-Geneseo*
*Geneseo, NY 14454, USA*

**J.R. MORROW**
*Department of Earth Sciences*
*University of Northern Colorado*
*Greeley, CO 80639, USA*

**P.B. WIGNALL**
*School of Earth and Environment*
*University of Leeds, Leeds LS2 9JT, UK*

ELSEVIER

Amsterdam - Boston - Heidelberg - London - New York - Oxford
Paris - San Diego - San Francisco - Singapore - Sydney - Tokyo

**Elsevier BV**
**Radarweg 29,**
**PO Box 211,**
**1000 AE Amsterdam**
**The Netherlands**

Elsevier Inc
525 B Street, Suite 1900
San Diego
CA 92101-4495
USA

Elsevier Ltd
The Boulevard
Langford Lane, Kidlington
Oxford OX5 1GB
UK

Elsevier Ltd
84 Theobald's Road,
London WC1Z 8RR
UK



First edition 2005

ISBN-13: 978-0-444-52127-9
ISBN-10: 0-444-52127-5

♾ The paper used in this publication meets the requirements of ANSI/NISO Z39.48-1992 (Permanence of Paper).
Transferred to digital print 2007
Printed and bound by CPI Antony Rowe, Eastbourne

05 06 07 08 09 10 10 9 8 7 6 5 4 3 2 1

# Contents

Understanding Late Devonian and Permian-Triassic Biotic and Climatic Events:
Towards an Integrated Approach
edited by D.J. Over, J.R. Morrow and P.B. Wignall



# *Chapter 1*

# *Introduction*

*Paul B. Wignall*

*School of Earth and Environment, University of Leeds, Leeds LS2 9JT, UK.*

The intense interest in the nature and origin of mass extinctions shown by the scientific community and the media has now lasted for two and a half decades. Initially focussed on the end-Cretaceous catastrophe, the last decade has seen a surge of interest in the other mass extinction events of the geological record, of which the Late Devonian and end-Permian events are two of the "big 5". Numerous research programmes are currently engaged in trying to understand these events and they share many features in common, for example, in the loss of equatorial reef communities, an association with oceanic anoxia, climate change and voluminous volcanism. This last factor is clearly the case with the end-Permian event, which coincides with the voluminous outpourings of the Siberian Traps, while similar volcanism is inferred rather than proved for the Late Devonian event. This volume draws together several studies on these key times in Earth history and reflects the diverse approaches that are necessarily required to understand such critical moments. The majority of the contributions originally formed part of a thematic symposium at the annual meeting of the Geological Society of America at Seattle, Washington in November 2003.

The Devonian contributions begin with Racki's review of the current state-of-play regarding the mass extinction at the mid-Late Devonian Frasnian-Famennian boundary. Despite much recent effort, he shows that there are still many outstanding problems that need resolving. No doubt some will prove more tractable than others. Thanks to the first order highstand of sea level in the Late Devonian, marine boundary sections are plentiful, but it is a quirk of palaeogeography that most are concentrated within a narrow equatorial belt. Thus, any latitudinal selectivity to extinction losses is difficult to demonstrate. This is unfortunate because temperature selectivity is a key factor in many extinction models. As McGhee discusses in the following chapter (and see also Stock's contribution) there is substantial palaeontological evidence that the extinction coincided with a cooling event. Joachimski et al.'s (2004) recent oxygen isotope data have provided some quantification of this cooling event, although the reported excursion of 5–7°C in sea-surface temperatures from a pre-excursion high of 32°C does not really seem like the sort of cooling liable to create an environmental catastrophe. There are no indications of glaciation at this time; for example no contemporaneous tillites are known.

As the study of individual mass extinction events matures it is notable that new advances are often made as the time resolution improves. Thus, the better events are constrained within a high-resolution time frame the more it will prove possible to determine cause-and-effect relationships. For Late Devonian studies, absolute dating has improved considerably in recent years – for example, Kaufmann et al. (2004) is an important contribution – and this, combined with improvements to the biostratigraphic resolution, will help improve extinction models. In this regard Girard et al.'s contribution to this volume provides valuable data and improved resolution to the *linguiformis* Zone, the terminal Frasnian conodont biozone and a key interval of the Frasnian-Famennian extinction crisis.

With the exception of the end-Ordovician event, meteorite impact has been invoked as the causal mechanism for all the major, mass extinctions and several minor extinction events too. However, with the obvious exception of the end-Cretaceous event, this extinction cause has not been widely favoured by the scientific community. Only for the Late Devonian event could impact be reasonably regarded as a serious contender among the range of proposed extinction mechanisms. This is due to the occurrence of several large impact craters of Late Devonian age, and also some smaller ones. Thus, Schieber and Over present data here on the beautifully preserved Flynn Creek crater of Tennessee. Whether any of these impacts was responsible for the extinction events depends essentially on demonstrating a coincidence of timing and here again recent improvements of radiometric dating are crucial. Currently, it appears that the impacts are not temporally related to extinctions, although here McGhee considers the possibility that the impacts triggered long-term climatic effects that may ultimately have had disastrous consequences.

Amongst the many victims of the extinctions, the stromatoporoids are particularly noteworthy because, prior to the Late Devonian, they were the dominant framework builders of Palaeozoic reefs while by the end of this interval they had gone extinct. Stock provides a thorough review of this group and shows how their fortunes reflect the controlling interplay of changing sea level and vicariance events. The Frasnian-Famennian crisis removed much of the stromatoporoids diversity and abundance, although they persisted throughout the Famennian. However, as Stock shows, these last survivors failed to show any evolutionary exuberance prior to their ultimate demise at the end of the Devonian, a phenomenon memorably named "dead clade walking" by Jablonski (2004). Rode and Lieberman model the geographic and stratigraphic range of brachiopod and bivalve taxa from the Middle and Upper Devonian of the northern Appalachian Basin with interesting results. They show that species which increase their range during the crisis interval in the *linguiformis* Zone, showed preferential survival and that consequently broad geographic range during this crisis interval (but not before it) is seen to favour survival. It is tempting to view this range expansion as a consequence of taxa moving into vacant ecospace as incumbent taxa are lost.

The end-Frasnian interval has long been known to be associated with the development of two anoxic events named after the Kellwasser Horizons developed in the classic German sections. Riquier et al. and Bond and Wignall provide new data on the geochemical aspects of these anoxic events and their geographic extent respectively. Riquier et al.'s multielement data demonstrate the intensity of anoxia in sections from Morocco, France and Germany and they argue, using evidence from fluctuations in barium concentrations, for elevated primary productivity during the anoxic events. The Kellwasser events have been primarily documented from North African and European sections whereas it has been unclear if they are also developed in the well-known Great Basin sequences of the western United States. Using a combination of facies analysis and pyrite framboid assay, Bond and Wignall show that Upper Kellwasser Event of Europe corresponds to the intensification and expansion of anoxia within the Great Basin. This *linguiformis* interval therefore witnessed a peak of anoxia, both in terms of intensity and geographic spread, and it is undoubtedly significant that this time coincides with the end-Frasnian mass extinction.

The final contributions to the volume look at the end-Permian mass extinction and its aftermath. Racki and Wignall concentrate on the significance of the Siberian Traps and their high latitude locus of eruption. Such a location is unique for Phanerozoic large igneous province eruption events, and it may have contributed to the apparent severity of the environmental consequences as these authors argue. The following contributions by Fraiser and Bottjer and Pruss et al. focus on the uniquely prolonged aftermath of the end-Permian mass extinction. This interval, which spanned the entire Early Triassic, was marked by the loss of many taxa that ultimately reappeared in the Middle Triassic. It has often been suggested that the absence of such Lazarus taxa reflects the poor preservation of Lower Triassic faunas and particularly the absence of silicified specimens. Fraiser and Bottjer successfully debunk this proposition by demonstrating the presence of appreciable numbers of silicified fossils in Lower Triassic strata. In the same vein, Wheeley and Twitchett (2005) have recently shown that Lazarus gastropods are not present in an unusually diverse earliest Triassic assemblage from Oman that includes silicified specimens. The absence of fossils in the Early Triassic is therefore a real phenomenon rather than an artefact of preservation. Pruss et al. further demonstrate the unusual nature of Lower Triassic rocks by highlighting the appearance of anachronistic microbial structures more typical of Proterozoic and Cambrian strata and by recording the presence of an impoverished trace fossil assemblage. These final studies demonstrate how much there is to learn about conditions in the aftermath of mass extinctions and both show that "normal" conditions were a long time in returning. For the end-Permian event at least, the post-apocalyptic world appears to have been as stressed as that which caused the extinction. In this regard the term "aftermath" may be something of a misnomer. During much of the Early Triassic the stressful conditions that began at the end of the Permian may have persisted in a prolongation of the crisis interval.

## References

Jablonski, D., 2004. The evolutionary role of mass extinctions: disaster, recovery and something in-between. In: Taylor, P. (Ed.), Extinctions in the History of Life, Cambridge University Press, Cambridge, pp. 151–177.

Joachimski, .M.M., van Geldern, R., Breisig, S., Buggisch, W., Day, J., 2004. Oxygen isotope evolution of biogenic calcite and apatite during the Middle and Late Devonian. Int. J. Earth Sci. 93, 542–553.

Kaufmann, B., Trapp, E., Mezger, K., 2004. The numerical age of the Upper Frasnian (Upper Devonian) Kellwasser horizons: a new U–Pb zircon date from Steinbruch Schmidt (Kellwerwald, Germany). J. Geol. 112, 495–501.

Wheeley, J.R., Twitchett, R.J., 2005. Palaeoecological significance of a new Griesbachian (Early Triassic) gastropod assemblage from Oman. Lethaia 38, 1–9.

Understanding Late Devonian and Permian-Triassic Biotic and Climatic Events:
Towards an Integrated Approach
edited by D.J. Over, J.R. Morrow and P.B. Wignall



## *Chapter 2*

# *Toward understanding Late Devonian global events: few answers, many questions*

*Grzegorz Racki**

*Department of Earth Sciences, Silesian University, 41-200 Sosnowiec, Poland*

> *"As one grows older, one realizes how little one knows: 'the more you learn, the more ignorant you become'. The joy of being a scientist is to discover this (...) As one accumulated information about how the Earth works, all the simple questions would be answered. Then the questions would have to become more intricate and harder to solve"* (Benton, 2003, p. 304)

**Abstract**

The Late Devonian was an epoch of dramatic evolutionary and environmental changes linked primarily with the Frasnian-Famennian (F-F) mass extinction. Current data and ideas support a prolonged, multi-causal nature of the biodiversity crisis, which favor Earth-bound mechanisms rather than a global cosmic catastrophe. The better understanding of the Late Devonian ocean-climate-biosphere system leads to several questions, and provides an agenda for future research. (1) *Magnitude and rank of biotic changes*: more detailed biodiversity studies are needed to place the end-Frasnian extinction in its Late Devonian context. In particular, the emerging severity of the end-Givetian and end-Famennian extinctions contrasts with the current overemphasis on the stepwise F–F crisis. (2) *Timing of the key boundaries*: a lack of radioisotopic dates hampers any estimation of true biodiversity dynamics, and the integrated comparison with reported ages of impact craters and magmatic events. (3) *Marine vs. terrestrial events*: insight into global ecosystem changes and correlation should be strengthened by chemostratigraphy, exemplified by the carbon isotope link between marine- and land-derived organic materials. (4) *High-resolution (bio)geochemical patterns*: isotope secular trends are poorly known at the intra-zonal and inter-basin scales, exemplified by prominent carbon isotope shifts across the Lower-Middle Frasnian passage. Further evidence is also awaited to verify cooling (vs. anoxia) pulses as the main stress factor in the F–F and end-Famennian marine settings, as well as climatic feedback with evolving weathering regimes on land and nutrient dynamic in marine realm. (5) *Near-equatorial vs. high-latitude domains*: refined data from extratropical successions, e.g. from the Kolyma Block, are still awaited. (6) *Tectonic and volcanic activity*: an integrated analysis of tectonic and igneous events, possibly triggered by superplume activity, will serve to evaluate any possible link with the Late Devonian biospheric perturbations. (7) Cyclostratigraphical perspective: includes growing research on refined magnetosusceptibility (MS) and various sea-level signatures to test whether they result from variation in Milankovitch frequency orbital variability. In addition, eustatic sea-level trends and their assumed glacioeustatic forcing have only recently been subject to discussion.

*Keywords:* Late Devonian; global events; biodiversity; geochronology; geochemistry; anoxia; palaeoclimate; volcanism

## 1. Introduction

The Late Devonian was an epoch of dramatic evolutionary and environmental changes, but focused primarily during one of the severest Phanerozoic extinctions close to the

*Corresponding author. Fax: +48-32-291-58-65. *E-mail address:* racki@us.edu.pl (G. Racki).

Frasnian-Famennian (F-F) boundary (Figs. 1–3), extensively summarized by McGhee (1996). In particular, a rapid decline of the low latitude, stromatoporoid-coral reefs (with a carbonate production loss of >90% according to Flügel and Kiessling, 2002) reflects a crucial ecosystem perturbation. This spectacular change in the stratigraphical record inspired introduction of the cosmic-accident concept by McLaren (1970) as the prime trigger of all global bioevents, and thus the Late Devonian successions have been studied intensively since 1970s. From the outset, problems of correlation and stage terminology seriously confused matters, but these were clarified in 1993, when the International Union of Geological Sciences approved a definition of the F–F boundary at the base of the conodont *Palmatolepis triangularis* Zone (Ziegler and Sandberg, 1990, 2001), i.e. immediately above the main anoxic Kellwasser Event and at a level of major extinction (Klapper et al., 1993;

*Figure 1.* Field photos of Late Devonian black shale horizons recording global anoxic events (see summary in Walliser, 1996): the Famennian Hangenberg (A) and *annulata* (B) events at Kowala, Poland (see also Fig. 9), and the two Late Frasnian Kellwasser events at Steinbruch Schmidt, Germany (C). Photos taken by P. Filipiak (A–B) and E. Schindler (C–D).

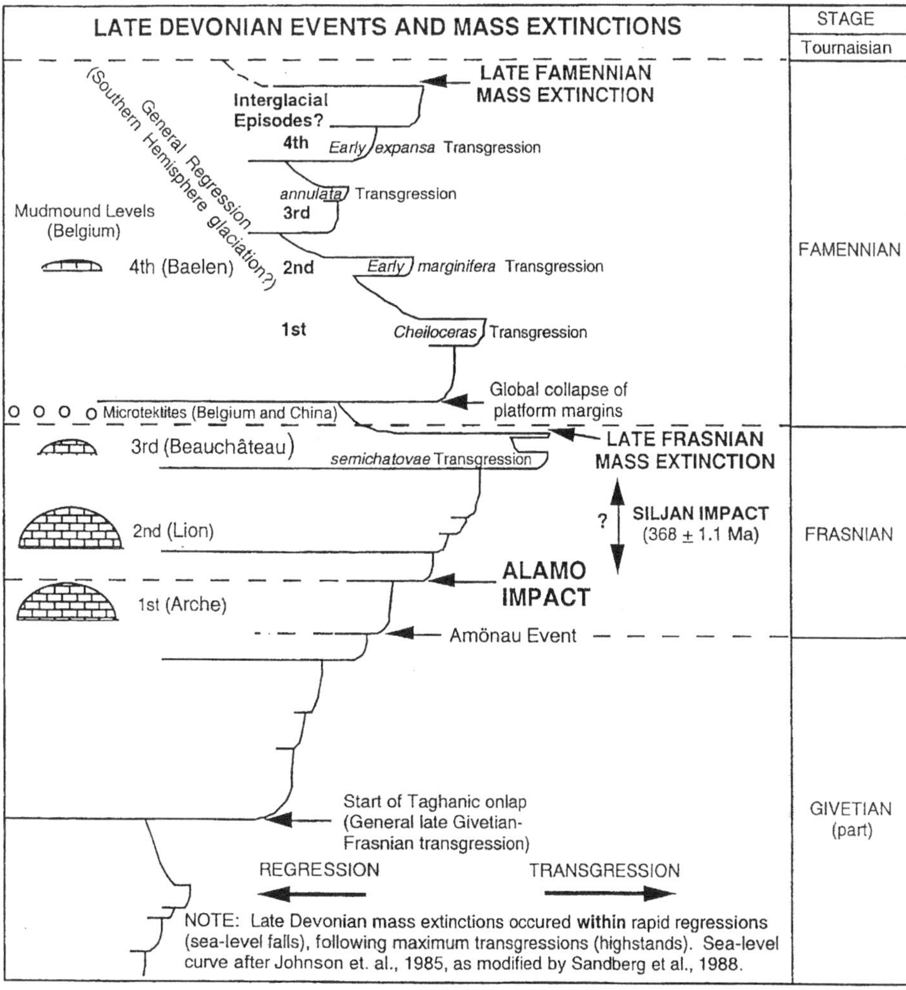


*Figure 2.* Late Devonian event stratigraphy, with reference to sea-level eustatic cyclicity (after Sandberg et al., 2000).

Walliser, 1996; House et al., 2000). In fact, studies of the eventful Late Devonian sedimentary and biotic record began as early as the 1980s as a paradigm of global changes and event stratigraphy, when Walliser (1981) and Eder and Franke (1982) linked widespread marker facies such as black shales (see Fig. 2) and reef extinction, respectively, with a global oceanographic trigger.

Three comprehensive summaries of the 'state-of-art' were published in the 1990s: McGhee (1996), Walliser (1996) and Hallam and Wignall (1997). Further progress is notable over the millennium crossroad, shown in two thematic 2002 issues of *Palaeogeography, Palaeoclimatology, Palaeoecology* (Racki and House, 2002) and *Acta Palaeontologica Polonica* (Baliński et al., 2002), which display the multidisciplinary efforts of international cooperation in Late Devonian studies. Other programs resulted in a series

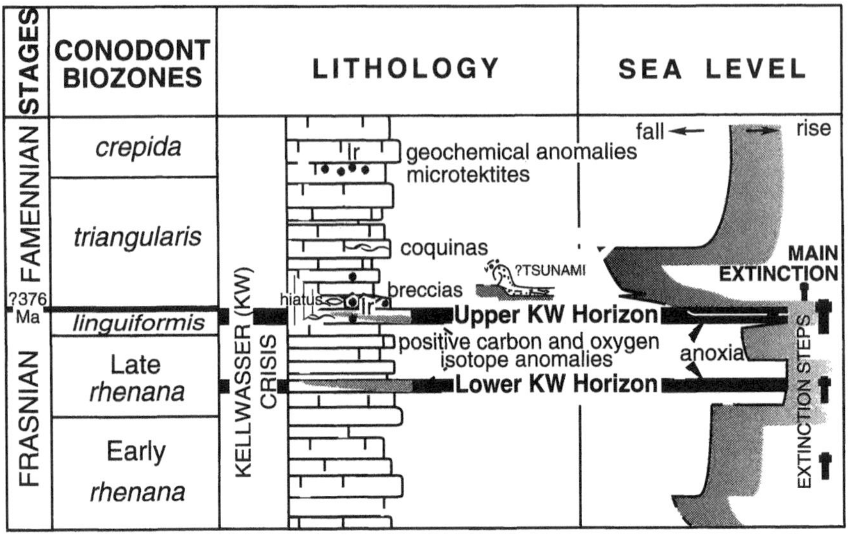


*Figure 3.* Diagram showing composite sedimentary and geochemical record across the F–F transition, and major eustatic and biotic events (modified Fig. 1 from Racki, 1999b; based on Fig. 3 in Joachimski and Buggisch, 1993; see also Figs 1 and 2).

of important papers by mainly Chinese-British (e.g. Chen et al., 2001, 2002, 2005; Chen and Tucker, 2003, 2004) and American (e.g. Murphy et al., 2000; Sageman et al., 2003) research groups. Some recent works have emphasized a gradual stepdown decline of many faunal groups, and the term "Kellwassser (KW) Crisis" that merges the effects of both anoxic KW events (Schindler, 1993), has entered the literature (Figs 2c and 3). Thus, more and more evidence has been accumulated for a prolonged nature of the F–F biosphere perturbation, and for an Earth-bound biodiversity crisis instead of a worldwide cosmic cataclysm resembling the end-Cretaceous event (see summary in Racki, 1999b; Copper, 2002).

Several new questions and strategic matters for future research emerged simultaneously with growing insight into the convoluted terrestrial processes. A better understanding of the Late Devonian ocean–climate–biosphere system leads to seven key frontiers, still waiting for extensive exploration, that were also raised in many presentations during the Geological Society of America Technical Session "Understanding Late Devonian and Permian–Triassic Biotic and Climatic Events: Towards an Integrated Approach" in Seattle in 2003. A subjective account of their coverage is a basic goal of this introductory article, with some references to continued polemics on primary extraterrestrial vs. Earth-bound causes as applied to the Late Devonian records (e.g. Alvarez, 2003; McGhee, this volume). Several joint research matters with the Permian–Triassic (P–Tr) boundary crisis are also outlined.

## 2. Magnitude and rank of biotic changes

The Late Devonian mass extinction was not a single instantaneous ("bedding-plane") killing event, as originally proposed by McLaren (1970) in his catastrophic impact scenario, but instead it consisted of a series of 'extinction pulses' (McGhee, 1996, 2001, this volume) occurring over several million years (ca. 5 Ma according to timescale of Kaufmann et al., 2004). As noted by McGhee (1996, pp. 44–46) and stressed recently by Bambach et al. (2004), *lowered origination* contributed more than elevated extinction to the F–F mass depletion of marine diversity that occurred during times of generally high extermination (see crustacean example in Rode and Lieberman, 2002). This is a different situation when compared to other globally distinct, 'true' mass extinctions, i.e. end-Ordovician, end-Permian, and end-Cretaceous. Alternatively, statistical analysis of pure taxonomic numbers is criticized as a straightforward proxy of the magnitude of ecological rebuilding in the ranking of mass extinction by McGhee et al. (2004). In fact, carbonate production by reefs appears to be not stabilized by their diversity over the long timescales (Kiessling, 2005).

The evolutionary dynamics across the F–F extinction horizon is overall still far from reliably documented (see summaries in Bultynck, 2000). Many palaeontological studies have provided a picture of biodiversity and ecologic changes across the major stage boundary in several regions and fossil groups, as exemplified particularly by the comprehensive work undertaken on the ostracods by Lethiers and Casier (summarized in Casier and Lethiers, 2001; also Groos-Uffenorde and Schindler, 1990; Olempska, 2002; Casier, 2003). However, lack of the high-resolution (bed-by-bed) documentation in reference or other biostratigraphically complete successions usually precludes a precise chronostratigraphical control needed to determine timing of the events occurring within the boundary interval itself (House, 1996; see examples in Blieck et al., 2000 and Streel et al., 2000). In fact, it is difficult to quantitatively assess of the biodiversity change during the F–F bioevent (and all other mass extinctions) so long as ancestor-descendant and biogeographical relationships in many fossil groups remain, to some extent, vague (MacLeod et al., 1997; Lieberman, 2002; Wood, 2004). Nevertheless, a progress is exemplified by Rode and Lieberman (2004), and increased resistance is postulated for shelly faunas that populated extensive offshore habitats.

Another hard matter is related to the poorly known survival-zone biota in the aftermath the F–F extinction that might potentially yield clues to the Late Palaeozoic biodiversity recovery (e.g. Erwin, 2001; Jablonski, 2002; Wood, 2004). A significant unbalanced excess of bacterially controlled productivity (Joachimski et al., 2001; Gong et al., 2002; see also Ormiston and Oglesby, 1995), coupled with frequent opportunistic marine algal blooms and suppression of major pelagic consumers (conodonts, ammonoids, sharks), is a remarkable feature in marine habitats (Paris et al., 1996; Ginter, 2002; Dzik, 2002; Chen and Tucker, 2003); however, a stronger effect in phytoplankton is not observed (Streel et al., 2000; see also Casier, 2003). These signatures from the photic zone could indicate a highly disorganized 'biological pump', seen during other mass extinction events, particularly the end-Cretaceous crisis (D'Hondt et al., 1998). Twitchett (2001) has suggested that the Lilliput effect – the small size of survivors in the immediate post-extinction aftermath – is a principal character of all recoveries after mass extinctions, confirmed by the F–F conodont and brachiopod response (Schülke, 1998; Renaud and Girard, 1999; Baliński, 2002). He relates this to a temporary decrease in food supply, with a resultant drop in the biomass and size in extreme oligotrophic settings. Thus, the major interruption episode of the trophic web due to large-scale destabilization events awaits more serious survey (Racki et al., 2002; Vermeij, 2004), especially that evidence to eutrophication is far more common (*cf.* Racki, 1999a; Joachimski et al., 2001, 2002; Tribovillard et al., 2004; Averbuch et al.,

2005). A post-extinction acme of microbial reef communities (see below) could be expression of fertility pulses, related to oceanic anoxia, on carbonate platform biota (as shown in the Aptian Tethyan margin; Immenhauser et al., 2005).

More importantly, it appears that detailed study is required for a longer interval than just across the Frasnian-Famennian boundary. The emerging severity of the end-Givetian and end-Famennian extinctions contrasts with a possibly overvalued significance, at least in *relative* terms, of the end-Frasnian biotic events (House, 2002), as demonstrated for echinoderms, bryozoans and radiolarians (Afanaseeva and Morozova, 1995; Webster et al., 1998; Głuchowski, 2002; Wang et al., 2003; Waters et al., 2003) and, probably, gastropods (Amler and Heidelberger, 2003). This demand is clear especially from biodiversity compilations of Copper (2002; in Brice et al., 2000) for Devonian atrypids and corals (see also terebratulid data; Garcia-Alcalde in Brice et al., 2000). Accordingly, the mid-Palaeozoic reef maximum in size and biodiversity corresponds to early-middle Givetian, and, by contrast, the Frasnian represents a late stage in a 'dying' episode that extinguishes development of the lower-diversity stromatoporoid-coral reef ecosystems in the calcite epeiric seas (Copper, 2002; Tapanila, 2005). Nevertheless, the end-Givetian biotic turnover remains conjectural in important aspects, and its detailed correlation with well-proved events in pelagic successions, which record a complex sequence of sedimentary events and faunal changes known as the Taghanic Biocrisis (see summary in Aboussalam and Becker, 2001), has not been achieved. In fact, a two-step collapse of Givetian stromatoporoid-coral bank biota is reported from the southern Laurussian carbonate shelf (Racki, 1993). In addition, Marshall et al. (2003) considered the aridity-driven changes in epeiric ecosystems as a trigger for the very significant Taghanic extinctions in both the marine and terrestrial realms. Hence, a greatly refined and time-constrained study of taxon stratigraphical ranges is required, particularly related to the many deepening pulses and reef decimations known during the Frasnian (House, 2002).

Even if metazoan reef ecosystems were impacted on a global scale (Copper, 2002), Webb (1996) claimed that the cyclical Middle–Late Devonian reef construction was continuous through the F–F extinction boundary, and without evident consequence for framework rigidity until the Devonian-Carboniferous (D–C) boundary (or Hangenberg) extinction event (*cf.* Shen and Webb, 2004). In north-western Australia, the most dramatic ecological changes caused by the F–F extinction are limited to back-reef habitats, while complex post-extinction microbial-sponge reef biotas represent a continuation of novel ecologies established already in the latest Frasnian (Wood, 2004; see also Reitner et al., 2001). Shallow-water buildups, marked by a calcimicrobial and stromatolitic framework with a few skeleton-dominated (stromatoporoid) examples, were widely distributed during the Famennian (Shen and Webb, 2004), but greatly constricted geographically mostly owing to a fall of northern hemisphere reefs after the D–C boundary (Kiessling, 2001); in fact, low-diversity reefs were particularly prone to environmental perturbations (Kiessling, 2005). A decoupling between reef construction and carbonate-platform development is emphasized by Kiessling et al. (2003, Fig. 15 therein) for the mid-Palaeozoic, because non-reefal carbonate production was particularly prolific during times of depressed reef growth (see the Famennian example in Peterhansel and Pratt, 2001).

In contrast to evolutionary pattern of several benthic groups (e.g. bivalves and gastropods; Amler, 1999; Amler and Heidelberger, 2003), the importance of Hangenberg environmental degradation is emphasized currently in radiolarians (Umeda, 2001) and plants

(Streel et al., 2000). In any case, the D–C extinction event could be more profound than previously thought (see summary in Walliser, 1996 and Caplan and Bustin, 1999).

## 3. Timing of the key boundaries

The second prominent uncertainty in Late Devonian event stratigraphy is tied to doubtful timing of the key boundaries. Almost all the Devonian ages are in flux, and appropriate time calibration is urgently needed. This is a principal goal of the Subcommission on Devonian Stratigraphy, but the progress is very slow. As shown in Figure 4, the numerical age of the F–F boundary remains highly controversial and has ranged from 376.5 Ma (Tucker et al., 1998) to 364 Ma (Compston, 2000) to 374.5 Ma (Gradstein et al., 2004; see also Gehmlich et al., 2000). However, a date around 376 Ma appears more probable after new U-Pb zircon analysis from a bentonite layer, intercalated between the two KW horizons at Steinbruch Schmidt, provided a date of 376.1 ± 1.6 Ma (Kauffmann et al., 2004).

A lack of consistent numerical dates hampers any estimation of true rates of biodiversity changes across the key intervals, as e.g. estimated ages for the Famennian Stage still range from 5.3 to 14.7 Ma (Fig. 4). This hindrance also precludes a definitive acceptance or rejection of the impact vs. volcanism models for extinction. For example, the central point in the impact discussion remains the timing of the Siljan crater, determined as 368 ± 1 Ma (see McGhee, 1996, Table 8.3). As stressed by Racki (1999b, p. 620): "although this crater is real, we cannot say exactly whether the documented impact occurred near the F–F

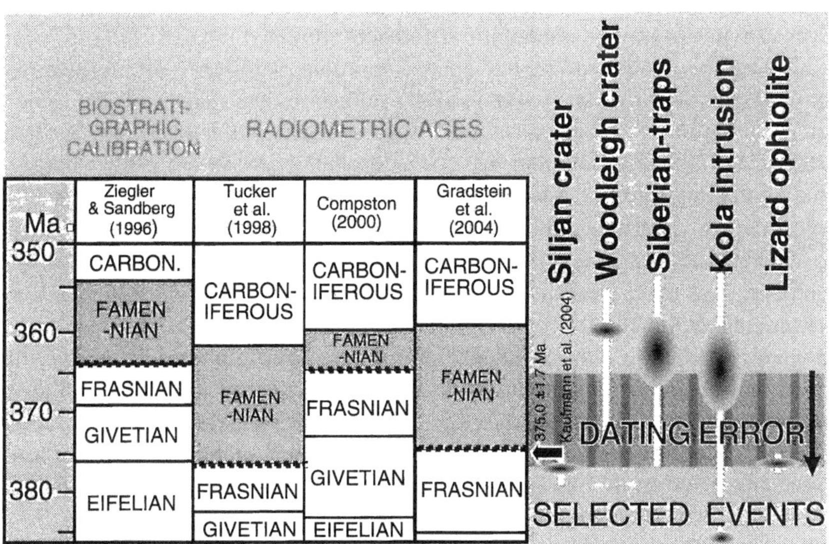


*Figure 4.* Comparison of four most recent Devonian time scales, and the selected Earth-bound and extraterrestrial event signatures to show their ambiguous absolute timing within established dating errors of the F–F boundary. Ages compiled from Kramm et al. (1993), Beard et al. (1996), Kravchinsky et al. (2002), Courtillot and Renne (2003), Vaughan and Scarrow (2003), Pervov et al. (2005), Reimold et al. (2005) and Uysal et al. (2005).

boundary, in the late Famennian or in the earlier Frasnian". However, the recently revised Siljan date (372 ± 2 Ma; Reimold et al., 2005) is again close to the currently proposed age for the F-F boundary, and only biostratigraphical dating of undoubted Siljan impact ejecta will be decisive. On the other hand, catastrophic volcanic paroxysm of the Devonian Siberian (Viluy) traps (see below), as highlighted by Courtillot and Renne (2003) for causation of the KW biotic crisis, could be temporally coupled both with the F–F as well as D–C extinction boundaries in the known range of dating errors (see Fig. 4).

Noteworthy, even in the case of P-Tr boundary and potentially coeval Siberian Traps, serious discrepancy and uncertainties in range of 3 Ma are evident between U-Pb and/or $^{40}Ar/^{39}Ar$ system dates (Wignall, 2001; Courtillot and Renne, 2003). In Late Devonian studies, similar doubts pertain even to the stratigraphical position of zircon-bearing ash levels sampled by Tucker et al. (1998) (see Streel et al., 2000; Over, 2002). The limiting factor in the geochronological calibration, based on high-resolution ion microprobe (SHRIMP) zircon $^{206}Pb/^{238}U$ ages, remain the origin and nature of the dated zircons; a crucial question of inherited grains is demonstrated by reallocation of a 381.1 Ma age to 367.6 Ma age for the Frasnian Little War Gap bentonite (Comptson, 2000, 2004).

## 4. Marine vs. terrestrial events

Ecosystem changes and event correlation between the marine and terrestrial realms are poorly known for the F–F extinction event, although it is assumed to affect both marine and terrestrial ecosystems; this is demonstrated by the diversity record of marine invertebrates, land-plant spores and macroflora (e.g. McGhee, 1996; Raymond, 2003; Raymond and Metz, 2004; but see more careful conclusion in Streel et al., 2000). A positive feedback scenario envisages the coupling of vegetation-promoted intensification of continental weathering, i.e. spread of rooted upland vascular plants and soil formation, with episodic increases of marine bio-productivity and promoted anoxic events (Kasig and Wilder, 1983; Wilder, 1994; Algeo et al., 1995; Ormiston and Oglesby, 1995; Algeo and Scheckler, 2003). Furthermore, the rise of trees probably had consequently a major effect on rapidly dropping levels of atmospheric $CO_2$ at that time, causing global cooling that led to South American glaciation (Berner, 2003; also Cox et al., 2001 and Raymond and Metz, 2004). On the other hand, the climatic response to mountain building-enhanced continental weathering and organic carbon burial is emphasized as well (Joachimski and Buggisch, 2002; Averbuch et al., 2005).

Pedogenic weathering rates are indeed a primary control on nutrient availability and, hence, may underscore an important influence on marine biotas, especially during anomalous greenhouse intervals (*cf.* Bambach, 1999). It is questionable, however, whether the palynological record supports any "vegetational revolutions" and land ecosystem overturns (see summary in Edwards et al., 2000 and Streel et al., 2000), rather than just a gradual, global increase of weathering rate, nutrient levels and bioproductivity during this epoch (Bambach, 1999; Racki, 1999b, p. 623; Martin, 2003; Sageman et al., 2003, p. 256). A scenario of river-borne nutrient flux is also generally disputed for eutrophication pulses beyond nearshore, especially near-estuarine, domains (*cf.* Sageman et al., 2003; Erba, 2004), although the surface waters may have been fertilized by solubilized nutrients from distant land sources (Tribovillard et al., 2004) and/or aeolian input (see Hladil, 2002). The

Wilder-Algeo F–F extinction model has only a limited role for upwelling (Ormiston and Oglesby, 1995; Algeo, 2003), but distal-shelf and oceanic environments must have been affected by other oceanographic nutrification phenomena (as discussed below). Then again, Sr-isotopic values and other geochemical signatures should be measured to refine continental input into the ocean (see modeling in Jones and Jenkyns, 2001), and a possible scale of subsequent eutrophication (also phosphogenesis; Fröhlich, 2004). The nutrient excess and eutrophic stress probably increased during the KW crisis (e.g. Racki, 1999a; Chen et al., 2002; Matsumoto et al., 2002; Racki et al., 2002; Chen and Tucker, 2004), and some other global events (e.g. Hangenberg Event; Caplan and Bustin, 1999). The destructive effects were at least partly accelerated by precipitation topographically controlled by tectonic activity (*cf.* Ormiston and Oglesby, 1995; Racki, 1998, 1999b; Averbuch et al., 2005). It is demonstrated by bioerosion-enhancing mesotrophic conditions within the Famennian Palliser platform, promoted by convergent movements and uplift during the Ellesmerian orogeny in western Laurussia (Peterhansel and Pratt, 2001). At general, Sr isotopes exhibit increasing values during Late Devonian (Denison et al., 1997), what reflect elevating input of continental Sr; a marked exception during late Frasnian (Fig. 6 in Veizer et al., 1999), casually linked with the both KW events (Chen et al., 2005), seems to correlate with some magmatic events (see the dates below and Fig. 4).

An interdisciplinary approach to land–sea interactions requires a common timescale. Potentially, a chemostratigraphic carbon isotopic correlation scheme developed from marine and land-derived organic matter could prove highly valuable, but this has not yet been accomplished. Such an integrative event-stratigraphic study has been realized for the expanded P-Tr boundary section in Greenland (Twitchett et al., 2001), where the fossil record contains both a well-preserved marine fauna as well as terrestrial palynomorphs, which have enabled comparison between the marine and terrestrial biotic crises using the same samples.

## 5. High-resolution (bio)geochemical patterns

A fourth problem is connected with major climatic and oceanographic processes. Substantial new data on the interplay between eutrophication, productivity and anoxia have recently been developed (see comprehensive summaries in Sageman et al., 2003 and Bond et al., 2004), culminating in a recent issue of *Chemical Geology* (Schultz and Rimmer, 2004). In the light of available evidence, the term "anoxic event" should be replaced rather by "intermittent or seasonal" anoxic event according to many recent authors (Murphy et al., 2000; Hudson, 2001; Schieber, 2001; Racki et al., 2002; see also evidence for episodic only euxinia during the P-Tr superanoxic event, Grice et al., 2005). Seasonal instability of the water column and episodic mixing are central to the productivity-anoxia feedback proposed in shelf anoxia model of Murphy et al. (2000) and Sageman et al. (2003). Further multiproxy geochemical signatures of pulses of anoxia or cooling are needed to determine the *primary* stress factor in the Late Devonian seas, especially given that they are auto-cyclically linked in the above model; the issue was highlighted in Joachimski and Buggisch (2003) in which they biogeochemically coupled the KW anoxic events with cooling pulses (see also discussion in Copper, 2002). Determination of the character and diverse nutrient sources should be thoroughly investigated in context of evolving primary producers (a diatom enigma? Schieber et al., 2000) and climatic setting, especially the presumed Frasnian cooling trend (Sageman et al., 2003), which is not confirmed by oxygen isotopic data (Joachimski et al., 2004).

In particular, benthic phosphor regeneration under anoxic conditions (Wallmann, 2003; Nederbragt et al., 2004; see also Fröhlich, 2004), paired with possible volcanogenic fertilization (Racki, 1999a; Siegmund et al., 2002; Racki and Wignall, this volume), are a step toward explaining the enigmatic phosphorus deficiency displayed in the short-term KW carbon isotope excursions modeled by Godderis and Joachimski (2004). As hypothesized for Mesozoic oceanic anoxic events (e.g. Jones and Jenkyns, 2001; Erba, 2004), a shift to meso- and eutrophic conditions in the remote parts of large oceans, paired with drastically reducing calcification, was induced by submarine igneous events that provided huge quantities of biolimiting elements, especially iron and $CO_2$. Such a vigorous hydrothermal fluid supply by low-temperature submarine springs and/or seeps is indeed assumed for growth of Devonian sponge mud-mounds and microbialites, such as some stromatolitic buildups (Playford and Wallace, 2001; Reitner et al., 2001).

As noted above, the global climate switched from a Frasnian greenhouse to an icehouse mode culminating in continental glaciations in the end-Famennian and Carboniferous (Streel et al., 2000). The numerical modeling by Ormiston and Oglesby (1995), however, undoubtedly precludes the F–F cooling onset, although a motive of lethally cold temperatures has been extensively explored for this boundary extinction since the work of Copper nearly three decades ago (Copper, 1977; see critical summary in Hallam and Wignall, 1997, p. 88). Devonian carbonate oxygen isotopes appear to be surprisingly unreliable in the climatic context (Joachimski et al., 2004), although useful in tracing of other Palaeozoic glaciations, including the Hangenberg Event (Brand et al., 2004). Two primary but independent research directions are obvious:

- There is a vital need for comparative isotopic analysis of marine carbonate and more reliable phosphate biogenic materials in many regions to assess the incipient icehouse F–F scenario as a consequence of enhanced organic carbon burial rate during the KW episodes (Buggisch, 1991; Joachimski and Buggisch, 2002, 2003).
- A geochemical and mineralogical investigation for evolving weathering regimes and consequent terrigenous fluxes is required for the evaluation of the greenhouse to icehouse evolution. This could be especially recorded in the degree of chemical alteration and maturity of sediments, as well as in clay mineralogy (e.g. Han et al., 2000; Pay et al., 2000; Devleeschouwer et al., 2002; Hladil, 2002; Matsumoto et al., 2002; Mahmudy Gharaie et al., 2004; Brand et al., 2004).

Joachimski et al. (2004) have recently provided such a refined palaeotemperature curve after oxygen isotope study of conodont apatites (Fig. 5). The complex fluctuations during the F–F interval appear to be succeeded by a surprisingly warm early Famennian (30–33°C; *cf.* Ormiston and Oglesby, 1995), while palynological data have been interpreted as evidence for cold and dry regimes in the Famennian (Streel et al., 2000). On the other hand, isotopic and clay mineral data from South China and Iran still suggest runaway greenhouse-type temperatures and enhanced weathering rates (Chen et al., 2002, 2005; Matsumoto et al., 2002; Mahmudy Gharaie et al., 2004), confirmed also by radiolarian record (Umeda, 2001). In the puzzling context, Hladil (2002, p. 244) concluded, with reference to collapsing carbonate complexes: “A stronger increase of the Th/U is typical for

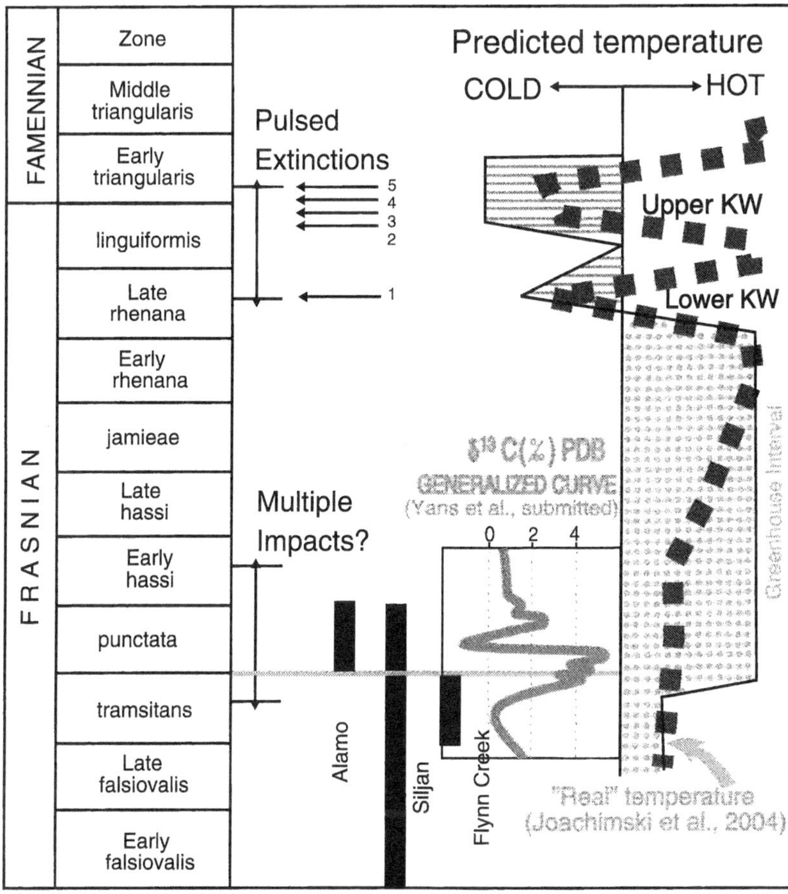


*Figure 5.* A lag-time multiple impact scenario of McGhee (2001, Fig. 3), proposed by analogy to the mid-Cenozoic feedback: an extraterrestrial factor triggered an anomalous greenhouse interval that collapsed during the KW Crisis toward the long-term cooling term (see also McGhee, this volume). This hypothesis is confronted both with a large carbon isotopic disturbance in Ardennes (Belgium) in the *punctata* Zone Yans et al. (submitted), and the overall trend of a hopefully real palaeotemperature curve, calculated from $\delta^{18}O$ in biogenic apatites by Joachimski et al. (2004). Note a gradual long-term warming tendency extremely disturbed in the broad F–F transition timespan (*cf.* Chen et al., 2005). Thus, the term "destabilized greenhouse" (Racki, 1998, p. 192) or "punctuated greenhouse" (Racki et al., 2002, p. 288) for the crucial crisis interval seems to be more reasonable than a "collapsed greenhouse" climate setting (McGhee, this volume).

the F/F and early Famennian intervals. It indicates a humid, but still hot climate in the Late Devonian tropics, in spite of the beginning of a glacioeustatic regime" (see also discussion in House et al., 2000).

Despite a large number of studies, the refined trends in biogeochemical cycling are poorly known at the intra-zonal and inter-basinal scales (e.g. Veizer et al., 1999; van Geldern and Joachimski, 2001). In the latter approach, as shown by Holmden et al. (1998), some carbon isotope excursions in epicontinental successions may originate from migration of isotopically dissimilar (i.e. temperature-salinity defined) water masses, and not from global-scale oceanographic changes (*cf.* also Dopieralska et al., 2003). Therefore, for example, primary vs. diagenetic meaning of negative $\delta^{13}C$ values, particularly in organic-rich sediments, needs comprehensive explanation (*cf.* Racki, 1999b; Joachimski et al., 2001, 2002). In addition, massive methane outbursts from dissociated hydrates could be involved in the C-isotopic features during regression and/or igneous episodes (Bratton, 1999; Chen et al., 2002; Matsumoto et al., 2002), as commonly accepted today for several biotic crises, including the P-Tr ecosystem collapse (Benton, 2003).

From a secular perspective, Devonian C-isotopic curves, based mostly on extremely reliable but limited measurements from low-Mg calcite brachiopods from separate continents, reveal only a broader supra-zonal pattern (Fig. 6). As discussed for a potentially biased whole-rock $\delta^{13}C$ values by Buggisch (2001), however, the reliable test for determining global vs. regional (also purely diagenetic) controls on isotope signals is comparative analysis of secular curves from distantly separated sedimentary basins located in different palaeogeographic and facies settings. Confidence of the results is finally verified by comparative organic and carbonate $\delta^{13}C$ values (Joachimski et al., 2002). Notably, the distinctive (nonetheless far from easily explainable; see Godderis and Joachimski, 2004) F–F inorganic isotopic pattern is preserved even in sparry bioclastic facies (Bond et al., 2004).

Of further, generally unappreciated significance is the fact that significant geochemical anomalies are not always associated with documented biotic events; they can still be found during 'background' intervals (e.g. Holser et al., 1996; Veizer et al., 1999). For example, the Silurian–Devonian boundary $\delta^{13}C_{carb}$ excursion of $\geq +5‰$ appears to be among the largest, well-documented, isotopic excursions in the Palaeozoic (Saltzman, 2002). Another large shift is discovered recently in the Lower-Middle Frasnian transition (*sensu* Ziegler and Sandberg, 2001), considered until recently as a biogeochemically and evolutionarily 'quiescent' interval (*cf.* Holser et al., 1996; Sepkoski, 1996; Prokoph and Veizer, 1999; but see McGhee, 2001). A high stratigraphic resolution of the isotopic record at intra-zonal scale, derived from a few, well-dated, biostratigraphically continuous sections in Ardennes, allowed Preat's group to detect a prominent positive-negative $\delta^{13}C_{carb}$ excursion (from

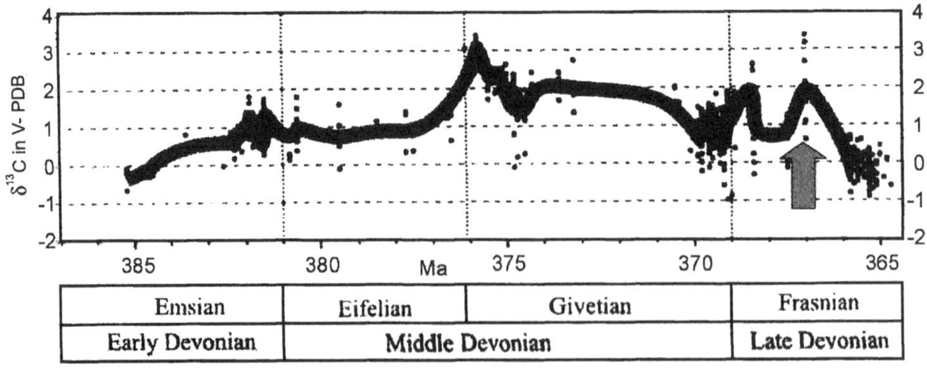


*Figure 6.* Devonian secular pattern of carbon isotopic ratio based on well-preserved brachiopod calcites after van Geldern and Joachimski (2001, Fig. 6); arrow shows increased $\delta^{13}C$ variation in the *Palmatolepis punctata* zonal timespan for comparison with the large carbon isotopic disturbance presented in Fig. 5.

5.85‰ to −1.20‰ ) in the *Palmatolepis punctata* Zone (Fig. 6); several data sets suggest it is supra-regional, and possibly even global (Yans et al., submitted). The large-scale changes in carbon cycling in the timespan are of higher-amplitude than the celebrated biogeochemical turnovers related to the F–F mass extinction. A link with the Alamo Icy Impact Event and/or other bolide strikes (Fig. 6), promoting a massive dissociation of oceanic methane hydrate, remains an attractive, but perhaps disputable, explanation, as preliminarily announced by Morrow et al. (2003).

## 6. Low- vs. high-latitude domains

A fundamental problem in Late Devonian investigations is caused by the palaeogeographical bias of studies, shown on the global reconstruction in Figure 7, which are almost entirely conducted on material from tropical palaeolatitudes (see summary in Kalvoda, 2002). Refined results from northern extratropical marine successions, e.g. from the Kolyma block (see Gagiev, 1985, 1997), would be especially desirable, as well as from diversity of the southern subpolar domains (e.g. Copper, 1977; Cotter, 2000; Streel et al., 2000; Melo and Loboziak, 2003). Useful, but still generalized, higher-latitude information from Siberia is now available (e.g. Dubotalov and Krasnov, 2000; Umeda et al., 2001; Stone et al., 2003; Yolkin et al., 2003). Also important is abundant data from poorly known near-equatorial regions, such as North China (Tarim microcontinent; Hao et al., 2003; Fig. 8), Iran (Matsumoto et al., 2002; Wendt et al., 2002; Mahmudy Gharaie et al., 2004), Turkey (Çapkinoğlu and Gedik, 2000; Higgs et al., 2002), Kazakhstan (Cook et al., 2002), Timan and the Polar Urals (House et al., 2000; Yudina et al., 2002), and partly even from

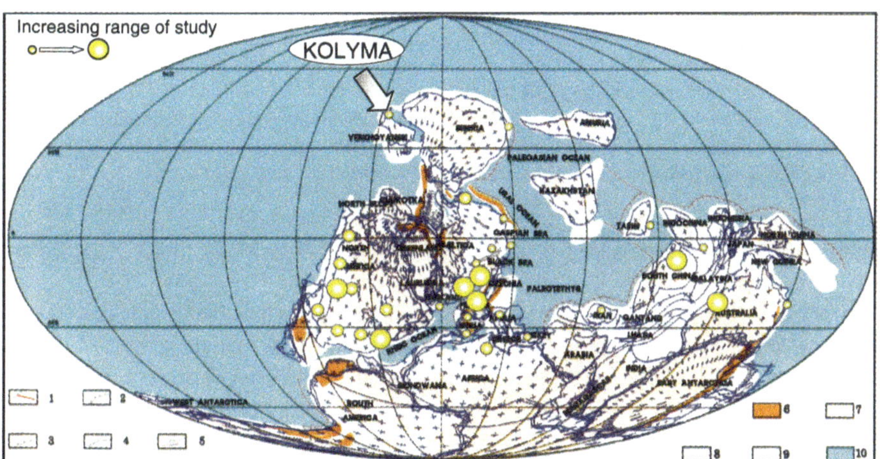


*Figure 7.* Palaeogeographical setting of relative progress in the study (size of the circle as a subjective expression of number and thematic scope of the works) of Late Devonian global events (reconstruction based on Golonka, 2000). Kolyma plate is arrowed as a key area to study of high-latitude topics. 1 – Oceanic spreading center and transform faults, 2 – subduction zone, 3, thrust fault, 4 – normal fault; 5, transform fault, 6 – mountains, 7 – landmass, 8 – ice sheet, 9 – shallow sea and slope, 10 – deep ocean basin.

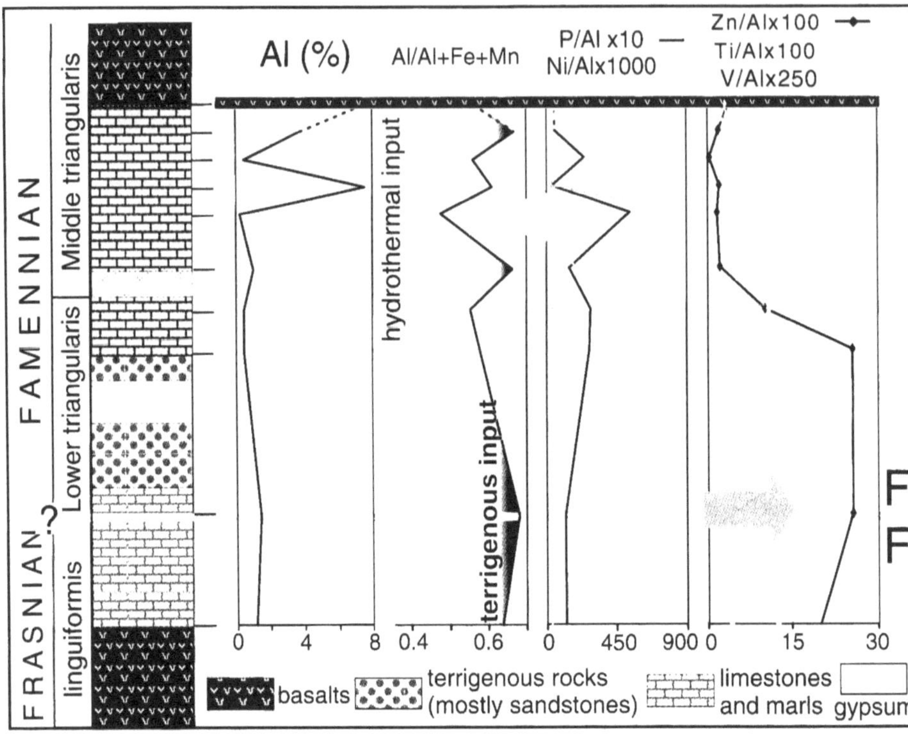


*Figure 8.* Simplified stratigraphic column of the F–F transition at the Bachu section in the Tarim basin, northern China, and trends of selected event-geochemical proxies (see discussion in Racki et al., 2002), based on Fig. 2 and analytical data from Table 1 in Hao et al. (2003). The position of the F–F boundary is approximated by a combined biostratigraphical–chemostratigraphical approach. Note an interruption of the mafic extrusive activity in the crucial interval, but also two white gypsum layers and a differentiated geochemical signature of two other events, interpreted by Hao et al. (2003) as a manifestation of large-scale rifting-hydrothermal processes in the Tarim (see also Han and Zhao, 2003), Kazakhstan and even South China basins (e.g. Ma and Bai, 2002). The weak Ni-anomaly near the F–F boundary, however, is obviously overwhelmed by continental input possibly paired with eutrophication and spread of anoxia (the very high V/Al ratio, but in one sample only), similar to Iranian and S-Chinese successions (Mahmudy Gharaie et al., 2004; Chen et al., 2005).

northern Canada (e.g. Patchett et al., 1999; Levman and Von Bitter, 2002; Klapper et al., 2004).

Thanks to broad international cooperation, the most comprehensively studied F–F section from event-stratigraphical, palaeobiological and geochemical perspectives is located in Poland, at the active Kowala quarry near Kielce, Holy Cross Mountains (see the array of articles in Racki and House, 2002, and Baliński et al., 2002; also Joachimski et al., 2001; Bond and Zatoń , 2003, and Bond et al., 2004; Fig. 9). The Kowala deep shelf-basin succession is distinguished by uniquely immature character of the organic matter (burial temperatures did not exceed 75°C; Belka, 1990), a key to modern geochemistry. Important results include:

- Discovery of isorenieratane and related organic compounds, diagnostic for green sulfur bacteria (Chlorobiaceae), an evidence for photic-zone anoxia (Joachimski et al., 2001; see also Bond et al., 2004).

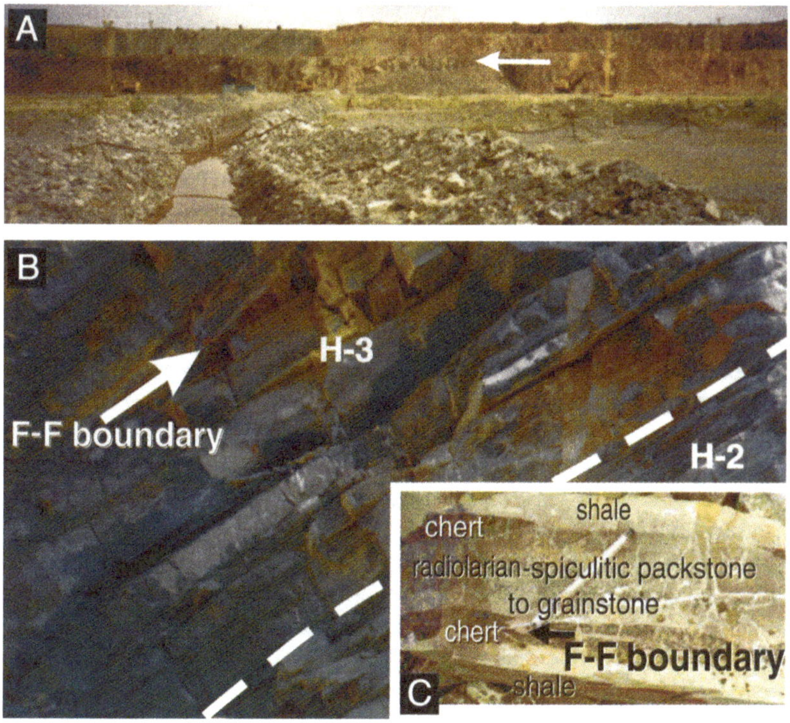


*Figure 9.* General view of active quarry at Kowala near Kielce in 1999 to show the section Kx (arrowed) presented in Racki et al. (2002) (A), and close-ups of dark-colored F–F boundary beds including marly unit H-2 and calcareous-cherty unit H-3 (B), and chert-bearing F–F boundary layer (in the bleached appearance only available) (C).

- Co-occurrence of distinctive positive organic $\delta^{13}C$ (Joachimski et al., 2001) and biogenic apatite $\delta^{18}O$ (Joachimski, unpubl.) excursions.
- a record of the F–F worldwide biosiliceous event (Fig. 9C), that indicates at least moderate eutrophication levels in the key boundary interval (Vishnevskaya et al., 2002).

Such a collective effort is just realized during study of other global events perfectly recorded at this succession (e.g. Famennian *annulata* and Hangenberg events – see Fig. 1A and B).

## 7. Tectonic and volcanic activity

A crudely established feature of the Late Devonian interval remains the accurate timing and understanding of the geodynamic nature of tectonic events, as reflected in the range of orogenic events, magmatic activity (see Fig. 4) and related hydrothermal and geothermal phenomena (see summary in Racki, 1998 and Averbuch et al., 2005). Nevertheless, the convoluted geotectonic Late Devonian history of the Palaeoasian and Palaeotethys oceans in Central Eurasia, between the East European, Siberian and Tarim cratons, and of the

Kazakhstan and other microcontinental blocks, has been better temporally constrained in recent years (Filippova et al., 2001; Bykadorov et al., 2003; Xiao et al., 2004; several articles in Jahn et al., 2004), together with the timing of Palaeotethys opening and spreading in East and Southeast Asia (e.g. Metcalfe and Allen, 2000; Guo et al., 2004).

An integrated analysis of Eovariscan orogeny and igneous events is required to evaluate the possible link between Devonian biotic events and processes, including tectonoeustasy (model of Cathles and Hallam, 1991; Racki, 1998), oceanic anoxia (Eder and Franke, 1982; Racki, 1999b; Peterhansel and Pratt, 2001; Tribovillard et al., 2004) and stromatoporoid-coral reef decline (Racki, 1998, 1999b; Reitner et al., 2001). Provenance and geochemical studies are particularly promising to establish causal correlation between tectonic and sedimentary events, even in volcano-sedimentary suites (Boulter et al., 2004). For example, evidence is provided for sedimentary-source changes in the well-studied Western Canada sedimentary basin (Patchett et al., 1999; Savoy et al., 2000; Stevenson et al., 2000). For this domain, Nelson et al. (2002) proposed also that carbonate-hosted Zn-Pb deposits and secondary coarse dolomites were record of the same age (362±9 Ma; *cf.* Dusel-Bacon et al., 2004), back-arc exhalative activity as far-field effects of subduction along the western Laurussian margin.

Despite the above announced uncertainty in timing (377–350 Ma; Kravchinsky et al., 2002; Courtillot and Renne, 2003), a key region for solving this problem is certainly the Siberian Viluy igneous province (Fig. 10), regarded as a precursor of the Siberian P-Tr ca aclysmic superplume eruption (Courtillot and Renne, 2003; Dobretsov, 2003; Kiselev et al., 2004). Recognized swarm feeder dikes, up to 200 m thick (Tomshin and Koroleva, 1990), are seen as a proxy of one of the largest Phanerozoic mantle superplume eruption (event 7 of Abbott and Isley, 2002), with the predicted surface area of flood basalt magmatism above 6 million $km^2$. In this model, the prolonged (18 Ma) tectonomagmatic reactivation by associated plume clusters occurred in both the eastern Laurussian (in particular the Pripyat-Dnieper-Donets rift graben and Kola province; Wilson and Lyashkevich, 1996) and the Siberian continents (see also Ernst and Buchan, 2001; http://www.largeigneousprovinces.org/record.html). Massive alkaline (carbonatitic) magmatism in the Kola igneous province (NW Russia), dated 360–380 Ma, is especially noteworthy (area covered *ca.* 100,000 $km^2$; Kramm et al., 1993; Beard et al., 1996; Arzamastsev et al., 2000), as well because of its potential for rapidly injecting substantial amounts of $CO_2$ and $SO_2$ into the atmosphere (Ray and Pande, 1999). Notable also for the volcanic catastrophe models (*cf.* Phipps Morgan et al., 2004), Late Devonian kimberlite magmatism is well known in Yakutia (376–350 Ma, Heaman et al., 2003, p. 177; Courtillot and Renne, 2003; Kiselev et al., 2004) as well as in several domains of East European Platform, including Kola and Pripyat regions (Wilson and Lyashkevich, 1996; Mahotkin et al., 2000; Bogatikov et al., 2001; Yutkina et al., 2004; Pervov et al., 2005).

Studies of the Palaeozoic igneous provinces are still in an early stage and replete with questions concerning petrological, geochemical, geochronological, volcanological and geophysical elements. In truth, exclusively direct biostratigraphical dating of tectono-volcanic phenomena in sedimentary records is decisive age determination, as exemplified by late Frasnian timing of the Pripyat-Dnieper-Donets magmatism, summarized in Racki (1998, p. 180, Fig. 3 therein). This concerns also, undervalued in the global event analysis, mineralization episodes, as shown by Iberian Pyrite Belt, Portugal (Oliveira

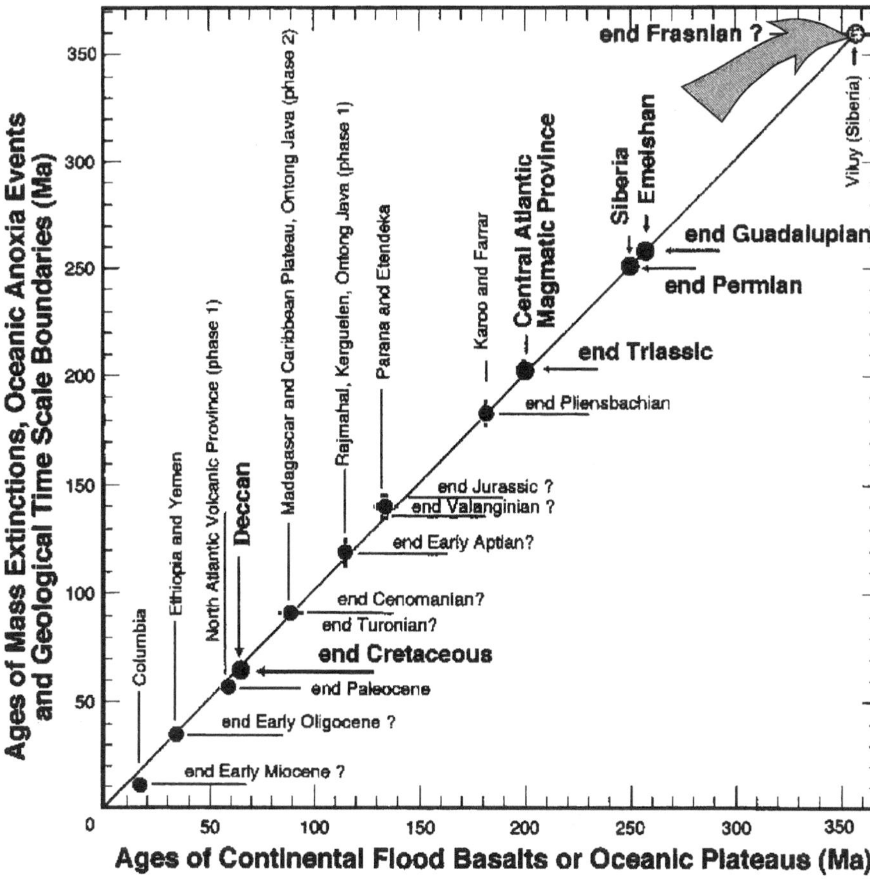


*Figure 10.* Correlation between the ages of large igneous provinces and mass extinctions and oceanic anoxia events, according to Courtillot and Renne (2003, Fig. 1), with emphasize on a firm arrangement of the F–F mass extinction and Siberian (Viluy) trap volcanism in this correlative scheme.

et al., 2004) and Zn-Pb-Ba ores in British Columbia, Canada (Paradis et al., 1998; Nelson et al., 2002).

According to Courtillot and Renne (2003), the high-latitude Siberian volcanism could be a crucial trigger for greenhouse feedback and a collapse of the thermohaline system for both the F–F and P-Tr biocrises (Fig. 10; see Racki and Wignall, this volume). There are basic differences, however, in the major climate-ocean system responses:

- F–F event: two pulses of cooling and anoxia in the highly destabilized greenhouse climate (and the biosiliceous acme; Racki, 1999a).
- P-Tr event: an apocalyptic greenhouse due to methane dissociation, superanoxia in the stagnant Panthalassan Ocean for ca. 10 Ma (and Chert Gap, Racki, 1999a; Beauchamp and Baud, 2002; see also Benton, 2003; Racki and Wignall, this volume).

## 8. Cyclostratigraphical perspective

Last, but not least, the advance includes continued search on magnetosusceptibility (MS) and variety of sea-level signatures to test whether their cause may lie in energy stimuli resulting from changes in the Earth's orbital parameters. This is a leading idea of House (1985), who speculated on an orbitally forced, climatically driven primary cause for many Devonian biotic events. Two examples of the highest temporal resolution at the millennial scale should be quoted:

- Seven-order MS scheme proposed by Crick et al. (2002), with the seventh order encompassing intervals less than 1 Ka. This cyclicity is forced by variation in the original rate of supply of iron-bearing materials into the marine system due either to climate-induced erosion cycles, or non-cyclic eustatic (Zhang et al., 2000; Averbuch et al., 2005) and/or epeirogenic sea-level changes (House, 2002). Understanding of this intriguing record awaits further investigation.
- Five-order cyclostratigraphy pattern of Chen and Tucker (2003), where depositional response to orbital precession (16–18 Ka in duration) and eccentricity (100 Ka) is reliably recognized. In the basinal cycles, however, even smaller-scale rhythmic stratifications were likely promoted by millennial-scale climatic forcing. On the basis of this orbital time calibration, the entire F–F biotic extinction took place over a period of 450 Ka, and the major crisis during a major sea-level fall was 400 Ka in duration (compare with previous calibration of Sandberg et al., 1988; see also Morrow and Sandberg, 2003).

This cyclostratigraphical approach can be full of pitfalls (e.g. Racki and Wignall, 2004), however, particularly when orbital frequencies, as recorded in the sedimentary archive, are severely modified by post-depositional alteration (Westphal et al., 2004). Nevertheless, independently refined conodont biocorrelation (Klapper, 1997; Gouwy and Bultynck, 2000), cyclostratigraphy, and integrated sequence and event stratigraphy are promising tools for high-resolution intra- and inter-basin correlation (e.g. Cotter, 2000), as well as estimation of more realistic accumulation rates (e.g. Chen et al., 2001; Morrow and Sandberg, 2003).

## 9. Two eustatic dilemmas

One of the recent issues open to debate concerns the Late Devonian eustatic cyclicity pattern, and the best example is a sea-level change during the key F–F interval. There are three competing hypotheses which were prominent at the GSA Session in Seattle 2003:

- 'Classic' transgression-regression couplets of Johnson et al. (1985), recently refined by Sandberg et al. (2002; see Fig. 1) and Morrow and Sandberg (2003);
- regressive change (Dopieralska et al., 2003); and
- transgressive change (Wignall and Bond, 2003; *cf.* Hallam and Wignall, 1999).

Sea-level rise scenario across the broad F–F transition may be more or less applicable to the western United States (*contra* Morrow and Sandberg, 2003), New York (Over, 2002; Sageman et al., 2003; but see cyclic pattern in Filer, 2002) and Morocco (Wendt and Belka,

1991; Fröhlich, 2004), in particular where synsedimentary tectonic movements were significant (e.g. Racki, 1998; Smith and Jacobi, 2001; Chow et al., 2004). However, this sea-level pattern seems difficult to demonstrate especially in shallow-water cratonic domains marked by an extensive F–F erosive hiatus and/or karst phenomena (see summary in Chen and Tucker, 2004; Chow et al., 2004), as shown particularly for the East European Platform (Alekseev et al., 1996). The typical case represents the Western Canada shelf, where the F–F sedimentary break is associated with surficial karstification, brecciation and neptunian dykes (Geldsetzer et al., 1993). A prominent palaeokarst within cyclic peritidal facies in Guilin, S China, similarly displays karren (scalloped surfaces), dissolution pits and pipes containing residual soil (*terra rossa*), indicators of a significant and long-lasting (>*ca.* 50 Ka) fall of sea level at the end of the Frasnian (Chen and Tucker, 2004).

Conversely, the geological evidence for evolving Late Devonian climate and lowstand regimes is broadly considered in the context of glacioeustatic control on the rapid sea-level fluctuations (Isaacson et al., 1999; Streel et al., 2000; Filer, 2002; Isaacson and Hladil, 2003; *contra* Ormiston and Oglesby, 1995). If so, glaciation seems the reasonable explanation for the major sea-level falls associated with the spread of Kellwasser-type anoxia over shelves (*cf.* Dopieralska et al., 2003). However, it is somewhat contrary to the autocyclic F–F scenario discussed by Buggisch (1991), where glacially forced regressions are associated essentially with the terminations of the KW intervals. The conodont apatite $\delta^{18}O$ record, in fact, reveals that both KW episodes are paired with cooling of tropical surface waters by 5–7°C (Fig. 5), due to lowered $CO_2$ partial pressures of the atmosphere caused by accelerated organic carbon burial (Joachimski and Buggisch, 2002, 2003). Are the KW signatures a response to eustatic lowering during two-step expansion of polar ice sheets (and not to a short-lived transgression, as traditionally thought; Johnson et al., 1985; Sandberg et al., 1988; Buggisch, 1991; Joachimski and Buggisch, 1993)? In fact, a temporal link between the sea-level change and anoxia spread is still somewhat differently understood (*cf.* Filer, 2002; Sandberg et al., 2002; Chen and Tucker, 2004; Averbuch et al., 2005). In summary, this is quite a different climatic record from that seen near the end of the Famennian, where eustatic fall is reliably interpreted to have resulted from a severe, terminal glacial episode (Johnson et al., 1985; Sandberg et al., 2002; Caplan and Bustin, 1999; Streel et al., 2000, pp. 143–145; Brand et al., 2004).

## 10. Implications and conclusions

Many aspects of Late Devonian (and other) global events remain conjectural, including timing and magnitude of ecosystem changes and mediated biodiversity dynamics, and especially their primary causes (see reviews in Walliser, 1996; Hallam and Wignall, 1997; Racki, 1999b; House, 2002; Vermeij, 2004). For example, the extraterrestrial motif is indeed still returning in Late Devonian studies, as demonstrated by Sandberg et al. (2002; Fig. 1) and Ellwood et al. (2003). The "impactors" regard this epoch as a timespan affected by bolides because "(…) substantial evidence of impact in that ca. 16 Myr interval" (Alvarez, 2003, p. 155 and Table 1), and this idea is in overall accordance with 'lag-time' multiple impacts hypothesis of McGhee (2001). A potential destabilizing role of impact bombardment in stressed Frasnian ecosystems is improbable on a global scale (Racki, 1999b) as considered for the 85 km diameter Siljan impact structure (Reimold et al., 2005). The Late Devonian record, especially craters and iridium anomalies, provides frustrating implications (McGhee,

*Table 1.* Late Devonian issues of our ignorance against succession features and selected regions to future investigation that would help to resolve the questions discussed in the text.

| Hot topical area | Succession characters that would help resolve the questions | Examples of regions to study that would help resolve the questions [Additional key references] |
|---|---|---|
| Magnitude and rank of biotic changes | Fossiliferous well-dated successions, preferably where pelagic and bank-to-reef facies interfinger | S China (Tsien and Fong, 1997; Chen et al., 2001); W Australia (Becker and House, 1997); Ardennes (Bultynck et al., 1998, 2000); Leon, Northern Spain (Van Loevezijn, 1987) [Copper, 2002] |
| Timing of the key boundaries | Well-dated sedimentary successions with volcanogenic intercalations | New York (Over, 2002); Southern Urals (Veimarn et al., 2004); Pripyat-Dnieper-Donets rift zone (Wilson and Lyashkevich, 1996) |
| Marine vs. terrestrial events | Well-dated terrestrial plant-bearing successions preferably with marine intercalations | Central and eastern Arctic (Copper, 2002, p. 38), Appalachians (Catskill delta; Scheckler, 1986); S Ireland (Jarvis, 2000), E China (Li and Wu, 1989) [Raymond and Metz, 1995; Edwards et al., 2000; Königshof et al., 2004] |
| Refined (bio)geochemical patterns | Conodont- and organic-rich successions with a minimal thermal and diagenetic overprint | S Holy Cross Mts, Poland (Belka, 1990; Joachimski et al., 2001); Rhenish Massif (Joachimski et al., 2002; Radke and Wilkes, 2004) [Joachimski et al., 2002, 2004] |
| Near-equatorial vs. high-latitude domains | Sections located in northern (1) and southern (2) high-latitudes | (1) Kolyma (Gagiev, 1985, 1997); Kuznetsk Basin (Yolkin et al., 1997, 2000)<br>(2) Brazil (Melo and Loboziak, 2003); South Africa (Cotter, 2000) [Copper, 1977] |
| Tectonic and volcanic activity | (1) Deep shelf-slope or oceanic facies that preferably records sea-floor volcanism and/or terrigenous input due to tectonic movements<br>(2) Volcanic suites, preferably with sedimentary intercalations | (1) Western Siberia (e.g. Rudnyi Altai; Umeda et al., 2001; see also Yolkin et al., 2003); Bavaria and Thuringia (Kiessling and Tragelehn, 1994; Gehmlich et al., 2000); Nevada and Canadian Cordillera (Savoy et al., 2000; Stevenson et al., 2000; Morrow and Sandberg, 2003)<br>(2) NW China (Hao et al., 2003; Fig. 8; Xiao et al., 2004); Iran (Mahmudy Gharaie et al., 2004); Siberian Viluy rift zone (Courtillot and Renne, 2003); New South Wales (Wright et al., 1990); SW Spain (Almaden; Pardo and Garcia-Alcalde, 1984) [Racki, 1998; Tribovillard et al., 2004] |
| Cyclo-stratigraphical perspective | Well-dated biostratigraphically complete successions with a minimal diagenetic overprint | [Crick et al., 2002; Chen and Tucker, 2003; Westphal et al., 2004] |

1996; Racki, 1999b; Hatsukawa et al., 2003), well exemplified by controversial timing of the Woodleigh structure (Renne et al., 2002). From other notable data, F–F whole-rock samples and conodonts reveal no obvious contribution of primitive Os supplied by extraterrestrial material, as would be expected when considering the impact theory (Brauns, 2001).

Seven topical areas of our Late Devonian ignorance, outlined above, are an agenda for future integrative, multi proxy research (Table 1), and some of these are addressed in the new IGCP Project 499 "Devonian land-sea interaction: evolution of ecosystems and climate" (Königshof et al., 2004). It is clear that a long series of environmental changes commenced in the late Middle Devonian and peaked at the F–F boundary; therefore, multicausal Earth-bound hypotheses are strongly supported (Racki and House, 2002). Repeated, possibly autocyclic co-occurrence of sea-level oscillations, eutrophic/anoxic conditions and rapid climatic shifts have been usually regarded as the most likely explanations (Joachimski and Buggisch, 2002; Bond et al., 2004; Averbuch et al., 2005). The key enigma is a threshold condition or trigger, different from all of the other background events that finally produce the biotic crisis. However, a promoting role for tectonic and magmatic phenomena in re-activated plate-tectonic settings has been explored (e.g. Averbuch et al., 2005), exemplified recently also by scenario of mantle-plume-induced, lithospheric gas explosions (Phipps Morgan et al., 2004). To definitively choose among the several model predictions, at least two critical data sets are especially required: accurate temperature curve and reliable intercalibrated chronostratigraphical scheme. In light of the most recent Late Devonian climate estimate of Joachimski et al. (2004), a variety of lag-time greenhouse-collapse model during the Kellwasser Crisis (*sensu* McGhee, this volume) should be combined with direct-effect greenhouse model in early Famennian; a tectono-volcanic stimuli (*cf.* Buggisch, 1991; Racki, 1999b; Averbuch et al., 2005; Chen et al., 2005) is at least so reasonable as multiple impacts of McGhee (this volume).

In light of so many questions raised in this chapter, a distillation of most well-documented geological facts is urgently needed as a first step in testing the diversity of proposed extinction models regardless of the triggering process, thoroughly discussed by McGhee (this volume). The complex F–F extinction scenario based on Earth-bound feedbacks demands an intricate explanation, but similar to the P-Tr boundary (and any geologic issue), with growing knowledge, the more and more questions have become convoluted and harder to solve simply (Benton, 2003). Although there is currently much confusion and many contradictory datasets and scenarios covering the Late Devonian timespan, a systematic and well-dated study of climatic proxies, building on the palaeotemperature pattern of Joachimski et al. (2004), seems to be the most urgent and fruitful next step on the path to understanding global events in the key interval of biosphere history.

## Acknowledgments

Discussions with many Devonian workers, in particular with George McGhee, Paul Copper, Tom Becker and Paul Wignall, are gratefully acknowledged. Formal reviews by two reviewers, John Bratton and Peter Harries, are much appreciated; their comments have greatly improved this review chapter. Ebi Schindler and Paweł Filipiak kindly supplied the field photos.

## References

Abbott, D.H., Isley, A.E., 2002. The intensity, occurrence, and duration of superplume events and eras over geological time. J. Geodyn. 34, 265–307.

Aboussalam, Z.S., Becker, R.T., 2001. Prospects for an upper Givetian substage. Mitt. Mus. Nat. Kd. Berl., Geowiss. Reihe 4, 83–99.

Afanaseeva, G.A., Morozova, I.P., 1995. Brachiopods and bryozoans in a crisis situation over the Frasnian-Famennian age boundary in Late Devonian [in Russian]. In: Rozanov, A.Y., Semikhatov, M.A. (Eds), Ekosistemnye Perestroiki i Evolutchya Biosfery, Vol. 2. PIN RAN, Moscow, pp. 45–48.

Alekseev, A.S., Kononova, L.I., Nikishin, A.M., 1996. The Devonian and Carboniferous of the Moscow Syneclise (Russian Platform): stratigraphy and sea-level changes. Tectonophysics 268, 149–168.

Algeo, T.J., 2003. The late Devonian marine P cycle and the origin of widespread black shales; enhanced rates of upwelling or chemical weathering? Geol. Soc. Am., Abstr. Pr. 35(6), 207–208.

Algeo, T.J., Berner, R.A., Maynard, J.B., Scheckler, S.E., 1995. Late Devonian oceanic anoxic events and biotic crises: 'rooted' in the evolution of vascular plants? GSA Today 5, 63–66.

Algeo, T.J., Scheckler, S.E., 2003. The role of soils as an interface between terrestrial and marine ecosystems; the Middle to Late Devonian. Geol. Soc. Am., Abstr. Pr. 35(6), 458–459.

Alvarez, W., 2003. Comparing the evidence relevant to impact and flood basalt at times of major mass extinctions. Astrobiology 3, 153–161.

Amler, M.R.W., 1999. Extinction and radiation of bivalves in the Late Devonian. In: Biology & Evolution of the Bivalvia, Meeting Volume and Abstracts, Malacological Society of London, http://www.sunderland.ac.uk/~es0mda/bivabst.htm.

Amler, M.R.W., Heidelberger, D., 2003. Late Famennian Gastropoda from south-west England. Palaeontology 46, 1151–1211.

Arzamastsev, A.A., Bea, F., Glaznev, V.N., Arzamastseva, L.V., Montero, P., 2000. Kola alkaline province in the Palaeozoic: evaluation of primary mantle magma composition and magma generation conditions. Russ. J. Earth Sci. 3, 1–21.

Averbuch, O., Tribovillard, N., Devleeschouwer, X., Riquier, L., Mistiaen, B., van Vliet-Lanoe, B., 2005. Mountain building-enhanced continental weathering and organic carbon burial as major causes for climatic cooling at the Frasnian–Famennian boundary (*c.* 376 Ma)? Terra Nova 17, 25–34.

Baliński, A., 2002. Frasnian-Famennian brachiopod extinction and recovery in southern Poland. Acta Palaeontol. Pol. 289–305.

Baliński, A., Olempska, E., Racki, G. (Eds), 2002. Biotic responses to the Late Devonian global events. Acta Palaeontol. Pol. 47, 1–404.

Bambach, R.K., 1999. Energetics in the global marine fauna: a connection between terrestrial diversification and change in the marine biosphere. Geobios 32, 131–144.

Bambach, R.K., Knoll, A.H., Wang, S.C., 2004. Origination, extinction, and mass depletions of marine diversity. Paleobiology 30, 522–542.

Beard, A.D., Downes, H., Vetrin, V., Kempton, P.D., Maluski, H., 1996. Petrogenesis of Devonian lamprophyre and carbonatite minor intrusions, Kandalaksha Gulf (Kola Peninsula, Russia). Lithos 39, 93–119.

Beauchamp, B., Baud, A., 2002. Growth and demise of Permian biogenic chert along northwest Pangea: evidence for end-Permian collapse of thermohaline circulation. Palaeogeogr. Palaeoclimatol. Palaeoecol. 184, 37–63.

Becker, R.T., House, M.R., 1997. Sea-level changes in the Upper Devonian of the Canning Basin, Western Australia. Cour. Forsch.–Inst. Senckenberg 199, 129–146.

Belka, Z., 1990. Thermal maturation and burial history from Conodont colour alternation data, Holy Cross Mountains, Poland. Cour. Forsch.–Inst. Senckenberg 118, 241–251.

Benton, M.J., 2003. When Life Nearly Died: The Greatest Mass Extinction of All Time. Thames & Hudson, London.

Berner, R.A., 2003. The rise of trees and their effects on Paleozoic atmospheric $CO_2$ and $O_2$. C.R. Geosci. 335, 1173–1177.

Blieck, A., Turner, S., Young, G.C., Luksevics, E., Mark-Kurik, E., Talimaa, V.N., Valiukevicius, J.J., 2000. Devonian vertebrate biochronology and global marine/non-marine correlation. Stuttgart, E. Schweizerbart'sche Verlagsbuchhandlung (Nägele u. Obermiller). In: Bultynck, P. (Ed.), Subcommission on Devonian Stratigraphy. Fossil Groups Important for Boundary Definition, Vol. 220. Cour. Forsch.–Inst. Senckenberg, pp. 161–193.

Bogatikov, O.A., Kononova, V.A., Pervov, V.A., Zhuravlev, D.Z., 2001. Sources, geodynamic setting of formation, and diamond-bearing potential of kimberlites from the northern margin of the Russian Plate: a Sr-Nd Isotopic and ICP-MS geochemical study. Petrology 9, 191–203.
Bond, D., Wignall, P.B., Racki, G., 2004. Extent and duration of marine anoxia during the Frasnian-Famennian (Late Devonian) mass extinction in Poland, Germany, Austria and France. Geol. Mag. 141, 173–193.
Bond, D., Zatoń, M., 2003. Gamma-ray spectrometry across the Upper Devonian basin succession at Kowala in the Holy Cross Mountains (Poland). Acta Geol. Pol. 53, 93–99.
Boulter, C.A., Hopkinson, L.J., Ineson, M.G., Brockwell, J.S., 2004. Provenance and geochemistry of sedimentary components in the Volcano-Sedimentary Complex, Iberian Pyrite Belt: discrimination between the sill–sediment-complex and volcanic-pile models. J. Geol. Soc. Lond. 161, 103–115.
Brand, U., Legrand-Blain, M., Streel, M., 2004. Biochemostratigraphy of the Devonian–Carboniferous boundary Global Stratotype Section and Point, Griotte Formation, La Serre, Montagne Noire, France. Palaeogeogr. Palaeoclimatol. Palaeoecol. 205, 337–357.
Bratton, J.F., 1999. Clathrate eustasy: methane hydrate melting as a mechanism for geologically rapid sea-level fall. Geology 27, 915–918.
Brauns, M., 2001. Osmium isotopes and the Upper Devonian "Kellwasser" event. American Geophysical Union, Fall Meeting 2001, abst. IP11A-0663.
Brice, D., Carls, P., Cocks, L.R.M., Copper, P., Garcia-Alcalde, J.L., Godefroid, J., Racheboeuf, P.R., 2000. Brachiopoda. In: Bultynck, P. (Ed.), Subcommission on Devonian Stratigraphy. Fossil Groups Important for Boundary Definition Vol. 220. Cour. Forsch.–Inst. Senckenberg, pp. 65–86.
Buggisch, W., 1991. The global Frasnian-Famennian "Kellwasser Event". Geol. Rundsch. 80, 49–72.
Buggisch, W., 2001. Whole-rock carbon isotope analysis ($\delta^{13}C$) of Devonian to Permian sediments. Terra Nostra 4, 12–15.
Bultynck, P. (Ed.), 2000. Subcommission on Devonian Stratigraphy. Fossil Groups Important for Boundary Definition,Vol. 220. Cour. Forsch.–Inst., Senckenberg, pp. 1–204.
Bultynck, P., Coen, A.M., Godefroid, J., 2000. Summary of the state of correlation in the Devonian of the Ardennes (Belgium-NE France) resulting from the decisions of the SDS. Cour. Forsch.–Inst. Senckenberg 225, 91–114.
Bultynck, P., Helsen, S., Hayduckiewich, J., 1998. Conodont succession and biofacies in upper Frasnian formations (Devonian) from the southern and central parts of the Dinant Synclinorium (Belgium): timing of facies shifting and correlation with late Frasnian events. Bull. Inst. Roy. Sci. Nat. Belg., Sci. Terre 68, 25–75.
Bykadorov, V.A., Bush, V.A., Fedorenko, O.A., Filippova, I.B., Miletenko, N.V., Puchkov, V.N., Smirnov, A.V., Uzhkenov, B.S., Volozh, Y.A., 2003. Ordovician–Permian palaeogeography of Central Eurasia: development of Palaeozoic petroleum-bearing basins. J. Petrol. Geol. 26, 325–350.
Çapkinoğlu, Ş., Gedik, I., 2000. Late Devonian conodont fauna of the Gümüşali Formation, the Eastern Taurides, Turkey. Turk. J. Earth Sci. 9, 69–89.
Caplan, M.L., Bustin, R.M., 1999. Devonian-Carboniferous Hangenberg mass extinction event, widespread organic-rich mudrock and anoxia: causes and consequences. Palaeogeogr. Palaeoclimatol. Palaeoecol. 148, 187–207.
Casier, J.G., 2003. Ostracods from the late Frasnian of the Neuville railway section (Dinant Synclinorium, Belgium): relation to the Kellwasser Event. Bull. Soc. Géol. France 174, 49–57.
Casier, J.G., Lethiers, F., 2001. Ostracods prove that the Frasnian/Famennian boundary mass extinction was a major and abrupt crisis. In: Buffetaut, E., Koeberl, C. (Eds), Geological and Biological Effects of Impact Events. Springer, Berlin, pp. 1–10.
Cathles, L.M., Hallam, A., 1991. Stress-induced changes in plate density, Vail sequences, epeirogeny, and short-lived global sea level fluctuations. Tectonics 10, 659–671.
Chen, D., Tucker, M.E., 2003. The Frasnian-Famennian mass extinction: insights from high-resolution sequence stratigraphy and cyclostratigraphy in South China. Palaeogeogr. Palaeoclimatol. Palaeoecol. 193, 87–111.
Chen, D., Tucker, M.E., 2004. Palaeokarst and its implication for the extinction event at the Frasnian-Famennian boundary (Guilin, South China). J. Geol. Soc. Lond. 161, 895–898.
Chen, D.Z., Qing, H.R., Li, R.W., 2005. The Late Devonian Frasnian–Famennian (F/F) biotic crisis: insights from $\delta^{13}C_{carb}$, $\delta^{13}C_{org}$ and $^{87}Sr$ / $^{86}Sr$ isotopic systematics. Earth Planet. Sci. Lett. 235, 151–166.
Chen, D.Z., Tucker, M.E., Jiang, M.S., Zhu, J.Q., 2001. Long-distance correlation between tectonic-controlled, isolated carbonate platforms by cyclostratigraphy and sequence stratigraphy in the Devonian of South China. Sedimentology 48, 57–78.

Chen, D., Tucker, M.E., Shen, Y., Yans, J., Preat, A., 2002. Carbon isotope excursions and sea-level change: implications for the Frasnian-Famennian biotic crisis. J. Geol. Soc. Lond. 159, 623–526.
Chow, N., George, A.D., Trinajstic, K.M., 2004. Tectonic control on development of a Frasnian–Famennian (Late Devonian) palaeokarst surface, Canning Basin reef complexes, northwestern Australia. Austr. J. Earth Sci. 51, 911–917.
Compston, W., 2000. Interpretation of SHRIMP and isotope dilution zircon ages for the Palaeozoic time-scale: II. Silurian to Devonian. Mineral. Mag. 64, 1127–1146.
Compston, W., 2004. SIMS U–Pb zircon ages for the Upper Devonian Snobs Creek and Cerberean Volcanics from Victoria, with age uncertainty based on $UO_2$/UO v. UO/U precision. J. Geol. Soc. Lond. 161, 223–228.
Cook, H.E., Zhemchuzhnikov, V.G., Zempolich, W.G., Zhaimina, V.Y., Buvtyshkin, V.M., Kotova, E.A., Golub, L.Y., Zorin, A.Y., Lehmann, P.J., Alexeiev, D.V., Giovannelli, A., Viaggi, M., Fretwell, N., LaPointe, P.A., Corboy, J.J., 2002. Devonian and Carboniferous carbonate platform facies in the Bolshoi Karatau, southern Kazakhstan; outcrop analogs for coeval carbonate oil and gas fields in the North Caspian Basin, western Kazakhstan. In: Zempolich, W.G., Cook, H.E. (Eds), Paleozoic Carbonates of the Commonwealth of Independent States (CIS): Subsurface Reservoirs and Outcrop Analogs, Vol. 74. Soc. Sediment. Geol., Spec. Publ., pp. 81–122.
Copper, P., 1977. Paleolatitudes in the Devonian of Brazil and the Frasnian-Famennian mass extinction. Palaeogeogr. Palaeoclimatol. Palaeoecol. 21, 165–207.
Copper, P., 2002. Silurian and Devonian reefs; 80 million years of global greenhouse between two ice ages. In: Kiessling W., Flügel, E., Golonka, J. (Eds), Phanerozoic Reef Patterns, Vol. 72. Soc. Econ. Paleontol. Mineral. Spec. Publ., pp. 181–238.
Cotter, E., 2000. Depositional setting and cyclic development of the lower part of the Witteberg group (Mid- to Upper Devonian), Cape Supergroup, Western Cape, South Africa. S. Afr. J. Geol. 103, 1–14.
Courtillot, V.E., Renne, P.R., 2003. On the ages of flood basalt events. C.R. Geosci. 335, 113–140.
Cox, J.E., Railsback, L.B., Gordon, E.A., 2001. Evidence from Catskill pedogenic carbonates for a rapid Late Devonian decrease in atmospheric carbon dioxide concentrations. Northeastern Geol. Environ. Sci. 23, 91–102.
Crick, R.E., Ellwood, B.B., Feist, R., El Hassani, A., Schindler, E., Dreesen, R., Over, D.J., Girard, C., 2002. Magnetostratigraphy susceptibility of the Frasnian/Famennian boundary. Palaeogeogr. Palaeoclimatol. Palaeoecol. 181, 67–90.
Denison, R.E., Koepnick, R.B., Burke, W.H., Hetherington, E.A., Fletcher, A., 1997. Construction of the Silurian and Devonian seawater $^{87}Sr/^{86}Sr$ curve. Chem. Geol. 140, 109–121.
Devleeschouwer, X., Herbosch, A., Préat, A., 2002. Microfacies, sequence stratigraphy and clay mineralogy of a condensed deep-water section around the Frasnian/Famennian boundary (Steinbruch Schmidt, Germany). Palaeogeogr. Palaeoclimatol. Palaeoecol. 181, 171–193.
D'Hondt, S., Donaghay, P., Zachos, J.C., Luttenberg, D., Lindinger, M., 1998. Organic carbon fluxes and ecological recovery from the Cretaceous-Tertiary mass extinction. Science 282, 276–279.
Dobretsov, N.L., 2003. Evolution of structures of the Urals, Kazakhstan, Tien Shan, and Altai-Sayan region within the Ural-Mongolian fold belt (Paleoasian Ocean). Russian Geol. Geophys. 44, 3–26.
Dopieralska, J., Belka, Z., Haack, U., 2003. Nd isotopic evidence for sea-level fluctuations and decoupling between the ocean and epeiric seas during the late Devonian. Geol. Soc. Am., Abstr. Pr. 35(6), 208.
Dubotalov, V.N., Krasnov, V.I., 2000. Middle Devonian and Frasnian seas of Siberia. Stratigr. Geol. Correl. 8, 557–578.
Dusel-Bacon, C., Wooden, J.L., Hopkins, M.J., 2004. U-Pb zircon and geochemical evidence for bimodal mid-Paleozoic magmatism and syngenetic base-metal mineralization in the Yukon-Tanana terrane, Alaska. Geol. Soc. Amer. Bull. 116, 989–1015.
Dzik, J., 2002. Emergence and collapse of the Frasnian conodont and ammonoid communities in the Holy Cross Mountains. Acta Palaeontol. Pol. 47, 565–650.
Eder, W., Franke, W., 1982. Death of Devonian reefs. Neues Jahrb. Geol. Paläont. Abh. 163, 241–243.
Edwards, D., Fairon-Demaret, M., Berry, C.M., 2000. Plant megafossils in Devonian stratigraphy: a progress report. In: Bultynck, P. (Ed.), Subcommission on Devonian Stratigraphy. Fossil Groups Important for Boundary Definition, Vol. 220. Cour. Forsch.–Inst. Senckenberg, pp. 25–37.
Erba, E., 2004. Calcareous nannofossils and Mesozoic oceanic anoxic events. Marine Micropaleont. 52, 85–106.
Ernst, R.E., Buchan, K.L., 2001. Large mafic magmatic events through time and links to mantle-plume heads. In: Ernst, R.E., Buchan, K.L. (Eds), Mantle Plumes: Their Identification through Time, Vol. 352. Geol. Soc. Am., Spec. Pap., pp. 483–575.

Erwin, D.H., 2001. Lessons from the past: biotic recoveries from mass extinctions. Proc. Natl. Acad. Sci. USA. 98, 5399–5403.

Filer, J.K., 2002. Late Frasnian sedimentation cycles in the Appalachian basin – possible evidence for high frequency eustatic sea-level changes. Sediment. Geol. 154, 31–52.

Filippova, I.B., Bush, V.A., Didenko, A.N., 2001. Middle Paleozoic subduction belts: the leading factor in the formation of the Central Asian fold-and-thrust belt. Russ. J. Earth Sci. 3, 405–426.

Flügel, E., Kiessling, W., 2002. Patterns of Phanerozoic reef crises. In: Kiessling, W., Flügel, E., Golonka, J. (Eds), Phanerozoic Reef Patterns, Vol. 72. Soc. Econ. Paleontol. Mineral. Spec. Publ., pp. 691–733.

Fröhlich, S., 2004. Phosphatic black pebbles and nodules on a Devonian carbonate shelf (Anti-Atlas, Morocco). J. Afr. Earth Sci. 38, 243–254.

Gagiev, M.H., 1985. Subdivision and correlation of Frasnian-Famennian boundary deposits (in terms of conodonts), USSR North-East (Fore-Kolyma Uplift), Vol. 75. Cour. Forsch.–Inst. Senckenberg, pp. 53–64.

Gagiev, M.H., 1997. Sedimentary evolution and sea-level fluctuations in the Devonian of North-east Asia. Cour. Forsch.–Inst. Senckenberg 199, 75–82.

Gehmlich, M., Linnemann, U., Tichomirowa, M., Gaitzsch, B., Kroner, U., Bombach, K., 2000. Geochronologie oberdevonischer bis unterkarbonischer Magmatite der Thüringischen und Bayerischen Faziesreihe sowie variszischer Deckenkomplexe und der Frühmolasse von Borna-Hainichen (Saxothuringisches Terrane). Z. Dt. Geol. Ges. 151, 337–363.

Geldsetzer, H.H.J., Goodfellow, W.D., McLaren, D.J., 1993. The Frasnian-Famennian extinction event in a stable cratonic shelf setting: Trout River, Northwest Territories, Canada. Palaeogeogr. Palaeoclimatol. Palaeoecol. 104, 81–95.

Ginter, M., 2002. Chondrichthyan fauna of the Frasnian-Famennian boundary beds in Poland. Acta Palaeontol. Pol. 47, 329–338.

Głuchowski, E., 2002. Famennian crinoids from the Holy Cross Mountains. Acta Palaeontol. Pol. 47, 319–328.

Godderis, Y., Joachimski, M.M., 2004. Global change in the late Devonian: modelling the Frasnian-Famennian short-term carbon isotope excursions. Palaeogeogr. Palaeoclimatol. Palaeoecol. 202, 309–329.

Golonka, J., 2000. Cambrian-Neogene Plate Tectonic Maps. Wyd. Uniw. Jagiellońskiego, Cracow.

Gong, Y., Li, B., Si, Y., Wu, Y., 2002. Late Devonian red tide and mass extinction. Chin. Sci. Bull. 47, 1138–1144.

Gouwy, S., Bultynck, P., 2000. Graphic correlation of Frasnian sections (Upper Devonian) in the Ardennes, Belgium. Bull. Inst. Roy. Sci. Nat. Belg., Sci. Terre 70, 25–52.

Gradstein, F., Ogg, J., Smith, A., 2004. A Geological Time Scale 2004. Cambridge Univ. Press., Cambridge (see International Stratigraphic Chart, International Commission of Stratigraphy; http://www.stratigraphy.org/).

Grice, K., Cao, C.Q., Love, G.D., Böttcher, M.E., Twitchett, R.J., Grosjean, E., Summons, R.E., Turgeon, S.C., Dunning, W., Jin, Y.G., 2005. Photic zone euxinia during the Permian-Triassic superanoxic event. Science 307, 706–708.

Groos-Uffenorde, H., Schindler, E., 1990. The effect of global events on entomozoacean Ostracoda. In: Whatley, R.C., Maybury, C. (Eds), Ostracoda and Global Events. International Symposium on Ostracoda, Vol. 10. Chapman and Hall, London, pp. 101–112.

Guo, F., Fan, W.M., Wang, Y.J., Li, C.W., 2004. Upper Paleozoic basalts in the southern Yangtze block: geochemical and Sr-Nd isotopic evidence for asthenosphere-lithosphere interaction and opening of the Paleo-Tethyan Ocean. Int. Geol. Rev. 46, 332–346.

Hallam, A., Wignall, P.B., 1997. Mass Extinctions and Their Aftermath. Oxford Univ. Press, Oxford.

Hallam, A., Wignall, P.B., 1999. Mass extinctions and sea-level changes. Earth-Sci. Rev. 48, 217–250.

Han, C., Zhao, G., 2003. Major types and characteristics of Late Paleozoic ore deposits, East Tianshan, Northwest China. Int. Geol. Rev. 45, 798–813.

Han, G., Preat, A., Chamley, H., Deconinck, J.F., Mansy, J.L., 2000. Palaeozoic clay mineral sedimentation and diagenesis in the Dinant and Avesnes Basins (Belgium, France): relationships with Variscan tectonism. Sediment. Geol. 136, 217–238.

Hao, W.C., Jianh, D.Y., Yao, J.X., Bai, S.L., Wang, X.P., 2003. Frasnian-Famennian boundary events in Tarim Basin, Xinjiang, Northwest China. Sci. China D 46, 865–871.

Hatsukawa, Y., Mahmudy Gharaie, M.H., Matsumoto, R., Toh, Y., Oshima, M., Kimura, A., Noguchi, T., Goto, T., Kakuwa, Y., 2003. Ir anomalies in marine sediments: case study for the Late Devonian mass extinction event. Geochim. Cosmochim. Acta 67 (18 Suppl.), A138–A138.

Heaman, L.M., Kjarsgaard, B.A., Creaser, R.A., 2003. The timing of kimberlite magmatism in North America: implications for global kimberlite genesis and diamond exploration. Lithos 71, 153–184.

Higgs, K.T., Finucane, D., Tunbridge, I.P., 2002. Late Devonian and early Carboniferous microfloras from the Hakkari Province of southeastern Turkey. Rev. Palaeobot. Palynol. 118, 141–156.
Hladil, J., 2002. Geophysical records of dispersed weathering products on the Frasnian carbonate platform and early Famennian ramps in Moravia, Czech Republic: proxies for eustasy and palaeoclimate. Palaeogeogr. Palaeoclimatol. Palaeoecol. 181, 213–250.
Holmden, C., Creaser, R.A., Muehlenbachs, K., Leslie, S.A., Bergström, S.M., 1998. Isotopic evidence for geochemical decoupling between ancient epeiric seas and bordering oceans: implications for secular curves. Geology 26, 567–570.
Holser, T., Margaritz, M., Ripperdan, R.L., 1996. Global isotopic events. In: Walliser, O.H. (Ed.), Global Events and Event Stratigraphy in the Phanerozoic. Berlin, Springer, pp. 63–88.
House, M.R., 1985. Correlation of mid-Palaeozoic ammonoid evolutionary events with global sedimentary perturbations. Nature 313, 17–22.
House, M.R., 1996. Juvenile goniatite survival strategies following Devonian extinction events. In: Hart, M.B. (Ed.), Biotic Recovery from Mass Extinction Events, Vol. 102. Geol. Soc. Lond., Spec. Publ., pp. 163–185.
House, M.R., 2002. Strength, timing, setting and cause of mid-Palaeozoic extinctions. Palaeogeogr. Palaeoclimatol. Palaeoecol. 181, 5–25.
House, M.R., Becker, R.T., Feist, R., Girard, C., Klapper, G., 2000. The Frasnian/Famennian boundary GSSP at Coumiac, southern France. Cour. Forsch.–Inst. Senckenberg 225, 59–75.
House, M.R., Menner, V.V., Becker, R.T., Klapper, G., Ovnatanova, N.S., Kuzmin, V., 2000. Reef episodes, anoxia and sea-level changes in the Frasnian of the southern Timan (NE Russian Platform). In: Insalaco, E., Skelton, P.W., Palmer, T.J. (Eds), Carbonate Platform Systems: Components and Interactions, Vol. 178. Geol, Soc. Lond., Spec. Publ., pp. 147–176.
Hudson, J.D., 2001. Advances in the study of 'black shales': source rocks for bricks and hydrocarbons. Proc. Yorskshire Geol. Soc. 53, 231–236.
Immenhauser, A., Hillgärtner, H., Van Bentum, E., 2005. Microbial-foraminiferal episodes in the Early Aptian of the southern Tethyan margin: ecological significance and possible relation to oceanic anoxic event 1a. Sedimentology 52, 77–99.
Isaacson, P., Hladil, J., 2003. Devonian glaciation in Gondwana: establishing Carboniferous eustasy. Geol. Soc. Am., Abstr. Pr. 35(6), 209.
Isaacson, P.E., Hladil, J., Shen, J.W., Kalvoda, J., Grader, G., 1999. Late Devonian (Famennian) glaciation in South America and marine offlap on other continents. In: Feist, R., Talent, J.A., Daurer, A. (Eds), IGCP 421 North Gondwana: Mid-Palaeozoic Terranes, Stratigraphy and Biota, Vol. 54. Abh. Geol. B.-A., pp. 239–257.
Jablonski, D., 2002. Survival without recovery after mass extinctions. Proc. Natl. Acad. Sci. USA 99, 8139–8144.
Jahn, B.M., Windley, B., Natal'in, B., Dobretsov, N. (Eds), 2004. Phanerozoic continental growth in Central Asia. J. Asian Earth Sci. 23(5), 599–815.
Jarvis, D.E., 2000. Palaeoenvironment of the plant bearing horizons of the Devonian-Carboniferous Kiltorcan Formation, Kiltorcan Hill, Co. Kilkenny, Ireland. In: Friend, P.F., Williams, B.P.J. (Eds), New Perspectives on the Old Red Sandstone, Vol. 180. Geol. Soc. Lond., Spec. Publ., pp. 333–341.
Joachimski, M., Buggisch, W., 1993. Anoxic events in the late Frasnian – causes of the Frasnian-Famennian faunal crisis? Geology 21, 675–678.
Joachimski, M.M., Buggisch, W., 2002. Conodont apatite $\delta^{18}O$ signatures indicate climatic cooling as a trigger of the Late Devonian mass extinction. Geology 30, 711–714.
Joachimski, M.M., Buggisch, W., 2003. Conodont apatite $\delta^{18}O$ signatures indicate climatic cooling as a trigger of the Late Devonian mass extinction – Reply. Geology 31, 384–384.
Joachimski, M.M., Ostertag-Henning, C., Pancost, R.D., Strauss, H., Freeman, K.H., Littke, R., Damsté, J.S., Racki, G., 2001. Water column anoxia, enhanced productivity and concomitant changes in $\delta^{13}C$ and $\delta^{34}S$ across the Frasnian-Famennian boundary (Kowala – Holy Cross Mountains/Poland). Chem. Geol. 175, 109–131.
Joachimski, M.M., Pancost, R.D., Freeman, K.H., Ostertag-Henning, C., Buggisch, W., 2002. Carbon isotope geochemistry of the Frasnian-Famennian transition. Palaeogeogr. Palaeoclimatol. Palaeoecol. 181, 91–109.
Joachimski, M.M., Van Geldern, R., Breisig, S., Buggisch, W., Day, J., 2004. Oxygen isotope evolution of biogenic calcite and apatite during the Middle and Late Devonian. Int. J. Earth Sci., 93, 542–553
Johnson, J.G., Klapper, G., Sandberg, C.A., 1985. Devonian eustatic fluctuations in Euramerica. Geol. Soc. Amer. Bull. 96, 567–587.
Jones, C.E., Jenkyns, H.C., 2001. Seawater strontium isotopes, oceanic anoxic events, and seafloor hydrothermal activity in the Jurassic and Cretaceous. Am. J. Sci. 301, 112–149.

Kalvoda, J., 2002. Late Devonian–Early Carboniferous foraminiferal fauna: zonations, evolutionary events, paleobiogeography and tectonic implications. Folia Facultatis Scientiarium Naturalium Universitatis Masarykianae Brunensis, Geologia 30, 1–213.
Kasig, W, Wilder, H., 1983. The sedimentary development of the Western Rheinisches Schiefergebirge and the Ardennes (Germany/Belgium). In: Martin, H., Eder, F.W. (Eds), Intercontinental Fold Belts. Springer, Berlin, pp. 185–209.
Kaufmann, B., Trapp, E., Mezger, K., 2004. The numerical age of the late Frasnian (Late Devonian) Kellwasser horizons – a new U-Pb zircon date from Steinbruch Schmidt (Kellerwald, Germany). J. Geol. 112, 495–501, http://www.uni-tuebingen.de/geo/gpi/mitarbeiter/kaufmann/devoniantimescale.pdf.
Kiessling, W., 2001. Paleoclimatic significance of Phanerozoic reefs. Geology 29, 751–754.
Kiessling, W., 2005. Long-term relationships between ecological stability and biodiversity in Phanerozoic reefs. Nature 433, 410–412.
Kiessling, W., Tragelehn, H., 1994. Devonian radiolarian faunas of conodont-dated localities in the Frankenwald (Northern Bavaria, Germany). Abh. Geol. B.-A. 50, 219–255.
Kiessling, W., Flügel, E., Golonka, J., 2003. Patterns of Phanerozoic carbonate platform sedimentation. Lethaia 36, 195–225.
Kiselev, A.I., Yarmolyuk, V.V., Egorov, K.N., Maslovskaya, M.N., Chernyshov, R.A., Kovalenko, V.I., 2004. Composition and sources of Middle Paleozoic basalts of the Vilyui Paleorift. Doklady Earth Sci. 397, 641–646.
Klapper, G., 1997. Graphic correlation of Frasnian (Upper Devonian) sequences in Montagne Noire, France, and Western Canada. In: Klapper, G., Murphy, M.A., Talent, J.A. (Eds), Paleozoic Sequence Stratigraphy, Biostratigraphy, and Biogeography: Studies in Honor of J. Granville ("Jess") Johnson, Vol. 321. Geol. Soc. Am. Spec. Pap., pp. 113–129.
Klapper, G., Feist, R., Becker, R.T., House, M.R., 1993. Definition of the Frasnian/Famennian stage boundary. Episodes 16, 433–441.
Klapper, G., Uyeno, T.T., Armstrong, D.K., Telford, P.G., 2004. Conodonts of the Williams Island and Long Rapids Formations (Upper Devonian, Frasnian-Famennian) of the Onakawana B Drillhole, Moose River Basin, Northern Ontario, with a revision of Lower Famennian species. J. Paleontol. 78, 371–387.
Königshof, P., Lazauskiene, J., Schindler, E., Wilde, V., Yalçin, M.N., 2004. New IGCP Project 499: "Devonian land-sea interaction: evolution of ecosystems and climate" (DEVEC). Facies 50, 347–348.
Kramm, U., Kogarko, L.N., Kononova, V.A., Vartiainen, H., 1993. The Kola Alkaline Province of the CIS and Finland: precise Rb-Sr ages define 380–360 Ma age range for all magmatism. Lithos 30, 33–44.
Kravchinsky, V.A., Konstantinov, K.M., Courtillot, V., Savrasov, J.I., Valet, J.P., Cherniy, S.D., Mishenin, S.G., Parasotka, B.S., 2002. Palaeomagnetism of East Siberian traps and kimberlites: two new poles and palaeogeographic reconstructions at about 360 and 250 Ma. Geophys. J. Int. 148, 1–33.
Levman, B.G., Von Bitter, P.H., 2002. The Frasnian-Famennian (mid-Late Devonian) boundary in the type section of the Long Rapids Formation, James Bay Lowlands, northern Ontario, Canada. Can. J. Earth Sci. 39, 1795–1818.
Li, X.X., Wu, X.Y., 1989. The succession of late Palaeozoic and Triassic plant assemblages of eastern China. J. Southeast Asian Earth-Sci. 3, 187–200.
Lieberman, B.S., 2002. Phylogenetic biogeography with and without the fossil record: gauging the effects of extinction and paleontological incompleteness. Palaeogeogr. Palaeoclimatol. Palaeoecol. 178, 39–52.
Ma, X.P., Bai, S.L., 2002. Biological, depositional, microspherule, and geochemical records of the Frasnian/Famennian boundary beds, South China. Palaeogeogr. Palaeoclimatol. Palaeoecol., 181, 325–346.
MacLeod, N., Rawson, P.F., Forey, P.L., Banner, F.T., Boudagher-Fadel, M.K., Bown, P.R., Burnett, J.A., Chambers, P., Culver, S., Evans, S.E., Jeffery, C., Kaminski, M.A., Lord, A.R., Milner, A.C., Milner, A.R., Morris, N., Owen, E., Rosen, B.R., Smith, A.B., Taylor, P.D., Urquhart, E., Young, R., 1997. The Cretaceous-Tertiary biotic transition. J. Geol. Soc. Lond. 154, 265–292.
Mahmudy Gharaie, M.H., Matsumoto, R., Kakuwa, Y., Milroy, P.G., 2004. Late Devonian facies variety in Iran: volcanism as a possible trigger of the environmental perturbation near the Frasnian-Famennian boundary. Geol. Quart. 48, 323–332
Mahotkin, I.L., Gibson, S.A., Thompson, R.N., Zhuravlev, D.Z., Zherdev, P.U., 2000. Late Devonian diamondiferous kimberlite and alkaline picrite (proto-kimberlite?) magmatism in the Arkhangelsk region, NW Russia. J. Petrol. 41, 201–227.
Marshall, J.E.A., Brown, J., Astin, T., 2003. The Taghanic Event: a late Middle Devonian aridity crisis. Geol. Soc. Am., Abstr. Pr. 35(6), p. 208.

Martin, R.E., 2003. The fossil record of biodiversity: nutrients, productivity, habitat area and differential preservation. Lethaia 36, 179–193.
Matsumoto, R., Mahmudy Gharaie, M.H., Kakuwa, Y., 2002. Was the Late Devonian mass extinction caused by massive dissociation of gas hydrate? Proceedings of the Fourth International Conference on Gas Hydrates, Vol. 1, Yokohama, Yokohama Symposia, pp. 75–79.
McGhee, G.R., 1996. The Late Devonian Mass Extinction. The Frasnian-Famennian Crisis. Columbia Univ. Press, New York.
McGhee, G.R., 2001. The "multiple impacts hypothesis" for mass extinction: a comparison of the Late Devonian and the late Eocene. Palaeogeogr. Palaeoclimatol. Palaeoecol. 176, 47–58.
McGhee, G.R., 2005. Testing Late Devonian extinction hypotheses. In: Morrow, J., Over, J., Wignall, P. (Eds), Developments in Paleontology and Straitigraphy, Understanding Late Devonian and Permian-Triassic Biotic and Climatic Events. Elsevier. Vol. 20, pp. 37–50.
McGhee, G.R., Sheehan, P.M., Bottjer, D.J., Droser, M.L., 2004. Ecological ranking of Phanerozoic biodiversity crises: ecological and taxonomic severities are decoupled. Palaeogeogr. Palaeoclimatol. Palaeoecol. 211, 289–297.
McLaren, D.J., 1970. Presidential address: time, life and boundaries. J. Paleontol. 48, 801–815.
Melo, J.H.G., Loboziak, S., 2003. Devonian–Early Carboniferous miospore biostratigraphy of the Amazon Basin, Northern Brazil. Rev. Palaeobot. Palynol. 124, 131–202.
Metcalfe, I., Allen, M.B. (Eds), 2000. Suture zones of east and southeast Asia. J. Asian Earth Sci. 18(6), 635–808.
Morrow, J.R., Racki, G., Małkowski, K., 2003. Timing of Frasnian (Early Late Devonian) carbon isotopic and Alamo impact events, Southern Nevada. Geol. Soc. Am., Abstr. Pr. 35(6), 209.
Morrow, J.R., Sandberg, C.A., 2003. Late Devonian sequence and event stratigraphy across the Frasnian-Famennian (F-F) boundary, Utah and Nevada. In: Harries, P.J. (Ed.), High-Resolution Approaches in Stratigraphic Paleontology. Kluwer Academic Publishers, Dordecht, pp. 351–419.
Murphy, A.E., Sageman, B.B., Hollander, D.J., 2000. Eutrophication by decoupling of marine biogeochemical cycles of C, N, and P: a mechanism for the Late Devonian mass extinction. Geology 28, 427–430.
Nederbragt, A.J., Thurow, J., Vonhof, H., Brumsack, H.J., 2004. Modelling oceanic carbon and phosphorus fluxes: implications for the cause of the late Cenomanian Oceanic Anoxic Event (OAE 2). J. Geol. Soc. Lond. 161, 721–728.
Nelson, J., Paradis, S., Christensen, J., Gabites, J., 2002. Canadian Cordilleran Mississippi Valley-type deposits: a case for Devonian-Mississippian back-arc hydrothermal origin. Econ. Geol. 97, 1013–1036.
Olempska, E., 2002. The Late Devonian Upper Kellwasser Event and entomozoacean ostracods in the Holy Cross Mountains, Poland. Acta Palaeont. Pol. 47, 247–266.
Oliveira, J.T., Pereira, Z., Carvalho, P., Pacheco, N., Korn, D., 2004. Stratigraphy of the tectonically imbricated lithological succession of the Neves Corvo mine area, Iberian Pyrite Belt. Portugal. Mineral. Deposita 39, 422–436.
Ormiston, A.R., Oglesby, R.J., 1995. Effect of the Late Devonian climate on source rock quality and location. In: Huc, A.Y. (Ed.), Paleogeography, Paleoclimate, and Source Rocks, Vol. 40. Am. Assoc. Pet. Geol. Stud. Geol., pp. 105–132.
Over, J., 2002. The Frasnian/Famennian boundary in central and eastern United States. Palaeogeogr. Palaeoclimatol. Palaeoecol. 181, 153–169.
Paradis, S., Nelson, J.L., Irwin, S.E.B., 1998. Age constraints on the Devonian shale-hosted Zn-Pb-Ba deposits, Gataga District, Northeastern British Columbia, Canada. Econ. Geol. 93, 184–200.
Pardo, M.V., Garcia-Alcalde, J.L., 1984. El Devonico de la region de Almaden (Ciudad Real, Espana). 1 Congr. Esp. Geol. I, 473–482.
Paris, F., Girard, C., Feist, R., Winchester-Seeto, T., 1996. Chitinozoan bio-event in the Frasnian-Famennian boundary beds at La Serre (Montagne Noire, Southern France). Palaeogeogr. Palaeoclimatol. Palaeoecol. 121, 131–145.
Patchett, P.J., Roth, M.A., Canale, B.S., de Freitas, T.A., Harrison, A.F., Embry, A.F., Ross, G.M., 1999. Nd isotopes, geochemistry, and constraints on sources of sediments in the Franklinian Mobile Belt, Arctic Canada. Geol. Soc. Amer. Bull. 111, 578–589.
Pay, M.D., Astin, T.R., Parker, A., 2000. Clay mineral distribution in the Devonian-Carboniferous sandstones of the Clair Field, west of Shetland, and its significance for reservoir quality. Clay Miner. 35, 151–151.
Pervov, V.A., Bogomolov, E.S., Larchenko, V.A., Levskii, L.K., Minchenko, G.V., Sablukov, S.M., Sergeev, S.A., Stepanov, V.P., 2005. Rb–Sr Age of kimberlites of the Pionerskaya Pipe, Arkhangel'sk Diamondiferous Province. Dokl. Earth Sci. 400, 67–71.

Peterhansel, A., Pratt, B.R., 2001. Nutrient-triggered bioerosion on a giant carbonate platform masking the postextinction Famennian benthic community. Geology 29, 1079–1082

Phipps Morgan, J., Reston, T.J., Ranero, C.R., 2004. Contemporaneous mass extinctions, continental flood basalts, and 'impact signals': are mantle plume-induced lithospheric gas explosions the causal link? Earth Planet. Sci. Lett. 217, 263–284.

Playford, P.E., Wallace, M.W., 2001. Exhalative mineralization in Devonian reef complexes of the Canning Basin, Western Australia. Econ. Geol. 96, 1595–1610.

Prokoph, A., Veizer, J., 1999. Trends, cycles and nonstationarities in isotope signals of Phanerozoic seawater. Chem. Geol. 161, 225–240.

Racki, G., 1993. Evolution of the bank to reef complex in the Devonian of the Holy Cross Mountains. Acta Palaeont. Pol. 37(1992), 87–182.

Racki, G., 1998. Frasnian-Famennian biotic crisis: undervalued tectonic control? Palaeogeogr. Palaeoclimatol. Palaeoecol. 141, 177–198.

Racki, G., 1999a. Silica-secreting biota and mass extinctions: survival patterns and processes. Palaeogeogr. Palaeoclimatol. Palaeoecol. 154, 107–132.

Racki, G., 1999b. The Frasnian-Famennian biotic crisis: how many (if any) bolide impacts? Geol. Rundsch. 87, 617–632.

Racki, G., House, M.R. (Eds), 2002. Late Devonian biotic crisis: ecological, depositional and geochemical Records. Palaeogeogr. Palaeoclimatol. Palaeoecol. 181(1–3), 1–374.

Racki, G., Racka, M., Matyja, H., Devleeschouwer, X., 2002. The Frasnian/Famennian boundary interval in the South Polish-Moravian shelf basins: integrated event-stratigraphical approach. Palaeogeogr. Palaeoclimatol. Palaeoecol. 181, 251–297.

Racki, G., Wignall, P.B., 2004. Comment on: orbital cyclostratigraphy of the Devonian Frasnian–Famennian transition in South China, by Gong, Y.M., Li, B.H., Wang, Y., Wu, Y., Palaeogeogr. Palaeoclimatol. Palaeoecol. 168(2001), 237–248. Palaeogeogr. Palaeoclimatol. Palaeoecol. 205, 169–170.

Racki G., Wignall, P.B., 2005. Late Permian double-phased mass extinction and volcanism: an oceanographic perspective. In: Morrow, J., Over, J., Wignall, P.B. (Eds), Developments, in Paleontology and Stratigraphy, Understanding Late Devonian and Permian-Triassic Biotic and Climatic Events. Elsevier. Vol. 20, 263–297.

Radke, M., Wilkes, H., 2004. Organic-geochemical characterisation of upper Famennian sediments from the Refrath 1 Borehole (Bergisch Gladbach-Paffrath Syncline; Ardennes-Rhenish Massif, Germany). Cour. Forsch.-Inst. Senckenberg 251, 43–61.

Ray, J.S., Pande, K., 1999. Carbonatite alkaline magmatism associated with continental flood basalts at stratigraphic boundaries: cause for mass extinctions. Geophys. Res. Lett. 26, 1917–1920.

Raymond, A., 2003. Terrestrial-marine links in the Frasnian-Famennian extinction event. Geol. Soc. Am., Abstr. Pr., 35(6), 459.

Raymond, A., Metz, C., 2004. Ice and its consequences: glaciation in the Late Ordovician, Late Devonian, Pennsylvanian-Permian, and Cenozoic compared. J. Geol. 112, 655–670.

Raymond, A., Metz, C., 1995. Laurussian land-plant diversity during the Silurian and Devonian mass extinction, sampling bias, or both. Paleobiology 21, 74–91.

Reimold, W.U., Kelley, S.P., Sherlock, S.C., Henkel, H., Koeberl, C., 2005. Laser argon dating of melt breccias from the Siljan impact structure, Sweden: implications for a possible relationship to Late Devonian extinction events. Meteoritics Planet. Sci. 40, 591–607.

Reitner, J., Hühne, C., Thiel, V., 2001. Porifera-rich mud mounds and microbialites as indicators of environmental changes within the Devonian/Carboniferous critical interval. Terra Nostra 4, 60–65.

Renaud, S., Girard, C., 1999. Strategies of survival during extreme environmental perturbations: evolution of conodonts in response to the Kellwasser Crisis (Upper Devonian). Palaeogeogr. Palaeoclimatol. Palaeoecol 146, 19–32.

Renne, P.R., Reimold, W.U., Koeberl, C., Hough, R., Claeys, P., 2002. Comment on: K-Ar evidence from illitic clays of a Late Devonian age for the 120 km diameter Woodleigh impact structure, Southern Carnarvon Basin, Western Australia, by I.T. Uysal, S.D. Golding, A.Y. Glikson, A.J. Mory and M. Glikson [Earth Planet. Sci. Lett. 192(2001), 218–289]. Earth Planet. Sci. Lett. 201, 247–252.

Rode, A.L., Lieberman, B.S., 2002. Phylogenetic and biogeographic analysis of Devonian phyllocarid crustaceans. J. Paleontol. 76, 271–286.

Rode, A.L., Lieberman, B.S., 2004. Using GIS to unlock the interactions between biogeography, environment, and evolution in Middle and Late Devonian brachiopods and bivalves. Palaeogeogr. Palaeoclimatol. Palaeoecol. 211, 345–359.

Sageman, B.B., Murphy, A.E., Verne, J.P., Ver Straaten, C.A., Hollander, D.J., Lyons, T.W., 2003. A tale of shales: the relative roles of production, decomposition, and dilution in the accumulation of organic-rich strata, Middle–Upper Devonian, Appalachian basin. Chem. Geol. 195, 229–273.

Saltzman, M.R., 2002. Carbon isotope ($\delta^{13}C$) stratigraphy across the Silurian–Devonian transition in North America: evidence for a perturbation of the global carbon cycle. Palaeogeogr. Palaeoclimatol. Palaeoecol. 187, 83–100.

Sandberg, C.A., Morrow, J.R., Ziegler, W., 2000. Late Devonian events and mass extinctions. In: Catastrophic Events and Mass Extinctions: Impacts and Beyond, Vol. 1053. Houston, Texas, Lunar and Planetary Institute Contribution, pp. 188–189.

Sandberg, C.A., Morrow, J.R., Ziegler, W., 2002. Late Devonian sea-level changes, catastrophic events, and mass extinctions. In: Koeberl, C., MacLeod, K.G. (Eds), Catastrophic Events and Mass Extinctions: Impacts and Beyond, Vol. 356, Boulder, Colorado, Geol. Soc. Am., Spec. Pap., pp. 473–487.

Sandberg, C.A., Ziegler, W., 1996. Devonian conodont biochronology in geologic time calibration. Senckenberg. Lethaea 76, 259–265.

Sandberg, C.A., Ziegler, W., Dreesen, R., Butler, J.L., 1988. Late Frasnian mass extinction: conodont event stratigraphy, global changes, and possible causes. Cour. Forsch.–Inst. Senckenberg 102, 263–307.

Savoy, L.E., Stevenson, R.K., Mountjoy, E.W., 2000. Provenance of Upper Devonian-Lower Carboniferous miogeoclinal strata, southeastern Canadian Cordillera: link between tectonics and sedimentation. J. Sediment. Res. 70, 181–193.

Scheckler, S.E., 1986. Geology, floristics and paleoecology of Late Devonian coal swamps from Appalachian Laurentia (U.S.A.). Ann. Soc. Géol. Belg. 109, 209–222.

Schieber, J., 2001. A role for organic petrology in integrated studies of mudrocks: examples from Devonian black shales of the eastern US. Int. J. Coal Geol. 47, 171–187.

Schieber, J., Krinsley, D., Riciputi, L., 2000. Diagenetic origin of quartz silt in mudstones and implications for silica cycling. Nature 406, 981–985.

Schindler, E., 1993. Event-stratigraphic markers within the Kellwasser Crisis near the Frasnian/Famennian boundary (Upper Devonian) in Germany. Palaeogeogr. Palaeoclimatol. Palaeoecol. 104, 115–125.

Schülke, I., 1998. Conodont community structure around the "Kellwasser mass extinction event" (Frasnian/Famennian boundary interval). Senckenberg. Lethaea 77, 243–261.

Schultz, R.B., Rimmer, S.M. (Eds), 2004. Geochemistry of organic-rich shales: new perspectives. Chem. Geol. 206(2–4), 163–426.

Sepkoski, J.J., 1996. Patterns of Phanerozoic extinction: a perspective from global data bases. In: Walliser, O.H. (Ed.), Global events and event stratigraphy in the Phanerozoic. Berlin, Springer, pp. 35–51.

Shen, J.W., Webb, G.E., 2004. Famennian (Upper Devonian) calcimicrobial (*Renalcis*) reef at Miaomen, Guilin, Guangxi, South China. Palaeogeogr. Palaeoclimatol. Palaeoecol. 204, 373–394.

Siegmund, H., Trappe, J., Oschmann, W., 2002. Sequence stratigraphic and genetic aspects of the Tournaisian "Liegender Alaunschiefer" and adjacent beds. Int. J. Earth Sci. 91, 934–949.

Smith, G.J., Jacobi, R.D., 2001. Tectonic and eustatic signals in the sequence stratigraphy of the Upper Devonian Canadaway Group, New York State. Am. Assoc. Pet. Geol. Bull. 85, 325–357.

Stevenson, R.K., Whittaker, S., Mountjoy, E.W., 2000. Geochemical and Nd isotopic evidence for sedimentary-source changes in the Devonian miogeocline of the southern Canadian Cordillera. Geol. Soc. Amer. Bull. 112, 531–539.

Stone, D.B., Minyuk, P., Kolosev, E., 2003. New paleomagnetic paleolatitudes for the Omulevka terrane of northeast Russia: a comparison with the Omolon terrane and the eastern Siberian platform. Tectonophysics 377, 55–82.

Streel, M., Caputo, M.V., Loboziak, S., Melo, J.H.G., 2000. Late Frasnian–Famennian climates based on palynomorph analyses and the question of the Late Devonian glaciations. Earth-Sci. Rev. 52, 121–173.

Tapanila, L., 2005. Palaeoecology and diversity of endosymbionts in Palaeozoic marine invertebrates: trace fossil evidence. Lethaia 38, 89–99.

Tomshin, M.D., Koroleva, O.V., 1990. Composite dykes of the Vilyuisk paleorift system, Siberian Platform, Yakutia, USSR. In: Parker, A.J. (Ed.), Mafic Dykes and Emplacement Mechanisms. Rotterdam, Balkema, pp. 535–540.

Tribovillard, N., Averbuch, O., Devleeschouwer, X., Racki, G., Riboulleau, A., 2004. Deep-water anoxia over the Frasnian–Famennian boundary (La Serre, France): a tectonically induced oceanic anoxic event? Terra Nova 16, 288–295.

Tsien, H.H., Fong, C.C.K., 1997. Sea-level fluctuations in South China. Cour. Forsch.–Inst. Senckenberg 199, 103–115.

Tucker, R.D., Bradley, D.C., Ver Straeten, C.A., Harris, A.G., Ebert, J.R., McCutcheon, S.R., 1998. New U-Pb zircon ages and the duration and division of Devonian time. Earth Planet. Sci. Lett. 158, 175–186.

Twitchett, R.J., 2001. Incompleteness of the Permian-Triassic fossil record: a consequence of productivity decline? Geol. J. 36, 341–353.

Twitchett, R.J., Looy, C.V., Morante, R., Visscher, H., Wignall, P.B., 2001. Rapid and synchronous collapse of marine and terrestrial ecosystems during the end-Permian biotic crisis. Geology 29, 351–354.

Umeda, M., 2001. Late Devonian pelagic cherts and extinction event. Geol. Soc. Am., Abstr. Pr. 33(6), 13.

Umeda, M., Sennikov, N.V., Obut, O.T., Ivata, K., Izokh, N.G., 2001. Frasnian-Famennian boundary radiolarian complexes of Rudny Altai siliceous formations [in Russian]. In: Vladimirov, A.G. (Ed.), Nauchno-Prakticheskaya Konferentsiya – Aktual'nyye Voprosy Geologii i Minerageniya Yuga Sibiri, pp. 178–183. Novosibirsk, Rossiyskaya Akademiya Nauk, Sibirskoye Otdeleniye, IGiL.

Uysal, I.T., Golding, S.D., Glikson, A.Y., Mory, A.J., Glikson, M., 2001. K-Ar evidence from illitic clays of a Late Devonian age for the 120 km diameter Woodleigh impact structure, Southern Carnarvon Basin, Western Australia. Earth Planet. Sci. Lett. 192, 218–289.

Uysal, I.T., Mory, J.A., Golding, S.D., Bolhar, R., Collerson, K.D., 2005. Clay mineralogical, geochemical and isotopic tracing of the evolution of the Woodleigh impact structure, Southern Carnarvon Basin, Western Australia. Contrib. Mineral. Petrol. 149, 576–590.

Van Geldern, R., Joachimski, M.M., 2001. The stable isotopic composition of Devonian brachiopods. Terra Nostra 2001(4), 76–80.

Van Loevezijn, G.B.S., 1987. Development and termination of the carbonate sedimentation on intracratonic Late Devonian platforms in the Cantrabrian Mountains (Spain). Z. Dt. Geol. Ges. 138, 197–209.

Vaughan, A.P.M., Scarrow, J.H., 2003. Ophiolite obduction pulses as a proxy indicator of superplume events? Earth Planet. Sci. Lett. 213, 407–416.

Veimarn, A.B., Puchkov, V.N., Abramova, A.N., Artyushkova, O.V., Baryshev, V.N., Degtyaryov, K.E., Kononova, L.I., Maslov, V.A., Mosejchuk, V.M., Pazukhin, V.N., Pravikova, N.V., Tevelev, A.V., Yarkova, A.V., 2004. Stratigraphy and geological events at the Frasnian-Famennian boundary in the Southern Urals. Geol. Quart. 48, 233–244.

Veizer, J., Ala, D., Azmy, K., Bruckschen, P., Buhl, D., Bruhn, F., Carden, G.A.F., Diener, A., Ebneth, S., Goddéris, Y., Jasper, T., Korte, C., Pawellek, F., Podlaha, O.G., Strauss, H., 1999. $^{87}Sr/^{86}Sr$, $\delta^{13}C$ and $\delta^{18}O$ evolution of Phanerozoic seawater. Chem. Geol. 161, 59–88.

Vermeij, G.J., 2004. Ecological avalanches and the two kinds of extinction. Evolut. Ecol. Res. 6, 315–337.

Vishnevskaya, V.S., Pisera, A., Racki, G., 2002. Siliceous biota and the Late Devonian biotic crisis: the Polish reference. Acta Palaeont. Pol. 47, 211–226.

Walliser, O.H., 1981. The geosynclinal development of the Rheinische Schiefergenirge (Rhenohercynian Zone of the Variscides; Germany). Geol. Mijnbouw 60, 89–96.

Walliser, O.H., 1996. Global events in the Devonian and Carboniferous. In: Walliser, O.H. (Ed.), Global Events and Event Stratigraphy in the Phanerozoic. Berlin, Springer, pp. 225–250.

Wallmann, K., 2003. Feedbacks between oceanic redox states and marine productivity: a model perspective focused on benthic phosphorus cycling. Glob. Biogeochem. Cy. 17, 1084, DOI 10.1029/2002GB001968.

Wang, Y., Aitchison, J.C., Luo, H., 2003. Devonian radiolarian faunas from South China. Micropaleontology 49, 127–145.

Waters, J.A., Maples, C.A., Lane, N.G., Marcus, S., Liao, Z.T., Liu, L., Hou, H.F., Wang, J.X., 2003. A quadrupling of Famennian pelmatozoan diversity: new Late Devonian blastoids and crinoids from Northwest China. J. Paleontol. 77, 922–948.

Webb, G.E., 1996. Was Phanerozoic reef history controlled by the distribution of non-enzymatically secreted reef carbonates (microbial carbonate and biologically induced cement)? Sedimentology 43, 947–971.

Webster, G.D., Lane, N.G., Maples, C.G., Waters, J.A.and Horowitz, A.B.S., 1998. Frasnian-Famennian extinction was a non-event for crinoids, blastoids and bryozoans. Geol. Soc. Am., Abstr. Pr. 30 (7), 30–31.

Wendt, J., Belka, Z., 1991. Age and depositional environment of Upper Devonian (early Frasnian to early Famennian) black shales and limestones (Kellwasser facies) in the Eastern Anti-Atlas. Facies 25, 51–90.

Wendt, J., Kaufmann, B., Belka, Z., Farsan, N., Bavandpur, A., 2002. Devonian/Lower Carboniferous stratigraphy, facies patterns and palaeogeography of Iran. Part I. Southeastern Iran. Acta Geol. Pol. 52, 129–168.

Westphal, H., Böhm, F., Bornholdt, F., 2004. Orbital frequencies in the carbonate sedimentary record: distorted by diagenesis? Facies 50, 3–11.

Wignall, P.B., 2001. Large igneous provinces and mass extinctions. Earth-Sci. Rev. 53, 1–33.

Wignall, P.B., Bond, D., 2003. Critical review of sea-level change during the Frasnian/Famennian mass extinction. Geol. Soc. Am., Abstr. Pr. 35(6), 208.

Wilder, H., 1994. Death of Devonian reefs – implications and further investigations. Cour. Forsch.—Inst. Senckenberg 172, 241–247.

Wilson, M., Lyashkevich, Z.M., 1996. Magmatism and the geodynamics of rifting of the Pripyat-Dnieper-Donets Rift, East European Platform. Tectonophysics 268, 65–81.

Wood, R., 2004. Palaeoecology of a post-extinction reef: Famennian (Late Devonian) of the Canning Basin, north-western Australia. Palaeontology 47, 415–445.

Wright, A.J., Pickett, J.W., Sewell, D., Roberts, J., Jenkins, T.B.H., 1990. Corals and conodonts from the Late Devonian Mostyn Vale Formation, Keepit, New South Wales. Mem. Ass. Australas. Paleont. 10, 211–254.

Xiao, W.J., Zhang, L.C., Qin, K.Z., Sun, S., Li, J.L., 2004. Paleozoic accretionary and collisional tectonics of the easter Tianshan (China): implications for the continental growth of central Asia. Am. J. Sci. 304, 870–895.

Xu, B., Gu, Z., Liu, Q., Wang, C., Li, Z., 2003. Carbon isotopic record from Upper Devonian carbonates at Dongcun in Guilin, southern China, supporting the worldwide pattern of carbon isotope excursions during Frasnian-Famennian transition. Chin. Sci. Bull.48, 1259–1264.

Yans, J., Corfield, R.M., Racki, G., Préat, A., submitted. Evidence for a major perturbation of the carbon cycle in the Middle Frasnian *punctata* conodont Zone. Terra Nova, Palaeogeogr. Palaeoclimatol. Palaeoecol.

Yolkin, E.A., Bakharev, N.K., Izokh, N.G., Kirda, N.P., Klets, A.G., Kontorovich, V.A., Obut, O.T., Timokhina, I.G., 2003. Late Devonian paleogeography of the West Siberian Plain and its mountainous surroundings. Russ. Geol. Geophys. 44, 169–180.

Yolkin, E.A., Gratsianova, R.T., Izokh, N.G., Yazikov, A.Y., Bakharev, N.K., 1997. Devonian sea-level fluctuations on the south-western margin of the Siberian continent. Cour. Forsch.–Inst. Senckenberg 199, 83–98.

Yolkin, E.A., Gratsianova, R.T., Izokh, N.G., Yazikov, A.Y., Bakharev, N.K., Alekseeva, R.E., Erina, M.V., Kim, A.I., Shishkina, G.R., 2000. Devonian standard boundaries within the shelf belt of the Siberian Old Continent (southern part of western Siberia, Mongolia, Russian Far East) and in the South Tien Shan. Cour. Forsch.–Inst. Senckenberg 225, 305–320.

Yudina, A.B., Racki, G., Savage, N.S., Racka, M., Małkowski, K., 2002. The Frasnian-Famennian events in a deep-shelf succession, Subpolar Urals: biotic, depositional and geochemical records. Acta Palaeontol. Pol. 47, 355–372.

Yutkina, E.V., Kononova, V.A., Bogatikov, O.A., Knyazkov, A.P., Kozar, N.A., Ovchinnikova, G.V., Levsky, L.K., 2004. Kimberlites of Eastern Priazov'e (Ukraine) and geochemical characteristics of their sources. Petrology 12, 134–148.

Zhang, S., Wang, X., Zhu, H., 2000. Magnetic susceptibility variations of carbonates controlled by sea-level changes: examples in Devonian to Carboniferous strata in southern Guizhou Province, China. Sci. China, D 43, 266–276.

Ziegler, W., Sandberg, C.A., 1990. The Late Devonian standard conodont zonation. Cour. Forsch.–Inst. Senckenberg 121, 1–115.

Ziegler, W., Sandberg, C.A., 2001. Utility of palmatolepids and icriodontids in recognizing Upper Devonian series, stage, and possible substage boundaries. Courier Forsch.-Inst. Senckenberg 225, 335–347.

Understanding Late Devonian and Permian-Triassic Biotic and Climatic Events:
Towards an Integrated Approach
edited by D.J. Over, J.R. Morrow and P.B. Wignall



# *Chapter 3*
# *Modelling Late Devonian extinction hypotheses*

*George R. McGhee Jr.**

*Department of Geological Sciences, Wright-Rieman Laboratories, Rutgers University, New Brunswick, NJ 08903, USA*

**Abstract**

Modelling the conditions of extinction hypotheses can lead to the discernment of the critical observations that must be made in order to definitively test those hypotheses. Models considered here are based upon the beginning assumption that the end-Frasnian biodiversity crisis was triggered in large part by pulses of global cooling. Both classes of models are catastrophic: large-igneous-province volcanism and asteroid or comet impact. A single impact model can be ruled out, as its predicted effects do not match those of the extinction pulses seen in the Late Frasnian. An episodic global-winter model would, but those effects can be produced by both the catastrophic volcanism and multiple impacts models. Alternative models suggest that both catastrophic volcanism and multiple impacts may trigger a global greenhouse-induced temperature increase, a predicted effect that does not match the empirical data. However, if both of those mechanisms are shifted back in time to produce lag-time models of short-term global temperature increase and then rapid temperature drops as the anomalous greenhouse interval collapsed, then both produce effects that do match the empirical data.

To definitively choose among the several model predictions two critical data are needed: (1) an accurate radiometric date on the Frasnian/Famennian (F/F) boundary, and (2) an accurate Late Devonian global temperature curve. Recent research results suggesting a 376 Ma age for the F/F boundary would indicate that several known impact craters post-date the F/F extinctions, and make the impact-induced extinction models less likely. High-research priority should thus be given to obtaining a biostratigraphic date or an accurate radiometric date on the Viluy Traps, as a test to corroborate the volcanic-induced extinction models.

*Keywords*: mass extinctions; multiple impacts hypotheses; catastrophic volcanism hypotheses; global cooling; Late Devonian

## 1. Introduction: modelling Late Devonian climates and extinctions

Over three decades have passed since McLaren (1970) proposed that the Frasnian/Famennian (F/F) mass extinction was triggered by an asteroid impact with the Earth, yet still no consensus exists as to the actual cause of the biodiversity crisis (see the reviews of McGhee, 1996; Walliser, 1996). It is generally agreed that the end-Devonian biodiversity crisis, also known as the Devonian/Carboniferous extinctions, arrow "D/C" in Figure 1, was triggered by the ecological effects associated with atmospheric $CO_2$ downdraw: lethal global cooling and climatic instability, marine regression produced by ice cap formation, and anoxic bottom waters produced by marine eutrophication (Walliser, 1996; Streel et al.,

*Corresponding author. Fax: +1-732-445-3374. *E-mail address:* mcghee@rci.rutgers.edu (G.R. McGhee).

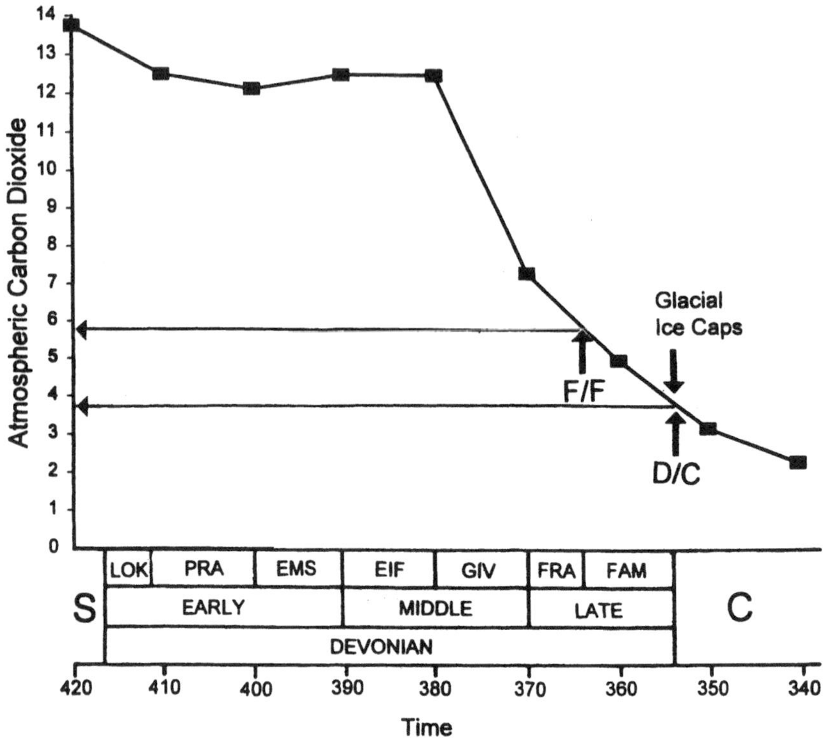


*Figure 1.* Carbon dioxide levels in the Earth's atmosphere during the interval of time from the latest Silurian to the early Carboniferous, from Berner (2001). Carbon dioxide levels are shown in factors of the $CO_2$ partial pressure of the preindustrial Earth's atmosphere (300 ppm $CO_2$), which has a value of 1.0 in the graph. Timescale after Gradstein and Ogg (1996).

2000; Algeo et al., 2001). Phanerozoic climatic models position the Late Devonian Epoch in the period of transition of the Earth's climate from a hot, 'greenhouse' world of the Early Palaeozoic to a cold 'icehouse' world of the Late Palaeozoic (Fischer and Arthur, 1977; Frakes et al., 1992). The Late Devonian cooling trend is proposed to have been driven by a dramatic decline in the partial pressure of $CO_2$ in the atmosphere from the Middle Devonian to the Early Carboniferous, a decline driven by the establishment of extensive vascular land plant cover and associated soil weathering (Wilder, 1989; Retallack, 1997; Berner, 2001; Algeo et al., 2001). It is clear that the dramatic decline of $CO_2$ partial pressures in the atmosphere during the 40 million years from the Givetian to the Early Carboniferous (Fig. 1) must have made the atmosphere more transparent to heat loss. The formation of glacial ice caps in Gondwana in the latest Famennian (Fig. 1), major ice caps that reached down to sea level (Streel et al., 2000) is concrete evidence of global cooling in this interval of geologic time.

Although the Frasnian/Famennian biodiversity crisis, arrow "F/F" in Figure 1, has also been argued to be triggered by lethal global cooling (Copper, 1977; see the review of the evidence in McGhee, 1996), the glacial ice caps arrived 10 million years too late to explain this crisis (Fig. 1). All attempts to find geologic evidence of glacial ice caps in the Late

Frasnian to Early Famennian interval of time have failed (Streel et al., 2000). Global sea level was at an all-time high for the Devonian, indicating there was no major build-up of glacial ice in this interval of time (although apparently rapid oscillations in sea level during the Late Frasnian and Early Famennian may indicate the formation of small- and high-altitude alpine-type glaciers; see Filer, 2002; Hladil, 2002). Atmospheric $CO_2$ concentrations in the Late Frasnian were six times higher than at present day, and twice as high as those present during the formation of the end-Devonian glacial ice caps (Fig. 1). Palaeobotanical evidence shows that the Frasnian world was quite hot, with extensive shallow seas covering much of the continental landmasses (Streel et al., 2000). In contrast to Frasnian climates, the Early Famennian world appears to have been cold and arid (Streel et al., 2000).

What could trigger lethal temperature drops in a world with atmospheric $CO_2$ partial pressures six times higher than at present (arrow "F/F" in Fig. 1)? That question is the focus of the models to be developed in this study.

## 2. Direct-effect global winter models

The first class of models to be considered is direct-effect models: the hypothetical cause produces an immediate effect. The effect to be produced is a rapid, and global, drop in temperature. For the purpose of modelling, the general shape of the Berner (2001) atmospheric $CO_2$ curve (Fig. 1) will be taken as a first-approximation of the general shape of the global temperature curve for this same interval of time (Fig. 2). At the very least, the actual pattern of global temperature change during the Devonian should agree with this first-approximation

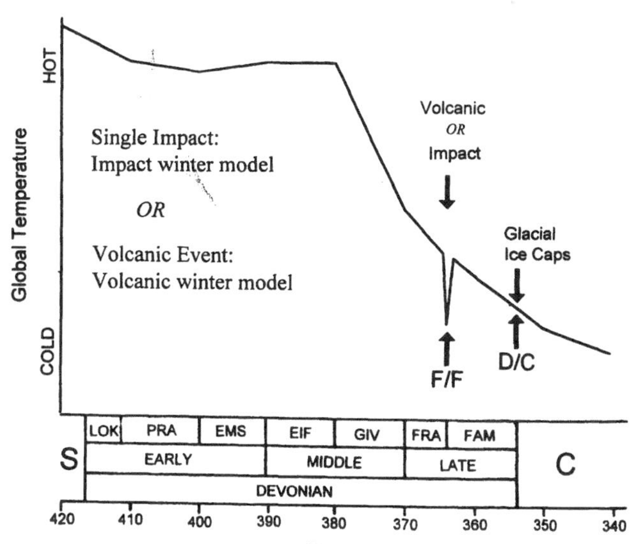


*Figure 2.* Direct-effect global winter model produced by a single impact of an asteroid or comet with the Earth, or by a catastrophic volcanic event.

curve (Fig. 2) to the extent that the Middle Devonian should be hotter than the Late Devonian, and the Frasnian should be hotter than the Famennian, based upon atmospheric $CO_2$ levels (Fig. 1).

McLaren's (1970) suggestion that the F/F extinctions were the direct effect of an asteroid or comet impact on the Earth was not taken seriously until the Alvarez et al. (1980) study, giving the initial geochemical evidence that the end-Cretaceous extinction was triggered by an asteroid impact (McGhee, 1981). The primary kill mechanism proposed for the end-Cretaceous asteroidal strike was lethal global cooling, the "impact winter" model (Alvarez et al., 1980; Bailey et al., 1994; Toon et al., 1994).

Applying the impact winter model to a single impact event produces the scenario shown in Figure 2. Impact of an asteroid or comet with the Earth is modelled to produce an immediate, rapid drop in global temperature in the Late Frasnian. Global temperatures then recover and rise back to the overall declining temperature trend in the Early Famennian as the atmospheric effects of the vaporized bolide and target rock diminish. The model predicts that a single impact should cause a single extinction event (Fig. 2).

At this point in the modelling a dilemma arises that lies at the heart of the current lack of consensus concernig the actual cause of the F/F biodiversity crisis: the exact same climatic effect has been proposed to be caused by catastrophic volcanism, the "volcanic winter" scenario, wherein rapid global cooling is produced by the vast amount of debris, ash, and gasses injected into the Earth's atmosphere during a catastrophic volcanic event (Racki et al., 2002). A volcanic-produced global dust and gas cloud would block light from the sun from reaching the Earth's surface, triggering planet-wide lethally cold temperatures even at the equator. As will become apparent in the models to be discussed, many of the climatic effects that have been proposed to be produced by extraterrestrial impacts and by Earth-based catastrophic volcanism are identical (Courtillot, 1999; Alvarez, 2003; Morgan et al., 2004), which makes differentiating the two potential causes of extinction problematical. This fundamental problem will be reconsidered in the conclusion of this paper.

The direct-effect model shown in Figure 2 can be rejected, whether it be impact or volcanically induced. The model prediction shown in Figure 2 does not match the actual temporal structure of the F/F extinctions: rather than a single instantaneous event (as proposed by McLaren, 1970), subsequent analyses of the past three decades indicate that the F/F biodiversity crisis consisted of a series of extinction pulses, spread over approximately one million years (see the reviews of Walliser, 1996; McGhee, 1996, 2001).

Rather than a single impact, it was later proposed that the Earth was impacted by several asteroids or comets over a significant span of geologic time (McGhee, 1982, p. 498), and that each impact event triggered its own pulse of global cooling and associated species extinctions (McGhee, 1996, pp. 172–173). Applying the impact winter model to multiple impact events produces the direct-effect scenario shown in Figure 3. The model predicts several sharp drops in global temperature in the Late Frasnian, and several discrete extinction pulses. Thus this model does produce effects that do match the temporal pattern of several discrete extinction pulses seen in the F/F biodiversity crisis (Walliser, 1996; McGhee, 1996).

However, two problems exist for the multiple impacts model (Fig. 3). First, it has been proposed that the exact same climatic effect could be produced by an extended period of catastrophic, and pulsed, flood-basalt volcanism (Courtillot, 1999; Alvarez, 2003). Second, the underlying global winter model itself has been called into question, both for impacts

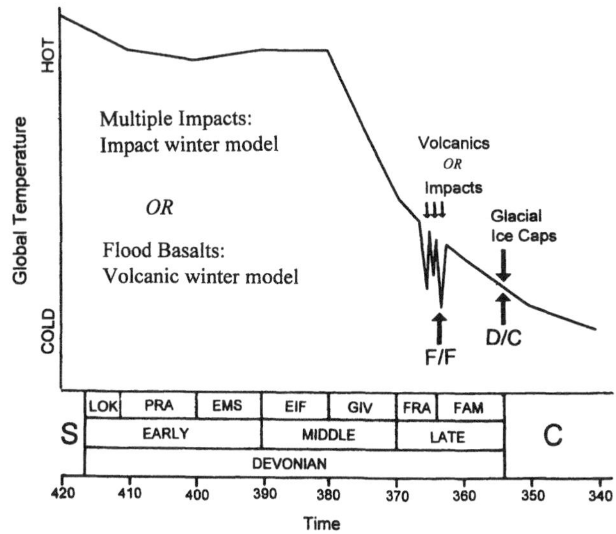


*Figure 3.* Direct-effect global winter model produced by multiple impacts of asteroids or comets with the Earth, or by an extended period of catastrophic flood-basalt volcanism.

and for volcanics. If the target rock of multiple-impact sites includes carbonates, then their subsequent vaporization upon impact would inject large quantities of $CO_2$ into the atmosphere, producing a greenhouse-induced period of global warming, not cooling. Actual empirical evidence for multiple-impact-produced global warming is present in the Late Eocene, when the Earth was impacted by at least three large bolides (Poag, 1997; Poag et al., 2003). For volcanics, more recent analyses have indicated that the initial period of "volcanic winter" would be quite brief, on the order of months, compared to a long-term period ($10^5$ years) of global warming produced by volcanic injection of large amounts of $CO_2$ into the atmosphere (Wignall, 2001).

## 3. Direct-effect greenhouse models

The direct-effect greenhouse climatic model for single events, either impact or volcanic, is given in Figure 4 and for multiple events in Figure 5. Both models can be rejected for the following reasons. First, the predicted temperature curves do not match the palaeobotanical data for the Late Frasnian and Early Famennian. Streel et al. (2000) have summarized the past three decades of palaeobotanical research that indicates that the Late Frasnian climate was warm, whereas the Early Famennian climate was cool and arid. Both direct-effect greenhouse models predict an Early Famennian that is hotter than the Late Frasnian (Figs. 4 and 5).

Second, the models produce only a brief initial period of global cooling, followed by an extended period of global warming, as indeed appears to have happened following the late Eocene multiple impacts (Poag, 1997; Poag et al., 2003), and appears to have happened during the Late Cretaceous period of catastrophic flood-basalt volcanism (Keller, 2001; Wilf et al., 2003; Nordt et al., 2003). The past three decades of palaeoecological research

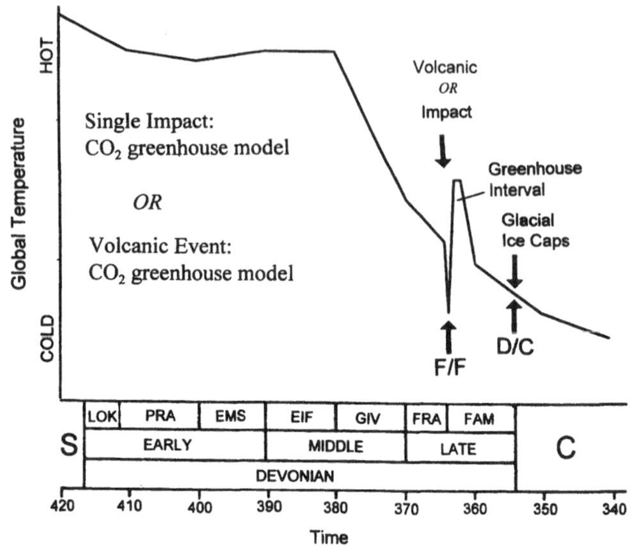


*Figure 4.* Direct-effect greenhouse climatic model produced by a single impact of an asteroid or comet with the Earth, or by a catastrophic volcanic event.

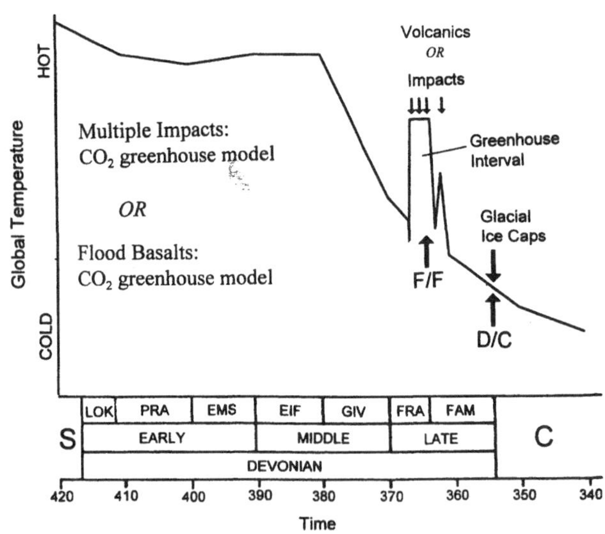


*Figure 5.* Direct-effect greenhouse climatic model produced by multiple impacts of asteroids or comets with the Earth, or by an extended period of catastrophic flood-basalt volcanism.

indicate that the Late Frasnian extinctions were triggered by global cooling, not warming (Copper, 1977, 2002; McGhee, 1996, 2001; Joachimski and Buggisch, 2002).

In summary of the direct-effect models section of this study, only the multiple-event, impact or volcanic, global cooling model (Fig. 3) produces a climatic scenario that matches the multiple-pulsed, cooling-related, extinction pattern seen in the F/F biodiversity crisis. However, both theoretical arguments (Wignall, 2001) and empirical data (Poag, 1997; Poag et al., 2003) exist to suggest that the direct-effect of both multiple impacts and long-term volcanism is global warming (Fig. 5) and not global cooling (Fig. 3), as will be discussed in the next section.

## 4. Lag-time greenhouse-collapse models

In lag-time models, the immediate effect of a hypothetical cause does not produce extinction, but rather creates temporary climatic conditions that do produce extinction when these conditions cease to exist. That is, the extinction effect occurs a significant period of time later, or lag-time, after the initial hypothetical cause. The "lag-time multiple impacts" hypothesis was first proposed by Poag (1997) for the Early Oligocene extinctions, and extended to the F/F extinctions by McGhee (2001).

The predicted effects of the lag-time multiple impacts hypothesis are illustrated in Figure 6. An extended period of impact bombardment is hypothesized to occur early in the Frasnian, producing an anomalous period of global warming. This interval of global warming interrupts the long-term global-cooling trend produced by the dramatic decline of $CO_2$ partial pressures in the atmosphere during the 40-million-year interval from the Givetian to the Early Carboniferous (Fig. 1), a decline driven by the establishment of extensive vascular land plant cover and associated soil weathering. As the period of bombardment begins to wane, the anomalous greenhouse condition begins to break up, and ultimately collapses. Global temperatures fall episodically as the greenhouse breaks up,

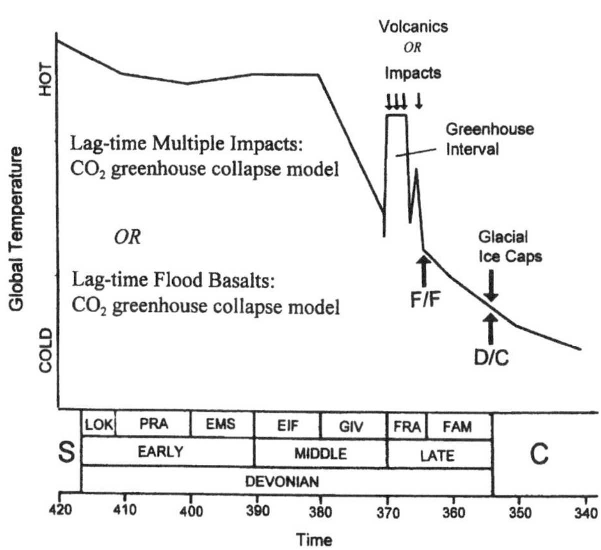


*Figure 6.* Lag-time greenhouse collapse model produced by multiple impacts of asteroids or comets with the Earth, or by an extended period of catastrophic flood-basalt volcanism.

and eventually plummet as the long-term global-cooling trend resumes (Fig. 6), producing a climatic scenario that matches the multiple-pulsed, cooling-related, extinction pattern seen in the F/F biodiversity crisis.

Once again, however, it can be argued that the exact same climatic scenario can be generated by an extended period of flood-basalt volcanism (Buggisch, 1991; Becker and House, 1994; Wilson and Lyashkevich, 1996; Racki, 1998, 1999; Streel et al., 2000). As for the lag-time multiple impacts hypothesis, the volcanism itself is not the direct killing mechanism, it instead creates an anomalous greenhouse interruption of the long-term global-cooling trend, the collapse of which does produce the thermal shocks necessary to trigger the F/F extinctions (Fig. 6).

## 5. Testing the models?

In summary, there exist two models that match the observed multiple-pulsed, cooling-related, extinction pattern seen in the F/F biodiversity crisis: the direct-effect multiple-events global-cooling model (Fig. 3), and the lag-time greenhouse-collapse global-cooling model (Fig. 6). A fundamental problem exists in that the climatic conditions of both of these models can be produced either by extraterrestrial impacts or by Earth-based catastrophic volcanism, a problem that not only exists for the F/F biodiversity crisis, but for all mass extinction events in general (Courtillot, 1999; Alvarez, 2003; Morgan et al., 2004). Thus how can we test these models in geologic time?

An interval of catastrophic flood-basalt volcanism should produce a "large igneous province" (LIP), whereas an interval of multiple impact events should produce a series of impact craters. Thus we can search the geologic record for these two key indicators: LIPs versus craters.

### *5.1. Evidence for catastrophic volcanism models*

The direct-effect catastrophic volcanism model (Fig. 3) predicts a LIP that has an age in the Late Frasnian, specifically in the Late *rhenana* to *linguiformis* Zonal interval (see McGhee, 1996 for discussion), whereas the lag-time model (Fig. 6) predicts a LIP that is much older, in the Early to Middle Frasnian.

The Pripyat-Dnieper-Donet rift, a major continental rift-system, is now known to have been active in the Ukraine region of Europe during the Late Devonian (Wilson and Lyashkevich, 1996; Racki, 1998, 1999). However, the volume of volcanic material erupted in the Pripyat-Dnieper-Donet rift appears to be relatively small (less than 10 000 $km^3$), thus the intensity of the volcanism associated with the rift was not near the magnitude associated with the Siberian or Deccan flood basalt fissure eruptions, both of which produced well over 1 000 000 $km^3$ of volcanic material (Wignall, 2001; Courtillot and Renne, 2003), in association with the end-Permian and end-Cretaceous mass extinctions (Courtillot, 1999). The possibility exists, however, that this volcanic activity was part of a series of plume clusters active in eastern Laurussia and Siberia (Abbott and Isley, 2002).

A more promising candidate for a major Late Devonian LIP, perhaps related to superplume activity (Abbott and Isley, 2002), are the Viluy Traps in eastern Siberia, which Courtillot and Renne (2003) suggest may have produced a volume of volcanic material that was originally equivalent to that of the Central Atlantic Magmatic Province (2 000 000 $km^3$). This volumetric estimate remains to be proved, however.

The initiation of the Pripyat-Dnieper-Donet magmatics is biostratigraphically dated in the Middle Frasnian (Racki, 1998), thus these volcanics are a prime candidate for an extinction trigger in the volcanic-induced lag-time model (Fig. 6). Outside of the East European Platform, volcanic activity in this time interval is also known from North America (Rotondo and Over, 2000; Over, 2002) and several eastern Laurussian regions (Racki et al., 2002; Yudina et al., 2002).

The Viluy Traps have no biostratigraphic date, but have been radiometrically dated to have an age in the range from 377 to 350 Ma (Courtillot and Renne, 2003). A major problem now arises in model testing (e.g. Fig. 3 versus Fig. 6) using radiometric ages rather than biostratigraphy: we do not precisely know the age of the F/F boundary itself: current estimates range from 364 to 376 Ma (Williams et al., 2000) – a range in age estimates of some 12 million years! This very important point will be considered in more detail in the discussion section of this paper.

### *5.2. Evidence for impact models*

The direct-effect multiple impacts model (Fig. 3) predicts impact craters that are Late Frasnian in age, whereas the lag-time model (Fig. 6) predicts craters that are Early to Middle Frasnian in age. Three decades of research have produced very little hard evidence for multiple impacts of asteroids or comets with the Earth during the Late Frasnian (reviewed in McGhee, 1996; Hallam and Wignall, 1997), although the search continues to the present day (e.g. Ma and Bai, 2002; Ellwood et al., 2003).

Much more solid evidence exists for major impacts in the Early to Middle Frasnian, and thus for the lag-time multiple impacts hypothesis (Fig. 6). The Alamo impact event in Nevada (USA) has been biostratigraphically dated to have occurred in the Middle Frasnian *punctata* Zone, and the Flynn Creek impact in Tennessee (USA) has been dated to have occurred in the Early Frasnian *transitans* Zone (Morrow et al., 1998; McGhee, 2001). Tentative evidence of a third impact event in eastern North America, older than the Flynn Creek impact, is currently being investigated (Schieber, 2003). The Alamo impactor in particular was quite large, producing ejecta debris that covers over 19 000 $km^2$ of southern Nevada and estimated to have produced a crater 70 km in diameter (Morrow et al., 1998; Sandberg et al., 2002). In addition, the Alamo impactor struck carbonate target rock and would have produced a massive injection of $CO_2$ into the atmosphere (McGhee, 2001), as predicted in the lag-time multiple impacts scenario (Fig. 6).

Unfortunately, only these two impact craters have reliable biostratigraphic dates. Several very large craters are known to have a general radiometric age in the Late Devonian time range (Grieve et al., 1995), such as the 52 km-diameter Siljan Crater in Sweden (368.0 $\pm$ 1.1 Ma), the 54 km-diameter Charlevoix Crater in Canada (357 $\pm$ 15 Ma), and huge 120 km-diameter Woodleigh Crater in Australia (359 $\pm$ 4 Ma; Uysal et al., 2001).

## 6. Discussion of recent research results

This paper was originally presented at the 2003 Paleontological Society symposium "Understanding Late Devonian and Permian-Triassic Biotic and Climatic Events: Towards an Integrated Approach," and concluded that two critical data were needed in order to definitively chose among the several extinction models: (1) an accurate radiometric date on the F/F boundary, and (2) an accurate Late Devonian global temperature curve (McGhee, 2003). In the past year, significant progress has been made on both of these questions, the consequences of which are discussed here.

### *6.1. An accurate radiometric date on the Frasnian/Famennian boundary*

Kaufmann et al. (2004) have published a 377.2 ± 1.7 Ma date for a bentonite bed precisely assigned to the Late *rhenana* Zone in the intensely studied Steinbruch Schmidt section in Germany (see discussion in McGhee, 1996), and have suggested an extrapolated date of 376.1 ± 1.6 Ma for the F/F boundary itself. This date is significantly older than the recently estimated dates of 364.9 Ma (Compston, 2000), 364.7 Ma (Young, 1996) and 364 Ma (Gradstein and Ogg, 1996), but close to the 376.5 Ma estimate of Tucker et al. (1998). The biostratigraphy accuracy of the Tucker et al. (1998) 381.1 ± 1.3 Ma date for the bentonite beds in the basal Chattanooga Shale has been questioned by Streel et al. (2000), who argue that these beds are Eifelian in age, and not Frasnian as reported by Tucker et al. (1998). However, subsequent conodont biostratigraphic analyses have confirmed the Frasnian age of the basal Chattanooga Shale (Rotondo and Over, 2000), and hence have corroborated the Tucker et al. (1998) 376.5 Ma extrapolated estimate for the age of the F/F boundary itself. Recently, Gradstein and Ogg (2004) themselves have abandoned their earlier estimate of 364 Ma for the F/F boundary for the 10 million year older estimate of 374.5 ± 2.6 Ma, an estimate whose ranges include the 376.5 Ma estimate of Tucker et al. (1998) and the 376.1 ± 1.6 Ma estimate of Kaufmann et al. (2004).

The immediate consequence of an age of the F/F boundary around 376 Ma is that the Siljan, Charlevoix, and Woodleigh impact craters are all much younger than the F/F boundary, and all post-date the extinctions in the Late Frasnian. This in turn makes the catastrophic-impact models less likely for two reasons: (1) it removes these impact events from the critical time intervals, and (2) it suggests that when these major impacts did occur in the Famennian (or later) they were associated with no major biotic crisis.

### *6.2. An accurate Late Devonian global temperature curve*

One possible way around the critical problem of the uncertain age of the F/F boundary is to construct an accurate Late Devonian global temperature curve, which then could be compared with the model predictions shown in Figures 3 and 6. Such an approach cannot definitively prove or disprove the multiple impacts versus catastrophic volcanics causal question, but it could provide a definitive test of direct-effect (Fig. 3) versus lag-time (Fig. 6) models. Early attempts to construct a Late Devonian temperature curve using oxygen isotope analyses using brachiopod shells yielded temperature measurements that were clearly too high, in the lethal range for marine organisms (see discussion in McGhee, 1996, pp. 139–140). More promising results are currently being generated using conodonts (Joachimski and Buggisch, 2002; Joachimski et al., 2004).

The recently published Middle and Late Devonian oxygen-isotope temperature curve of Joachimski et al. (2004) suggests that global sea-surface temperatures in the Middle Devonian and Early Frasnian ranged from 22 to 25°C, but then increased markedly in the Middle to Late Frasnian to 32°C, dropped sharply by 5–7°C in two abrupt pulses in the Late Frasnian Late *rhenana* and *linguiformis* Zones, and then increased again in the Famennian to a range of 30–33°C. At first glance, such a temperature profile better matches the predictions of the lag-time model (Fig. 6), which predicts a marked increase in temperature in the earlier Frasnian, followed by sharp drops of temperature in the later Frasnian.

The Joachimski et al. (2004) analyses indicate that global sea-surface temperatures dropped from 34 to 26°C in two sharp pulses in the Late Frasnian Late *rhenana* and *linguiformis* Zones, coincident with the F/F extinction pulses. As such, these analyses corroborate the three decades of palaeoecological analyses that have indicated that the F/F extinction pulses were associated with cooling events (Copper, 1977, 2002; McGhee, 1996, 2001; Streel et al., 2000).

What does not match are the surprisingly high temperature values for the Famennian, which do not follow the predictions of the lag-time model (Fig. 6) nor the palaeobotanical data, which indicate that the Early Famennian was cool and dry (Streel et al., 2000). Both the palaeobotanical data, and the Berner (2001) $CO_2$ curve (Fig. 1), predict that the Famennian should be cooler than the Givetian, not vice versa. The Joachimski et al. (2004) Famennian temperature analyses are also puzzling in that the $CO_2$ models of Goddéris and Joachimski (2004) predict a drop in atmospheric $CO_2$ composition from 2925 ppm in the Late *rhenana* Zone (Late Frasnian) to 1560 ppm in the Late *triangularis* Zone (three zones up into the Famennian), a decline in global $CO_2$ partial pressures consistent with the Berner (2001) $CO_2$ curve (Fig. 1). It is difficult to resolve these low $CO_2$ values in the Early Famennian with a temperature that is as high as that in the Frasnian, with its much more $CO_2$-rich atmosphere. Clearly, much more work needs be done in constructing an accurate Late Devonian temperature curve.

## 7. Conclusions

In order to definitively test the impact-induced versus volcanic-induced extinction models, or to definitively test direct-effect versus lag-time extinction models, we must be able to accurately date these two key indicators in geologic time, relative to the F/F boundary: LIPs versus craters. Recent research results suggesting a 376 Ma age for the F/F boundary would indicate that several known impact craters post-date the F/F extinctions, and make the impact-induced extinction models less likely. High research priority should thus be given to obtaining a biostratigraphic date or an accurate radiometric date on the Viluy Traps, as a test to corroborate the volcanic-induced extinction models. Failing this, we may be able at least to test direct-effect versus lag-time extinction models if we can construct an accurate Late Devonian global temperature curve.

## Acknowledgements

I thank J. Morrow, J. Over, and P. Wignall, organizers of the 2003 Paleontological Society symposium "Understanding Late Devonian and Permian-Triassic Biotic and Climatic Events: Towards an Integrated Approach" for their invitation to speak. I also thank G. Racki, J. Over, and J. Day for constructive comments on the original manuscript.

## References

Abbot, D.H., Isley, A.E., 2002. The intensity, occurrence, and duration of superplume events and eras over geological time. J. Geodyn. 34, 265–307.

Algeo, T.J., Scheckler, S.E., Maynard, J.B., 2001. Effects of the Middle to Late Devonian spread of vascular land plants on weathering regimes, marine biotas, and global climate. In: Gensel, P.G., Edwards, D. (Eds), Plants Invade the Land. Columbia University Press, New York, pp. 213–236.

Alvarez, W., 2003. Comparing the evidence relevant to impact and flood basalt at times of major mass extinctions. Astrobiology 3, 153–161.

Alvarez, L.W., Alvarez, W., Asaro, F., Michel, H.V., 1980. Extraterrestrial cause for the Cretaceous-Tertiary extinction. Science 208, 1095–1108.

Bailey, M.E., Clube, S.V.M., Hahn, G., Napier, W.M., Valsechi, G.B., 1994. Hazards due to giant comets: climate and short-term catastrophism. In: Gehrels, T. (Ed.), Hazards due to Comets and Asteroids. University of Arizona Press, Tucson, pp. 479–536.

Becker, R.T., House, M.R., 1994. Kellwasser events and goniatite successions in the Devonian of the Montagne Noire with comments on possible causations. Cour. Forsch. Inst. Senckenberg 169, 45–77.

Berner, R.A., 2001. The effect of the rise of land plants on atmospheric $CO_2$ during the Paleozoic. In: Gensel, P.G., Edwards, D. (Eds), Plants Invade the Land. Columbia University Press, New York, pp. 173–178.

Buggisch, W., 1991. The global Frasnian-Famennian "Kellwasser Event." Geol. Rundsh. 80, 49–72.

Compston, W., 2000. Interpretation of SHRIMP and isotope dilution zircon ages for the Palaeozoic time-scale: II. Silurian to Devonian. Mineral. Mag. 64(6), 1127–1146.

Copper, P., 1977. Paleolatitudes in the Devonian of Brazil and the Frasnian-Famennian mass extinction. Palaeogeogr., Palaeoclimatol., Palaeoecol. 21, 165–207.

Copper, P., 2002. Reef development at the Frasnian/Famennian mass extinction boundary. Palaeogeogr., Palaeoclimatol., Palaeoecol. 181, 27–65.

Courtillot, V.E., 1999. Evolutionary Catastrophes: The Science of Mass Extinction. Columbia University Press, New York.

Courtillot, V.E., Renne, P.R., 2003. On the ages of flood basalt events. Comptes Rendus Geosci. 335, 113–140.

Ellwood, B.B., Benoist, S.L., El Hassani, A., Crick, R.E., 2003. Possible multiple bolide impacts at the Frasnian-Famennian boundary: evidence from two sections, Bou Tchrafine and Jebel Amelane, located in the Anti-Atlas of Morocco. Geological Society of America, Abstracts with Programs 35(6), 209.

Filer, J.K., 2002. Late Frasnian sedimentation cycles in the Appalachian basin – possible evidence for high frequency eustatic sea-level changes. Sed. Geol. 154, 31–52.

Fischer, A.G., Arthur, M.A., 1977. Secular variation in the pelagic realm. Soc. Econ. Paleontol. Mineral. Spec. Publ. 25, 19–50.

Frakes, L.A., Francis, J.E., Syktus, J.I., 1992. Climate Models of the Phanerozoic. Cambridge University Press, Cambridge.

Goddéris, Y., Joachimski, M.M., 2004. Global change in the Late Devonian: modelling the Frasnian-Famennian short-term carbon isotope excursions. Palaeogeogr., Palaeoclimatol., Palaeoecol. 202, 309–329.

Gradstein, F.M., Ogg, J., 1996. A Phanerozoic time scale. Episodes 19, 3–5.

Gradstein, F.M., Ogg, J., 2004. Geologic time scale 2004 – why, how, and where next! Lethaia 37, 175–181.

Grieve, R.A.F., Rupert, J., Smith, J., Therriault, A., 1995. The record of terrestrial impact cratering. GSA Today 5(189), 194–196.

Hallam, A., Wignall, P.B., 1997. Mass Extinctions and Their Aftermath. Oxford University Press, Oxford.

Hladil, J., 2002. Geophysical records of dispersed weathering products on the Frasnian carbonate platform and early Famennian ramps in Moravia, Czech Republic: proxies for eustacy and palaeoclimate. Palaeogeogr., Palaeoclimatol., Palaeoecol. 181, 213–250.
Joachimski, M.M., Buggisch, W., 2002. Conodont apatite $\delta^{18}O$ signatures indicate climate cooling as a trigger of the Late Devonian (F-F) mass extinction. Geology 30, 711–714.
Joachimski, M.M., van Geldern, R., Breisig, S., Buggisch, W., Day, J., 2004. Oxygen isotope evolution of biogenic calcite and apatite during the Middle and Late Devonian. Internat. J. Earth Sci. 93, 542–553.
Kaufmann, B., Trapp, E., Mezger, K., 2004. The numerical age of the Upper Frasnian (Upper Devonian) Kellwasser horizons: a new U-Pb zircon date from Steinbruch Schmidt (Kellerwald, Germany). J. Geol. 112, 495–501.
Keller, G., 2001. The end-Cretaceous mass extinction in the marine realm: year 2000 assessment. Planet. Space Sci. 49, 817–830.
Ma, X.P., Bai, S.L., 2002. Biological, depositional, microspherule, and geochemical records of the Frasnian/Famennian boundary beds, South China. Palaeogeogr., Palaeoclimatol., Palaeoecol. 181, 325–346.
McGhee, G.R., 1981. The Frasnian-Famennian extinctions: a search for extraterrestrial causes. Bull. Field Mus. Nat. Hist. 52(7), 3–5.
McGhee, G. R., 1982. The Frasnian-Famennian extinction event: a preliminary analysis of Appalachian marine ecosystems. In: Silver, L. T., Schultz, P. H. (Eds), Geological Implications of Impacts of Large Asteroids and Comets on the Earth. Geological Society America, Special Paper No. 190, 491–500.
McGhee, G.R., 1996. The Late Devonian Mass Extinction. Columbia University Press, New York.
McGhee, G.R., 2001. The 'multiple impacts hypothesis' for mass extinction: a comparison of the Late Devonian and the late Eocene. Palaeogeogr., Palaeoclimatol., Palaeoecol. 176, 47–58.
McGhee, G.R., 2003. Testing Late Devonian extinction hypotheses. Geological Society of America, Abstracts with Programs 35(6), 207.
McLaren, D.J., 1970. Time, life, and boundaries. J. Paleontol. 44, 801–815.
Morgan, J.P., Reston, T.J., Ranero, C.R., 2004. Contemporaneous mass extinctions, continental flood basalts, and 'impact signals': are mantle plume-induced lithospheric gas explosions the causal link? Earth Planet. Sci. Lett. 217, 263–284.
Morrow, J.R., Sandberg, C.A., Warme, J.E., Kuehner, H.-C., 1998. Regional and possible global effects of sub-critical Late Devonian Alamo impact event, southern Nevada, USA. J. Br. Interplanet. Soc. 51, 451–460.
Nordt, L., Atchley, S., Dworkin, S., 2003. Terrestrial evidence for two greenhouse events in the latest Cretaceous. GSA Today 13(12), 4–9.
Over, D.J., 2002. The Frasnian/Famennian boundary in central and eastern United States. Palaeogeogr., Palaeoclimatol., Palaeoecol. 181, 153–169.
Poag, C.W., 1997. Roadblocks on the kill curve: testing the Raup hypothesis. Palaios 12, 582–590.
Poag, C.W., Mankinen, E., Norris, R. D., 2003. Late Eocene impacts: geologic record, correlation, and paleoenvironmental consequences. In: Prothero, D.R., Ivany, L.C., Nesbitt, E.A. (Eds), From Greenhouse to Icehouse: The Marine Eocene-Oligocene Transition. Columbia University Press, New York, pp. 495–510.
Racki, G., 1998. Frasnian-Famennian biotic crisis: undervalued tectonic control? Palaeogeogr., Palaeoclimatol., Palaeoecol. 141, 177–198.
Racki, G., 1999. The Frasnian-Famennian biotic crisis: how many (if any) bolide impacts? Geol. Rundsch. 87, 617–632.
Racki, G., Racka, M., Matyja, H., Devleeschouwer, X., 2002. The Frasnian/Famennian boundary interval in the South Polish-Moravian shelf basins: integrated event-stratigraphic approach. Palaeogeogr., Palaeoclimatol., Palaeoecol. 181, 251–297.
Retallack, G.J., 1997. Early forest soils and their role in Devonian global change. Science 276, 583–585.
Rotondo, K.A., Over, D.J., 2000. Biostratigraphic age of the Belpre Ash (Frasnian) in the Chattanooga and Rhinestreet Shales of the Appalachian Basin. Geological Society of America, Abstracts with Programs 32(7), A225.
Sandberg, C.A., Morrow, J.R., Ziegler, W., 2002. Late Devonian sea-level changes, catastrophic events, and mass extinctions. In: Koeberl, C., MacLeod, K.G. (Eds), Catastrophic Events and Mass Extinctions: Impacts and Beyond. Geological Society of America, Special Paper No. 356, 473–487.
Schieber, J., 2003. Disturbed beds in lower Frasnian black shales of Tennessee and Kentucky: their possible connection to impacts. Geological Society of America, Abstracts with Programs 35(6), 209.
Streel, M., Caputo, M.V., Loboziak, S., Melo, J.H.G., 2000. Late Frasnian-Famennian climates based on palynomorph analyses and the question of the Late Devonian glaciations. Earth-Sci. Rev. 52, 121–173.

Toon, O.B., Zahnle, K., Turco, R.P., Covey, C., 1994. Environmental perturbations caused by asteroid impacts. In: Gehrels, T. (Ed.), Hazards due to Comets and Asteroids. University of Arizona Press, Tucson, pp. 791–826.

Tucker, R.D., Bradley, D.C., Ver Straeten, C.A., Harris, A.G., Ebert, J.R., McCutcheon, S.R., 1998. New U-Pb zircon ages and the duration and division of Devonian time. Earth Planet. Sci. Lett. 158, 175–186.

Uysal, I.T., Goldring, S.D., Glikson, A.Y., Mory, A.J., Glikson, M., 2001. K-Ar evidence from illitic clays of a Late Devonian age for the 120 km diameter Woodleigh impact structure, southern Carnarvon Basis, Western Australia. Earth Planet. Sci. Lett. 192, 281–289.

Walliser, O.H., 1996. Global events in the Devonian and Carboniferous. In: Walliser, O.H. (Ed.), Global Events and Event Stratigraphy. Springer-Verlag, Berlin, pp. 225–250.

Wignall, P.B., 2001. Large igneous provinces and mass extinctions. Earth-Sci. Rev. 53, 1–33.

Wilder, H., 1989. Neue Ergebnisse zum oberdevonischen Riffsterben am Nordrand des mitteleuropäischen Variscikums. Fortschr. Geol. Rheinl. Westfal. (Geol. Landesamt Nordrh.-Westfal.) 35, 57–74.

Wilf, P., Johnson, K.R., Huber, B.T., 2003. Correlated terrestrial and marine evidence for global climate changes before mass extinction at the Cretaceous-Paleogene boundary. Proc. Natl. Acad. Sci. USA 100, 599–604.

Williams, E.A., Friend, P.F., Williams, B.P.J., 2000. A review of Devonian time scales: databases, construction and new data. In: Friend, P.F., Williams, B.P.J. (Eds), New Perspectives on the Old Red Sandstone. Geological Society of London, Special Publications No. 180, pp. 1–21.

Wilson, M., Lyashkevich, Z.M., 1996. Magmatism and the geodynamics of rifting of the Pripyat-Dnieper-Donets rift, East European Platform. Tectonophysics 268, 65–81.

Young, G.C., 1996. Devonian (Chart 4). In: Young, G.C., Laurie, J.R. (Eds), An Australian Phanerozoic Timescale. Oxford University Press, Oxford.

Yudina, A.B., Racki, G., Savage, N.M., Racka, M., Malkowski, K., 2002. The Frasnian-Famennian events in a deep-shelf succession, Subpolar Urals: biotic, depositional, and geochemical events. Acta Palaeont. Pol. 47, 355–372.

Understanding Late Devonian and Permian-Triassic Biotic and Climatic Events:
Towards an Integrated Approach
edited by D.J. Over, J.R. Morrow and P.B. Wignall



## *Chapter 4*

# *Sedimentary fill of the Late Devonian Flynn Creek crater: a hard target marine impact*

*Jürgen Schieber*[a,*], *D. Jeffrey Over*[b]

[a]*Department of Geological Sciences, Indiana University, Bloomington, IN 47405, USA*
[b]*Department of Geological Sciences, SUNY-Geneseo, Geneseo, NY 14454, USA*

## Abstract

The 3.6 km wide Flynn Creek crater in north-central Tennessee was produced by an asteroid that struck a flat lying succession of Ordovician carbonates. The crater is filled by a basal breccia and Late Devonian Chattanooga black shale. Conodonts in shallow water lag deposits that overlie the Ordovician succession in the region indicate lower Frasnian flooding of the area. The continuous stratigraphic record in the crater spans impact and post-impact deposits; the recovery of shallow water components and lower Frasnian conodonts in initial marine deposits above the crater fill breccia indicate that marine sedimentation commenced immediately after impact and that the impact occurred in shallow water. Most recent radiometric calibrations of Devonian conodont zones suggest that the impact occurred around 382 Ma.

Stratigraphy and sedimentary features suggest the following sequence of events: (1) shallow water impact during the early Frasnian; (2) formation of basal breccia as a fallback deposit; (3) deposition of graded breccia as displaced water rushed back into the crater; (4) ejecta washed back into the crater by storm-induced waves and currents; (5) accumulation and preservation of black shale first in the crater, and also as sea level rose outside the crater. Because the target rocks were lithified carbonates, the Flynn Creek crater has the morphologic characteristics of a subaerial impact. The sediment fill, however, reflects the shallow marine setting of the impact site.

*Keywords*: wet impact; impact age; sedimentary processes; black shale; preservation

## 1. Introduction

After years of controversy, the field of impact geology is rapidly merging with the geological mainstream as geologists from various specializations increasingly investigate the effect of impacts on the preserved geologic record. More and more impacts are uncovered by non-specialists during the conventional geological investigations, such as the Chesapeake Bay structure (Poag, 1997), the Alamo Breccia (Warme and Sandberg, 1996; Sandberg et al., 2002), and spherule layers (Simonson and Harnik, 2000). While in earlier impact studies the focus was on verifying the impact and determining the processes and events at impact time, the study of post-impact sediments is rapidly gaining in importance because they preserve a record of the transition between impact-related effects and post-impact geological history. In some instances crater-fill sediments are the only preserved

*Corresponding author. Fax: +1-812-855-7899. *E-mail address:* jschiebe@indiana.edu (J. Schieber).

long-term record of post-impact geological and environmental processes that originally affected a much larger area (Beales and Lozej, 1975; Partridge et al., 1993; Grieve, 1997).

As documented by Roddy (1968), the Flynn Creek structure of north-central Tennessee resulted from a Devonian asteroid impact that produced a crater of 3.6 km diameter and a 150 m depth in the flat-lying Ordovician carbonates (Figs. 1, and 2). Petrography of brecciated rocks shows that only Ordovician strata were affected (Roddy, 1968). Silurian and Middle Devonian strata, which potentially occurred in the area, had been eroded prior to the Late Devonian (Conant and Swanson, 1961). The impact deformed underlying strata to a depth of approximately 500 m and brecciated more than 2 $km^3$ of rock, 1.3 $km^3$ of which was ejected (Roddy, 1968). Deformation includes folding and faulting of underlying strata, uplift along the crater rim, and formation of a central high due to rebound of the crater floor (Fig. 2). Within the resulting annular trough, deformed Ordovician carbonates were covered by a breccia layer of 40 m average thickness (Roddy, 1968). Elsewhere in central Tennessee, Ordovician carbonates are overlain by the Upper Devonian Chattanooga Shale, a marine black shale unit of less than 10 m thickness (Conant and Swanson, 1961). In contrast, at Flynn Creek up to 55 m of black shale are preserved (Roddy, 1968).

The Flynn Creek structure, once the subject of controversy (Boon and Albritton, 1936; Dietz, 1960; Conant and Swanson, 1961; Shoemaker and Eggleton, 1961; Bucher, 1963), has an impact origin (Roddy, 1966, 1968). Supporting observations are structural features that match other impact craters, shatter cones, and remnants of an ejecta blanket in a rim graben (Roddy, 1968). Because the crater is filled by Late Devonian black shales a mid-Devonian impact on a low relief land surface, followed by a period of subaerial erosion, was proposed by Roddy (1968). Marine flooding of the crater during the early Late

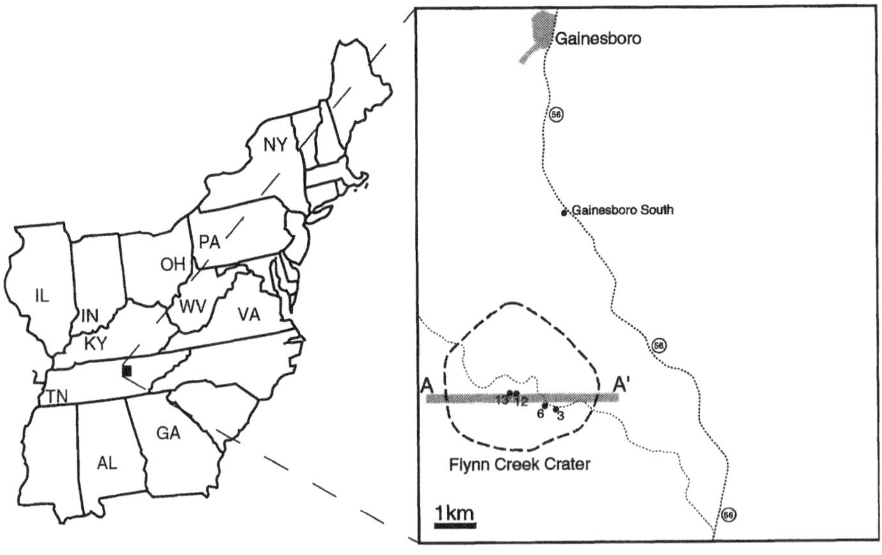


*Figure 1.* Location of study area in central Tennessee and enlarged detail map of study area with crater location and outline. The gray horizontal bar A-A' notes location of the crater cross section from Figure 2. The numbers within the crater indicate location of drill cores examined for this study. Roads are marked by short dashed lines.

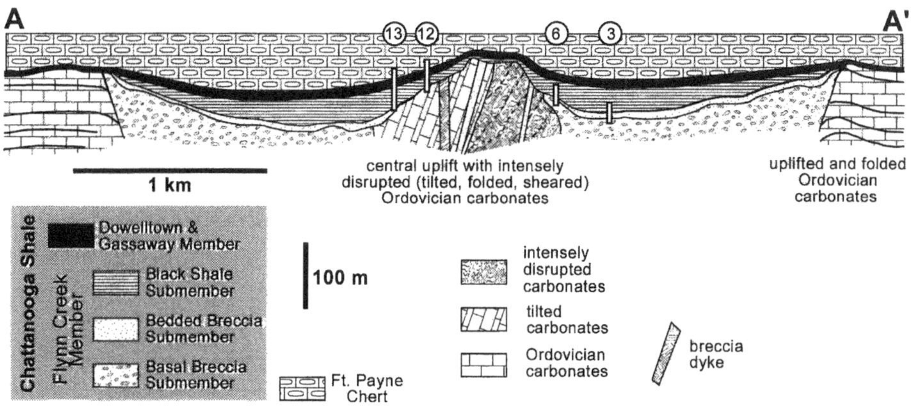


*Figure 2.* Stratigraphic overview and schematic presentation of stratigraphic relationships in the crater (crater topography from Roddy, 1966, 3 times vertical exaggeration). Vertical white bars and circled numbers mark projections of drill cores studied.

Devonian is indicated by conodonts from basal bedded crater deposits (Huddle, 1963). Roddy (1977) considered the possibility of a shallow marine impact in a later publication, but his data did not allow a clear determination.

Conversations by the senior author with David Roddy prior to his death in 2002 provided the impetus to conduct this study. Dave was aware that several critical questions still required an answer. Most pressing among these were: (1) whether the impact occurred on land or in water; (2) whether there was a temporal gap between the impact breccias and the subsequent post-impact deposits; and (3) how much constraint on the impact age could be obtained from the conodonts in the crater fill sediments. Cores examined in the summer of 1999 and 2000 lead to the observations that resolve these questions.

Sedimentologic and petrographic examination of the crater fill shows conclusively that the impact must have occurred when the area was covered by a shallow sea. Marine crater sedimentation commenced immediately after impact, and conodonts from the crater fill allow determination of impact time within the constraints of Devonian conodont stratigraphy and radiometric age determination (Klapper, 1997; Tucker et al., 1998).

The Chattanooga Shale, as defined by Conant and Swanson (1961), comprises only a small portion of the black shale fill in the crater. The bulk of the black shale is part of an earlier deposited member of the Chattanooga Shale, largely absent elsewhere, that extends the record of Devonian black shale deposition in central Tennessee.

## 2. Description of crater fill

We propose the name Flynn Creek Member for the portion of the crater fill that underlies the Dowelltown Member of the Chattanooga Shale of earlier definition (Conant and Swanson, 1961). The Flynn Creek Member consists, in ascending order, of three distinct units: (1) basal breccia; (2) bedded breccia; and (3) black shale submembers.

### *2.1. Basal breccia submember*

The basal breccia consists of a very poorly sorted, chaotic mixture of boulder to granule size angular carbonate clasts derived from underlying strata. Coarser particles are set in a matrix of sand- to silt-size carbonate detritus and dolomite cement; deformation fabrics are common at the base. This unit is on average 40 m thick and passes without visible break into an uppermost division with clear normal grading. No other sedimentary structures are apparent. The unit is capped by a thin (2 cm) carbonaceous shale drape (Fig. 3).

### *2.2. Bedded breccia submember*

The bedded breccia begins at the lowest shale drape that contains approximately 25% carbonate clasts, quartz and chert grains, silicified fossil debris, and well-rounded phosphate granules in a fine-grained matrix of organic matter, dolomicrite, and clays (Fig. 4). The shale is overlain by dolomite-cemented beds of gravel, granule, sand, and silt-size carbonate debris that range from 50 to 150 cm in thickness (Figs. 3–5). Beds vary in number from place to place, are separated from each other by shale drapes, are massive to crudely wavy-parallel bedded (centimeter to decimeter scale), and fine upwards in places. Roddy (1968) distinguished poorly bedded breccia, bedded dolomitic breccia, and bedded dolomite subdivisions that reflect the relative proportions of coarse versus fine carbonate debris. Poorly bedded breccia is dominated by carbonate clasts ranging in size from a few millimeters to several centimeters (Fig. 5); bedded dolomitic breccia contains a substantial proportion of sand- to silt-size carbonate debris; and bedded dolomite is dominated by sand- to silt-size grains and contains essentially no coarser clasts (Fig. 6). Poorly bedded breccia is more prominent along the margins of the crater; bedded dolomitic breccia and bedded dolomite dominate the interior portions of the crater (Fig. 2). The bedded breccia contains rounded carbonate clasts, several percent of scattered quartz grains in which no shock lamellae, chert grains, and rounded phosphate granules were observed (Figs. 5, and 6). Although bedded dolomite shows considerable recrystallization, many carbonate grains still have fine crystalline cores with impurities that indicate detrital origins (Fig. 7). Within bedded dolomite layers occur thinner (2–5 cm), graded dolomite beds that have horizontal lamination, water escape structures, and fading ripples (as defined by Stow and Shanmugam, 1980) in the basal portions. Dolomite beds may also show contorted laminae and, where overlain by a shale drape, have an irregular bumpy surface, probably a result of water escape (Fig. 3).

### *2.3. Black shale submember*

Depending on position within the crater, the bedded breccia/dolomite interval may be directly overlain by Devonian black shales, or the contact may be marked by a layer of coarse sandstone. This sandstone, commonly near the crater margin, consists of 75% carbonate clasts, subordinate quartz and chert grains, silicified fossil debris, and rounded phosphate granules (Fig. 8). Thin sandstone layers, a few millimeters to 3 cm thick; of the same general composition occur throughout the basal 13 m of the black shale succession.

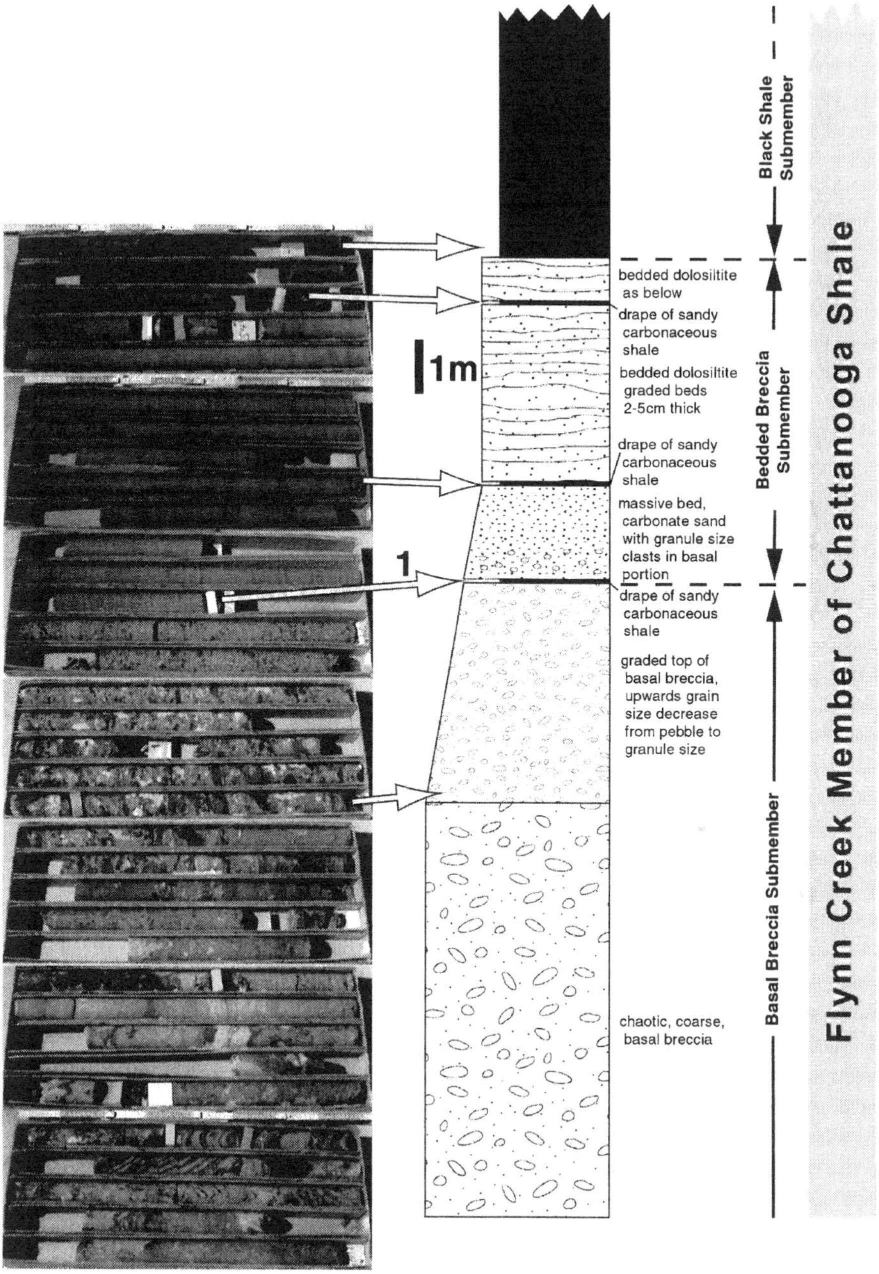


*Figure 3.* Core and corresponding stratigraphic section from drill hole 3 in the area between eastern crater rim and central uplift (see Fig. 1). Shown are the basal breccia submember, bedded breccia submember, and the bottom of the black shale submember of the Flynn Creek Member of the Chattanooga Shale. The first carbonaceous shale drape following breccia deposition is marked with arrow 1. Core boxes are approximately 82 cm long.

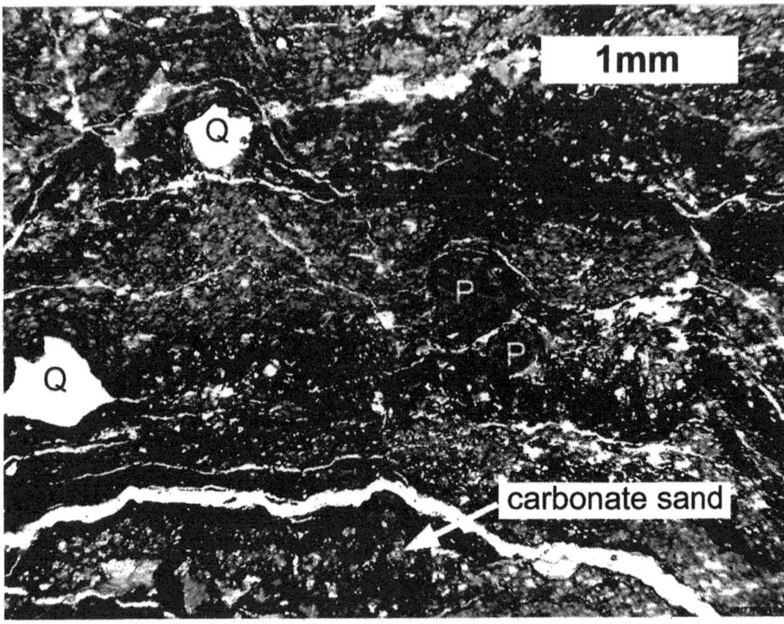


*Figure 4.* Photomicrograph of shale drape above basal breccia (arrow 1 in Fig. 3). Grains of quartz and chert are marked with Q, rounded phosphatic grains are marked with P. Note stringer of sand size carbonate debris (carbonate sand, white arrow). Wavy-crenulated nature of laminae due to compression of shale drapes over an irregular surface (e.g. projecting clasts).

Sandstone layers have sharp contacts and basal scour features on the underlying black shale and contain black shale rip-up clasts. Larger clasts can project above the upper surface of sandstone layers. The black shales are faintly banded, darker bands of clay, quartz silt, dolomite, and abundant organic matter alternate with lighter bands enriched in dolomite grains (Fig. 9). Silt- to sand-size carbonate debris at the base of lighter bands indicates that these constitute graded couplets (Fig. 9). Thin beds containing sand-sized quartz and pyrite grains, usually with diffuse lower and upper boundaries, carry the imprint of early diagenetic infilling of cysts of the marine alga *Tasmanites* (Fig. 10; Schieber, 1996, 1998a). Basal portions of the black shale succession also contain 10–30 cm thick intervals that have intense soft sediment deformation. The upper portion of the black shale succession is essentially identical to the Dowelltown Member.

## 3. Stratigraphic relationships

Within the bedded breccia submember, beds of bedded breccia are dominant near the crater margins, and bedded dolomite dominates the crater interior. Core logs and gamma ray profiles allow reconstruction of stratigraphic relationships between the black shale submember and the overlying Dowelltown Member of the Chattanooga Shale (Fig. 11). Only the upper portion of the crater fill succession is a direct match for the Dowelltown and Gassaway members at the reference section 20 km to the south at Hurricane Bridge.

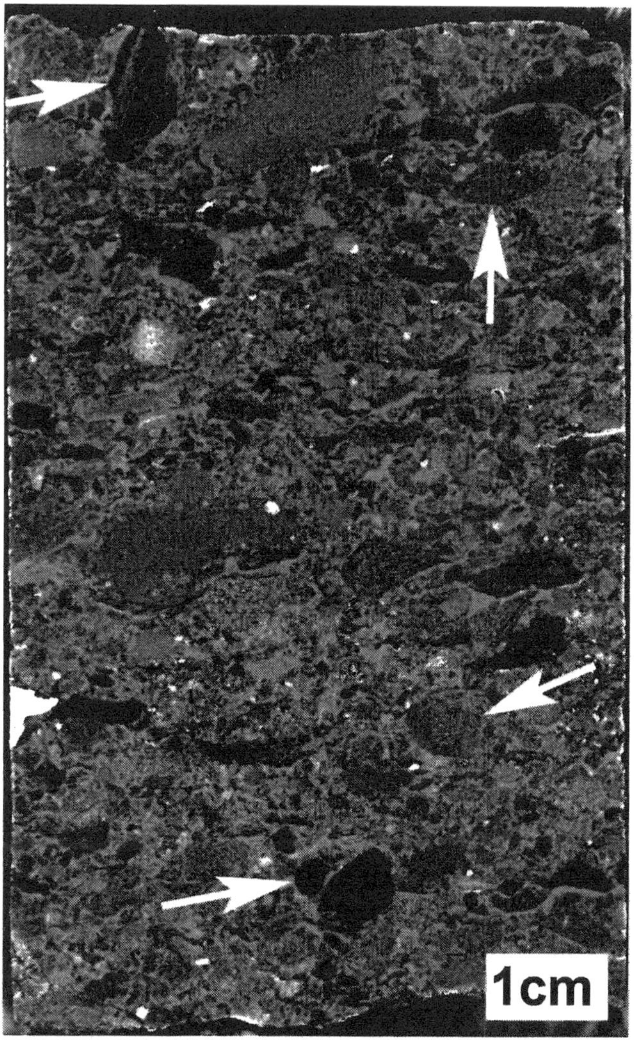


*Figure 5.* Photo of core sample of bedded breccia (core # 3, bedded breccia submember). Several well-rounded carbonate clasts are clearly visible (white arrows).

The black shale of the Flynn Creek Member forms a thick succession with comparatively low gamma ray response and lacks an obvious equivalent outside the crater. The shale also contains disseminated dolomite, a constituent that is absent in the overlying Dowelltown and Gassaway members. Gamma ray and core logs indicate several traceable horizons within the black shale submember of the Flynn Creek Member. These horizons differ in dip from the overlying Dowelltown Member and indicate that the Flynn Creek Member suffered erosional truncation prior to Dowelltown deposition (Fig. 11). The contact between the Flynn Creek Member and the base of the overlying Dowelltown Member is marked by a lenticular, coarse sandy lag with quartz and chert grains, silicified fossil debris, and rounded phosphate granules (Fig. 12).

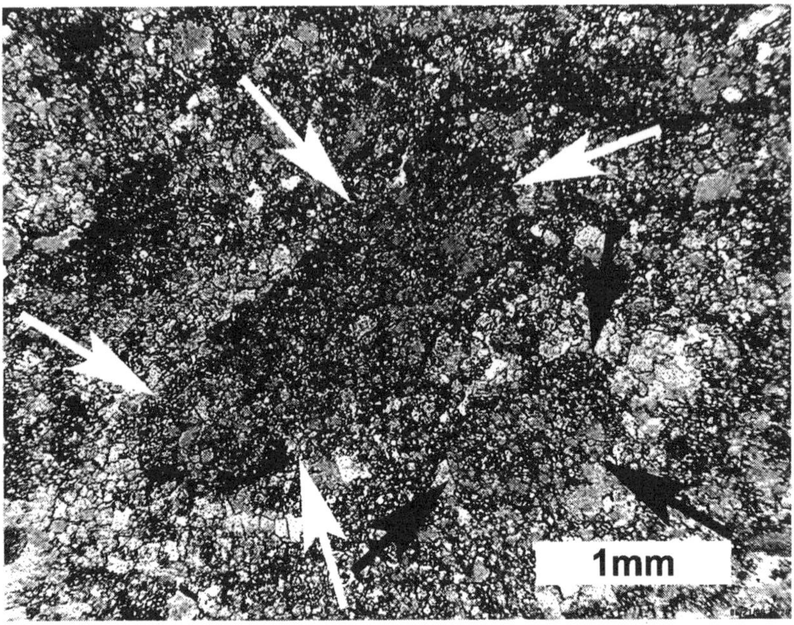


*Figure 6.* Photomicrograph of small rounded carbonate clasts in bedded dolomite (marked with white and black arrows).

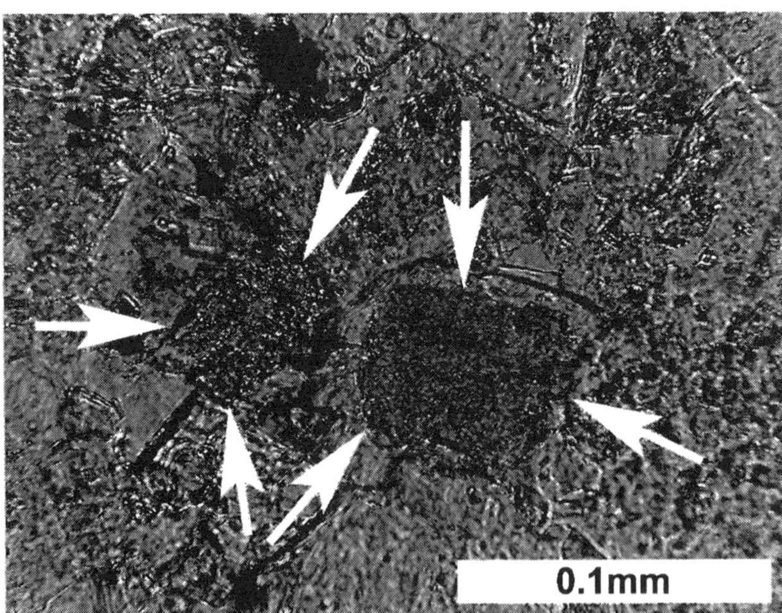


*Figure 7.* Photomicrograph of bedded dolomite. White arrows point out detrital cores in two dolomite grains.

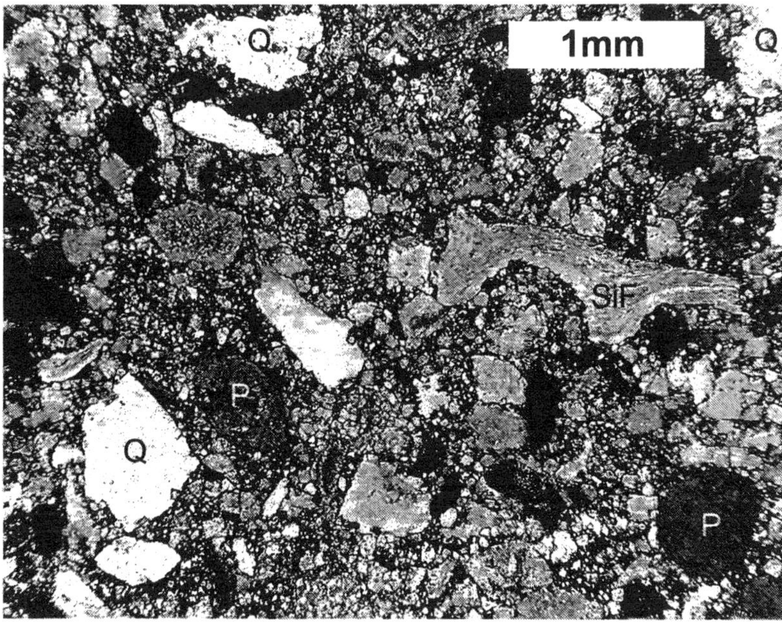


*Figure 8.* Photomicrograph of sandstone bed at base of Dowelltown Member. Grains marked Q are mainly pieces of chert, SiF identifies a silicified brachiopod shell, and P marks rounded phosphatic grains.

## 4. Reconstruction of impact-related processes and depositional history

In comparison to other wet target impact craters (e.g. Poag, 1997), the chaotic basal breccia is best interpreted as a fall-back deposit that formed immediately after the impact. The graded top portion suggests deposition controlled by settling velocity of particles, commonly observed where particles settle through a turbulent fluid/sediment mixture (Boggs, 1995). Thus, impact occurred while the area was covered by water. Impact-displaced water rushed back into the void and carried freshly ejected material back into the crater. The turbulence associated with such a scenario is extreme and allows for short-term suspension transport of pebble-size particles (Nichols, 1999). As turbulence abated the coarsest particles settled first, and sand- to silt-size material settled last. The absence of a depositional break between the coarse chaotic and the graded top portion of the basal breccia are indicative of a depositional continuum, typical of other graded breccias at marine impact sites (e.g. Poag and Aubry, 1995; Poag, 1997; Dypvik and Jansa, 2003). The basal breccia submember, including the graded top portion, probably represents a time interval measurable in hours.

The shale drape over the basal breccia indicates low energy conditions after impact-related turbulence had subsided (Figs. 3, and 4). Petrographic equivalents of the quartz and chert grains, silicified fossil debris, conodonts, and rounded phosphate granules do not occur in breccia samples. Outside the crater, however, these grain types comprise the basal Dowelltown Member lag that overlies Ordovician strata in many localities (Conant and Swanson, 1961; Schieber, 1998b). Conodonts from the basal Dowelltown lag range in age from upper Givetian to lower Frasnian and suggest that shallow water conditions persisted

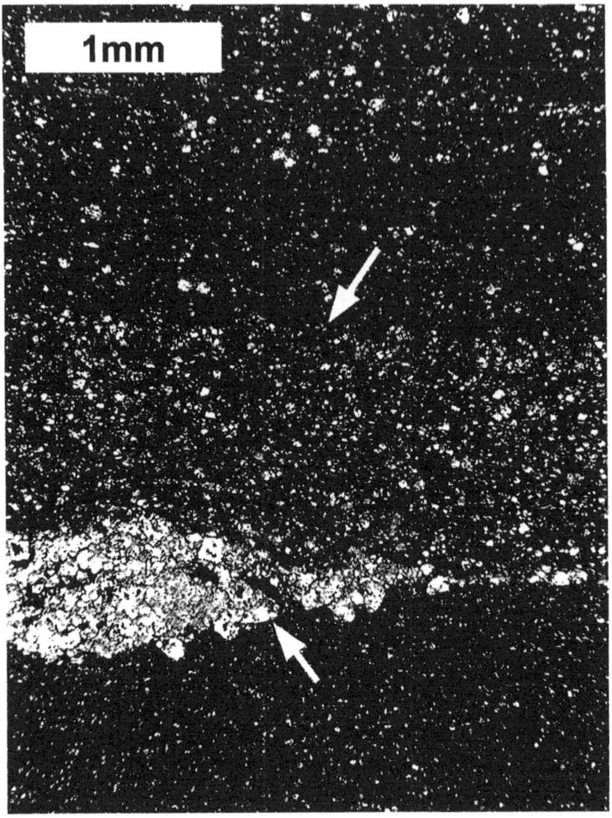


*Figure 9.* Photomicrograph of banded black shale. Light colored grains are mainly dolomite. Arrows point to base and top of a lighter colored layer, which contains more dolomite than darker layers above and below. In places, as shown here, these light colored bands may have lenses of silt- to sand-size carbonate detritus at the base.

for a long time period in the region and prevented accumulation of fine-grained sediments (Ettensohn et al., 1989), or discontinuous deposition where earlier deposited material was winnowed and resistant particles accumulated. The lack of mudstone intercalations in the lag suggests deposition above fair weather wave base. The epicontinental setting of the Devonian inland sea and water depth estimates for shale deposition in the Chattanooga Shale suggest a water depth of 10 m or less (Conant and Swanson, 1961; Schieber, 1998a). The composition of the shale drape that covers the basal breccia implies that the carbonate particles were derived from an ejecta blanket outside the crater, were washed across the crater rim during storm events, and settled in the deeper, mud-dominated portions of the crater. Considering the overall shallow water conditions in the area this should have been a frequent occurrence. Abundant coarse material in this shale drape suggests rapid accumulation, possibly representing only a few hundreds to thousands of years.

Thick beds of bedded breccia/dolomite indicate pulses of coarse clastic deposition. Rounded carbonate clasts, together with scattered quartz, chert, and phosphate grains, indicate reworking and mixing of ejecta with lags in the crater vicinity (Fig. 6). Storms were

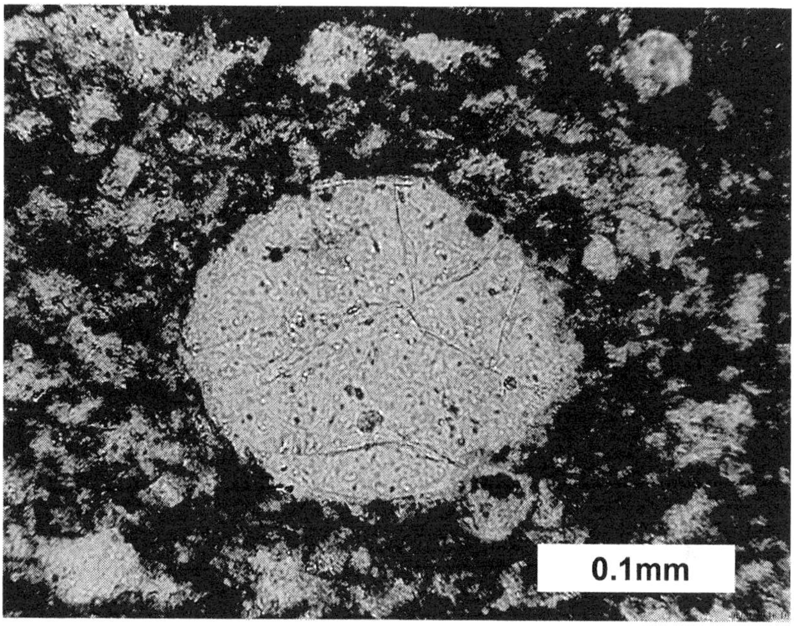


*Figure 10.* Photomicrograph of chalcedonic quartz grain (center) in banded black shale interpreted as early diagenetic (pre-compaction) infill of *Tasmanites* cyst.

probably the main agent that moved these materials back into the crater to form depositional aprons (Roddy, 1968). Gravity flow processes dominated sediment transport on the steep slopes, up to 30 degrees, along the crater rim. The typically massive appearance of bedded breccia beds indicates grainflow as the most likely transport mechanism (Middleton and Southard, 1984). Fine-grained turbidite flow is indicated for thin, graded dolomite beds that have horizontal laminae and fading ripples (Stow and Shanmugam, 1980; Stow and Piper, 1984). Water escape structures, attesting to rapid deposition, are consistent with such an interpretation. Within the bedded breccia submember, shale drapes amount to 1% or less of the accumulated sediment (Fig. 3). Because the deeper water in the crater favors mud accumulation, rapid emplacement of the entire interval is indicated.

The coarse sandstone at the base of the black shale succession in the crater is a mixture of carbonate ejecta with basal Chattanooga lag. The relative increase of lag-related chert grains, silicified fossil debris, conodonts, and phosphate granules indicates a dwindling supply of carbonate debris through gradual removal of ejecta. The occurrence of sand layers of this type through the basal 13 m of the black shale section indicates, for that stratigraphic interval, that the water outside the crater was still shallow enough to allow wave reworking and lag formation. Steep slopes, black shale rip-ups, clasts projecting above the bed surface, and basal scours suggest that these sandstone layers were emplaced by grainflow. Because the bedded breccia and black shale submembers span several conodont zones (see section 6), this suggests an initial time interval of several hundred thousand years when black shale deposition occurred only within the crater, while shallow water conditions and lag formation persisted outside.

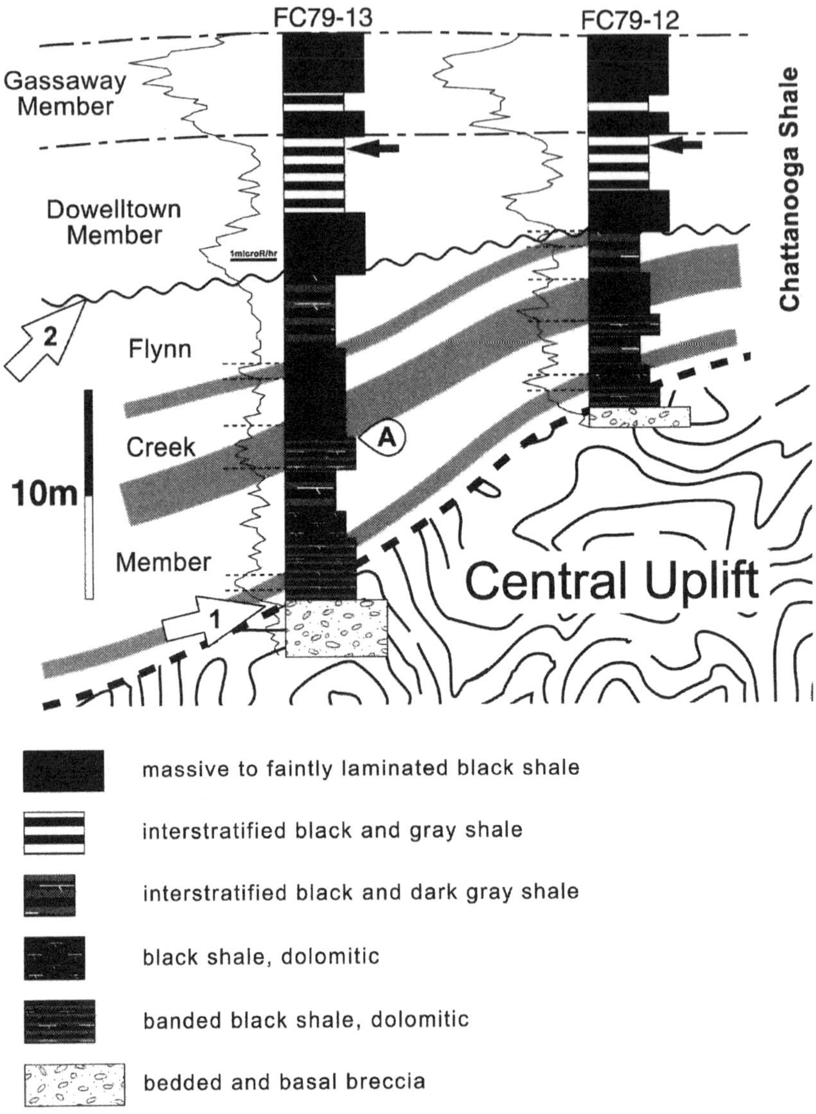


*Figure 11.* Black shale stratigraphy from drill cores at the western flank of the central uplift within the crater (the last two numbers of the core labels correspond to core numbers given in Fig. 1). The gamma ray profiles were constructed from scintillometer measurements made in 10 cm intervals on drill cores. Gamma ray profiles typical for the Dowelltown and Gassaway members only match the upper portion of the black shale succession in the crater (gamma ray counts increase toward the left). The underlying black shale submember of the Flynn Creek Member has an overall lower gamma ray activity. The black arrows indicate the position of the Center Hill Ash Bed. Gray bands follow horizons marked by increased gamma ray response. Toward the right, the uppermost band is truncated at the Dowelltown contact, indicating an angular unconformity between the black shales. The black shale submember consists of several different facies types that are stacked in two depositional cycles. Marker A is a reference point for conodont samples collected from the black shale submember. Conodont samples 7/4/2000–1, 7/4/2000–2, and 7/4/2000–3 were collected with even spacing from 21 m to 13.5 m above marker A. Arrow 2 marks approximate projection of location of 7/4/2000–3. Arrow 1 marks stratigraphic location of FC base conodont sample.

*Figure 12.* Photo of core sample from the crater fill that was collected from the base of the Dowelltown Member. Scours and pyrite cemented sandy lag that marks the base of the Dowelltown Member of the Chattanooga Shale are visible at the top of the sample. An uncompressed burrow tube, pointed out with white arrow, is filled with sand from the lag. The black shale shows differential compaction around the burrow (binocular examination), the amount of which suggests a water content/porosity of about 35% at the time the burrow was emplaced.

Graded bedding in black shales of the Flynn Creek Member and steep depositional slopes suggest a fine-grained turbidite component (Stow and Shanmugam, 1980), consistent with intervals of soft sediment deformation that are attributable to downslope sediment creep and slumping. Darker carbonaceous bands probably reflect "pelagic" background sedimentation (Fig. 9). Silica-filled *Tasmanites* cysts in these shales are due to in situ silica precipitation during early diagenesis (Schieber, 1996). Abundant cysts probably represent sediment-starved intervals when biogenic components such as algal matter and radiolaria dominated sedimentation (Schieber, 1998a).

The Flynn Creek Member was truncated prior to Dowelltown deposition (Fig. 11). Within the context of erosion in the Chattanooga Shale this implies regression and partial removal of the Flynn Creek Member, followed by transgression and deposition of the Dowelltown Member. In several places in central Tennessee the Chattanooga Shale has a thin, usually less than 1 m thick, interval of dolomite bearing black shale at the base (Schieber, 1998b). This interval is laterally discontinuous and separated from overlying shale by an erosion surface. Previously considered part of the Dowelltown Member, petrographic characteristics, location beneath an erosion surface, and low gamma ray intensity indicate that this stratum is equivalent to the Flynn Creek Member. Preservation outside of the crater implies that sea level rose sufficiently during deposition of the Flynn

Creek Member to allow mud accumulation outside of the crater, and that the region was once covered by a pre-Dowelltown black shale blanket. Water depth may have increased from around 10 m to as much as 50 m (Schieber, 1994, 1998a).

At the time of Dowelltown deposition, black shales of the Flynn Creek Member still had about 35% porosity (Fig. 12). Thus, as these shales compacted further the crater continued to influence structure, distribution, and thickness of post-impact deposits. Continued compaction of the Flynn Creek Member caused the Dowelltown Member to be 40% thicker in the crater than in the type section, 7 versus 5 m (Fig. 11).

The disconformity that separates the Dowelltown and Gassaway members is of regional extent (Schieber, 1998b). It approximates the Frasnian–Famennian boundary (Over, 2002) and is representative of the boundary between transgressive–regressive cycles IId and IIe of Johnson et al. (1985). The Gassaway Member in the crater is only marginally thicker than in surrounding areas, an indication that by Famennian time the crater fill had compacted sufficiently to produce a more or less level surface.

## 5. Comparison to other marine impacts

The presence of water and the properties of the target rock determine associated sedimentary processes and the sedimentary signature that a marine impact leaves in the rock record (Dypvik and Jansa, 2003). If the target consists of unconsolidated or poorly lithified sediments or sedimentary rocks, the resulting crater may lack a continuous uplifted rim and show resurge gullies and rimwall collapse (Dypvik and Jansa, 2003). The Flynn Creek crater shows an uplifted rim that was not significantly beveled by post-impact erosion and was not dissected by resurge gullies (Roddy, 1968), an indication that the Ordovician target rocks were lithified by that time. The crater morphology is a close match to that expected of a subaerially produced crater. The smooth-flattened top of the central high/uplift (Roddy, 1968), in contrast to the sharp central peaks of terrestrial impact craters, suggests post-depositional modification by wave action, the result of shallow water cover (Dypvik and Jansa, 2003).

## 6. Time of impact

Conodonts from the fill of the Flynn Creek structure clearly constrain the relative age of the Flynn Creek Member basal breccia, bedded breccia, and black shale submembers, as well as the overlying Dowelltown Member of the Chattanooga Shale. The Frasnian conodonts are readily placed within the Montagne Noire (MN) zonation proposed by Klapper (1989, 1997). The basal and bedded breccia submembers contain a mixed fauna of Late Ordovician and Devonian conodonts (Fig. 13). The Ordovician taxa, including *Yaoxianognathus* and *Plectodina* are typical of the Nashville Group that floors the crater (S. Leslie, pers. comm., 2004). *Ancyrodella alata* and *Polygnathus pennatus* from the breccia beds are limited to the early Frasnian, *A. alata* occurs only within MN Zones 3 and 4. The black shale interbeds within the basal and bedded breccias, in conjunction with the occurrence of *Ancyrodella rotundiloba* and *P. pennatus* that were tabulated but not illustrated by Huddle (1963), from the base and top of the basal breccia, as well as from the bedded breccia and base of the black shale submember, indicate impact and infill of the

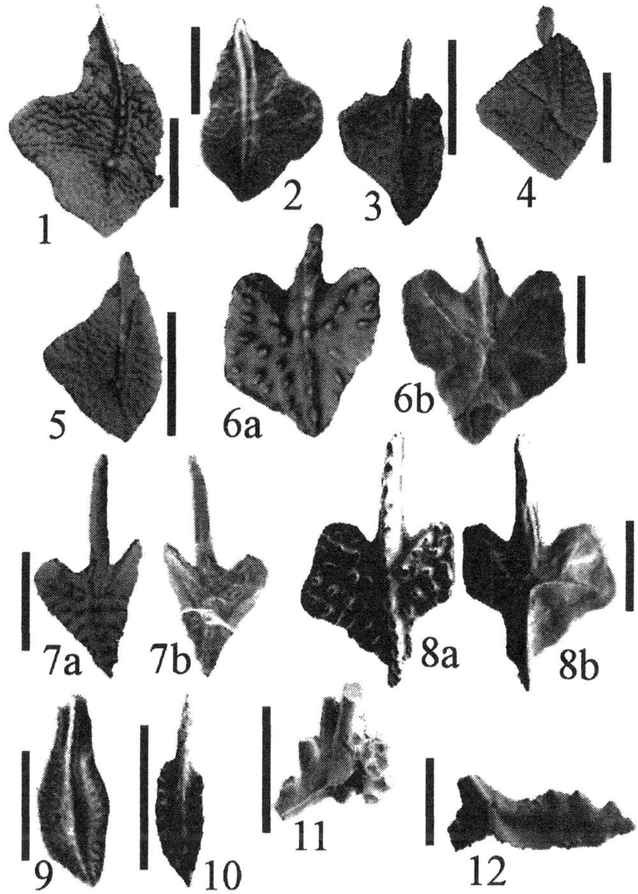


*Figure 13.* SEM digital images of conodonts from the Dowelltown and Flynn Creek members of the Chattanooga Shale. Specimens are reposited in the Orton Museum, Ohio State University, Columbus, Ohio. Scale bars are 0.5 mm. 13-1 – *Palmatolepis punctata* (Hinde, 1879) late form, upper view, inverse image of mold, OSU 52225, Dowelltown Member, TN70-09. 13-2 – *Palmatolepis bohemica* Klapper and Foster, 1993, upper view, OSU 52226, Dowelltown Member, HC2-02. 13-3 – *Palmatolepis* aff. *P. bohemica,* upper view, inverse image of mold, OSU 52227, black shale submember of Flynn Creek Member, FC 7/4/2000–02. 13-4 - *Palmatolepis* aff. *P. punctata*, upper view, inverse image of mold, OSU 52228, black shale submember of Flynn Creek Member, FC 7/4/2000–02. 13-5 – *Palmatolepis punctata* early form, upper view, inverse image of mold, OSU 52229, black shale submember of Flynn Creek Member, FC 7/4/2000–02. 13-6 – *Ancyrodella curvata* Branson and Mehl (1934b) early form of Feist and Klapper (1985), upper inverse image of mold and lower view, OSU 52230, black shale submember of Flynn Creek Member, FC 7/4/2000–02. 13-7 – *Ancyrodella gigas* Youngquist, 1947, upper inverse image of mold and lower view, OSU 52231, black shale submember of Flynn Creek Member, FC 7/4/2000–02. 13-8 – *Ancyrodella alata* Glenister and Klapper, 1966, upper and lower view, OSU 52232, basal breccia submember of Flynn Creek Member, FC base. 13-9 – *Polygnathus dubius* Hinde (1879), upper view, OSU 52233, basal breccia submember of Flynn Creek Member, FC base. 13-10 – *Polygnathus pennatus* Hinde (1879), upper view, OSU 52235, basal breccia submember of Flynn Creek Member, FC base. 13-11 – *Plectodina* sp., inner oblique view, OSU 52235, basal breccia submember of Flynn Creek Member, FC base. 13-12 – *Yaoxianognathus abruptus* (Branson and Mehl, 1934a), outer view, OSU 52236, basal breccia submember of Flynn Creek Member, FC base.

structure within MN Zone 3 to 4. Although these conodonts could have been transported into the crater from the neighboring platform, samples from the black shale submember, as well as 4524SD of Huddle (1963), did not contain any Ordovician conodonts, an indication that Ordovician strata in the area of the crater were buried by this time. The black shale submember yielded diverse early middle Frasnian conodonts, including *Ancyrodella curvata* early form of Feist and Klapper (1985), the early morphotype of *Palmatolepis punctata*, as well as other palmatolepids, that are typical of MN Zones 5 and 6, but do range higher. The Dowelltown Member, marked by a disconformity and basal lag, overlies the Flynn Creek Member and Ordovician strata. Regionally this lag contains Ordovician through Late Devonian conodonts. *Palmatolepis bohemica* and the late morphotype of *P. punctata* from the lowest black shale beds of the Dowelltown, just above the basal lag, indicate the middle of MN Zone 6 into MN Zone 8. This supports the correlation of the basal Dowelltown Member with the basal Rhinestreet Shale indicated by de Witt et al. (1993).

With some limitations, and acknowledging analytical error ranges of ±2 m.y. for published radiometric dates, as well as competing geochronological schemes by Sandberg and Ziegler (1996) and Streel et al. (2000), an approximate absolute age of the impact was calculated by calibrating the (MN) zonation of Klapper (1989) with Devonian radiometric ages compiled by Tucker et al. (1998). The approach is based on the Frasnian Composite Standard of Klapper (1997), where one Frasnian composite standard unit equals 0.174 m.y. if a duration of 6 m.y. for the Frasnian is used (Tucker et al., 1998). This yields an absolute age of ~382.24 Ma. for the base of MN 3 and 381.82 Ma. for the top of MN 4. With an MN Zone 3 through 4 age for the oldest post-impact sediments, the 0.42 m.y. time interval from 382.24 to 381.82 Ma. thus brackets the time of impact. Obviously, such an estimate represents a simplification and has inherent limitations, such as the unproven assumption that Frasnian composite standard units are of uniform duration, the small number of radiometric ages available, and the broad range of analytical error. Nonetheless, the approximate 382 Ma. impact age allows comparison to other Devonian impacts that have been dated by radiometric means and lack conodont data.

Because there are numerous Late Devonian impacts in the rock record, such as Flynn Creek, the Alamo Impact (southern Nevada), and Siljan (Sweden), Sandberg et al. (2002) have suggested that all of these may be related to a Late Devonian comet shower. Conodont data indicate that the Alamo Impact in southern Nevada occurred during the early middle Frasnian *punctata* Zone (Sandberg et al., 2002), an age assignment that is equivalent to MN Zones 5 and 6, close to the time of the Flynn Creek impact. Argon–argon age determination on rocks from the Siljan crater in Sweden yielded an impact age of 377 Ma. (Reimold et al., 2005).

Based on conodont ages, the MN Zone 3 to 4 Flynn Creek impact pre-dates the MN Zone 5 to 6 Alamo Impact. The "calculated" age of 381.6 Ma. for the top of MN 5 clearly predates 377 Ma. for the Siljan impact, making it unlikely that there was any causal relationship between Flynn Creek, the Alamo Impact, and the Siljan Impact.

## 7. Conclusions

The asteroid that produced the Flynn Creek crater struck a series of flat-lying and lithified Ordovician carbonates that were covered by a shallow sea. The impact occurred during the

Lower Frasnian, approximately 382 million years ago, and the marine crater fill sedimentation commenced immediately after the impact. The Flynn Creek Member of the Chattanooga Shale represents strata deposited after the impact and before deposition of the widespread Dowelltown Member.

The regional water depth at the time of impact was on the order of 10 m or less, and gradually increased after impact due to a general sea level rise observed during the Frasnian. The Flynn Creek crater is older than the Alamo Impact of southern Nevada and the Siljan Impact of Sweden. The sedimentary history recorded by the crater fill indicates repeated regressions and transgressions overprinted on a gradual rise in sea level during the early and middle Frasnian.

## Acknowledgments

Reviews of early drafts of the manuscript by Gil Klapper and John Warme were greatly appreciated. The senior author acknowledges the Donors of the Petroleum Research Fund (Grants 28236-AC2 and 30744-AC8), administered by the American Chemical Society, for funding of this study. The late Dave Roddy generously shared his understanding of the Flynn Creek Structure and provided access to drill cores and sample materials. Dave was able to comment on the first draft of this manuscript, but his untimely death in 2002 prevented him from seeing it go into print. Flynn Creek was one of Dave's favorite impact structures, and this paper is dedicated to his memory.

## Locality information

HC2-02 – Tennessee Highway 56, Hurricane Bridge, DeKalb County, near Smithville, Tennessee, Center Hill Dam Quad, UTM 16SFQ125(E)884(N) 0.05 m above base of the Dowelltown Member of the Chattanooga Shale.

FC base – Flynn Creek crater fill, approximately 400 m NE of Antioch, Jackson County, in creek bed along Cub Hollow Rd., Gainesboro Quad, UTM 16SFR208(E)154(N), base of breccia.

FC 7/4/2000-02 – Flynn Creek crater fill, approximately 450 m SW of Antioch, Jackson County, along Shady Grove Rd., Gainesboro Quad, UTM 16SFR202(E)148(N), 10 m below the top of the Flynn Creek Member of the Chattanooga Shale.

TN70-09 – Tennessee Highway 70, Little War Gap, Hawkins County, near Striggersville, Tennessee, Kyle's Ford Quad, UTM 17SLL194(E)418(N), 0.5 m above base of Dowelltown Member of Chattanooga Shale.

Type Section for Flynn Creek Member (composite section) – (a) Lower portion of member (basal and bedded breccia) outcrops in creek bed along Cub Hollow Road. Base of section approximately 640 m (0.4 miles) ENE of intersection of Cub Hollow Road and Flynn Creek Road, top of section approximately at intersection. (b) Upper portion of member (black shale) outcrops along Shady Grove Road. Base of section at intersection of Shady Grove Road and Flynn Creek Road. Top of section approximately 500 m (0.3 miles) SW of intersection marked by base of Dowelltown Member.

## References


Beales, F. W., Lozej, G. P., 1975. Ordovician tidalites in the unmetamorphosed sedimentary fill of the Brent Meteorite Crater, Ontario. In: Ginsburg, R. N. (Ed.), Tidal Deposits, A Casebook of Recent Examples and Fossil Counterparts, Springer, New York, 315–323.

Boggs, S., 1995. Principles of Sedimentology and Stratigraphy. Prentice-Hall, Englewood Cliffs, NJ, 774 pp.

Boon, J.D., Albritton, C.C., 1936. Meteorite scars in ancient rocks. Field and Lab. 5, 53–64.

Branson, E.B., Mehl, M.G., 1934a. Conodonts from the Joachim (middle Ordovician) of Missouri. University of Missouri Studies 8(2), 77–100, Plates 5–7 [imprint 1933].

Branson, E.B., Mehl, M.G., 1934b. Conodonts from the grassy creek shale of Missouri. University of Missouri Studies 8(3), 171–259, Plates 13–21 [imprint 1933].

Bucher, W.H., 1963. Cryptoexplosion structures caused from without or from within the earth? (astroblemes or geoblemes?). Am. J. Sci. 261, 597–649.

Conant, L.C., Swanson, V.E., 1961. Chattanooga shale and related rocks of central Tennessee and nearby areas. United States Geological Survey Professional Paper 357, 91 pp.

de Witt, Jr., Roen, W.J.B., Wallace, L.G., 1993. Stratigraphy of Devonian black shales and associated rocks in the Appalachian Basin. U.S. Geological Survey Bulletin 1909, B1–B57.

Dietz, R.S., 1960. Meteorite impact suggested by shattercones in rocks. Science 131, 1781–1784.

Dypvik, H., Jansa, L.F., 2003. Sedimentary signatures and processes during marine bolide impacts: a review. Sediment. Geol. 161, 309–337.

Ettensohn, F.R., Goodman, P.T., Norby, R., Shaw, T.H., 1989. Stratigraphy and biostratigraphy of the Devonian–Mississippian black shales in west-central Kentucky and adjacent parts of Indiana and Tennessee, Proceedings of the 1988 Eastern Oil Shale Symposium, 237–245.

Feist, R., Klapper, G., 1985. Stratigraphy and conodonts in pelagic sequences across the middle-upper Devonian boundary, Montagne Noire, France. Palaeontographica Abt. A. 188, 1–18.

Glenister, B.F., Klapper, G., 1966. Upper Devonian conodonts from the Canning Basin, Western Australia. J. Paleontol. 40, 777–842.

Grieve, R.A.F., 1997. Extraterrestrial impact events: the record in the rocks and the stratigraphic column. Paleogeogr. Paleoclimatol. Paleoecol. 132, 5–23.

Hinde, G.J., 1879. On conodonts from the Chazy and Cincinnati group of the Cambro-Silurian and from the Hamilton and Genesee shale divisions of the Devonian in Canada and the United States. Geol. Soc. Lond. Quart. J. 35, 351–369, Plates 15–17.

Huddle, J.W., 1963. Conodonts from the Flynn Creek cryptoexplosion structure, Tennessee. USGS Prof. Paper 475C, C55–C57.

Johnson, J.G., Klapper, G., Sandberg, C.A., 1985. Devonian eustatic fluctuations in Euramerica. GSA Bulletin, 96, 567–587.

Klapper, G., 1989. The Montagne Noire Frasnian (Upper Devonian) conodont succession. In: McMillan, N.J., Embry, A.F., Glass, D.J. (Eds), Devonian of the World. Can. Soc. Petrol. Geol. Mem. 14(3), 449–478.

Klapper, G., 1997. Graphic correlation of Frasnian (Upper Devonian) sequences in Montagne Noire, France, and western Canada. In: Klapper, G., Murphy, M.A., Talent, J.A. (Eds), Paleozoic Sequence Stratigraphy, Biostratigraphy, and Biogeography: Studies in Honor of J. Granville ("Jess") Johnson: Boulder, Colorado, Geological Society of America Special Paper 321, 113–184.

Klapper, G., Foster, C.T., Jr., 1993. Shape analysis of Frasnian species of the Late Devonian conodonts Genus *Palmatolepis*: the Paleontological Society. Mem. 32, 1–35.

Middleton, G.V., Southard, J.B., 1984. Mechanics of Sediment Movement. SEPM Short Course No. 3, p. 401.

Nichols, G., 1999. Sedimentology and Stratigraphy. Blackwell, Oxford, 355.

Over, D.J., 2002. The Frasnian/Famennian boundary in central and eastern United States. Paleogeogr. Paleoclimatol. Paleoecol. 181, 153–169.

Partridge, T. C., Kerr, S. J., Metcalfe, S. E., Scott, L., Talma, A. S., Vogel, J. C., 1993. The Pretoria Saltpan: a 200,000 year Southern African lacustrine sequence. Paleogeogr. Paleoclimatol. Paleoecol. 101, 317–337.

Poag, C.W., 1997. The Chesapeake bay bolide impact: a convulsive event in Atlantic coastal plain evolution. Sediment. Geol. 108, 45–90.

Poag, C.W., Aubry, M.-P., 1995. Upper Eocene impactites of the U.S. East Coast: depositional origins, biostratigraphic framework, and correlation. Palaios 10, 16–43.

Reimhold, W.U., Kelley, S.P., Sherlock, S.C., Henkel, H., Koeberl, C., 2005. Laser argon dating of melt breccias from the Siljan impact structure, Sweden: Implications for a possible relationship to Late Devonian extinction events. Meteoritics & Planetary Science 40(4), 591-607.

Roddy, D.J., 1966. The Paleozoic Crater at Flynn Creek, Tennessee. Unpublished Ph.D. thesis, California Institute of Technology, 232.

Roddy, D.J., 1968. The Flynn Creek Crater, Tennessee. In: French, B.V., Short, N.M. (Eds), Shock Metamorphism of Natural Materials, Mono Book Corp., Baltimore, 291–322.

Roddy, D.J., 1977. Pre-impact conditions and cratering processes at the Flynn Creek Crater, Tennessee. In: Roddy, D.J., Pepin, R.O., Merrill, R.B. (Eds), Impact and Explosion Cratering, Pergamon Press, New York, pp. 277–308.

Sandberg, C. A., Ziegler, W., 1996. Devonian conodont biochronology in geologic time calibration. Senckenbergiana Lethaea 76, 259–265.

Sandberg, C.A., Morrow, J.R., Ziegler, W., 2002. Late Devonian sea level changes, catastrophic events, and mass extinctions. In: Koeberl, C., MacLeod, K.G. (Eds), Catastrophic Events and Mass Extinctions: Impacts and Beyond: Boulder, Colorado, Geological Society of America Special Paper 356, 473–487.

Schieber, J., 1994. Evidence for high-energy events and shallow water deposition in the Chattanooga Shale, Devonian, central Tennessee, USA. Sediment. Geol. 93, 193–208.

Schieber, J., 1996. Early diagenetic silica deposition in algal cysts and spores: a source of sand in black shales? J. Sediment. Res. 66, 175–183.

Schieber, J., 1998a. Sedimentary features indicating erosion, condensation, and hiatuses in the Chattanooga Shale of Central Tennessee: relevance for sedimentary and stratigraphic evolution. In: Mudstones and Shales 1: Basin Studies, Sedimentology, and Paleontology, Schieber, J., Zimmerle, W., Sethi, P. (Eds), Stuttgart, Schweizerbartsche Verlagsbuchhandlung, 187–215.

Schieber, J., 1998b. Developing a sequence stratigraphic framework for the Late Devonian Chattanooga Shale of the southeastern US: relevance for the Bakken Shale. In: Christopher, J.E., Gilboy, C.F., Paterson, D.F., Bend, S.L. (Eds), Eight International Williston Basin Symposium, Saskatchewan Geological Society, Special Publication No. 13, 58–68.

Shoemaker, E.M., Eggleton, R.E., 1961. Terrestrial features of impact origin. In: Proc. Geophys. Lab./Lawrence Radiation Lab. Cratering Symp., Lawrence Radiation Lab Report UCRL-6438, pt. 1, paper A, 27.

Simonson, B.M., Harnik, P., 2000. Have distal impact ejecta changed through geologic time? Geol. 28, 975–978.

Stow, D.A.V., Piper, D.J.W., 1984. Deep-water fine-grained sediments: facies models: In: Stow, D.A.V., Piper, D.J.W. (Eds), Fine-Grained Sediments: Deep-Water Processes and Facies, Geological Society of London Special Publication, 611–645.

Stow, D.A.V., Shanmugam, G., 1980. Sequence of structures in fine-grained turbidites: comparison of recent deep-sea and ancient flysch sediments. Sediment. Geol. 25, 23–42.

Streel, M., Caputo, M.V., Lovboziak, S., Melo, J.H.G., 2000. Late Frasnian-Famennian climate: based on palynomorph analyses and the question of the Late Devonian glaciation. Earth-Science Rev. 52, 121–173.

Tucker, R.D., Bradley, D.C., Ver Straeten, C.A., Harris, A.G., Ebert, J.R., McCutcheon, S.R., 1998. New U-Pb zircon ages and the duration and division of Devonian time. Earth Planet. Sci. Lett. 158, 175–186.

Warme, J. E., Sandberg, C. A., 1996. The Alamo megabreccia: record of a Late Devonian impact in southern Nevada. GSA Today 6, 1–7.

Youngquist, W., 1947. A new upper Devonian conodont fauna from Iowa. J. Paleontol. 21, 95–112.

Understanding Late Devonian and Permian-Triassic Biotic and Climatic Events:
Towards an Integrated Approach
edited by D.J. Over, J.R. Morrow and P.B. Wignall



## Chapter 5

# Devonian stromatoporoid originations, extinctions, and paleobiogeography: how they relate to the Frasnian-Famennian extinction

Carl W. Stock*

*Department of Geological Sciences, University of Alabama, Box 870338, Tuscaloosa, AL 35487, USA*

**Abstract**

Origination and extinction of stromatoporoid genera show that the rate of origination exceeded that of extinction from the early Early Devonian (Lochkovian) through the early Middle Devonian (Eifelian). For the remainder of the Devonian (Givetian-Famennian), extinction outpaced origination. During the Early and Middle Devonian, tropical to subtropical marine habitats were divided into the Eastern Americas Realm and the Old World Realm – in North America these were separated by the Transcontinental Arch and the Canadian Shield. Eustatic sea level, which had been rising since the Emsian, reached a point late in the Givetian (Taghanic Onlap) where the land barrier was breached, and Old World taxa invaded the Eastern Americas. Sea level continued to rise episodically through most of the Frasnian, followed by an abrupt fall in conjunction with the Frasnian-Famennian extinction. It is concluded that the decrease in stromatoporoid genus diversity, which occurred as Old World and Eastern Americas faunas came into competition, continued with further sea-level rise during the Frasnian. This lowered diversity put the stromatoporoids in a vulnerable position for extinction when climatic disturbances occurred at the end of the Frasnian.

## 1. Introduction

A great many publications dealing with the Frasnian-Famennian (F-F) extinction focus on events that occurred in close proximity to the extinction. Relatively few studies take the longer view of what happened well before the F-F extinction to see if any long-term factors contributed. Here the history of stromatoporoid genus originations, extinctions, and paleobiogeography throughout the Devonian are examined.

Stromatoporoids, a class of calcified sponges that existed from the Early Ordovician through the Late Devonian, lived in shallow, tropical to subtropical, normal marine environments, and were the major constructors of reefs in the Silurian and Devonian. They have long been recognized as one of the taxa that were profoundly affected by the F-F extinction (e.g. Stearn, 1987), as were other groups, such as rugose corals (Sorauf and Pedder, 1986), brachiopods (Racki, 1998), conodonts (Aldridge, 1988), trilobites (Briggs et al., 1988), amm-onoid cephalopods (House, 2002), as well as additional taxa noted by Schindler (1993). The entire reef ecosystem, of which stromatoporoids were an integral part, was wiped out during the F-F extinction (Copper, 2002a). Regarding the details of what

*Corresponding author. Fax: 1-205-348-0818. *E-mail address:* cstock@wgs.geo.ua.edu (C.W. Stock).

happened to the stromatoporoids during and after this extinction, it is known that: (1) global stromatoporoid biomass plummeted at the end of the Frasnian, but there was some recovery during the Famennian; (2) labechiid stromatoporoids, which had played a minor role in stromatoporoid communities since the Late Ordovician mass extinction (e.g. Nestor and Stock, 2001), returned to a prominent role; (3) a few non-labechiid, mostly clathrodictyid, stromatoporoids also did well; and (4) total extinction of stromatoporoids occurred at the close of the Famennian – geologically younger late Paleozoic and Mesozoic "stromatoporoids" are the product of convergent evolution within the Porifera (e.g. Wood, 1991). However, much less has been published about Devonian stromatoporoid evolution prior to the F-F extinction. Their near-extinction at the close of the Frasnian does not appear to represent a sharp decline in diversity, but rather the culmination of a long-term trend.

## 2. Originations and extinctions

Information on origination and extinction of stromatoporoid genera was taken from Stearn et al. (1999), with a few exceptions. Geologic ranges of several stromatoporoid genera that were reported at the series level were refined to stage-level data (e.g. *Rosenella* was changed from Lower Devonian to Lochkovian and Emsian; *Oslodictyon* was changed from Lower Devonian to Lochkovian). To this was added the Famennian *Vietnamostroma* (Nguyen and Mistiaen, 1998), which had been called "New Genus Hung" by Stearn et al. (1999, p. 18). McGhee (1996) pointed out that the overall loss of diversity associated with an extinction event is as likely due to decreased originations as it is to increased extinctions. Examination of available data on origination and extinction of stromatoporoid genera during the Devonian reveals that for the first four ages (Lochkovian-Eifelian), origination exceeded extinction (Table 1; Fig. 1). From the Late Middle Devonian (Givetian) through the close of the Devonian, extinction outpaced origination. Not only did the proportions reverse, but also the number of originations decreased, with a low of four in the Frasnian, and the number of extinctions increased. The preponderance of extinction relative to origination was expected for the Frasnian, especially because the data are combined within each age (i.e. they give the impression that all genus ranges terminate at the end of an age); however, an as nearly strong, similar relationship was not anticipated for the Givetian.

In terms of overall occurrences, stromatoporoids reached peak levels during the Givetian and Frasnian (Stock, 1990). Genus diversity was highest in the Eifelian, and decreased only slightly through the Frasnian (Table 1; Fig. 2). These high levels of ecological success and diversity do not suggest evolutionary difficulties.

## 3. Paleobiogeography and changing sea level

### *3.1. Introduction*

Many causes for the F-F extinction have been suggested, including: (1) global cooling; (2) global warming; (3) too much or too little seasonality; (4) marine regression and resultant loss of habitat; (5) widespread oceanic anoxia; and (6) bolide impact (McGhee, 1996). Most of these have been presented as short-term phenomena in close temporal proximity

*Table 1.* Genera analyzed in time and space.

| ORDER/Genus | Prev | Loch | Prag | Ems | Eif | Giv | Fras | Fam |
|---|---|---|---|---|---|---|---|---|
| LABECHIIDA | | | | | | | | |
| *Rosenella* | x | O | ? | O | O | ? | O | O |
| *Cystostroma* | x | ? | ? | O | ? | ? | O | O |
| *Labechia* | x | O | ? | O | ? | ? | O | O |
| *Labechiella* | x | ? | O | O | O | O | O | O |
| *Stromatocerium* | x | ? | ? | O | | | | |
| *Platiferostroma* | | | | | | | | O |
| *Parastylostroma* | | | | | | | O | O |
| *Stylostroma* | x | O | ? | O | ? | ? | O | O |
| *Pachystylostroma* | x | ? | ? | ? | ? | ? | ? | O |
| *Pennastroma* | | | | | | | | O |
| *Spinostroma* | | | | | | | | O |
| *Pararosenella* | | | | | | | | O |
| *Lophiostroma* | x | ? | ? | ? | ? | ? | O | |
| *Vietnamostroma* | | | | | | | | O |
| CLATHRODICTYIDA | | | | | | | | |
| *Clathrodictyon* | x | ? | ? | O | OE | OE | | |
| *Bullulodictyon* | | | | | | | O | |
| *?Coenellostroma* | | | | O | O | | | |
| *?Kyklopora* | | | | | | | | O |
| *Oslodictyon* | x | O | | | | | | |
| *Stelodictyon* | x | E | | | | | | |
| *Yabeodictyon* | x | ? | ? | O | | | | |
| *Gerronostroma* | x | ? | O | O | OE | OE | O | O |
| *Petridiostroma* | x | OE | ? | ? | OE | OE | | |
| *Atelodictyon* | | O | O | O | OE | OE | OE | O |
| *Coenostelodictyon* | | O | | | | | | |
| *?Cubodictyon* | | | | | O | | | |
| *Intexodictyon* | x | E | O | O | | | | |
| *Tienodictyon* | | | | O | O | | | |
| *Anostylostroma* | | | | | OE | OE | O | O |
| *Belemnostroma* | | O | | | | | | |
| *Hammatostroma* | | | | | | O | OE | |
| *Nexililamina* | | | | O | O | | | |
| *Pseudoactinodictyon* | | | | OE | OE | OE | O | |
| *Schistodictyon* | x | ? | O | O | OE | OE | O | |
| ACTINOSTROMATIDA | | | | | | | | |
| *Actinostroma* | | O | O | O | O | OE | OE | |
| *Bifariostroma* | | | | O | O | O | O | |
| *Plectostroma* | x | O | O | O | O | O | O | |
| *Actinostromella* | x | O | | | | | | |
| *Araneosustroma* | x | O | | | | | | |
| STROMATOPORELLIDA | | | | | | | | |
| *Stromatoporella* | | | O | OE | OE | OE | | |
| *Clathrocoilona* | | | | O | O | OE | OE | |

*Table 1.* (*Continued*)

| ORDER/Genus | Prev | Loch | Prag | Ems | Eif | Giv | Fras | Fam |
|---|---|---|---|---|---|---|---|---|
| *Dendrostroma* | | | | | | **OE** | **O** | |
| *Simplexodictyon* | **x** | **?** | **?** | **O** | **O** | | | |
| *Trupetostroma* | | | | | **OE** | **OE** | **OE** | |
| *Stictostroma* | | | | **OE** | **OE** | **OE** | **OE** | |
| *Syringodictyon* | | | | | **E** | | | |
| *Styloporella* | | | | | | | **O** | |
| *Tubuliporella* | | | **O** | **?** | **O** | | | |
| *Hermatostroma* | | | | | **O** | **OE** | **OE** | |
| *Hermatoporella* | | | | | | **O** | **OE** | |
| *Hermatostromella* | **x** | **O** | **O** | **O** | | | | |
| *Synthetostroma* | | | | | | **O** | **O** | |
| *Idiostroma* | | | | | **O** | **OE** | **O** | |
| STROMATOPORIDA | | | | | | | | |
| *Stromatopora* | **x** | **O** | **O** | **O** | **O** | **OE** | **O** | **O** |
| *Climacostroma* | | | | | **O** | **O** | | |
| *Glyptostromoides* | | | | **O** | **O** | **O** | | |
| *Neosyringostroma* | | | | **O** | **O** | **OE** | | |
| *Pseudotrupetostroma* | | | | **O** | **O** | **O** | | |
| *Taleastroma* | | | | | **OE** | | | |
| *Ferestromatopora* | | | | | | **OE** | **O** | |
| *Arctostroma* | | | | | | **O** | **OE** | |
| *Syringostromella* | **x** | **OE** | **O** | **O** | | | | |
| *Salairella* | | | **O** | **O** | **O** | **O** | **O** | |
| SYRINGOSTROMATIDA | | | | | | | | |
| *Syringostroma* | | **O** | **?** | **OE** | **OE** | **E** | | |
| *Atopostroma* | | **OE** | **?** | **O** | | | | |
| *Columnostroma* | | **E** | **O** | **?** | **E** | **O** | | |
| *Coenostroma* | **x** | **E** | **?** | **O** | **OE** | **OE** | | |
| *Habrostroma* | **x** | **OE** | **O** | **OE** | **OE** | **OE** | | |
| *Parallelostroma* | **x** | **OE** | **O** | | | | | |
| *Stachyodes* | | | | | **O** | **OE** | **OE** | |
| AMPHIPORIDA | | | | | | | | |
| *Amphipora* | **x** | **O** | **O** | **O** | **OE** | **OE** | **OE** | **O** |
| *Euryamphipora* | | | | | | | **O** | |
| *Novitella* | | | | | | | **O** | |
| *Vacuustroma* | | | | **O** | **?** | **O** | **O** | |
| Total Genera | 27 | 24 | 17 | 36 | 38 | 36 | 35 | 18 |
| Originations | | 7 | 3 | 11 | 9 | 6 | 5 | 5 |
| Extinctions | | 6 | 1 | 6 | 8 | 11 | 24 | 18 |

Abbreviations: Prev=genera originating previous to the Devonian; Loch=Lochkovian; Prag=Pragian; Ems=Emsian; Eif=Eifelian; Giv=Givetian; Fras=Frasnian; Fam=Famennian; O=Old World Realm; E=Eastern Americas Realm. Use of O and/or E for the Frasnian and Famennian represent areas formerly in those realms.

*Source:* Most data from Stearn et al. (1999).

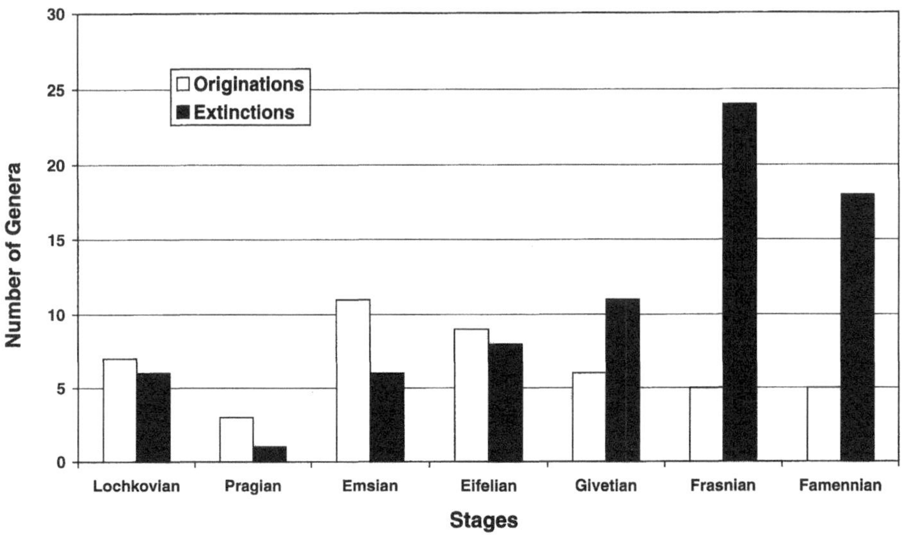


*Figure 1.* Histogram of originations and extinctions of Devonian stromatoporoid genera by stage.

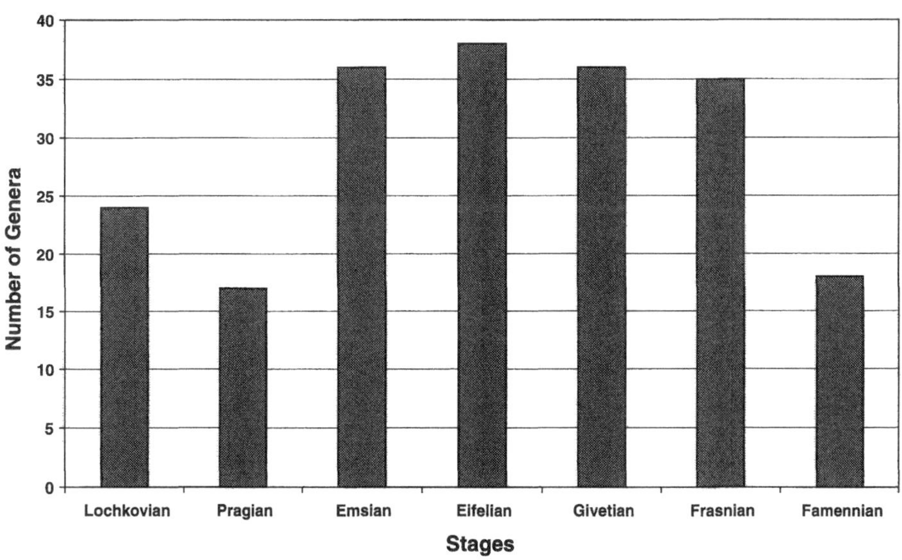


*Figure 2.* Histogram of diversity of Devonian stromatoporoid genera by stage. No inferred occurrences of genera, as indicated by question marks in Table 1, are included.

to the F-F boundary. Consequently, less attention has been paid to long-term trends affecting Devonian organisms well before the close of the Frasnian – these include changes in paleogeography that altered the geographic distribution of the organisms involved.

The geographic ranges of organisms are defined by barriers. In the case of marine organisms, the most effective barriers are landmasses; however, other factors can be important as well. Temperature – stromatoporoids were limited to tropical-to-subtropical latitudes, because they required warm temperatures in order to secrete their $CaCO_3$ skeletons; stromatoporoids played a prominent role in the determination of paleolatitude in the Devonian paleogeographic reconstructions of Heckel and Witzke (1979). Depth – they lived in only shallow water; Boucot (1975) found that stromatoporoids were characteristic of his Benthic Assemblages 2 and 3. Turbidity – the circulatory canals and flagellated chambers facilitated feeding and gas exchange within stromatoporoids, but these spaces could become clogged by fine-grained sediments, be they siliciclastic or carbonate; as nearly all stromatoporoid fossils are found in carbonate rocks, it is most likely due to the higher frequency of turbidity in siliciclastic, relative to carbonate, depositional environments. Salinity – nearly all stromatoporoids are found in rocks representing deposition in normally saline environments, although some beds containing concentrations of *Amphipora* are thought to have been deposited in hypersaline lagoons (Fischbuch, 1962).

The rise of geographic barriers that separate areas of otherwise similar ecological conditions typically results in an overall increase in biodiversity. For example, when a land barrier divides a particular marine habitat into two isolated regions, two sets of taxa may independently evolve from common ancestors. Initially this may happen at the species level, but as time goes by, progressively higher taxonomic levels (i.e. first genera, then families, etc.) may diverge. This phenomenon is known as vicariance biogeography (Lieberman, 2000; Cecca, 2002). When the barrier is removed, then the evolutionarily diversified taxa are brought into competition for the same ecological resources, including space. This leads to an overall drop in diversity, as the most fit species survive through natural selection.

Sea-level rise in the Devonian was episodic, with each episode representing a gradual increase followed by an abrupt decrease at the end (Fig. 3). A rise in sea level increases the portion of a continent that is flooded by epeiric seas; typically, the larger area provides habitats for a greater diversity of organisms. A fall in sea level reduces the portion flooded by epeiric seas; the reduced area brings species into competition for the remaining resources, resulting in extinction of the less adapted.

### *3.2. Historical background*

For most of the Devonian, the marine biotas of the world were divided into three realms (e.g. Blodgett et al., 1990): (1) Malvinokaffric Realm – southern high-latitude areas; (2) Eastern Americas Realm (EAR) – southeastern North America and northwestern South America; and (3) Old World Realm (OWR) – all other marine habitats. Seas of the Malvinokaffric Realm and South American portions of the Eastern Americas Realm were too cold for stromatoporoids. General recognition of these realms is based primarily on evidence presented for brachiopods (e.g. Johnson and Boucot, 1973) and rugose corals (e.g. Oliver, 1977; Fig. 4).

Throughout much of the Cambrian through the Devonian, marine biotas of eastern North America were separated from those of the western and northern parts of the continent by two conterminous physiographic features, the Canadian Shield and the Transcontinental Arch, the latter extending from Minnesota and Wisconsin to Arizona and New Mexico (Fig. 5). In the Devonian this landmass extended through Greenland and the Baltic Shield (e.g. Witzke, 1990).

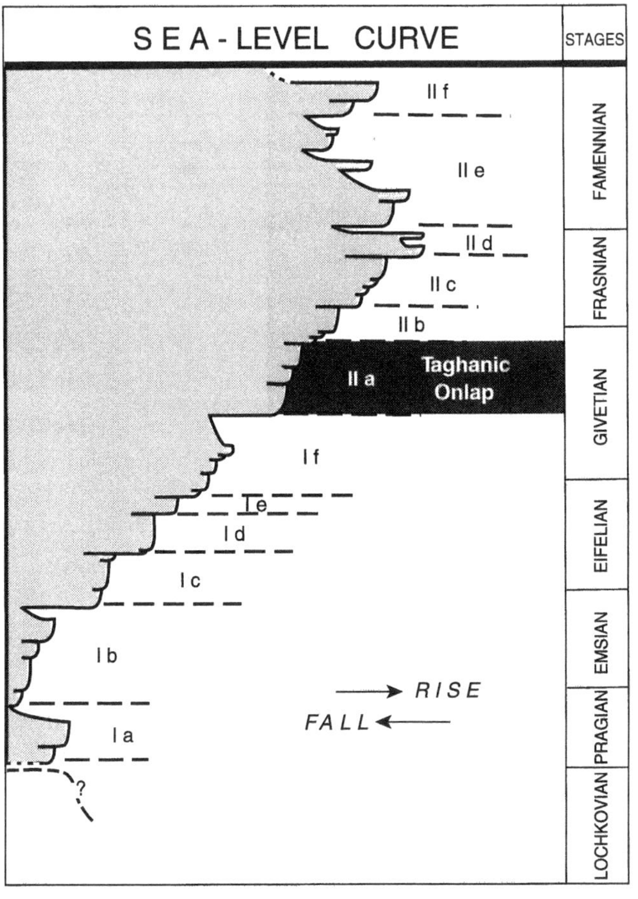


*Figure 3.* Devonian sea-level curve (after Johnson and Sandberg, 1988).

An overwhelming number of Devonian stromatoporoid genera lived in only the OWR, or Late Devonian areas that had been in the OWR. Of the 74 genera tabulated (Table 1), 40 were restricted to OWR areas, and only two (*Stelodictyon*; *Syringodictyon*) never left the EAR during the Devonian. This relationship should not be surprising as the OWR was much larger, and had a greater latitudinal range than did the EAR (Fig. 5).

### 3.3. *Effectiveness of the transcontinental barrier*

The degree to which the Canadian Shield and Transcontinental Arch precluded exchange between the OWR and EAR varied through time. A prime example of a breach in the Transcontinental Arch is thought to have occurred from Early Pragian through Early Emsian, when EAR taxa invaded the Great Basin of the southwestern U.S. (e.g. Nevada); this is coincident with the Johnson et al. (1985) T-R Cycle Ia (Johnson, 1990). Boucot et al. (1969, p. 40) stated that the brachiopod fauna of Nevada during that time interval, "... is

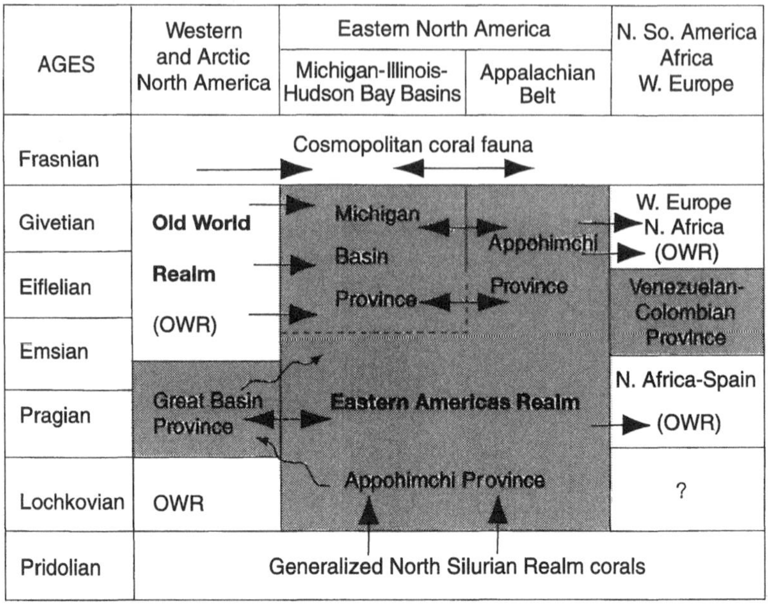


*Figure 4.* Distribution of the Old World and Eastern Americas Realms in North America, based on rugose corals. Arrows indicate direction of migration (modified after Oliver, 1990).

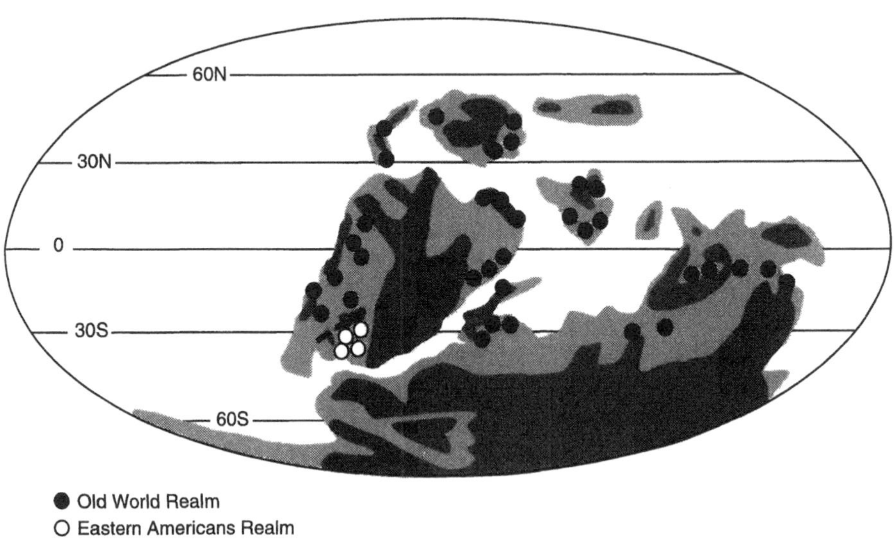


*Figure 5.* Paleogeographic map of the world during the Givetian, showing the locations of stromatoporoids. Dark gray areas are dry land, and light gray areas are epeiric seas. The dashed line marks the location of the Transcontinental Arch (modified after Copper, 2002b).

clearly and wholly of Appalachian [i.e. EAR] aspect …". Oliver and Pedder (1989; Fig. 4) indicated that rugose corals demonstrated the same relationship, where EAR forms inhabited an otherwise OWR area. Blodgett et al. (1988) saw that gastropods demonstrated the same biogeography. A reassessment of Nevada rugose corals by Pedder and Murphy (2004, p. 847) caused them to conclude that, "… there is no evidence that the pioneer corals of the Great Basin Province emigrated from the Appohimchi Province, or from any other part of the EAR," so apparently the barrier was effective, at least for the rugose corals. Boucot and Blodgett (2001) noted the presence of an anomalous OWR fauna in the upper Eifelian of the central Appalachians (EAR).

The transcontinental barrier has been depicted with different degrees of completeness for various times in the Devonian. Witzke (1990) illustrated an intact Transcontinental Arch-Canadian Shield barrier for the Early Devonian; however, for the Middle Devonian (Eifelian-Givetian, undifferentiated) he showed three breaks: (1) South Dakota–Nebraska–Kansas; (2) North Dakota–Minnesota; and (3) Manitoba–Hudson Bay–Ontario. Unfortunately, Witzke (1990) did not explain these breaks. In several Devonian paleogeographic maps, Copper (2002b) showed the Transcontinental Arch and Canadian Shield inundated by the sea to different extents. Other evidence for flooding of the Canadian Shield comes from the discovery of xenoliths of normal marine Devonian carbonates from Jurassic kimberlites (Cookenboo et al., 1998; McCracken et al., 2000).

Nearly all stromatoporoid genera that originated in the Devonian began their existence in the OWR, or seemingly simultaneously in the OWR and EAR (Table 1). These apparent simultaneous originations most likely represent the appearance of a genus in one realm, and the extension of its range into the second realm within the same geologic age. Following the total disappearance of stromatoporoids from the EAR and the North American parts of the OWR during the Pragian, first reported by Stock (1990), four genera entered the EAR from the OWR, and one originated in both realms. The Pragian hiatus in stromatoporoid occurrences in North America is most likely due to sea-level fall at the close of Sloss' (1963) Tippecanoe cratonic sequence, which reduced areas of marine deposition and exposed Pragian marine deposits to erosion. In the Eifelian six genera emigrated from the OWR, and three genera first appeared in both realms. The Givetian had eight OWR immigrants, the highest, plus one apparent simultaneous origination. Three additional genera entered the EAR from the OWR in the Frasnian. Clearly the barrier between the two realms was breached at least a few times from the Emsian through the Givetian, however briefly, to allow the passage of OWR genera into the Eastern Americas Realm.

Two cases of stromatoporoid species breaching the transcontinental barrier are known. Prosh and Stearn (1993) reported the occurrence of four species, one each in *Stromatoporella*, *Stictostroma*, *Habrostroma*, and *Parallelopora*, in both southwestern Ellesmere Island, Canadian Arctic (OWR), and Michigan-southern Ontario (EAR) in strata close to the Lower-Middle Devonian boundary. *Habrostroma centrotum* was demonstrated to have occurred in Lochkovian (lower Lower Devonian) rocks of Bathurst and Ellesmere Islands, Canadian Arctic (OWR), and Maine, New York, and Virginia (EAR) by Stock and Burry-Stock (2004).

Stock (1990; repeated in Stock, 1997) declared that stromatoporoid endemism between the OWR and the EAR ended with the Eifelian. Although not using the more up-to-date information on Devonian stromatoporoid systematics and paleobiogeography (Stearn et al., 1999), this claim is still true to some degree. In the Eifelian, 21 genera were endemic to the OWR, two were endemic to the EAR, and 15 occurred in both realms (Table 1). The latter represent 40 percent of Eifelian stromatoporoid genera – certainly not representative of a continuous, uniform realm, nor the result of separation by an impenetrable barrier.

At this point one may wonder if the transcontinental barrier in North America was an effective deterrent to the exchange of stromatoporoids between the OWR and EAR during the Devonian. One distribution clearly demonstrates that a group of stromatoporoids was prevented from crossing – no Devonian stromatoporoids in order Labechiida have been reported from the EAR (Table 1). In addition to the labechiids, 21 non-labechiid genera that originated in the OWR or carried over from the Silurian into the Lochkovian of the OWR never arrived in the EAR prior to the Taghanic Onlap (Table 1).

## 4. Givetian sea-level rise: the Taghanic Onlap

By Givetian time EAR stromatoporoids lived in what are now the states of Michigan, Indiana, Illinois, Iowa, and Missouri. Most Givetian stromatoporoid genera come from the OWR, including the western U.S. (e.g. Nevada), western and arctic Canada, Australia, northwestern Africa, and Eurasia, from Spain, France, and the United Kingdom to Siberia, China, and Vietnam.

Global eustatic sea level, which had been rising episodically since the Emsian (Fig. 3), reached a point in the Late Givetian where the inter-realm barrier was breached, a phenomenon known as the Taghanic Onlap (Johnson, 1970; Sandberg et al., 2002). Originally, Johnson (1970) saw the Taghanic Onlap occurring from the latest Givetian through the earliest Frasnian; however, Johnson et al. (1985) placed the Taghanic Onlap as beginning at about the middle of the Givetian, in the Middle *varcus* Zone, and ending in the Late Givetian, in the Lowermost *asymmetricus* Zone, coincident with their T-R Cycle IIa. Johnson and Sandberg (1988) changed the placement of the Givetian-Frasnian boundary from within the Lower *asymmetricus* Zone (Johnson et al., 1985) to within the Lowermost *asymmetricus* Zone, so that the Taghanic Onlap (Cycle IIa) ended just prior to the end of the Givetian. Boucot and Blodgett (2001) noted that the Malvinokaffric Realm ceased to exist after the Eifelian due to a decrease in the gradient of global climate (i.e. the earth became warmer); thus, it may be assumed that the Taghanic Onlap and subsequent sea-level rise in the Frasnian were accompanied by global warming. This rise in Frasnian ocean temperature is supported by Joachimski et al. (2004), based on oxygen isotope ratios in well preserved conodont apatite and brachiopod calcite.

The impact of the Taghanic Onlap on a few faunal groups is relatively well understood. Johnson and Boucot (1973) stated that the end result was a single cosmopolitan brachiopod fauna in Frasnian North America. Oliver (1977) claimed the same result for the rugose corals, seeing the Taghanic Onlap as a gradual mixing of OWR and EAR genera. His basic understanding of the Taghanic Onlap has not changed since then (e.g. Oliver and Pedder, 1989; Oliver, 1990, 1997). Boucot (1975) and Oliver (e.g. 1977) recognized the existence of an EAR until the end of the Givetian, even though it contained fewer endemic genera through time. According to Sorauf and Pedder (1986, p. 1277), "… there was no significant turnover of [rugose coral] genera at the Middle Devonian-Frasnian boundary; 24 of the 27 Early Frasnian genera were holdovers from previous faunas". Boucot (1990) contrasted the previous statement with one by Oliver and Pedder (1989, p. 233), who stated, "In the Late

Givetian every family and genus of rugose corals that originated as an [EAR] endemic disappeared;" however, if many rugose coral genera became extinct and were replaced by other genera during the Late Givetian, Sorauf and Peddar (1986) were correct. Boucot (1990) likened the extinction of Late Givetian EAR rugose corals to that experienced simultaneously by EAR brachiopods. Scrutton (1997) noted a strong decline in rugose coral diversity about mid-way through the Taghanic Onlap (Cycle IIb; Upper *varcus* Zone), followed by a moderate recovery that preceded the severe drop in diversity accompanying the F-F extinction.

One example of an OWR immigrant arriving in the EAR is *Actinostroma*. The earliest evidence of its arrival is the presence of one specimen of *A. tyrrelli* in the Logansport Limestone of northern Indiana (Galloway and St. Jean, 1957). According to Shaver et al. (1986), the Logansport Limestone is now more correctly called the Traverse Formation; the lower and middle portions are very fossiliferous, including abundant stromatoporoids. Shaver et al. (1986) assigned a Givetian age to the Traverse, so *A. tyrrelli* could very well have arrived with the Taghanic Onlap. By the Early Frasnian, *A. clathratum* was a common component of the stromatoporoid fauna in Iowa (Stock, 1982; Smith, 1994), the only state in the U.S. Midwest that has abundant Frasnian stromatoporoids.

There could be some difficulty determining which stromatoporoid genera were present in the EAR prior to the Taghanic Onlap, because nearly all lower Givetian strata in the North American mid-continent do not contain stromatoporoids; an exception is the small fauna described by Birkhead and Fraunfelter (1973) from the St. Laurent Limestone of southeastern Missouri and Lingle Limestone of southern Illinois. Sea level continued to rise episodically through most of the Frasnian (Fig. 3) with an abrupt fall at its close. There is but one tropical to subtropical realm in the Frasnian (Fig. 6). Johnson et al. (1986) recognized that the rise of sea level, and associated removal of barriers, contributed to extinction during the Frasnian.

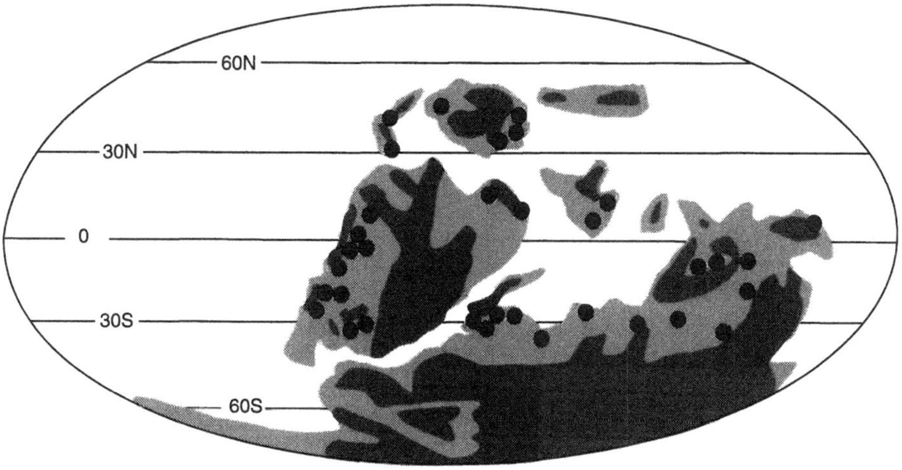


*Figure 6.* Paleogeographic map of the world during the Frasnian, showing the locations of stromatoporoids. Dark gray areas are dry land, and light gray areas are epeiric seas (modified after Copper, 2002b).

## 5. The Frasnian record

### *5.1. Introduction*

Stearn (1987) cited studies in former OWR areas Alberta, Canada (Stearn, 1975), Belgium (Lecompte, 1951, 1952), Czech Republic (Zukalova, 1971), Poland (Kazmierczak, 1971), and Afghanistan (Mistiaen, 1985a, b) that documented a decline in stromatoporoid species diversity through the Frasnian. Stearn's (1987) approach is repeated here, where trends involving genera are reported. Lithostratigraphic units are correlated with the latest information on conodont zonations and the Johnson and Sandberg (1988) sea-level cycles (Fig. 7). Where possible, generic revisions of Stearn et al. (1999) are applied. The former OWR area of Western Australia and the former EAR area of Iowa, USA are added.

### *5.2. Alberta*

Stearn (1975) described the stromatoporoids from the Ancient Wall reef complex, Mount Haultain area of Alberta. He delineated five assemblages, 1–5 in ascending order. Data on conodont zonations come from McLean and Klapper (1998).

Assemblage 1 came from the lower part of the Flume Member of the Cairn Formation, in which there are conodonts of Montagne Noire (MN) zones 1–3; it contains five genera of stromatoporoids. Assemblage 2, the upper part of the Flume Member and the upper member of the Cairn Formation, is in MN zones 4–8; it contains 12 genera. Assemblage

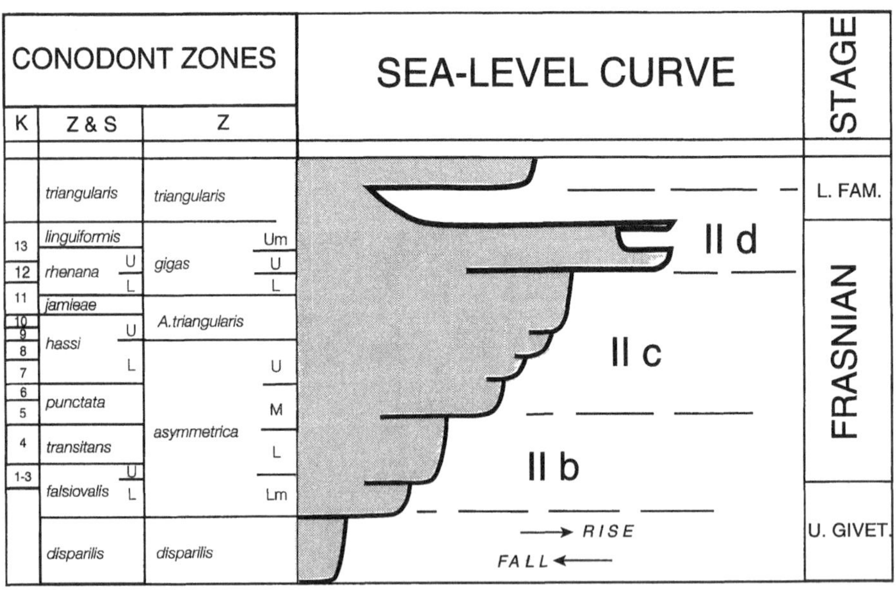


*Figure 7.* Frasnian sea-level curve correlated with three conodont zonations; see Figure 3 for context. K = Klapper (1988); Z & S = Ziegler & Sandberg (1990); Z = Ziegler (1962, 1971). Equivalence of Z and Z & S from Ziegler and Sandberg (1990); equivalence of Z & S and K from Klapper and Becker (1999).

3, the lower part of the Peechee Member of the Southesk Formation, is in MN zones 9–11; it contains six genera. Assemblage 4, the upper part of the Peechee Member, is in MN Zone 11; it contains four genera. Assemblage 5, the Arcs Member of the Southesk Formation, is in MN Zone 12; it contains 10 genera. In summary, the five assemblages contain 5, 12, 6, 4, and 10 genera in ascending order.

### *5.3. Belgium*

Stromatoporoids from the Dinant Basin of Belgium were described by Lecompte (1951, 1952), who reported occurrences of Frasnian stromatoporoids at six stratigraphic levels. Conodont zonations are from Boulvain et al. (1999).

Lecompte's (1951) assemblages are listed in ascending order; lithostratigraphic updates are in parentheses:

(1) F2b, Schistes à *Receptaculites neptuni* et *Spirifer bisinus* (Fm. de Nismes) – one genus; Lower *asymmetricus* Zone or upper Lower *falsiovalis* Zone (MN Zone 3).
(2) F2d, "Récifs" du marbre à *Disphyllum* (Fm. du Moulin Liénaux, Mem. de l'Arche) – two genera; Middle *asymmetricus* Zone or *punctata* Zone (MN zones 5–6).
(3) F2g, Calcaires stratifies (Fm. des Grands Breux, Mbr. Bieumont) – 17 genera; lower Upper *asymmetricus* Zone or Lower *hassi* Zone (MN Zone 7).
(4) F2h, "Récifs" de marbre à Stromatopores (Fm. des Grands Breux, Mbr. du Lion) – 11 genera; Upper *asymmetricus* Zone or *hassi* Zone (MN zones 7–10).
(5) F2i, Schistes à *Reticularia pachyrhyncha* (Fm. de Neuville) – three genera; assemblages 5 and 6 are in the Lower *rhenana* Zone (MN zones 11–12).
(6) F2j, "Récifs de marbre rouge (Fm. de Neuville) – four genera.

In summary, the six assemblages contain 1, 2, 17, 11, 3, and 4 genera in ascending order.

### *5.4. Czech Republic*

Zukalova (1971) described stromatoporoids from Moravia, Czech Republic, and delineated three stratigraphic assemblages. Data on conodont zonations are from Dvorak et al. (1984).

The lowest assemblage (lower Frasnian) contains seven genera; it is in the Lower *asymmetricus* Zone (MN zones 3–4). The middle assemblage (lower Frasnian b+) contains 13 genera; it is in the *asymmetricus* Zone (MN zones 4–8). The highest assemblage (upper Frasnian) contains 11 genera; it is in the *triangularis–gigas* zones (MN zones 9–13). In summary, the three assemblages contain 7, 13, and 11 genera in ascending order.

### *5.5. Poland*

Only two stromatoporoid assemblages were delineated by Kazmierczak (1971) for the Frasnian of the Holy Cross Mountains, Poland. Conodont zonation data are from Racki and Turnau (2000).

The lower assemblage, the Upper Sitkowka Beds, contains 16 genera; it is in the Upper *falsiovalis* Zone (MN Zone 3). The upper assemblage, the Kadzielnia Beds, contains six genera; it is in the *transitans* Zone, and possibly extends into the *punctata* Zone (MN zones 4–5).

### 5.6. *Afghanistan*

Mistiaen (1985a,b) described the stromatoporoids from the Central Mountains of Afghanistan, dividing the Frasnian into lower, middle, and upper parts. Conodont zonations for these rocks are unavailable. The lower Frasnian contains eight genera, the middle Frasnian two genera, and the upper Frasnian one genus.

### 5.7. *Western Australia*

Four stromatoporoid assemblages were delineated by Cockbain (1984) from the Frasnian of the Canning Basin, Western Australia. Some of the assemblages are lateral equivalents of others. Information on conodont zonations was provided by G. Klapper (2004, pers. commun.).

There are two marginal slope facies assemblages, in ascending order: (1) Sadler Limestone – 12 genera; MN zones 1–4; and (2) Virgin Hills Formation (part) – two genera; overall MN zones 6–13, but stromatoporoid-bearing portion about MN Zone 7. There are two platform facies assemblages: (1) Pillara Limestone – 12 genera; no zonally diagnostic conodonts known – Cockbain (1984) represented the Pillara Limestone as a lateral equivalent of the Sadler Limestone and lower Virgin Hills Formation, suggesting the Pillara is in MN zones 1–8; and (2) Napier Formation (part) – three genera; no conodont data are available – Cockbain represented the stromatoporoid-bearing portion of the Napier as a lateral equivalent of the stromatoporoid-bearing portion of the Virgin Hills, suggesting it is in MN Zone 7. In summary, at least in the marginal-slope facies, two assemblages contain 12 and 2 genera in ascending order.

### 5.8. *Iowa*

Data on the occurrence of stromatoporoids in the Frasnian of Iowa come from several sources. Smith (1994) described the stromatoporoids from the Idlewild Member of the Lithograph City Formation. The fauna from the Mason City Member of the Shell Rock Formation was described by Stock (1982). A brief summary of the stromatoporoid faunas of the Nora Member of the Shell Rock Formation, and Cerro Gordo and Owen Members of the Lime Creek Formation was presented by Stock (1984b); a few of these were described by Stock (1984a). Conodont zonation data are from Day (1996).

There are four stromatoporoid assemblages in the Frasnian of Iowa, in ascending order: (1) Idlewild Member – 14 genera; MN zones 3–4; (2) Mason City Member – 11 genera; MN Zone 6; (3) Nora Member – 11 genera; conodonts are undiagnostic, but probably equivalent to MN zones 7 and/or 8; and (4) Cerro Gordo and Owen Members – seven genera; MN zones upper 11–13. Day (1996) placed assemblage 1 in sea-level Cycle IIb-2, assemblage 2 in lower Cycle IIc, assemblage 3 in upper Cycle IIc, and assemblage 4 in Cycle IId. In summary, the four assemblages contain 14, 11, 11, and 7 genera in ascending order.

### 5.9. *Summary*

Many factors make a direct comparison of stromatoporoid genus diversity through the Frasnian a difficult undertaking. In each of the seven examples cited above, one person conducted all work – the thesis by Smith (1994) was supervised by Stock – so that the same approach to the research was used within each area; this is good. Because seven paleontologists (one per area) conducted the research, publishing the stromatoporoid systematics over a period of 43 years (1951–1994), and because the conodont biostratigraphy spans 20 years (1984–2004), differences in the data must be assumed to exist among the seven examples. The various authors reported stromatoporoid assemblages from stratigraphic intervals that are not uniform among the examples.

Stromatoporoid genus diversity through the Frasnian for six of the seven examples – there are no conodont data for Afghanistan – were plotted (Fig. 8) to compare diversity trends among the six areas, and to determine if Stearn's (1987) conclusion that stromatoporoid species diversity declined through the Frasnian, also holds true at the genus level. Almost nowhere on Figure 8 is there a place where intervals of equal duration can be compared. In four of the six areas, stromatoporoids did not appear until MN Zone 3, and in a different combination of four areas, stromatoporoids are absent from MN Zone 13. In Poland, there are none higher than MN Zone 5, and in Western Australia, none higher than MN Zone 7. In three of the areas (Alberta, Belgium, Czech Republic), genus diversity

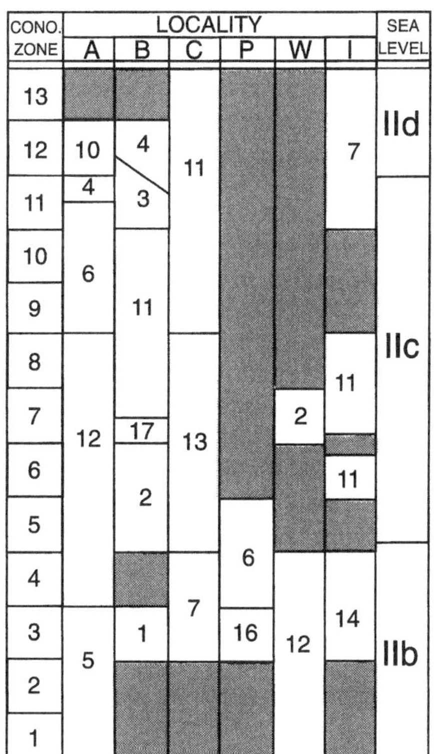


*Figure 8.* Stromatoporoid genus diversity through the Frasnian. Montagne Noire conodont zones are used to economize on space – equivalence with other conodont zonations is given in Figure 7. Localities: A = Alberta; B = Belgium; C = Czech Republic; P = Poland; W = Western Australia; I = Iowa. Gray areas represent unconformities and/or non-stromatoporoid-bearing strata.

starts low, increases to a peak about midway, and then decreases toward the close of the Frasnian; the late peak in genus diversity in Alberta is the only strong exception. The three other areas (Poland, Western Australia, Iowa) demonstrate the trend in decreasing diversity through time that was seen by Stearn (1987), although they include different spans of geologic time. Mistiaen (1985a, b) indicated a trend of genus diversity drop through the Frasnian in Afghanistan. No obvious correlation is seen between genus diversity and sea-level cycles (Fig. 8).

The obvious conclusion here is that much more work must be done to improve the bases for comparison of Frasnian stromatoporoid assemblages, before any meaningful world-scale correlations can be conducted among geographic areas and with global sea-level cycles.

## 6. The Frasnian-Famennian extinction

Global cooling appears to be the most favored factor invoked as a cause for the F-F extinction. Sorauf and Pedder (1986) noted that the rugose coral extinction in the latest Frasnian was marked by the abrupt disappearance of all shallow, warm-water species, whereas basin-dwelling cool-water species were only marginally affected. Very few species returned to shallow-water habitats in the Early Famennian – in only the Late Famennian was there an increase in diversity. They felt that sea-level change, oceanic cooling, or reduction in phytoplankton levels were of questionable importance in explaining the extinction. Sorauf (1989, p. 337) concluded that the F-F extinction of rugose corals was due to, "… the motion of shifting continents, resulting in changes in configuration of ocean basins and in changes for paths of oceanic currents and major gyres," resulting in changes in evaporite deposition, orogenic and epiorogenic activity, eustatic sea level, and climate.

Stearn (1987) contrasted the gradual decline in generic diversity of Frasnian stromatoporoids with the rapid decline of rugose coral species at the end of the Frasnian reported by Sorauf and Pedder (1986). He went on to explain that cooler-water stromatoporoids might have fared better than their warmer-water counterparts during the Famennian. Referring to Sorauf and Pedder (1986), Stearn (1987, p. 679) stated that, "They believe, however, that their results are consistent with a rapid decline of ocean temperatures as well as with a hypothesis of meteoric impact. The results presented here appear to support the first hypothesis." Stock (1990) concluded that the F-F extinction of stromatoporoids was brought on by the combined decreases of sea level and temperature, phenomena supported by Sandberg et al. (2002). Cooling was further supported by Joachimski and Buggisch (2002), who reported positive $\delta^{18}O$ excursions in samples of conodont apatite from two F-F boundary sections, noting a coincidence with positive excursions in carbonate $\delta^{13}C$ (Joachimski et al., 2002). Joachimski and Buggisch (2002, p. 714) concluded that, "… repeated and significant temperature changes during the latest Frasnian may be the key to the question concerning 'the inhibiting factor that caused the cessation of new species originations (McGhee, 1988, p. 256).'"

Stromatoporoids lived in very shallow water, possibly at times shallower than for most other marine macro-organisms, including the corals. This has been noted for onshore–offshore transects (e.g. Laporte, 1967; Stock, 1979, p. 318), and with depth variations at single localities (e.g. Lecompte, 1954; Walker and Alberstadt, 1975); Scrutton (1998) agreed that stromatoporoids tended to inhabit shallower environments than did corals, and that this relationship was more pronounced in the Devonian than in the Silurian. The shallower the sea, the less protection there is against the effects of short-term changes in temperature on benthic organisms. Shallow, warm-water, tropical to subtropical forms such as stromatoporoids, were ill-equipped to deal with global cooling, regardless of the rapidity of the onset of cooling. Deeper-water organisms were already tolerant of cooler temperatures, as they lived in habitats protected from major environmental disturbances.

## 7. The Famennian record

When McLaren (1970) first suggested that the F-F extinction could have been caused by an "astronomical catastrophe," he made note of the negative effect it had on the stromatoporoids; however, the only genus he mentioned as typical of the Famennian was *Labechia*. Later, McLaren (1983, p. 319) said that no stromatoporoids "... are known in the Early Famennian with the exception of *Labechia*, and at least six families disappeared". Today there are 18 genera known from the Famennian (Table 1; Fig. 2), 12 of which are labechiids, and four of which are clathrodictyids. Stromatoporoids of ostensibly Tournaisian (Early Mississippian) age were reported initially from the former USSR (e.g. Vasilyuk, 1966) and China (e.g. Wang, 1978a, b), but these strata are now known to be Famennian.

Stearn (1987, p. 677) summarized the recovery of stromatoporoids following the F-F extinction as a matter of "... survival of genera better adapted to cool water". He noted three distinctive stromatoporoid faunas for the Famennian: (1) almost exclusively labechiids; (2) labechiids mixed with clathrodictyids; and (3) dominated by clathrodictyids, to the exclusion of labechiids, and therefore more Frasnian-like. Based on the paleogeographic distribution of Famennian stromatoporoids, he suggested that labechiids might have been better adapted to cool water. If cool water was preferred by labechiids, one would expect them to have lived in the higher-latitude EAR (Fig. 5); however, they were absent (Table 1).

Stock (1990), on the other hand, concluded that depth could have been the controlling factor. In her summary of the geographic distribution of Famennian and Lower Carboniferous – now known to be upper Famennian – stromatoporoids, Bogoyavlenskaya (1982) described two communities. The first community includes the Western Slope of the Urals, Novaya Zemlya, the Donets Basin, and several other parts of Russia, and southeastern China (i.e. Guangxi, Guizhou, Hunan). The second community includes the Eastern Slope of the Urals, central Kazakhstan, and Western Europe (i.e. France, Belgium, Germany, Czech Republic). In all cases where an entire stromatoporoid fauna is described, Stearn's (1987) faunas 1 and 2 occurred in Bogoyavlenskaya's (1982) first community, whereas his fauna 3 occurred in her second community. According to Hamilton (1970), the Western Slope of the Urals, Novaya Zemlya, the Yogorsk Peninsula, Bolshaya Zelents Island, and Dolgi Island (all Russia) represent deposition in a miogeosyncline (i.e. shallower water), whereas the Eastern Slope of the Urals represents deposition in a eugeosyncline (i.e. deeper water). The implications here are that stromatoporoid faunas of dominantly non-labechiids were restricted to relatively deeper water, and dominantly labechiid and mixed faunas were restricted to shallower water. Perhaps the generally cooler conditions prevailing in association with Famennian glaciations (e.g. Crowell, 1999) continued to have had a negative impact on non-labechiid stromatoporoids in shallower water.

## 8. Unknowns

The geological record of Famennian stromatoporoids is better known than is that of their Frasnian counterparts, in spite of the much greater diversity (Table 1; Fig. 2) and geographic extent (Stock, 1990) of the latter. Attention should now be paid to elucidating the history of the Frasnian, and Givetian, stromatoporoids, so that we can better understand what may have contributed to their decimation in the F-F extinction.

Was the effect of the Taghanic Onlap a gradual or episodic encounter between OWR and EAR stromatoporoids? Was the apparent invasion of OWR forms into the EAR only a unidirectional phenomenon? Were the extinctions of stromatoporoid genera gradually distributed throughout the Frasnian, or were they clustered into step-wise events? Were Frasnian genus-level extinctions globally synchronous, or did certain areas serve as refugia? How do Givetian and Frasnian evolutionary events correlate with the Johnson et al. (1985) sea-level curve? The work of Stearn (2001) and Shapo (2003) represent beginnings toward answering these questions. Stearn (2001) divided the Devonian stromatoporoid succession of arctic and western Canada into 10 biostratigraphic assemblages, and indicated which species are characteristic of each assemblage. He also revised the names of 89 species. Shapo (2003) described the stromatoporoids from the middle Givetian Solon and Rapid members of the Little Cedar Formation of east-central Iowa – these strata were deposited early in the Taghanic Onlap. She described one species each for *Petridiostroma*, *Schistodictyon*, *Clathrocoilona*, *Stictostroma*, and *Coenostroma*.

## 9. Conclusions

(1) From the Early Devonian through the early Middle Devonian (Eifelian), stromatoporoid genus origination exceeded extinction.
(2) From the late Middle Devonian (Givetian) through the Late Devonian, stromatoporoid genus extinction exceeded origination.
(3) Far more genera were endemic to, and far more genera originated in, the Old World Realm, relative to the Eastern Americas Realm.
(4) A small number of genera entered the Eastern Americas Realm from the Old World Realm during the Emsian and Eifelian Ages.
(5) The Givetian reduction in total genus diversity was most likely caused by sea-level rise (Taghanic Onlap), which resulted in a merger of the Eastern Americas and Old World Realms.
(6) The Frasnian reduction in genus diversity was most likely caused by: (a) episodic sea-level rise, resulting in further reduction in endemism during the Frasnian; interrupted by (b) brief, but abrupt, sea-level falls, resulting in smaller epeiric seas in which species would have been forced to compete for resources; followed by (c) abrupt, greater than usual, sea-level drawdown and global cooling at the end of the Frasnian, possibly associated with the onset of Gondwana glaciations.
(7) Stromatoporoids made a partial recovery during the Famennian, including the appearance of several new labechiid genera, only to become extinct by the close of the Devonian.

## Acknowledgments

Many colleagues have contributed my understanding of Devonian stromatoporoid evolution and paleobiogeography, including, in approximately chronological order, J. E. Sorauf, J. St. Jean, J. M. Dennison, C. W. Stearn, W. A. Oliver, Jr., B. J. Witzke, C. A. Sandberg, and J. R. Morrow. Stearn, B. D. Webby, and H. Nestor helped with the refinement of the geologic ranges of several genera – this publication would have been impossible without their shared insights. Thanks also go to M. Coen-Aubert, J. Hladil, and G. Racki for suggesting conodont references, and G. Klapper for sharing information on conodonts of the Canning Basin. Careful reviews by Stearn and P. Copper did much to strengthen this paper.

## References

Aldridge, R.J., 1988. Extinction and survival in the Conodonta. In: Larwood, G.P., (Ed.), Extinction and Survival in the Fossil Record. Syst. Assoc. Sp. Vol. 34, Clarendon Press, Oxford, pp. 231–256.

Birkhead, P.K., Fraunfelter, G.H., 1973. Some Middle Devonian stromatoporoids from southern Illinois and southeastern Missouri. J. Paleontol. 47, 1069–1076.

Blodgett, R.B., Rohr, D.M., Boucot, A.J., 1988. Lower Devonian gastropod biogeography of the Western Hemisphere. In: McMillan, N.J., Embry, A.F., Glass, D.J., (Eds), Devonian of the World. Proceedings of the Second International Symposium on the Devonian System, Vol. III. Can. Soc. Petrol. Geol. Mem. 14, pp. 281–294.

Blodgett, R.B., Rohr, D.M., Boucot, A.J., 1990. Early and Middle Devonian gastropod biogeography. In: McKerrow, W.S., Scotese, C.R., (Eds), Palaeozoic Palaeogeography and Biogeography. Geol. Soc. Mem. No. 12, Geological Society Publishing House, Bath, pp. 277–284.

Bogoyavlenskaya, O.V., 1982. Stromatoporaty poznego devona-rannego karbona. Paleo. Zh., 1982(1), 33–38. [Transl. In Paleo. J. 15, 29–36]

Boucot, A.J., 1975. Evolution and Extinction Rate Controls. Developments in Palaeontology and Stratigraphy, No. 1, Elsevier, Amsterdam, 427pp.

Boucot, A.J., 1990. Silurian and pre-Upper Devonian bio-events. In: Walliser, O., Kauffman, E.G., (Eds), Abrupt Changes in the Global Biota, Springer-Verlag, Berlin, pp. 125–132.

Boucot, A.J., Blodgett, R.B., 2001. Silurian – Devonian biogeography. In: Brunton, C.H.C., Cocks, L.R.M., Long, S.L., (Eds), Brachiopods Past and Present. Syst. Assoc. Sp. Vol. 63, Clarendon Press, Oxford, pp. 335–344.

Boucot, A.J., Johnson, J.G., Talent, J.A., 1969. Early Devonian brachiopod zoogeography. Geol. Soc. Amer. Sp. Pap. 119, 113pp.

Boulvain, F., Bultynck, P., Coen, M., Coen-Aubert, M., Lacroix, D., Laloux, M., Casier, J.-C., Dejonghe, L., Dumoulin, V., Ghysel, P., Godefroid, J., Helsen, S., Mouravieff, N, A., Sartenaer, P., Tourneur, F., Vanguestaine, M., 1999. Les formations du Frasnien de la Belgique. Mem. Geol. Surv. Belgium No. 44, 125pp.

Briggs, D.E.G., Fortey, R.A., Clarkson, E.N.K., 1988. Extinction and the fossil record of the arthropods. In: Larwood, G.P., (Ed.), Extinction and Survival in the Fossil Record. Syst. Assoc. Sp. Vol. 34, Clarendon press, Oxford, pp. 171–209.

Cecca, F., 2002. Palaeobiogeography of Marine Fossil Invertebrates: Concepts and Methods. Taylor & Francis, London, 273pp.

Cockbain, A.E., 1984. Stromatoporoids from the Devonian reef complexes Canning Basin Western Australia. Geol. Surv. Western Australia, Bull. 129, 108pp.

Cookenboo, H.O., Orchard, M.J., Daoud, D.K., 1998. Remnants of Paleozoic cover on the Archean Canadian Shield: limestone zenoliths from kimberlite in the central Slave craton. Geology 26, 391–394.

Copper, P., 2002a. Reef development at the Frasnian/Famennian mass extinction boundary. Palaeogeog., Palaeoclimatol., Palaeoecol. 181, 27–65.

Copper, P., 2002b. Silurian and Devonian reefs: 80 million years of global greenhouse between two ice ages. In: Kiessling, W., Flügel, E., Golonka, J., (Eds), Phanerozoic Reef Patterns. SEPM Sp. Pub. No. 72, SEPM, Tulsa, pp. 181–238.

Crowell, J.C., 1999. Pre-Mesozoic ice ages: their bearing on understanding the climate system. Geol. Soc. Amer. Mem. 192, 106pp.

Day, J.E., 1996. Faunal signatures of Middle-Upper Devonian depositional sequences and sea level fluctuations in the Iowa Basin: U.S. Midcontinent. In: Witzke, B.J., Ludvigson, G.A., Day, J.E., (Eds), Paleozoic Sequence Stratigraphy: Views from the North American Craton. Geol. Soc. Amer. Sp. Pap. 306, Geological Society of America, Boulder, pp. 277–300.

Dvorak, J., Friakova, O., Galle, A., Hladil, J., Skocek, V., 1984. Correlation of the reef and basin facies of Frasnian age in the Krtiny HV-105 borehole in the Moravian Karst. Sb. Geol. Ved. Geol. 39, 71–98.

Fischbuch, N.R, 1962. Stromatoporoid zones of the Kaybob reef, Alberta. J. Alberta Soc. Petrol. Geol. 10, 62–72.

Galloway, J.J., St. Jean, J., Jr., 1957. Middle Devonian Stromatoporoidea of Indiana, Kentucky, and Ohio. Bulls. Amer. Paleo. 43, 25–308.

Hamilton, W., 1970. The Uralides and the motion of the Russian and Siberian Platforms. Geol. Soc. Amer. Bull. 81, 2553–2576.

Heckel, P.H., Witzke, B.J., 1979. Devonian world palaeogeography determined from distribution of carbonates and related lithic palaeoclimate indicators. In: House, M.R., Scrutton, C.T., Bassett, M.G., (Eds), The Devonian System. Palaeontological Association Special Papers in Palaeontology 23, Palaeontological Association, London, pp. 99–123.

House, M.R., 2002. Strength, timing, setting and cause of mid-Palaeozoic extinctions. Palaeogeogr., Palaeoclimatol., Palaeoecol. 181, 5–25.

Joachimski, M.M., Buggisch, W., 2002. Conodont apatite ?$^{18}$O signatures indicate climatic cooling as a trigger to the Late Devonian mass extinction. Geology 30, 711–714.

Joachimski, M.M., Pancost, R.D., Freeman, K.H., Ostertag-Henning, C., Buggisch, W., 2002. Carbon isotope geochemistry of the Frasnian-Famennian transition. Palaeogeogr., Palaeoclimatol., Palaeoecol. 181, 91–109.

Joachimski, M.M., van Geldern, R., Breisig, S., Buggisch, W., Day, J.E., 2004. Oxygen isotope evolution of biogenic calcite and apatite during the Middle and Late Devonian. Int. J. Earth Sci. (Geol. Rundsch.) 93, 542–553.

Johnson, J.G., 1970. Taghanic Onlap and the end of North American Devonian provinciality. Geol. Soc. America Bull. 81, 2077–2093.

Johnson, J.G., 1990. Lower and Middle Devonian brachiopod-dominated communities of Nevada, and their position in a biofacies-province-realm model. J. Paleontol. 64, 902–941.

Johnson, J.G., Boucot, A.J., 1973. Devonian brachiopods. In: Hallam, A., (Ed.), Atlas of Palaeobiogeography. Elsevier Scientific, Amsterdam, pp. 89–96.

Johnson, J.G., Klapper, G., Sandberg, C.A., 1985. Devonian eustatic fluctuations in Euramerica. Geol. Soc. America Bull. 96, 567–587.

Johnson, J.G., Klapper, G., Sandberg, C.A., 1986. Late Devonian eustatic cycles around margin of Old Red Continent. Ann. Soc. Géol. Belgique 109, 141–147.

Johnson, J.G., Sandberg, C.A., 1988. Devonian eustatic events in the western United States and their biostratigraphic responses. In: McMillan, N. J., Embry, A.F., Glass, D.J., (Eds), Devonian of the World, Vol. III. Can. Soc. Petrol. Geol. Mem. 14, Canadian Society of Petroleum Geologists, Calgary, pp. 171–178.

Kazmierczak, J., 1971. Morphogenesis and systematics of the Devonian Stromatoporoidea from the Holy Cross Mountains, Poland. Palaeontol. Pol. 26, 146pp.

Klapper, G., 1988. The Montagne Noire Frasnian (Upper Devonian) conodont succession. In: McMillan, N.J., Embry, A.F., Glass, D.J., (Eds), Devonian of the World, Vol. III. Can. Soc. Petrol. Geol. Mem. 14, Canadian Society of Petroleum Geologists, Calgary, pp. 449–468.

Klapper, G., Becker, R.T., 1999. Comparison of Frasnian (Upper Devonian) conodont zonations. Boll. Soc. Paleo. Ital. 37, 339–348.

Laporte, L.F., 1967. Carbonate deposition near mean sea-level and resultant facies mosaic: Manlius Formation (Lower Devonian) of New York State. Amer. Assoc. Petrol. Geol. Bull. 51, 73–101

Lecompte, M.J., 1951, 1952. Les stromatoporoids du Dévonien moyen et supérieur du Bassin de Dinant. Inst. Roy. Sci. Nat. Belgique Mem. 116, 117, 359pp.

Lecompte, M.J., 1954. Quelques données rélatives à la genèse et aux charactères écologiques des "récifs" du Frasnien de l'Ardenne. Inst. R. des Sci. Nat. de Belgique, Victor Van Straelen Jubilaire 1, 153–194. (Translated in Int. Geol. Rev. 1959, 1(7), 1–24)

Lieberman, B.S., 2000. Paleobiogeography: Using Fossils to Study Global Change, Plate Tectonics, and Evolution. Topics in Geobiology, Vol. 16. Kluwer Academic/Plenum, New York, 208pp.

McCracken, A.D., Armstrong, D.K., Bolton, T.E., 2000. Conodonts and corals in kimberlite zenoliths confirm a Devonian seaway in central Ontario and Quebec. Can. J. Earth. Sci. 37, 1651–1663.
McGhee, G.R., Jr., 1988. The Late Devonian extinction event: evidence for abrupt ecosystem collapse. Paleobiology 14, 250–257.
McGhee, G.R., Jr., 1996. The Late Devonian Mass Extinction: The Frasnian/Famennian Crisis. Columbia University Press, New York, 303pp.
McLaren, D.J., 1970. Presidential address: time, life, and boundaries. J. Paleontol. 44, 801–815.
McLaren, D.J., 1983. Bolides and biostratigraphy. Geol. Soc. America Bull. 94, 313–324.
McLean, R.A., Klapper, G., 1998. Biostratigraphy of Frasnian (Upper Devonian) strata in western Canada, based on conodonts and rugose corals. Bull. Can. Petrol. Geol. 46, 515–563.
Mistiaen, B., 1985a, 1985b. Phénomènes récifaux dans le Dévonien d'Afghanistan (Montagnes Centrales), analyse et systématique des Stromatopores. Soc. Géol. Nord., Pub. 11, 1/2, 381/245.
Nestor, H., Stock, C.W., 2001. Recovery of the stromatoporoid fauna after the Late Ordovician extinction. In: Ezaki, Y., Mori, K., Sugiyama, T., Sorauf, J.E., (Eds), Proceedings of the 8th International Symposium on Fossil Cnidaria and Porifera. Bull. Tohoku Univ. Mus. No. 1, pp. 329–337.
Nguyen, H.H., Mistiaen, B., 1998. Uppermost Famennian stromatoporoids of north central Viet Nam. J. Geol., Ser. B, No. 11–12/1998, 57–75.
Oliver, W.A., Jr., 1977. Biogeography of Late Silurian and Devonian rugose corals. Palaeogeogr., Palaeoclimatol., Palaeoecol. 22, 85–135.
Oliver, W.A., Jr., 1990. Extinctions and migrations of Devonian rugose corals in the Eastern Americas Realm. Lethaia 23, 167–178.
Oliver, W.A., Jr., 1997. Ptenophyllids in eastern North America. Coral Res. Bull., 5, 203–209.
Oliver, W.A., Jr., Pedder, A.E.H., 1989. Origins, migrations, and extinctions of Devonian Rugosa on the North American Plate. In: Jell, P.A., Pickett, J.W. (Eds), Fossil Cnidaria 5. Assoc. Austral. Palaeontol., Mem. No. 8, Association of Australasian Palaeontologists, Brisbane. pp. 231–237.
Pedder, A.E.H., Murphy, M.A., 2004. Emsian (Lower Devonian) Rugosa of Nevada: revision of systematics and stratigraphic ranges, and reassessment of faunal provincialism. J. Paleontol. 78, 838–865.
Prosh, E.C., Stearn, C.W., 1993. Early Devonian age of the Detroit River Group, inferred from Arctic stromatoporoids. Can. J. Earth Sci. 30, 2465–2474.
Racki, G., 1998. The Frasnian-Famennian brachiopod extinction events: a preliminary review. Acta Palaeontol. Pol. 43, 395–411.
Racki, G., Turnau, E., 2000. Devonian series and stage boundaries in Poland. Cour. Forsch.-Inst. Senckenberg 225, 145–158.
Sandberg, C.A., Morrow, J.R. Ziegler, W., 2002. Late Devonian sea-level changes, catastrophic events, and mass extinctions. In: Koeberl, C., MacLeod, K.G., (Eds), Catastrophic Events and Mass Extinctions: Impacts and Beyond. Geol. Soc. America Sp. Pap. 356, Geological Society of America, Boulder, pp. 473–487.
Schindler, E., 1993. Event-stratigraphic markers within the Kellwasser Crisis near the Frasnian/Famennian boundary (Upper Devonian) in Germany. Palaeogeog., Palaeoclimatol., Palaeoecol. 104, 115–125.
Scrutton, C.T., 1997. The Palaeozoic corals, I: origins and relationships. Proc. Yorkshire Geol. Soc. 51, 177–208.
Scrutton, C.T., 1998. The Palaeozoic corals, II: structure, variation and palaeoecology. Proc. Yorkshire Geol. Soc. 52, 1–57.
Shapo, D.E., 2003. Systematics and morphometric analysis of stromatoporoids from the Little Cedar Formation, Middle Devonian, east-central Iowa. MS Thesis, University of Iowa, Iowa City, IA, USA.
Shaver, R.H., Ault, C.H., Burger, A.M., Carr, D.D., Droste, J.B., Eggert, D.L., Gray, H.H., Harper, D., Hasenmueller, N.R., Hasenmueller, W.A., Horowitz, A.S., Hutchinson, H.C., Keith, B.D., Keller, S.J., Patton, J.B., Rexroad, C.B., Wier, C.E., 1986. Compendium of Paleozoic rock-unit stratigraphy in Indiana – a revision. Indiana Geol. Surv. Bull. 59, 203pp.
Sloss, L.L., 1963. Sequences in the cratonic interior of North America. Geol. Soc. America Bull. 74, 93–111.
Smith, M.J., 1994. Upper Devonian stromatoporoids from the Idlewild Member of the lithograph City Formation of north-central Iowa. MS Thesis, University of Alabama, Tuscaloosa, AL, USA.
Sorauf, J.E., 1989. Rugosa and the Frasnian-Famennian extinction event: a progress report. In: Jell, P.A., Pickett, J.W., (Eds), Fossil Cnidaria 5. Assoc. Australas. Paleontol. Mem. 8, Association of Australasian Palaeontologists; Brisbane, pp. 327–338.
Sorauf, J.E., Pedder, A.E. H., 1986. Late Devonian rugose corals and the Frasnian-Famennian crisis. Can. J. Earth Scis. 23, 1265–1287.

Stearn, C.W., 1975. Stromatoporoid assemblages, Ancient Wall reef complex (Devonian), Alberta. Can. J. Earth Sci. 12, 1631–1667.

Stearn, C.W., 1987. Effect of the Frasnian-Famennian extinction event on the stromatoporoids. Geology 15, 677–680.

Stearn, C.W., 2001. Biostratigraphy of Devonian stromatoporoid faunas of arctic and western Canada. J. Paleontol. 75, 9–23.

Stearn, C.W., Webby, B.D., Nestor, H., Stock, C.W., 1999. Revised classification and terminology of Palaeozoic stromatoporoids. Acta Palaeontol. Pol. 44, 1–70.

Stock, C.W., 1979. Upper Silurian (Pridoli) Stromatoporoidea of New York. Bulls. Amer. Paleontol. 76, 289–389.

Stock, C.W., 1982. Upper Devonian (Frasnian) Stromatoporoidea of north-central Iowa: Mason City Member of the Shell Rock Formation. J. Paleontol. 56, 998–1005.

Stock, C.W., 1984a. Upper Devonian (Frasnian) Stromatoporoidea of north-central Iowa: redescription of the type specimens of Hall and Whitfield (1873). J. Paleontol. 58, 773–788.

Stock, C.W., 1984b. The distribution of stromatoporoids in the Upper Devonian of north-central Iowa. In: Anderson, W.I., (Ed.), General Geology of North-Central Iowa. Guidebook for the 48th Annual Tri-State Geological Field Conference, October 6–7, 1984, University of Northern Iowa, Cedar Falls, pp. 125–129.

Stock, C.W., 1990. Biogeography of the Devonian stromatoporoids. In: McKerrow, W.S., Scotese, C.R., (Eds), Palaeozoic Palaeogeography and Biogeography. Geol. Soc. Mem. No. 12, Geological Society, Publishing House, Bath, pp. 257–265.

Stock, C.W., 1997. Paleobiogeographical range of North American Devonian stromatoporoids: the roles of global and regional controls. In: Perejón, A., Comas-Rengifo, M.J., (Eds), Proceedings of the VII International Symposium on Fossil Cnidaria and Porifera, Vol. II. Bol. R. Soc. Española Hist. Nat. (Sec. Geol.) 92, pp. 281–288.

Stock, C.W., Burry-Stock, J.A., 2004. New data reinforce the conclusion that the inter-realm barrier in North America was selectively breached by the stromatoporoid *Habrostroma centrotum* during the Lochkovian Age (Early Devonian). Geol. Soc. Amer. Abs. Prog. 36(5), 91–92.

Vasilyuk, N.P., 1966. Korally i stromatoporoidei. In: Aizenverg, D. E., (Ed.), Fauna nizov Turne (zony $C_1^ta$) Donetskogo basseina. Akademiya Nauk Ukrainskoi SSR, Institut Geologicheskikh Nauk, Kiev, 128pp.

Walker, K. R., Alberstadt, L. P., 1975. Ecological succession as an aspect of structure in fossil communities. Paleobiology 1, 238–257.

Wang, S.-B., 1978a. Order Stromatoporoidea. In: Atlas of Fossils of Southwest China, Sichuan Volume, Part II: From Carboniferous to Mesozoic. Chengdu Institute of Geology and Mineral Resources, Geological Publishing House, Peking, pp. 123–137, 616–618.

Wang S.-B., 1978b. Order Stromatoporoidea. In: Atlas of Fossils of Southwest China, Guizhou Volume, Part II: From Carboniferous to Mesozoic. Chengdu Institute of Geology and Mineral Resources, Geological Publishing House, Peking, pp. 98–106, 566–567.

Witzke, B. J., 1990. Palaeoclimatic constraints for Palaeozoic palaeolatitudes of Laurentia and Euramerica. In: McKerrow, W. S., Scotese, C. R., (Eds), Palaeozoic Palaeogeography and Biogeography. Geol. Soc. Mem. No. 12, Geological Society, Publishing House, Bath, pp. 57–73.

Wood, R. A., 1991. Problematic reef-building sponges. In: Simonetta, A., Conway Morris, S., (Eds), The Early Evolution of Metazoa and the Significance of Problematic Taxa. Cambridge University Press, Cambridge.

Ziegler, W., 1962. Taxonomie und Phylogenie Oberdevonischer Conodonten und ihre stratigraphische Bedeutung. Hess. Land. Bod. Ab. 38, 166pp.

Ziegler, W., 1971. Conodont stratigraphy of the European Devonian. In: Sweet, W. C., Bergström, S. M., (Eds), Symposium on Conodont Biostratigraphy. Geol. Soc. Amer. Mem. 127, Geological Society of America, Boulder, pp. 227–284.

Ziegler, W., Sandberg, C. A., 1990. The Late Devonian standard conodont zonation. Cour. Forsch.-Inst. Senckenberg 121, 115pp.

Zukalova, V., 1971. Stromatoporoidea from the Middle and Upper Devonian of the Moravian Karst. Czech. Ústred. Ústav. Geol., Rozpr. 37, 143pp.

Understanding Late Devonian and Permian-Triassic Biotic and Climatic Events:
Towards an Integrated Approach
edited by D.J. Over, J.R. Morrow and P.B. Wignall



## *Chapter 6*

# *Using environmental niche modeling to study the Late Devonian biodiversity crisis*

*Alycia L. Stigall Rode**

*Department of Geological Sciences, Ohio University, 316 Clippinger Laboratories, Athens, OH 45701, USA*

*Bruce S. Lieberman*

*Department of Ecology and Evolutionary Biology, University of Kansas, 1475 Jayhawk Blvd., Lawrence, KS 66045, USA*

### Abstract

Geographic ranges are estimated for brachiopod and bivalve species during the late Middle (mid-Givetian) to the middle Late (terminal Frasnian) Devonian to investigate range changes during the time leading up to and including the Late Devonian biodiversity crisis. Species ranges were predicted using GARP (Genetic Algorithm using Rule-set Prediction), a modeling program developed to predict fundamental niches of modern species. This method was applied to fossil species to examine changing ranges during a critical period of Earth's history. Comparisons of GARP species distribution predictions with historical understanding of species occurrences indicate that GARP models predict accurately the presence of common species in some depositional settings. In addition, comparison of GARP distribution predictions with species-range reconstructions from geographic information systems (GIS) analysis suggests that GARP modeling has the potential to predict species ranges more completely and tailor ranges more specifically to environmental parameters than GIS methods alone. Thus, GARP modeling is a potentially useful tool for predicting fossil species ranges and can be used to address a wide array of palaeontological problems.

The use of GARP models allows a statistical examination of the relationship of geographic range size with species survival during the Late Devonian. Large geographic range was statistically associated with species survivorship across the crisis interval for species examined in the *linguiformis* Zone but not for species modeled in the preceding Lower *varcus* or *punctata* zones. The enhanced survival benefit of having a large geographic range, therefore, appears to be restricted to the biodiversity crisis interval.

*Keywords:* prediction, invasion, extinction, survival, biogeography

## 1. Introduction

The geographic distribution of species is controlled by a variety of factors: biotic; environmental; and historical (Brown and Lomolino, 1998). The fundamental ecological niche of a species exerts a primary control on the geographic distribution of a species. Reconstructing species niches is an essential step in predicting the area a species could

*Corresponding author. Fax: +1-740-593-0486. *E-mail address:* stigall@ohio.edu (A.L. Stigall Rode).

inhabit (Peterson, 2001). The fundamental niche is the set of environmental tolerances and limits in multidimensional space that define where a species is potentially able to maintain populations (Grinnell, 1917; Hutchinson, 1957). Species, however, rarely occupy their entire fundamental niche due to historical contingencies (e.g. their ancestors never inhabited the regions) or biological factors (e.g. competitive exclusion) (Brooks and McLennan, 1991, 2002; Brown and Lomolino, 1998). Modeling species ranges based on the fundamental niche is a major research area in modern biology and can be used to provide insight into geographic range changes, to predict new regions where species could occur, and to predict the effects of climate change on geographic distribution. Ranges of species in the fossil record are also controlled by the same types of ecological variables, so understanding the interplay between species fundamental niches and realized ranges is also important for palaeontologists. Numerous methods exist for reconstructing species ranges, including simplistic models designed around one variable and sophisticated computer learning-based systems (e.g. Stockwell and Peters, 1999; Haltuch et al., 2000). The GARP (Genetic Algorithm using Rule-set Prediction) modeling system, a computer learning-based system, predicts species ranges based on the fundamental ecological requirements modeled using the environmental characteristics of a set of known occurrence sites (Stockwell and Peters, 1999). This method has been successful at predicting species ranges and as a tool for investigating ecological and evolutionary questions in the modern biota (e.g. Peterson et al., 1999, 2001, 2002a–c; Anderson et al., 2002; Feria and Peterson, 2002). In this paper, we will explore the use of GARP for reconstructing ranges of shallow-marine brachiopod and bivalve species during the Givetian and Frasnian Ages (late Middle and early Late Devonian).

The Late Devonian is an excellent time to examine changing geographic ranges for several reasons. Firstly, the Middle to Late Devonian transition involved a dramatic change from a highly endemic Middle Devonian fauna to a cosmopolitan Late Devonian biota (Oliver, 1976, 1990; McGhee, 1989, 1996). In addition, the Late Devonian was a time of major biodiversity decline associated with the Frasnian–Famennian biodiversity crisis (McGhee, 1988, 1996). This crisis event was characterized by elevated extinction levels, reduced speciation rates, and ecological reorganization (McGhee, 1988, 1990, 1996; Oliver and Pedder, 1994; Droser et al., 2000; Rode and Lieberman, 2002). Finally, changing patterns of geographic range, particularly range expansions or species invasions during the Middle to Late Devonian transition, have been implicated in species survival during the biodiversity crisis interval (McGhee, 1996; Rode and Lieberman, 2004).

Quantifying invertebrate fossil ranges is currently a promising area of palaeontological study. Recent geographic information systems (GIS) work with Palaeozoic invertebrates (e.g. Rode and Lieberman, 2003, 2004, 2005) has built on the earliest use of GIS methods in palaeontology (e.g. Juliusson and Graham, 1999; Graham, 2000; Ferguson et al., 2001). Computer learning-based genetic algorithms, however, have not been previously attempted in palaeontology, because most traditional range reconstruction methods are based on determining areas that surround known occurrence points, and do not allow simultaneous consideration of multiple variables under multiple rule sets. The use of the GARP algorithm to explore Palaeozoic species ranges, therefore, provides a potentially useful step quantifying species ranges and producing additional information for palaeoecological and macroevolutionary studies.

## 2. Methods

### *2.1. Geographic and stratigraphic intervals examined*

#### *2.1.1. Geographic extent*

The focus of this analysis is to reconstruct species ranges of brachiopod and bivalve species for three time intervals during the Givetian and Frasnian Ages. The geographic area of this study is restricted to the northern Appalachian Basin of eastern North America including the Devonian outcrop belt in the states of New York, Maryland, Pennsylvania, West Virginia, and Virginia (Fig. 1). This region was chosen for study because the area represents one of the most complete Devonian sequences in the world with an extensively studied, well-preserved fauna. Within the region, the area of interest was divided by a grid system into smaller areas of 0.5° latitude by 0.5° longitude (Fig. 1), which is a standard

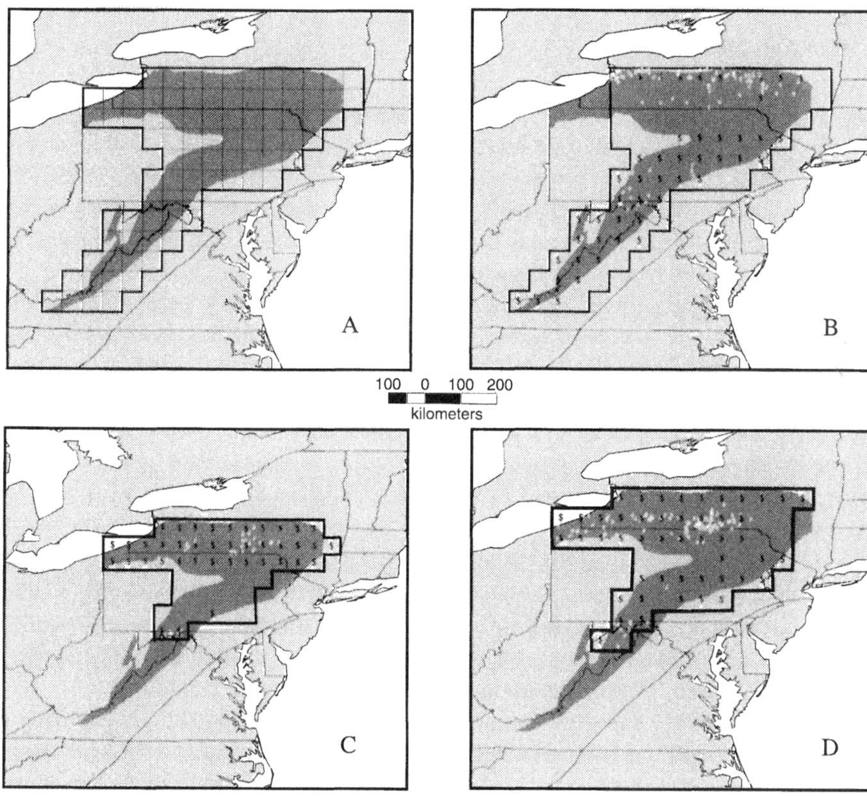


*Figure 1.* (A) Devonian outcrop belt of New York, Pennsylvania, West Virginia, Virginia, and Maryland with 0.5° × 0.5° grid overlain. Distribution of environmental data (black triangles), and species occurrence data (white circles), and geographic area considered within the modeling experiment (black outline) for the (B) Lower *varcus* Zone, (C) *punctata* Zone, and (D) *linguiformis* Zone.

procedure when using the GARP modeling system (R. Scachetti-Pereira pers. comm., 2002). Stratigraphic and environmental information (as discussed below) was obtained for each grid box individually.

### 2.1.2. *Temporal range*

To examine changes in species ranges through time, three time intervals were investigated, each approximating a conodont zone. The intervals examined were the Lower *varcus* (middle Givetian), *punctata* (middle Frasnian), and *linguiformis* (terminal Frasnian) conodont zones, which are estimated at 1.5, 0.6, and 0.4 million years, respectively [zone durations approximated based on the relative durations of conodont zones in Sandberg and Ziegler (1996) calibrated against the Devonian time scale of Tucker et al. (1998)]. Both the Lower *varcus* and *punctata* zones precede the Late Devonian biodiversity crisis, providing necessary baseline information, whereas the *linguiformis* Zone is within crisis interval (McGhee, 1996). Species survivorship through the biodiversity crisis interval is further examined by comparing the temporal range of species into the Famennian in statistical analyses.

### 2.1.3. *Stratigraphic framework*

During the Middle to Late Devonian, the depositional setting of the northern Appalachian Basin was dominated by a shallow siliciclastic ramp system, the Catskill Delta (Woodrow, 1985). The Catskill Delta is a progradational system derived from the weathering of the Acadian highlands formed by several progressive tectophases of the Early to Late Devonian Acadian orogeny (Ettensohn, 1985). Progressive weathering and subsidence produced thick and laterally extensive deposits throughout the foreland basin beginning in the Middle Devonian. The depositional setting during this time was a gently sloping siliciclastic shelf with storm processes dominating in the platform setting and tidal influences dominating in the nearshore (Brett and Baird, 1994; Prave and Duke, 1991). The shoreline migrated progressively westward as the foreland basin filled from the Middle to Late Devonian (Rickard, 1975; de Witt et al., 1993). Consequently, the area of marine deposition in the study area becomes progressively smaller in younger stratigraphic intervals. Hence, the areal extent of marine rocks available for study in the *punctata* Zone is smaller than in the Lower *varcus* Zone and so on (Fig. 1). The westward progression of facies does not pose a problem in this analysis, because the full complement of environments examined in the oldest time intervals remains present in the basin during the youngest intervals. Cross-environmental analyses, therefore, remain possible and the reduction in depositional area is accounted for by examining relative areas in statistical analyses. Table 1 lists the stratigraphic references used in determining palaeoenvironmental conditions.

Stratigraphic data for depositional environments within each conodont zone were collected from stratigraphic units interpreted to approximately correlate to the zone of interest. While the boundaries of these units may not precisely coincide with the temporal boundaries of the conodont zone of interest, they do represent the best approximation and provide the most accurate data available to reconstruct sedimentary conditions during the temporal intervals under investigation.

The Lower *varcus* Zone of the middle Givetian includes the well-characterized stratigraphic units of the Hamilton Group of New York and the Mahantango Formation of

*Table 1.* Stratigraphic information references by conodont zone.

| Lower *varcus* Zone | *punctata* Zone | *linguiformis* Zone |
|---|---|---|
| Batt, 1996 | *Adams et al., 1956 | Babcock and Wegweiser, 1998 |
| *Batt, 1999 | Applebaum, 1993 | *Dennison, 1985 |
| *Brett and Baird, 1994 | *Bishuk et al., 1991 | *Dennison et al., 1979 |
| *Brett et al., 1986 | *Bridge and Dingman, 1981 | *de Witt, 1960 |
| Dennison, 1985 | *Bowen et al., 1970 | *de Witt et al., 1993 |
| *Dennison and Hasson, 1976 | *Bowen et al., 1974 | Ehrets, 1981 |
| *Dennison et al., 1979 | *Colton and de Witt, 1958 | Frakes, 1963 |
| Ellison, 1963 | *Dennison, 1985 | *Frakes, 1964 |
| Ellison, 1965 | *Dennison et al., 1979 | *Jacobi and Smith, 1999 |
| Epstein, 1986 | *de Witt et al., 1993 | Karman, 1968 |
| *Faill et al., 1973 | *Fletcher, 1962 | *Kirchgasser et al., 1994 |
| *Hasson and Dennison, 1979 | Jacobi and Smith, 1999 | *Krajewski and Williams, 1971 |
| Lafferty et al., 1994 | Kirchgasser, 1965 | *Leighton, 2000 |
| Linsley, 1994 | *Kirchgasser, 1983 | *McGhee, 1976 |
| *Mayer, 1994 | *Kirchgasser et al., 1994 | *McGhee and Sutton, 1981 |
| *Mayer et al., 1994 | *Krajewski and Williams, 1971 | *McGhee and Sutton, 1983 |
| *Miller, 1986 | *Lundegard et al., 1985 | *McGhee and Sutton, 1985 |
| *Oliver and Klapper, 1981 | *McGhee and Sutton, 1985 | *Metzger et al., 1974 |
| *Prave and Duke, 1991 | *Oliver and Klapper, 1981 | *Oliver and Klapper, 1981 |
| *Prave et al., 1996 | *Over et al., 1999 | *Over, 1997 |
| *Rickard, 1975 | Patchen and Dugolinsky, 1979 | *Over et al., 1999 |
| Rodeheaver, 1992 | *Rickard, 1975 | Patchen and Dugolinsky, 1979 |
| *Savarese et al., 1986 | Sutton, 1963 | *Pepper and de Witt, 1950 |
| Sevon, 1985 | *Sutton and McGhee, 1985 | *Rahmanian, 1979 |
| *Ver Straeten and Brett, 1999 | *Sutton et al., 1970 | *Rickard, 1975 |
| Willard, 1935a | *Tesmer, 1966 | *Roe, 1976 |
| Willard, 1935c | Willard, 1934 | *Schultz, 1974 |
| Woodrow, 1985 | Willard, 1935b | *Smith and Jacobi, 2000 |
| *Wygart, 1986 | Woodrow, 1985 | Sutton, 1963 |
| | | *Sutton and McGhee, 1985 |
| | | *Tesmer, 1966 |
| | | *Tesmer, 1974 |
| | | Walker and Sutton, 1967 |
| | | Willard, 1934 |
| | | Willard, 1935b |
| | | *Williams and Slingerland, 1985 |
| | | Woodrow, 1985 |

*key citation.

Pennsylvania, Maryland, and West Virginia. The stratigraphic units used to estimate environmental parameters within the interval are the Ludlowville Formation of the Hamilton Group, the Panther Sandstone, the Plattekill Formation, the Millboro Shale Member of the Mahantango Formation, and the Mahantango Formation undivided (Appendix 1.1).

Environmental parameters during the *punctata* Zone of the middle Frasnian were estimated by using the characteristics of the Sonyea Group of New York, specifically the correlative Cashaqua, Rock Stream, Glen Aubrey, and lower Walton formations. Correlative portions of the Trimmers Rock Formation and Bralier Formation as well as the correlative strata of the Chemung, Portage, and Catskill magnafacies of Pennsylvania, Maryland, and West Virginia were also considered (Appendix 1.2).

The *linguiformis* Zone environment was estimated using the characteristics of the upper Java Formation (upper portion of the Hanover and Wiscoy members), Mansfield Shale, and Slide Mountain Formation of New York. Parts of the Trimmers Rock Formation, Foreknobs Formation, and other correlative strata of the Chemung, Portage, and Catskill magnafacies of Pennsylvania, Maryland, and West Virginia were also examined (Appendix 1.3).

### 2.2. Species occurrence information

Species geographic distribution data included in this analysis were assembled from examination of museum collections. Museums with extensive taxonomic and stratigraphic material from the northern Appalachian Basin of North America were visited and occurrence data collected for brachiopod and bivalve species. These occurrence data include species identifications (verified by A.L.S.), geographic location from which the fossil was collected, keyed to latitude and longitude values with maps, and the stratigraphic position of each specimen. Stratigraphic information was then converted to the approximate correlative conodont zone based on current literature (Rode and Lieberman, 2004). Only material with sufficient stratigraphic and locality information to identify a specimen's presence within a narrow geographic region and particular conodont zone were included within the database. Taxonomic identifications were based on comparison of specimens with the most up-to-date references available. Collection from the following museums were used: American Museum of Natural History; the Carnegie Museum of Natural History; the Museum of Comparative Zoology (Harvard University); the Peabody Museum of Natural History (Yale University); the University of Iowa Museum of Paleontology; the University of Michigan Museum of Paleontology; and the United States National Museum of Natural History. The entire database created from the museum data and further details on its assembly is published in Rode and Lieberman (2004).

This database was further culled, with only species represented by five or more occurrences during a conodont zone of interest retained for the present analysis. The GARP modeling algorithm (see discussion below), has been shown to be effective with sample sizes as small as five species occurrences (Peterson and Cohoon, 2002; Stockwell and Peterson, 2002), so five spatially distinct occurrence points within a conodont zone was used as the lower cutoff for species inclusion in this analysis. Included species are listed in Table 2, and species occurrence data used in this analysis are presented in Appendix 2.

### 2.3. Acquisition of environmental data and creation of base layers

The niche of shallow-marine species is controlled by a variety of environmental factors, such as water depth, wave energy, substrate type, oxygenation levels, and biotic interactions. The factors included in this study are primarily abiotic in nature and include the variables

*Table* 2. List of species and the conodont zones in which they were modeled using GARP.

| Species | Conodont Zone |
|---|---|
| *Ambocoelia gregaria* (Hall) | *linguiformis* |
| *Ambocoelia umbonata* (Conrad) | *linguiformis* |
| *Athyris angelica* (Hall) | *linguiformis* |
| *Athyris cora* (Hall) | *varcus* |
| *Athyris spiriferoides* (Eaton) | *varcus* |
| *Cariniferella carinata* (Hall) | *varcus, linguiformis* |
| *Cariniferella tioga* (Hall) | *linguiformis* |
| *Cupularostrum contracta* (Hall) | *linguiformis* |
| *Cupularostrum exima* (Hall) | *punctata, linguiformis* |
| *Cypricardella bellistriata* (Conrad) | *varcus* |
| *Cyrtospirifer chemungensis* (Hall) | *linguiformis* |
| *Douvillina cayuta* (Hall) | *linguiformis* |
| *Eoschizodus chemungensis* (Conrad) | *punctata* |
| *Floweria chemungensis* (Conrad) | *linguiformis* |
| *Floweria prava* (Hall) | *linguiformis* |
| *Goniophora chemungensis* (Vanuxem) | *punctata* |
| *Grammysia elliptica* Hall and Whitfield | *punctata* |
| *Leptodesma nitida* (Hall) | *punctata* |
| *Leptodesma spinerigum* (Conrad) | *varcus, linguiformis* |
| *Mucrospirifer mucronatus* (Conrad) | *varcus* |
| *Nervostrophia nervosa* (Hall) | *linguiformis* |
| *Palaeoneilo constricta* (Conrad) | *varcus, punctata* |
| *Paracyclas lirata* Conrad | *varcus* |
| *Praewaagenoconcha speciosa* (Hall) | *punctata, linguiformis* |
| *Productella rectispina* (Hall) | *linguiformis* |
| *Pseudatrypa devoniana* (Webster) | *linguiformis* |
| *Ptychopteria chemungensis* Conrad | *punctata* |
| *Schizophoria impressa* (Hall) | *linguiformis* |
| *Spinatrypa spinosa* (Hall) | *varcus, linguiformis* |
| *Spinocyrtia granulosa* (Conrad) | *varcus* |
| *Strophonella hybrida* Hall and Whitfield | *linguiformis* |
| *Tylothyris mesacostalis* (Hall) | *punctata, linguiformis* |

considered to be most important for determining habitable areas for benthic marine organisms (Brenchley and Harper, 1998) as well as the types of data typically compiled by sedimentary geologists interested in reconstructing depositional environments.

Eleven environmental factors were used to predict species ranges in this analysis (Table 3). Successful GARP analyses have been produced with as few as four and as many as 19 environmental variables (e.g. Feria and Peterson, 2002; Anderson et al., 2002; Peterson et al., 2002). Statistical analyses by Peterson and Cohoon (1999) have shown that although as few as five environmental variables can achieve nearly maximum accuracy in results, the inclusion of additional variables enhances detail and does not reduce accuracy. In addition, although some covariation is present within the environmental variables (e.g. water

*Table 3.* Explanation of coding strategy for variables used in construction of environmental base maps.

*Percent mud, silt, or sand:*
- Approximate fraction of each grain size within the sedimentary package

*Percent limestone:*
- Approximate percentage of limestone beds within the sedimentary package

*Bedding style:*
- Approximate thickness of sedimentary beds. Decimals indicate the relative abundance of each bedding type.

1. Thin: centimeter scale bedding
2. Moderate: decimeter scale bedding
3. Thick: meter scale bedding

*Substrate type:*
- Character of the substrate on which benthic organisms reside. Decimals indicate the relative abundance of each substrate type.

1. Muddy: fine grained, soupy sediment with abundant water in pore spaces for nutrients and deposit feeders
2. Silty: intermediate substrate type
3. Sandy: well sorted, coarser grained sediment

*Inferred water depth/energy zone:*
- Relative water depth with respect to storm and fair weather wave bases. Offshore settings within the Appalachian basin may have had water depths of 50 to 150 m (Prave et al., 1996). Decimals indicate the relative placement within the energy zone.

0. Offshore: below storm wave base
1. Subtidal: at storm wave base
2. Lower intertidal: at the lower boundary of the fair weather wave base (low tide waves)
3. Upper intertidal: at the lower boundary of high tide waves
4. Subaerial: above the high tide interval

*Depositional environment:*
- Inferred sedimentary environment of deposition. Decimals indicate the relative placement within depositional environments.

0. Basin
1. Outer shelf
2. Middle shelf
3. Inner shelf
4. Deltaic-estuarine
5. Coastal plain/alluvial setting

*Ichnofacies:*
- Representative icnofauna found within the stratigraphic unit. Decimals indicate the relative placement within ichnofacies.

0. Anoxic, traces absent
1. *Zoophycus* ichnofacies
2. *Cruziana* ichnofacies
3. *Skolithos* ichnofacies

*Table 3. (Continued)*

4. *Tubiphytes* ichnofacies
5. *Scoyenia* ichnofacies

*Oxygenation:*

- Inferred oxygen content of the water column at the water–substrate interface. Decimals indicate the relative placement within oxygenation zones.

1. Anaerobic
2. Dysaerobic
3. Normal marine
4. Subaerial

*Biofacies*

- Community of species present. Community names and associations follow Bowen et al. (1974), McGhee (1976), McGhee and Sutton (1981, 1983, 1985), and Sutton et al. (1970).

Lower *varcus* Zone:

1. Anoxic, fossils rare
2. Dysaerobic, *Ambocoelia*, *Palaeoneilo*, chonetids
3. Open marine, *Cypricaridella*, *Tropidoleptus*, *Athyris*, and *Ambocoelia*
4. Continental, root traces

*punctata* Zone:

1. Ammonites, conodonts
2. *Rhipidomella* fauna
3. *Cypricardella* fauna
4. Continental, root traces, plant material

*linguiformis* Zone:

1. Ammonites, cephalopods
2. *Ambocoelia-Cariniferella* fauna
3. *Tylothyris-Schizophoria* fauna
4. *Cyrtospirifer-Douvillina* fauna
5. Continental

depth and depositional environment), the GARP algorithm (discussed below) is designed to analyze poorly structured domains and is not sensitive to environmental covariation (Stockwell and Peters, 1999).

Environmental variables were coded for each grid box (Fig. 1) from published stratigraphic columns and descriptions. The raw data are presented in Appendices 1.1-1.3 and include all environmental data as well as the key references from which the data were derived. The raw data were converted to numerical values appropriate for use in the modeling program using the coding scheme presented in Table 3. The coded data are presented in Tables 4–6. Because each grid box encompasses an area of roughly 43 × 56 km, variability in environmental conditions occurs commonly within regions. Environmental variability also occurs temporally through the stratigraphic interval considered. This variability in environmental parameters within a single grid box was incorporated by coding the variable with a value intermediate between the end member states present within the region.

*Table 4*. Data used in reconstructing environmental base maps for the Lower *varcus* Zone.

| Longitude | Latitude | % Mud | % Silt | % Sand | % ls | Bedding | Substrate | Water depth | Environ | Oxygen | Biofacies |
|---|---|---|---|---|---|---|---|---|---|---|---|
| −78.75 | 42.75 | 55 | 20 | 0 | 25 | 1 | 1 | 1 | 1.5 | 1.5 | 3 |
| −78.25 | 42.75 | 50 | 30 | 0 | 20 | 1 | 1 | 1.5 | 1.5 | 1.5 | 3 |
| −77.75 | 42.75 | 70 | 0 | 0 | 30 | 1 | 1 | 1.5 | 0.5 | 1 | 3 |
| −77.25 | 42.75 | 70 | 15 | 0 | 15 | 1 | 1 | 0 | 0 | 1.5 | 2 |
| −76.75 | 42.75 | 80 | 10 | 0 | 10 | 1 | 1 | 0 | 0 | 1 | 2 |
| −76.25 | 42.75 | 60 | 20 | 0 | 20 | 1.5 | 2 | 1.5 | 2 | 1 | 3 |
| −75.75 | 42.75 | 30 | 60 | 0 | 10 | 1.5 | 2 | 1 | 2 | 2 | 3 |
| −75.25 | 42.75 | 15 | 10 | 75 | 0 | | | | | | |
| −74.75 | 42.75 | 20 | 0 | 80 | 0 | 3 | 3 | 3.5 | 4 | 4 | 4 |
| −74.25 | 42.75 | 20 | 10 | 70 | 0 | 3 | 3 | 4 | 5 | 5 | 4 |
| −75.25 | 42.25 | 20 | 10 | 70 | 0 | 3 | 3 | 4 | 5 | 5 | 4 |
| −77.25 | 41.25 | 90 | 10 | 0 | 0 | 1.5 | 1.5 | 1 | 0.5 | 2 | 2 |
| −76.75 | 41.25 | 80 | 20 | 0 | 0 | 1.5 | 1.5 | 1 | 0.5 | 2 | 2 |
| −76.25 | 41.25 | 63 | 35 | 2 | 0 | 1.5 | 1.5 | 1 | 1.5 | 2 | 3 |
| −75.75 | 41.25 | 43 | 47 | 10 | 0 | 1.25 | 1.5 | 1 | 1.5 | 2 | 3 |
| −75.25 | 41.25 | 52 | 44 | 4 | 0 | 1.25 | 2 | 1.5 | 2 | 2.5 | 3 |
| −74.75 | 41.25 | 20 | 28 | 52 | 0 | 2.5 | 2.5 | 2 | 2.5 | 2.5 | 3 |
| −78.25 | 40.75 | 80 | 20 | 0 | 0 | 1.5 | 1.5 | 1 | 0.5 | 2 | 2 |
| −77.75 | 40.75 | 57 | 40 | 3 | 0 | 1.5 | 1.5 | 1 | 1.5 | 2 | 3 |
| −77.25 | 40.75 | 62 | 25 | 13 | 0 | 1.25 | 1.5 | 1 | 1.5 | 2 | 3 |
| −76.75 | 40.75 | 27 | 40 | 33 | 0 | 1.25 | 2 | 1.5 | 2 | 2.5 | 3 |
| −76.25 | 40.75 | 21 | 47 | 32 | 0 | 1.25 | 2 | 1.5 | 2 | 2.5 | 3 |
| −75.75 | 40.75 | 14 | 46 | 40 | 0 | 1.25 | 2 | 1.5 | 2 | 2.5 | 3 |
| −75.25 | 40.75 | 10 | 53 | 37 | 0 | 1.25 | 2 | 1.5 | 2 | 2.5 | 3 |

| | | | | | | | | | | | |
|---|---|---|---|---|---|---|---|---|---|---|---|
| −78.75 | 40.25 | 48 | 52 | 0 | 0 | 1.25 | 1.5 | 1 | 0.5 | 2 | 2 |
| −78.25 | 40.25 | 36 | 50 | 14 | 0 | 1.25 | 1.5 | 1 | 1.5 | 2 | 3 |
| −77.75 | 40.25 | 28 | 50 | 22 | 0 | 1.25 | 2 | 1.5 | 2 | 2 | 3 |
| −77.25 | 40.25 | 15 | 41 | 44 | 0 | 2.5 | 3 | 3 | 3 | 3 | 3 |
| −76.75 | 40.25 | 0 | 6 | 94 | 0 | 2.5 | 3 | 2 | 3.5 | 3 | 3 |
| −76.25 | 40.25 | | | | | | | 4 | 5 | 5 | 4 |
| −75.75 | 40.25 | | | | | | | 4 | 5 | 5 | 4 |
| −78.75 | 39.75 | 33 | 50 | 17 | 0 | 1.25 | 1.5 | 1.5 | 1.5 | 2 | 3 |
| −78.25 | 39.75 | 25 | 50 | 25 | 0 | 1.25 | 2 | 1.5 | 2 | 2 | 3 |
| −77.75 | 39.75 | 15 | 43 | 42 | 0 | 2.5 | 2.5 | 2 | 3 | 2.5 | 3 |
| −79.75 | 39.25 | 100 | 0 | 0 | 0 | 1 | 1 | 0 | 0 | 0 | 1 |
| −79.25 | 39.25 | 75 | 25 | 0 | 0 | 1.5 | 1.5 | 1 | 0.5 | 2 | 2 |
| −78.75 | 39.25 | 31 | 50 | 19 | 0 | 1.25 | 1.5 | 1 | 1.5 | 2 | 3 |
| −78.25 | 39.25 | 25 | 50 | 25 | 0 | 1.25 | 2 | 1 | 1.5 | 2 | 3 |
| −79.75 | 38.75 | 100 | 0 | 0 | 0 | 1 | 1 | 0 | 0 | 0 | 1 |
| −79.25 | 38.75 | 77 | 23 | 0 | 0 | 1.5 | 1.5 | 1 | 0.5 | 2 | 2 |
| −78.75 | 38.75 | 35 | 50 | 15 | 0 | 1.25 | 1.5 | 1 | 2 | 2 | 3 |
| −78.25 | 38.75 | 21 | 50 | 29 | 0 | 1.25 | 2 | 1 | 1.5 | 2 | 3 |
| −80.25 | 38.25 | 100 | 0 | 0 | 0 | 1 | 1 | 0 | 0 | 0 | 1 |
| −79.75 | 38.25 | 87 | 13 | 0 | 0 | 1.5 | 1.5 | 1 | 0.5 | 2 | 2 |
| −79.25 | 38.25 | 46 | 40 | 14 | 0 | 1.5 | 2 | 1 | 2 | 2.5 | 3 |
| −80.25 | 37.75 | 66 | 32 | 2 | 0 | 1.5 | 1.5 | 1 | 0.5 | 2 | 2 |
| −79.75 | 37.75 | 34 | 50 | 16 | 0 | 1.25 | 2 | 1.5 | 2 | 2.5 | 3 |
| −81.25 | 37.25 | 75 | 25 | 0 | 0 | 1.5 | 1.5 | 1 | 0.5 | 2 | 2 |
| −80.75 | 37.25 | 38 | 62 | 0 | 0 | 1 | 1 | 0 | 1 | 2 | 3 |
| −80.25 | 37.25 | 39 | 50 | 11 | 0 | 1.25 | 1.5 | 1 | 1.5 | 2 | 3 |

*Table 5.* Data used in reconstructing environmental base maps for the *punctata* Zone.

| Longitude | Latitude | % Mud | % Silt | % Sand | % ls | Bedding | Substrate | Water depth | Environ | Oxygen | Biofacies |
|---|---|---|---|---|---|---|---|---|---|---|---|
| −78.75 | 42.75 | 90 | 0 | 0 | 10 | 1.00 | 1.00 | 0.00 | 1.00 | 2.00 | 1 |
| −78.25 | 42.75 | 80 | 0 | 0 | 20 | 1.00 | 1.00 | 0.00 | 1.00 | 2.00 | 2 |
| −77.75 | 42.75 | 70 | 5 | 5 | 20 | 1.00 | 1.00 | 0.50 | 1.00 | 2.00 | 2 |
| −77.25 | 42.75 | 50 | 33 | 10 | 7 | 1.50 | 1.50 | 1.00 | 2.00 | 3.00 | 2 |
| −76.75 | 42.75 | 50 | 29 | 26 | 5 | 1.50 | 2.00 | 1.00 | 2.00 | 3.00 | 2 |
| −76.25 | 42.75 | 69 | 8 | 22 | 0 | 1.50 | 2.50 | 2.00 | 3.00 | 3.00 | 3 |
| −75.75 | 42.75 | 69 | 8 | 22 | 0 | 1.50 | 2.50 | 2.00 | 3.00 | 3.00 | 3 |
| −75.25 | 42.75 | 35 | 15 | 50 | 0 | 1.50 | 2.50 | 2.00 | 3.00 | 3.50 | 3 |
| −74.75 | 42.75 | 15 | 10 | 75 | 0 | 3.00 | 3.00 | 3.50 | 5.00 | 4.00 | 4 |
| −74.25 | 42.75 | 10 | 10 | 80 | 0 | 3.00 | 3.00 | 3.50 | 5.00 | 4.00 | 4 |
| −80.25 | 42.25 | 90 | 0 | 0 | 10 | 1.00 | 1.00 | 0.00 | 1.00 | 2.00 | 1 |
| −79.75 | 42.25 | 90 | 0 | 0 | 10 | 1.00 | 1.00 | 0.00 | 1.00 | 2.00 | 1 |
| −79.25 | 42.25 | 90 | 0 | 0 | 10 | 1.00 | 1.00 | 0.00 | 1.00 | 2.00 | 1 |
| −78.75 | 42.25 | 90 | 0 | 0 | 10 | 1.00 | 1.00 | 0.00 | 1.00 | 2.00 | 1 |
| −78.25 | 42.25 | 80 | 0 | 0 | 20 | 1.00 | 1.00 | 0.00 | 1.00 | 3.00 | 2 |
| −77.75 | 42.25 | 70 | 5 | 5 | 20 | 1.00 | 1.00 | 0.50 | 1.00 | 3.00 | 2 |
| −77.25 | 42.25 | 40 | 45 | 12 | 3 | 1.50 | 2.00 | 1.00 | 2.00 | 3.00 | 2 |
| −76.75 | 42.25 | 40 | 45 | 20 | 5 | 1.50 | 2.25 | 1.50 | 2.00 | 3.00 | 2 |
| −76.25 | 42.25 | 85 | 0 | 15 | 0 | 1.50 | 1.50 | 2.00 | 2.50 | 3.00 | 3 |
| −75.75 | 42.25 | 75 | 20 | 10 | 0 | 2.00 | 1.50 | 2.50 | 3.50 | 3.00 | 3 |
| −75.25 | 42.25 | 35 | 15 | 50 | 0 | 1.50 | 2.50 | 2.50 | 4.00 | 3.50 | 3 |
| −74.75 | 42.25 | 20 | 10 | 70 | 0 | 3.00 | 3.00 | 3.50 | 4.50 | 4.00 | 4 |
| −74.25 | 42.25 | 10 | 10 | 80 | 0 | 3.00 | 3.00 | 4.00 | 5.00 | 4.00 | 4 |
| −73.75 | 42.25 | 5 | 5 | 90 | 0 | 3.00 | 3.00 | 4.00 | 5.00 | 4.00 | 4 |
| −80.25 | 41.75 | 90 | 0 | 0 | 10 | 1.00 | 1.00 | 0.00 | 1.00 | 2.00 | 1 |
| −79.75 | 41.75 | 90 | 0 | 0 | 10 | 1.00 | 1.00 | 0.00 | 1.00 | 2.00 | 1 |
| −79.25 | 41.75 | 90 | 0 | 0 | 10 | 1.00 | 1.00 | 0.00 | 1.00 | 2.00 | 1 |
| −78.75 | 41.75 | 90 | 0 | 0 | 10 | 1.00 | 1.00 | 0.00 | 1.00 | 3.00 | 1 |

| | | | | | | | | | | | |
|---|---|---|---|---|---|---|---|---|---|---|---|
| −78.25 | 41.75 | 80 | 0 | 0 | 20 | 1.00 | 1.00 | 0.00 | 1.00 | 3.00 | 2 |
| −77.75 | 41.75 | 70 | 5 | 5 | 20 | 1.00 | 1.00 | 0.50 | 1.00 | 3.00 | 2 |
| −77.25 | 41.75 | 70 | 4 | 26 | 0 | 2.00 | 2.50 | 1.00 | 2.00 | 3.00 | 2 |
| −76.75 | 41.75 | 70 | 4 | 26 | 0 | 2.00 | 2.50 | 1.00 | 2.00 | 3.00 | 2 |
| −76.25 | 41.75 | 69 | 8 | 22 | 0 | 1.50 | 2.50 | 2.00 | 3.00 | 3.00 | 3 |
| −75.75 | 41.75 | 69 | 8 | 22 | 0 | 1.50 | 2.50 | 2.00 | 3.00 | 3.00 | 3 |
| −75.25 | 41.75 | 20 | 10 | 70 | 0 | 3.00 | 3.00 | 3.50 | 4.00 | 4.00 | 4 |
| −74.75 | 41.75 | 10 | 10 | 80 | 0 | 3.00 | 3.00 | 4.00 | 5.00 | 4.00 | 4 |
| −74.25 | 41.75 | 5 | 5 | 90 | 0 | 3.00 | 3.00 | 4.00 | 5.00 | 4.00 | 4 |
| −77.25 | 40.25 | 50 | 35 | 15 | 0 | 1.50 | 1.00 | 2.00 | 2.00 | 3.00 | 2 |
| −78.75 | 39.75 | 20 | 75 | 5 | 0 | 1.00 | 2.00 | 1.00 | 2.00 | 3.00 | |
| −78.25 | 39.75 | 50 | 35 | 15 | 0 | | | | | 3.00 | |
| −79.75 | 39.25 | 85 | 10 | 0 | 5 | | | | | 1.00 | |
| −79.25 | 39.25 | 65 | 30 | 5 | 0 | 1.00 | 1.50 | 0.00 | 1.00 | 2.00 | |
| −78.75 | 39.25 | 57 | 38 | 5 | 0 | | | | | 3.00 | |
| −78.25 | 39.25 | 50 | 35 | 15 | 0 | | | | | 3.00 | |
| −79.75 | 38.75 | 65 | 30 | 5 | 0 | | | | 1.00 | 2.00 | |
| −79.25 | 38.75 | 57 | 38 | 5 | 0 | | | | 1.00 | 3.00 | |
| −78.75 | 38.75 | 50 | 35 | 15 | 0 | 1.10 | 1.50 | 0.00 | 1.00 | 3.00 | |
| −80.25 | 38.25 | 65 | 30 | 5 | 0 | | | | | 2.00 | |
| −79.75 | 38.25 | 57 | 38 | 5 | 0 | 1.50 | | | 1.00 | 3.00 | |
| −79.25 | 38.25 | 50 | 35 | 15 | 0 | | | | | 3.00 | |
| −80.75 | 37.75 | 85 | 10 | 0 | 5 | | | | | 1.00 | |
| −80.25 | 37.75 | 75 | 22 | 0 | 3 | | | | | 2.00 | |
| −79.75 | 37.75 | 57 | 38 | 5 | 0 | 2.00 | | | | 3.00 | |
| −79.25 | 37.75 | 50 | 35 | 15 | 0 | | | | | 4.00 | |
| −81.25 | 37.25 | 75 | 22 | 0 | 3 | | | | | 2.00 | |
| −80.75 | 37.25 | 57 | 38 | 5 | 0 | 2.00 | | | | 3.00 | |
| −80.25 | 37.25 | 20 | 40 | 40 | 0 | 2.50 | 2.25 | 0.00 | 2.00 | 3.00 | |
| −79.75 | 37.25 | 20 | 40 | 40 | 0 | | | | | 3.00 | |
| −81.75 | 36.75 | 57 | 38 | 5 | 0 | | | | | 3.00 | |

*Table 6.* Data used in reconstructing environmental base maps for the *linguiformis* Zone.

| Longitude | Latitude | % Mud | % Silt | % Sand | % ls | Bedding | Substrate | Water depth | Environ | Ichno | Oxygen | Biofacies |
|---|---|---|---|---|---|---|---|---|---|---|---|---|
| −78.75 | 42.75 | 73 | 13 | 4 | 10 | 1.00 | 1.50 | 0.00 | 0.0 | 1.0 | 2.0 | 1 |
| −78.25 | 42.75 | 30 | 60 | 5 | 5 | 1.00 | 2.00 | 1.00 | 2.0 | 2.0 | 2.5 | 2 |
| −77.75 | 42.75 | 25 | 60 | 20 | 5 | 2.20 | 2.25 | 1.25 | 3.5 | 3.0 | 2.5 | 2 |
| −77.25 | 42.75 | 24 | 28 | 52 | 0 | 2.00 | 2.50 | 1.50 | 2.0 | 2.0 | 3.0 | 3 |
| −76.75 | 42.75 | 69 | 3 | 43 | 0 | 2.00 | 2.50 | 1.50 | 1.0 | 1.0 | 3.0 | 4 |
| −76.25 | 42.75 | 40 | 0 | 60 | 0 | 2.00 | 2.50 | 2.50 | 4.0 | 4.0 | 3.0 | 4 |
| −75.75 | 42.75 | 15 | 10 | 75 | 0 | 3.00 | 3.00 | 4.00 | 5.0 | 5.0 | 4.0 | 5 |
| −75.25 | 42.75 | 10 | 10 | 80 | 0 | 3.00 | 3.00 | 4.00 | 5.0 | 5.0 | 4.0 | 5 |
| −74.75 | 42.75 | 5 | 5 | 90 | 0 | 3.00 | 3.00 | 4.00 | 5.0 | 5.0 | 4.0 | 5 |
| −74.25 | 42.75 | 3 | 2 | 95 | 0 | 3.00 | 3.00 | 4.00 | 5.0 | 5.0 | 4.0 | 5 |
| −80.25 | 42.25 | 85 | 7 | 0 | 8 | 1.00 | 1.00 | 0.00 | 0.0 | 1.0 | 2.0 | 1 |
| −79.75 | 42.25 | 85 | 7 | 0 | 8 | 1.00 | 1.00 | 0.00 | 0.0 | 1.0 | 2.0 | 1 |
| −79.25 | 42.25 | 85 | 7 | 0 | 8 | 1.00 | 1.00 | 0.00 | 0.0 | 1.0 | 2.0 | 1 |
| −78.75 | 42.25 | 73 | 13 | 4 | 10 | 1.10 | 1.00 | 0.00 | 1.0 | 2.0 | 2.0 | 2 |
| −78.25 | 42.25 | 70 | 10 | 5 | 15 | 1.05 | 1.00 | 1.00 | 1.5 | 2.0 | 2.0 | 2 |
| −77.75 | 42.25 | 40 | 35 | 25 | 0 | 1.50 | 1.50 | 1.25 | 2.0 | 2.5 | 2.5 | 2 |
| −77.25 | 42.25 | 20 | 40 | 40 | 0 | 2.00 | 2.25 | 1.50 | 2.5 | 2.0 | 3.0 | 3 |
| −76.75 | 42.25 | 40 | 5 | 55 | 0 | 2.00 | 2.50 | 1.50 | 2.0 | 2.0 | 3.0 | 4 |
| −76.25 | 42.25 | 40 | 0 | 60 | 0 | 2.00 | 2.50 | 2.50 | 4.0 | 4.0 | 4.0 | 4 |
| −75.75 | 42.25 | 15 | 10 | 75 | 0 | 3.00 | 3.00 | 4.00 | 5.0 | 5.0 | 4.0 | 5 |
| −80.25 | 41.75 | 85 | 7 | 0 | 8 | 1.00 | 1.00 | 0.00 | 0.0 | 1.0 | 2.0 | 1 |
| −79.75 | 41.75 | 85 | 7 | 0 | 8 | 1.00 | 1.00 | 0.00 | 0.0 | 1.0 | 2.0 | 1 |
| −79.25 | 41.75 | 85 | 7 | 0 | 8 | 1.00 | 1.00 | 0.00 | 0.0 | 1.0 | 2.0 | 1 |

| | | | | | | | | | | | | |
|---|---|---|---|---|---|---|---|---|---|---|---|---|
| −78.75 | 41.75 | 73 | 13 | 4 | 10 | 1.10 | 1.00 | 0.00 | 1.5 | 2.0 | 2.0 | 2 |
| −78.25 | 41.75 | 77 | 8 | 15 | 0 | 1.50 | 1.00 | 1.25 | 1.0 | 2.0 | 2.5 | 2 |
| −77.75 | 41.75 | 77 | 8 | 15 | 0 | 1.50 | 1.00 | 1.50 | 1.0 | 2.0 | 2.5 | 2 |
| −77.25 | 41.75 | 38 | 4 | 52 | 0 | 2.00 | 2.50 | 1.50 | 2.0 | 2.0 | 3.0 | 3 |
| −76.75 | 41.75 | 69 | 3 | 43 | 0 | 2.00 | 2.50 | 1.50 | 3.0 | 3.0 | 3.0 | 4 |
| −76.25 | 41.75 | 40 | 0 | 60 | 0 | 2.00 | 2.50 | 2.50 | 3.5 | 3.5 | 4.0 | 4 |
| −75.75 | 41.75 | 5 | 20 | 75 | 0 | 3.00 | 3.00 | 4.00 | 4.5 | 5.0 | 4.0 | 5 |
| −76.25 | 41.25 | 45 | 40 | 13 | 2 | 1.00 | 2.00 | 0.00 | 0.5 | 1.0 | 3.0 | |
| −75.25 | 41.25 | 15 | 82 | 3 | 0 | 1.50 | 2.00 | 0.00 | 1.0 | 1.0 | 2.5 | |
| −74.75 | 41.25 | 8 | 90 | 2 | 0 | 1.50 | 2.00 | 0.00 | 2.0 | 2.0 | 3.0 | |
| −78.25 | 40.75 | 30 | 45 | 25 | 2 | 1.00 | 2.25 | 1.00 | 3.0 | 3.0 | 3.0 | |
| −77.75 | 40.75 | 35 | 50 | 15 | 0 | 2.00 | 2.00 | 1.00 | 2.5 | 2.0 | 3.0 | |
| −77.25 | 40.75 | 39 | 51 | 10 | 0 | 1.50 | 2.50 | 0.00 | 0.5 | 1.0 | 3.5 | |
| −76.75 | 40.75 | 36 | 50 | 14 | 0 | 1.00 | 1.50 | 0.00 | 0.5 | 1.0 | 3.0 | |
| −76.25 | 40.75 | 15 | 79 | 5 | 1 | 1.00 | 1.50 | 0.00 | 0.5 | 1.0 | 3.5 | |
| −75.75 | 40.75 | 10 | 82 | 8 | 0 | 1.00 | 2.00 | 0.00 | 1.0 | 1.0 | 3.0 | |
| −75.25 | 40.75 | 8 | 82 | 10 | 0 | 1.50 | 2.00 | 0.00 | 1.0 | 1.0 | 3.0 | |
| −78.75 | 40.25 | 25 | 20 | 55 | 0 | 1.00 | 2.25 | 0.00 | 0.5 | 1.0 | 3.0 | |
| −78.25 | 40.25 | 25 | 20 | 55 | 0 | 1.00 | 2.50 | 0.00 | 0.5 | 1.0 | 3.5 | |
| −77.25 | 40.25 | 60 | 30 | 10 | 0 | 1.50 | 1.00 | 1.50 | 1.0 | 1.0 | 3.5 | |
| −76.75 | 40.25 | 70 | 20 | 1 | 0 | 1.50 | 1.00 | 0.00 | 2.0 | 2.0 | 3.5 | |
| −76.25 | 40.25 | 39 | 56 | 5 | 0 | 1.50 | 1.50 | 0.00 | 2.0 | 2.0 | 3.0 | |
| −78.75 | 39.75 | 80 | 18 | 2 | 0 | 1.00 | 1.50 | 1.00 | 2.5 | 3.0 | 3.0 | |
| −78.25 | 39.75 | 80 | 10 | 10 | 0 | 1.00 | 1.00 | 1.00 | 2.5 | 2.0 | 3.0 | |
| −79.25 | 39.25 | 5 | 25 | 70 | 0 | 1.00 | 3.00 | 2.00 | 4.0 | 3.0 | 3.0 | 4 |

For example, a grid box in which deposits attributable to middle-shelf (value 2) and inner-shelf (value 3) environments are present are coded with a value of 2.5 for that parameter. Intermediate coding is an effective coding strategy for incorporating environmental variability within the system in a repeatable and an objective manner. This method is analogous to analyses of the modern biota that utilize time-averaged data, such as mean annual temperature, as environmental variables (Anderson et al., 2003). Additionally, because contour interpolation was used to create grid surfaces for modeling, intermediate, noninteger values are useful data in the analysis.

Once the complete set of stratigraphic and environmental parameters was assembled and coded, the data were imported into ArcView 3.2 (ESRI, 1999) for creation of environmental coverages. The data were imported as an event theme and converted to an ArcView shapefile. Using the Spatial Analyst extension of ArcView 3.2 (ESRI, 1999), an interpolated grid surface was constructed for each environmental variable in each time slice. The interpolation was accomplished at a grid size of 0.03° under an inverse distance weight interpolation procedure using a fixed radius of 75 km at the second power. The radius value was set so that interpolations would include the center of all bounding grid boxes for increased continuity and follows standard contouring protocol (Davis, 2002). An example of an interpolated surface output is presented in Figure 2.

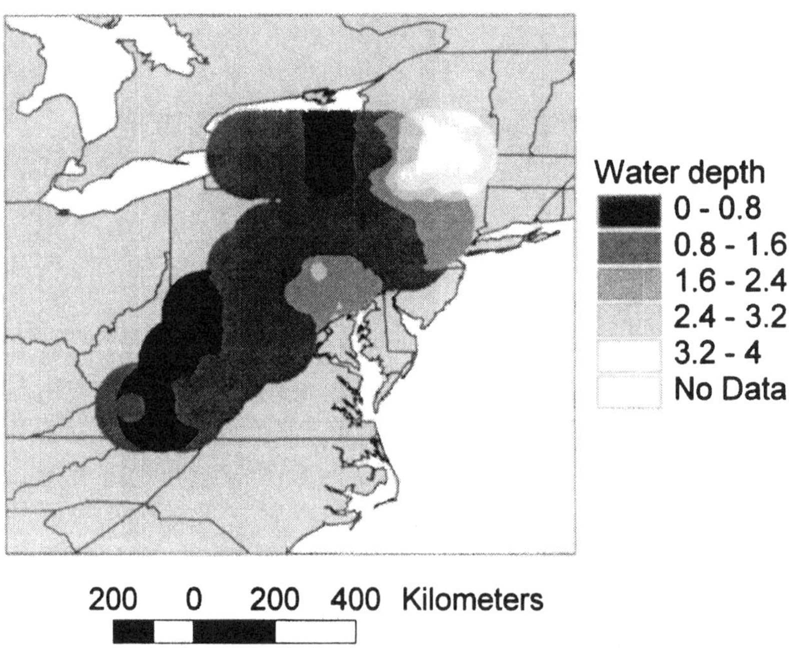


*Figure 2.* An example of an interpolated environmental layer surface showing water depth in the Lower *varcus* Zone. The numeric values pertain to the coded water depth values in Table 3. Note that water depth shallows to the east nearing the Acadian Highlands and deepens to the west into the Appalachian Foreland Basin. The deeper water region in western New York corresponds to a tectonic basin feature, the Rome Trough (Saverese et al., 1986).

### *2.4. Distribution modeling*

#### *2.4.1. Choice of modeling system*

Numerous statistical methods exist for analyzing and predicting the geographic distributions of species, including multiple regression, logistic regression, and genetic algorithms. Multiple regression analysis is a useful method for handling multifactor data and has been used widely in studies that attempt to predict percent cover (e.g. Haltuch et al., 2000). Multiple regression requires assumptions of multivariate normality and equal variance–covariance matrices, however, that are not likely to be met with the data available in palaeontological studies. Logistic regression is free from the requirement of multivariate normality and predicts a dichotomous dependent variable; logistic regression, however, requires that absence data represent true absences, not undersampling (Buchan and Padilla, 2000). This is an unacceptable assumption for palaeontological data where sampling is typically neither statistical nor uniform and species occurrences represent only a subset of a species range due to the limited availability of outcrops available for study. In addition, both multiple linear regression and logistic regression are associated with high error rates and limited ability to accurately predict occurrences (Goodwin et al., 1998; Haltuch et al., 2000; Chong et al., 2001).

Genetic algorithms provide an alternative to standard regression modeling by including several algorithms in an iterative, artificial-intelligence-based approach. This approach automates decision-making by repeatedly analyzing a series of local rules that combine categorical, range-type, and logistic rules to obtain higher significance levels than global rules, such as those applied in regression modeling (Stockwell and Peters, 1999; Stockwell and Peterson, 2002). In addition to maximizing the significance of the prediction, genetic algorithms also strive to achieve predictive accuracy, which is a weakness of the other methods mentioned above (Peterson and Vieglas, 2001). Genetic algorithms are particularly effective for analyzing museum data sets that are assembled by sampling that was neither uniform nor designed for statistical tests, and where environmental data consist of poorly structured domains (Stockwell and Peterson, 2002). The specific genetic algorithm that has been designed for use with biological occurrence data is GARP. GARP is designed to predict species ranges based on the fundamental niche, which is reconstructed from environmental data (Peterson and Vieglas, 2001). The GARP system has been tested extensively and has been shown to achieve high accuracy with low numbers of species occurrence data, even when there are as few as five environmental parameters (Peterson and Cohoon, 1999; Peterson, 2001; Stockwell and Peterson, 2002; Anderson et al., 2003).

#### *2.4.2. GARP implementation*

All modeling analyses in this study used DesktopGarp 1.1.4 developed by R. Scachetti-Pereira (www.lifemapper.org/desktopgarp). GARP works as in iterative process of rule selection, evaluation, testing, and incorporation or rejection (Stockwell and Peters, 1999). The species occurrence data points are divided equally into training and test points. The training data set is randomly sampled to create 1250 presence and background (absence) data points. A local rule is generated randomly from a set of possibilities (e.g. logistic regression, logit, atomic), applied to the training data, and tested with an internal test. In each iteration, predictive accuracy is assessed with 1250 points resampled from the test data set and 1250 points randomly sampled from the study region as a whole. The genetic component of the algorithm consists of mutating rules that include point mutations, deletions, and crossovers followed by an assessment of whether the mutation resulted in increased accuracy. The program uses the change in predictive accuracy from the prior iteration to determine whether a particular rule should be incorporated in the model or discarded (Stockwell and Peters, 1999). The algorithm continues until results converge on earlier models or after 1000 iterations.

Prior to running complete analyses on all species within the database, a jackknifing procedure was performed to determine the suitability of the 11 reconstructed environmental variables for species range prediction. This procedure has been used previously to test the efficacy of environmental layers (e.g. Peterson et al., 2002; Stockwell and Peterson, 2002). Environmental jackknife analysis was accomplished by implementing the "all combinations of selected layers" option within GARP. The environmental jackknife procedure was performed for all included species with greater than ten unique occurrence points within the Lower *varcus* Zone. These taxa are: *Athyris spiriferoides* (Eaton), *Cariniferella carinata* (Hall), *Palaeoneilo constricta* (Conrad), and *Spinatrypa spinosa* (Hall). The errors in terms of omission and commission values were assessed for each environmental layer using multiple linear regression in Minitab 14 (Minitab Inc., 2003). In this context, multiple linear regression is used to assess whether the inclusion of each environmental layer increases amount of error, measured as omission and commission, which is a distinctly different application than that discussed above (Davis, 2002). Multiple linear regression analysis was performed for each species individually and for each conodont zone with species pooled. Each species had a unique set of environmental factors that were significantly correlated with high omission and commission values; however, no factor was significantly associated with error in all four species. For *Athyris spiriferoides*, error was associated with limestone percent; error in *Cariniferella carinata* was associated with silt percent; for *Palaeoneilo constricta*, error was associated with limestone percent, mud percent, and oxygenation variables; and error in *Spinatrypa spinosa* was related to limestone percent. To characterize further how various environmental layers contributed to errors, the niche of each species was predicted using 100 replications when the layers significantly associated with error were removed and with all environmental layers included. The resulting error between these two experiments for each species was compared using a *t*-test and including a Dunn–Šidák correction to account for multiple comparisons (Sokal and Rolfe, 1995) in Minitab 14. In all cases, there was no significant difference in error between species models run with all environmental layers and those with the potentially error-inducing layers removed. Each of the environmental variables was therefore considered informative, and species niche for range predictions were reconstructed using all environmental variables.

Within this analysis, species ranges were predicted by running 200 replicate models of each species ecological niche at a convergence level of 0.01. All environmental variables (Table 3) were included within the analysis as justified above. The best subset selection option was invoked, and the ten best species predictions under an omission threshold of 10% and a commission threshold of 50% were retained following standard protocol (A.T. Peterson, pers. comm. 2004). Range prediction maps were output as Arc/Info Grids and imported into ArcView 3.2 for analysis. Within ArcView Spatial Analyst, the ten best subset outcomes were summed to provide a distribution model for each species (e.g. Fig. 3).

### *2.5. Determining robustness of GARP models*

The GARP models were compared with other predictions of species distribution to determine the robustness of GARP distribution predictions. GARP range predictions are based on a set of five to 50 rules developed using a variety of methods (e.g. logit, regression, etc.), and these rules are not included with the model output. The complexity of the GARP rule-set means that direct statistical comparison of predicted species ranges to environmental base layers is not readily available, and untangling the precise effect of a single environmental layer on the final species range prediction is a complex problem. An alternative approach was thus taken: predicted species ranges were compared with previously published descriptions of expected species occurrences (e.g. Bowen et al., 1974; McGhee, 1976; McGhee and Sutton, 1981, 1983; Sutton et al., 1970; Sutton and McGhee, 1985). Predicted ranges within the *linguiformis* Zone were compared with ranges expected based on the community palaeoecology studies of McGhee and Sutton (1981) and with those reconstructed by digitizing a polygon to enclose points by Rode and Lieberman (2004) (Table 7). Because the geographic range reconstruction undertaken by Rode and Lieberman (2004) encompasses a greater geographic area, comparisons between GIS- and

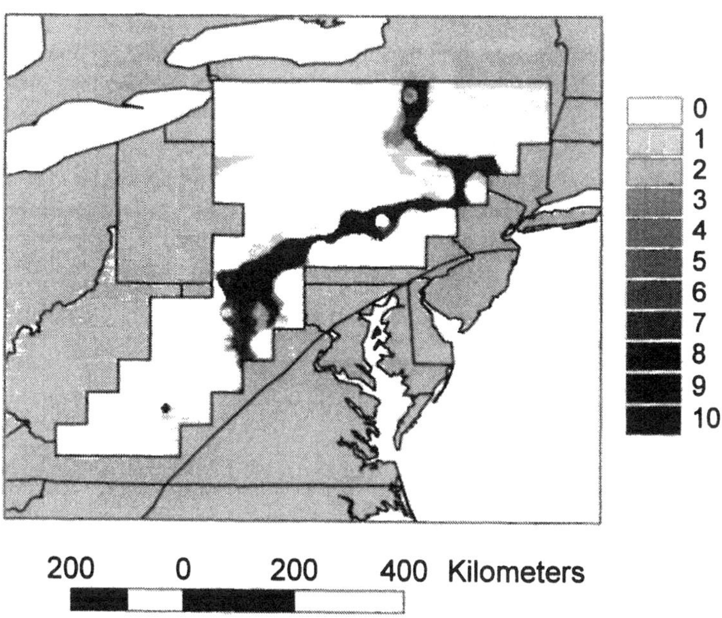


*Figure 3.* An example of a predicted species range map: *Leptodesma spinerigum* (Conrad) during the Lower *varcus* Zone. Darker shades indicate an increased number the ten best subset maps predict *L. spinerigum* to occur at a location. Comparison with Fig. 2 indicates that the range of *L. spinerigum* partially, though not precisely, follows the general trend of the Givetian shoreline.

*Table 7.* Comparison of differences in species ranges predicted by GIS bounding-polygon reconstruction and GARP.

| Species | GIS ($km^2$) | GARP ($km^2$) | Conodont zone |
|---|---|---|---|
| *Cariniferella carinata* | 8,700 | 39,700 | Lower *varcus* |
| *Cypricardella bellistriata* | 39,700 | 25,800 | Lower *varcus* |
| *Cyrtospirifer chemungensis* | 2,000 | 4,200 | *linguiformis* |
| *Praewaagenoconcha speciosa* | 17,000 | 53,500 | *linguiformis* |
| *Spinatrypa spinosa* | 5,900 | 20,300 | *linguiformis* |

GARP-modeled ranges were restricted to species endemic to roughly the northern part of the Appalachian Basin (Devonian outcrop belts in Maryland, New York, Pennsylvania, Virginia, and West Virginia).

### *2.6. Examination of species survival with environment*

Statistical analyses were conducted to assess the effects of distribution changes on species survival during the Late Devonian biodiversity crisis. The areal extent of each species range was calculated in ArcView by summing the areas common to six or more of the ten best subset predictions following the method of Peterson et al. (2002). Since the total geographic area modeled within each conodont zone was different (Fig. 1), these values are reported in Table 8 as both raw numbers and as percentages of the total geographic area within the modeling limits of each conodont zone. A comparison of species range and species survival suggested a relationship may exist, because species that survive tend to have larger geographic ranges (Table 9). The relationship between the area of species' ranges and species survival was investigated statistically using an ANOVA including a Dunn–Šidák correction to account for multiple comparisons (Tables 10–13). In addition, ranges of species that crossed two time intervals were compared to determine the amount of expansion or reduction in geographic distribution (Table 14).

## 3. Results and discussion

### *3.1. Comparison of outputs with other predictions*

The GARP models were developed for 10 species in the Lower *varcus* Zone, nine species in the *punctata* Zone, and 20 species in the *linguiformis* Zone (Table 2). Each of these models produced a unique species range based on the 11 environmental parameters considered. Comparison of the *linguiformis* Zone predictions with the community palaeoecology established by McGhee and Sutton (1981) provides the most direct comparison with previously published descriptions of expected species occurrences and is used as a case study to test the robustness of the GARP predictions as the data sets underlying the two distribution models are distinct.

McGhee and Sutton (1981) established three community types based on brachiopod and bivalve species in the Java Formation of New York and Foreknobs Formation of West

*Table 8.* Geographic ranges predicted from GARP modeling with status of species analyzed by relevant conodont zone. Species are designated survivors (Y) if they persisted through the biodiversity crisis and into the Famennian and victims (N) if they did not.

| Species | Area ($km^2$) | Coverage (%) | Survivor | Conodont zone |
|---|---|---|---|---|
| *Ambocoelia gregaria* | 27,100 | 19.4 | Y | *linguiformis* |
| *Ambocoelia umbonata* | 9,500 | 6.8 | Y | *linguiformis* |
| *Athyris angelica* | 42,400 | 30.4 | Y | *linguiformis* |
| *Athyris cora* | 21,600 | 11.6 | N | Lower *varcus* |
| *Athyris spiriferoides* | 39,700 | 21.3 | N | Lower *varcus* |
| *Cariniferella carinata* | 7,300 | 5.2 | N | *linguiformis* |
| *Cariniferella carinata* | 39,700 | 21.3 | N | Lower *varcus* |
| *Cariniferella tioga* | 6,000 | 4.4 | N | *linguiformis* |
| *Cupularostrum contracta* | 42,700 | 30.6 | Y | *linguiformis* |
| *Cupularostrum exima* | 24,000 | 17.2 | Y | *linguiformis* |
| *Cupularostrum exima* | 5,900 | 5.1 | Y | *punctata* |
| *Cypricardella bellistriata* | 25,700 | 13.8 | N | Lower *varcus* |
| *Cyrtospirifer chemungensis* | 7,200 | 5.1 | N | *linguiformis* |
| *Douvillina cayuta* | 6,600 | 4.7 | N | *linguiformis* |
| *Eoschizodus chemungensis* | 31,100 | 26.5 | N | *punctata* |
| *Floweria chemungensis* | 40,000 | 28.6 | Y | *linguiformis* |
| *Floweria parva* | 31,400 | 22.5 | N | *linguiformis* |
| *Goniophora chemungensis* | 7,300 | 6.2 | Y | *punctata* |
| *Grammysia elliptica* | 34,500 | 29.4 | Y | *punctata* |
| *Leptodesma nitida* | 14,100 | 12.0 | N | *punctata* |
| *Leptodesma spinerigum* | 39,200 | 28.1 | Y | *linguiformis* |
| *Leptodesma spinerigum* | 23,400 | 12.5 | Y | Lower *varcus* |
| *Mucrospirifer mucronatus* | 16,300 | 8.8 | N | Lower *varcus* |
| *Nervostrophia nervosa* | 4,800 | 3.5 | N | *linguiformis* |
| *Palaeoneilo constricta* | 5,300 | 4.5 | Y | *punctata* |
| *Paleoneilo constricta* | 39,300 | 21.1 | Y | Lower *varcus* |
| *Paracyclas lirata* | 31,500 | 16.9 | N | Lower *varcus* |
| *Praewaagenoconcha speciosa* | 53,500 | 38.3 | Y | *linguiformis* |
| *Praewaagenoconcha speciosa* | 8,500 | 7.3 | Y | *punctata* |
| *Productella rectispina* | 22,900 | 16.4 | Y | *linguiformis* |
| *Pseudatrypa devoniana* | 40,600 | 29.1 | N | *linguiformis* |
| *Ptychopteria chemungensis* | 1,800 | 1.5 | N | *punctata* |
| *Schizophoria impressa* | 40,200 | 28.9 | Y | *linguiformis* |
| *Spinatrypa spinosa* | 20,300 | 14.6 | N | *linguiformis* |
| *Spinatrypa spinosa* | 32,900 | 17.6 | N | Lower *varcus* |
| *Spinocyrtia granulosa* | 11,900 | 6.4 | N | Lower *varcus* |
| *Strophonella hybrida* | 26,600 | 19.1 | N | *linguiformis* |
| *Tylothyris mesacostalis* | 29,100 | 20.9 | Y | *linguiformis* |
| *Tylothyris mesacostalis* | 12,400 | 10.6 | Y | *punctata* |

Virginia. Five species were included both in McGhee and Sutton's (1981) community analysis and this analysis: *Ambocoelia gregaria* (Hall); *Athyris angelica* (Hall); *Floweria chemungensis* (Conrad); *Leptodesma spinerigum* (Conrad); and *Tylothyris mesacostalis* (Hall). McGhee and Sutton (1981) characterized *Ambocoelia gregaria* and *Athyris angelica* as

*Table 9.* Size of geographic range ($km^2$) versus species survival through the crisis interval for species during the conodont zones considered.

| | | Survivors | Victims |
|---|---|---|---|
| Lower *varcus* | $31.3 \times 10^3$ | $24.7 \times 10^3$ | |
| *punctata* | | $12.3 \times 10^3$ | $15.6 \times 10^3$ |
| *linguiformis* | | $33.7 \times 10^3$ | $16.7 \times 10^3$ |
| Total | | $26.7 \times 10^3$ | $20.9 \times 10^3$ |

*Table 10.* ANOVA table showing analysis of geographic range versus survival through the Late Devonian biodiversity crisis for Lower *varcus* Zone species.

| Source | Degrees of freedom | Sum of squares | Mean squares | $F$-value | $P$-value |
|---|---|---|---|---|---|
| Survival | 1 | 2.47E + 07 | 2.47E + 07 | 0.22 | 0.648 |
| Error | 8 | 8.79E + 08 | 1.10E + 08 | | |
| Total | 9 | 9.03E + 08 | | | |

*Table 11.* ANOVA table showing analysis of geographic range versus survival through the Late Devonian biodiversity crisis for *punctata* Zone species.

| Source | Degrees of freedom | Sum of squares | Mean squares | $F$-value | $P$-value |
|---|---|---|---|---|---|
| Survival | 1 | 2.17E + 07 | 2.17E + 07 | 0.14 | 0.715 |
| Error | 7 | 1.05E + 09 | 1.50E + 08 | | |
| Total | 8 | 1.08E + 08 | | | |

*Table 12.* ANOVA table showing analysis of geographic range versus survival through the Late Devonian biodiversity crisis for *linguiformis* Zone species.

| Source | Degrees of freedom | Sum of squares | Mean squares | $F$-value | $P$-value |
|---|---|---|---|---|---|
| Survival | 1 | 1.42E + 09 | 1.42E + 09 | 8.64 | 0.009 |
| Error | 16 | 2.95E + 09 | 1.64E + 08 | | |
| Total | 17 | 4.37E + 09 | | | |

*Table 13.* ANOVA table showing analysis of geographic range versus survival through the Late Devonian biodiversity crisis for all species.

| Source | Degrees of freedom | Sum of squares | Mean squares | $F$-value | $P$-value |
|---|---|---|---|---|---|
| Survival | 1 | 4.28E + 08 | 4.28E + 08 | 2.15 | 0.151 |
| Error | 35 | 6.97E + 09 | 1.99E + 08 | | |
| Total | 36 | 7.40E + 09 | | | |

*Table 14.* Relative change in geographic range size of species that occurred in two conodont zones based on GARP predicted modeling and its general association with species survival through the crisis interval.

| Species | Conodont zones of transition | Relative size of younger range (%) | Survival |
|---|---|---|---|
| *Cupularostrum exima* | *punctata* to *linguiformis* | 379 | Y |
| *Praewaagenoconcha speciosa* | *punctata* to *linguiformis* | 526 | Y |
| *Tylothyris mesacostalis* | *punctata* to *linguiformis* | 197 | Y |
| *Cariniferella carinata* | *varcus* to *linguiformis* | 24 | N |
| *Leptodesma spinerigum* | *varcus* to *linguiformis* | 224 | Y |
| *Spinatrypa spinosa* | *varcus* to *linguiformis* | 83 | N |
| *Palaeoneilo constricta* | *varcus* to *punctata* | 21 | Y |

dominant members of the open-shelf setting, *Tylothyris mesacostalis* and *Floweria chemungensis* as key members of the outer-platform community, and *Leptodesma spinerigum* as a dominant species within the inner-platform and nearshore settings. These predictions are borne out by the GARP predictions (Fig. 4): *Athyris angelica* and *Ambocoelia gregaria* occupy the most basinward positions, whereas *Leptodesma spinerigum* is predicted to occupy a more nearshore setting, and *Tylothyris mesacostalis* and *Floweria chemungensis* are most frequently predicted to occur in a central, middle-shelf setting. McGhee and Sutton (1981) noted that whereas these five species are characteristic of specific depositional settings, their ranges often extend into adjacent settings. This is also illustrated in the results from the GARP models. For example, although *Athyris angelica* occurs throughout the basinal setting, this species also commonly occurs in the outer shelf and in some areas of the middle- or inner-shelf setting (Fig. 4b). The GARP models also illustrate that whereas species ranges follow depositional setting in part, a one-to-one correlation does not exist, indicating the importance of additional environmental factors in determining the fundamental or realized niche of each species.

Comparison of species ranges predicted from community analyses of McGhee and Sutton (1981) with species distributions predicted by GARP shows that GARP modeling is a robust way in which to predict species ranges. GARP modeling shows a high level of predictive accuracy with known species occurrences and expected palaeoecology and appears to be a viable approach for reconstructing the ranges of fossil species in shallow marine ecosystems.

### *3.2. Comparison of GARP and GIS enclosure ranges*

Ranges of Devonian brachiopod and bivalve species have previously been modeled using GIS (Rode and Lieberman, 2004, 2005). This technique essentially involves enclosing known species occurrences during a conodont zone interval within a polygon. Detailed descriptions of this method are published elsewhere (Rode and Lieberman, 2004) and are not repeated herein. The method by which species occurrence points are enclosed within a minimum area polygon has the potential to both under- and over-predict species ranges. Under-prediction is expected typically, since all localities where a species lived will

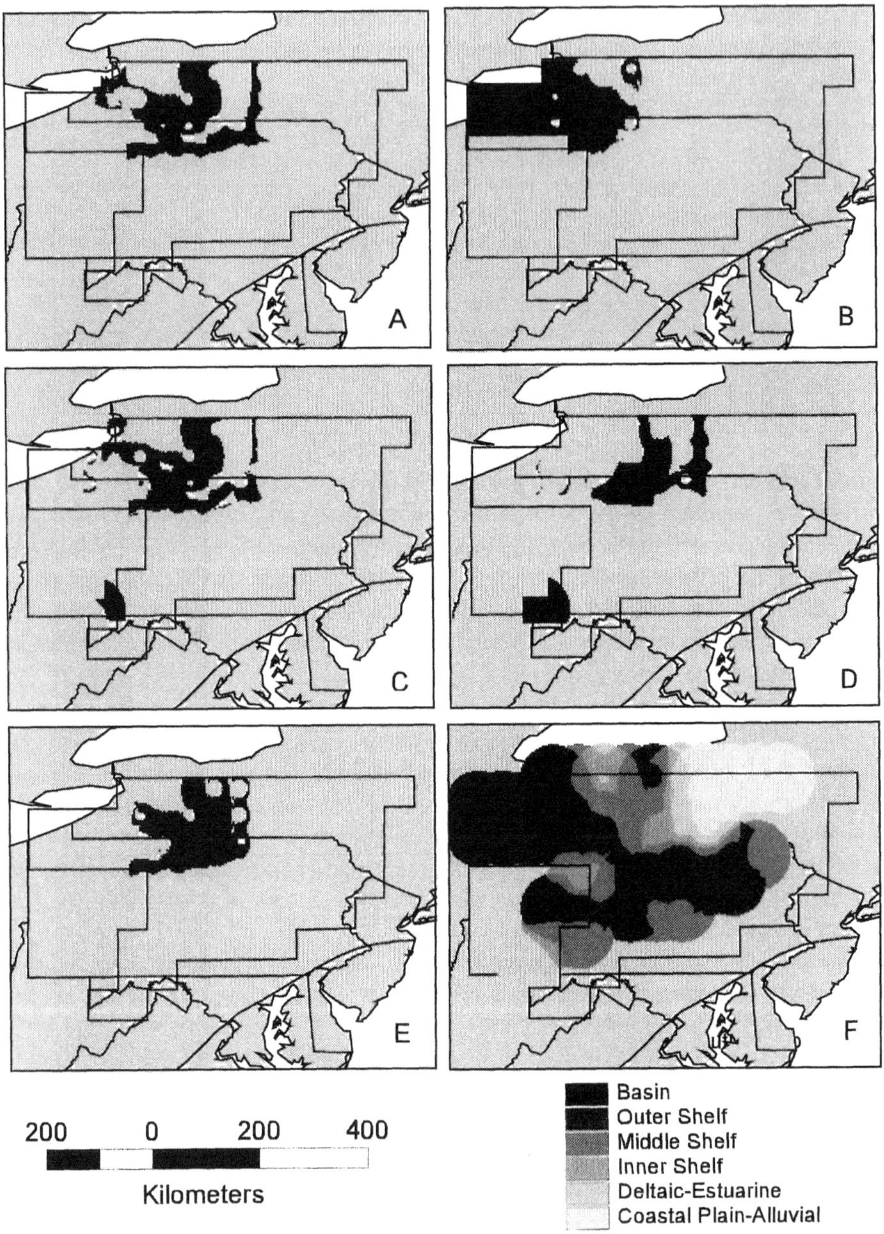


*Figure 4.* GARP prediction maps. The area shown in black indicates the part of the modeled region, bounded by the line, in which all of the best subset maps predict the species to occur. (A) *Ambocoelia gregaria* (Hall), (B) *Athyris angelica* (Hall), (C) *Floweria chemungensis* (Conrad), (D) *Leptodesma spinerigum* (Conrad), (E) *Tylothryris mesacostalis* (Hall). (F) Interpolated surface grid indicating depositional environment, the key parameter examined in McGhee and Sutton (1981), and one of the eleven parameters examined within this analysis. Circular holes within the reconstructed ranges indicate a lack of environmental data at a particular site, which prevents accurate prediction in that part of the range.

necessarily not be present within the data set. Over-prediction can occur when a bounding polygon includes areas that would have been inhospitable to the species due to local changes in shoreline, sediment influx, or other environmental conditions.

The GARP distribution models produced herein were compared with the minimum area polygon ranges reconstructed using GIS by Rode and Lieberman (2004) to further characterize the general utility of the GARP method for fossil invertebrates. Because the GIS data sets of Rode and Lieberman (2004) included some geographic areas outside the study area of this project, only species whose range occurs entirely within the northern Appalachian Basin were used for comparison. It is important to note, however, that the same data set discussed above is the data source for both these studies, so the two studies are directly comparable. Two species from the Lower *varcus* and three species from the *linguiformis* Zones were examined (Table 14) and side-by-side comparisons of species range predictions are shown in Figure 5.

In most pairwise comparisons of polygon enclosure and GARP prediction models, the ranges are roughly consistent between the two outputs. Commonly, though not always, the GARP-predicted range encompasses the entire polygon enclosure range and predicts species to occur in additional areas adjacent to the polygon range (i.e. Fig. 5A,B; E,F; G,H; I,J for: *Cariniferella carinata* (Hall), *Cyrtospirifer chemungensis* (Hall), *Praewaagenoconcha speciosa* (Hall), and *Spinatypra spinosa* (Hall), respectively). Often the predicted ranges compare quite closely with the polygon ranges, e.g. in *Cyrtospirifer chemungensis* (Fig. 5E,F), which may suggest both that GARP is accurately predicting known ranges, and that the method of collection of data for the GIS polygon enclosure ranges may be sufficient to reasonably capture the actual species range. The distribution of *Cypricardella bellistriata* (Conrad) differs from the general pattern described above; instead of the GARP prediction encompassing a greater area than the polygon reconstruction, it includes a smaller area (Fig. 5C,D). Comparison of the two range reconstructions indicates that the preferred habitat of *C. bellistriata* is discontinuous within the polygon enclosure. This indicates that when species ecological preferences result in discontinuous populations, polygon enclosure ranges will over-predict species ranges, whereas the predictive approach produces a more accurate representation of the area a species can successfully colonize. The GARP modeling algorithm seems to estimate species ranges successfully, and does not appear to suffer from significant under- or over-prediction errors when compared with GIS constructed species occurrence enclosure ranges. Because the GARP algorithm-predicted species ranges are based on a rule-set that has been trained on the data and contains both internal and external tests, it should be expected to produce more refined estimates of species range than GIS enclosure models. In fact, predicted ranges that exceed the known species occurrence-bounding polygon provide testable hypotheses for future work in assessing the boundaries of species ranges, predicting ranges of species groups and boundaries of community types, and also determining the quality of the fossil record.

### *3.3. Quantifying geographic range change and species survival*

The size of a species geographic range has previously been shown to be related to species survival during the Late Devonian biodiversity crisis interval (Rode and Lieberman, 2004). Rode and Lieberman (2004) determined this relationship using species ranges estimated

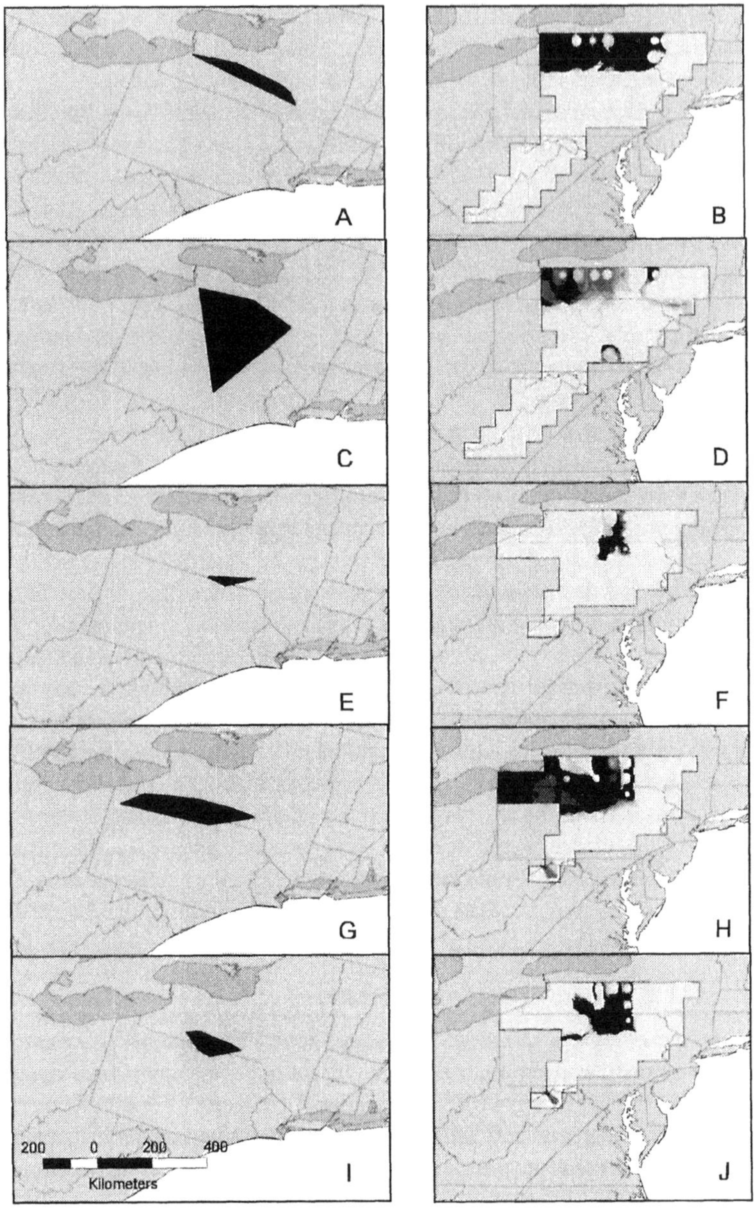


*Figure 5.* Comparison of GIS polygon enclosure range reconstructions and GARP distribution predictions. (A) Polygon enclosure range and (B) GARP prediction range for *Cariniferella carinata* (Hall) during the *varcus* Zone; (C) Polygon enclosure range and (D) GARP prediction range for *Cypricardella bellistriata* (Conrad) during the *varcus* Zone; (E) Polygon enclosure range and (F) GARP prediction range for *Cyrtospirifer chemungensis* (Hall) during the *linguiformis* Zone; (G) Polygon enclosure range and (H) GARP prediction range for *Praewaagenoconcha speciosa* (Hall) during the *linguiformis* Zone; (I) Polygon enclosure range and (J) GARP prediction range for *Spinatrypa spinosa* (Hall) during the *linguiformis* Zone. Circular holes within the reconstructed ranges indicate a lack of environmental data at a particular site, which prevents accurate prediction in that portion of the range.

by the bounding-polygon method. Because it was noted that GARP range predictions, although often congruent with GIS range reconstructions, can differ in size and geographic area from polygon inclusion reconstructions, it is worth investigating the further resiliency of these results. Thus, the areal changes within the range of a single species between conodont zones based on GARP modeling were quantified (Table 14). There are seven species whose ranges could be predicted for two conodont zones, and of these, four species exhibited range increases and three exhibited range decreases. Range expansions appear to be the result of species colonizing additional habitat, and are not just attributable to habitat tracking. For example, in the *linguiformis* Zone, the increased range of *Leptodesma spinerigum* (Fig. 6C,D) and *Praewaagenoconcha speciosa* (Fig. 6E,F) cannot be explained by habitat tracking, because the east-west breadth, hence the number of environments occupied, increased. Range contractions, when they occurred, resulted from a species becoming restricted to a subset of its prior range. An example of this for *Cariniferella carinata* is shown in Figure 6A,B. In this case, although the ancestral population of *C. carinata* occupied a wide array of environments in the Lower *varcus* Zone, *linguiformis* Zone populations were restricted to a very narrow geographic area.

Again, based on GARP modeling, and matching the results from Rode and Lieberman (2004) based on GIS, a correlation may exist between changes in species distribution and survival through the biodiversity crisis interval and into the Famennian (Table 14). Of the seven species whose ranges were predicted in two different time slices, all species that undergo range expansion into the *linguiformis* Zone persisted through the biodiversity crisis. None of the species with ranges contracting from the Lower *varcus* into the *linguiformis* Zones survive into the Famennian. Not only range expansion, but also the timing of range expansion and the areas that species expanded into, may have controlled species survival during the Late Devonian as the range of *Palaeoneilo constricta* contracts between the Lower *varcus* and *punctata* Zones, but it survives the biodiversity crisis. Perhaps expansion or contraction during the *linguiformis* Zone may be crucial, although there were too few data to consider this statistically. This may indicate that range size and range expansion may be parameters that were critical for surviving this particular biodiversity crisis, although not as critical for species survival during the adjacent background intervals.

To further examine the effect timing and geographic range had on survival, ANOVAs were computed to compare the mean ranges of species that persisted into the Famennian with species that became extinct by the end of the Frasnian. Each species was characterized as a survivor if it persisted into the Famennian or a victim if it did not; the geographic range was calculated for each species per stage; conodont zones were analyzed separately and together to determine whether species that survived the crisis had statistically larger ranges and whether the survival advantage was constant or varied across time. The differences between survivor and victim species ranges were not statistically different in either the *punctata* or the Lower *varcus* zones (Tables 10, 11); in fact, during the *punctata* zone, the average geographic range of species that became extinct was higher than for survivor species, although not significantly so (Tables 8, 10). In addition, the ANOVA in which species extant in all conodont zones were pooled together also did not indicate significant survival differences by geographic range (Table 13). A significant size difference was recovered for the *linguiformis* Zone, however ($p = 0.002$) (Table 12). The significance of this result remained even after compensating for multiple analyses using the Dunn–Šidák

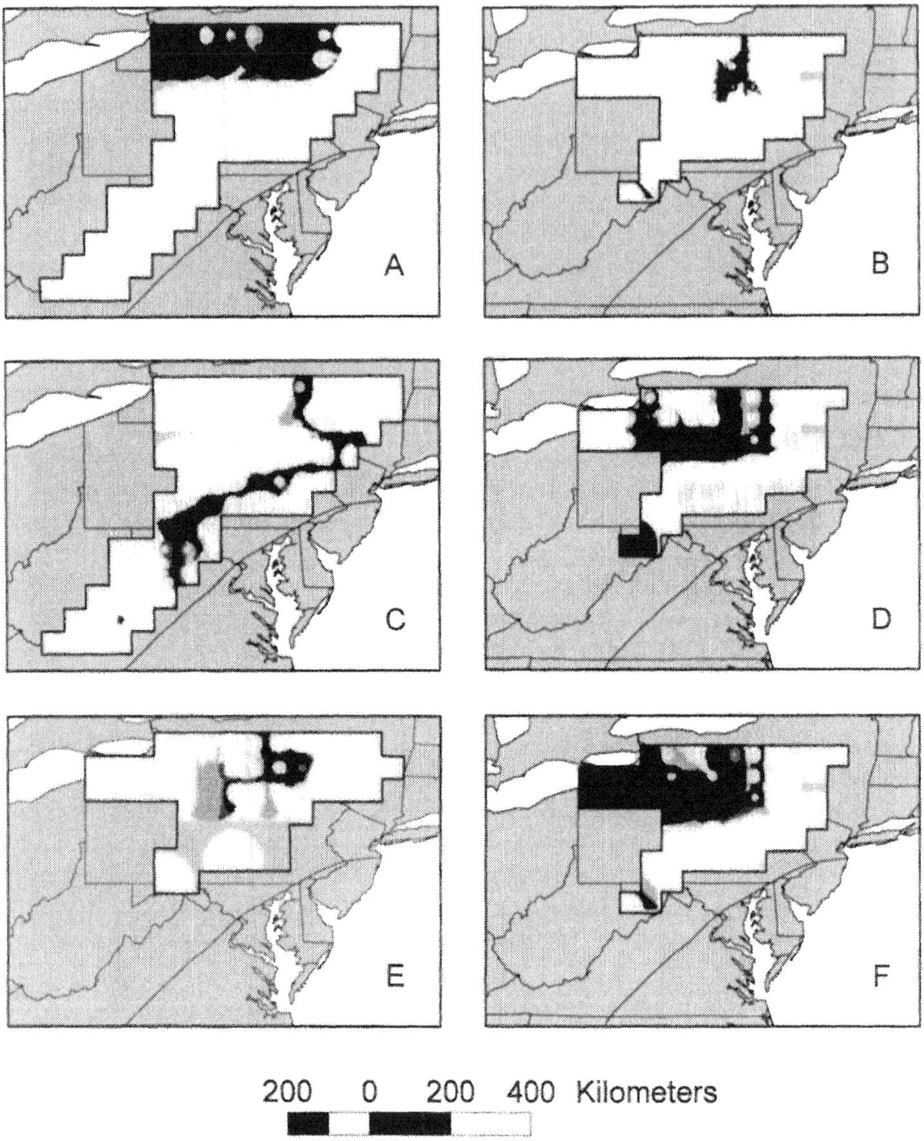


*Figure 6.* Comparison of the geographic ranges predicted by GARP modeling for species that occurred in two of the relevant conodont zones. (A) Lower *varcus* Zone and (B) *linguiformis* Zone distribution of *Cariniferella carinata* (Hall); (C) Lower *varcus* Zone and (D) *linguiformis* Zone distribution of *Leptodesma spinerigum* (Conrad); (E) *punctata* Zone and (F) *linguiformis* Zone distribution of *Praewaagenoconcha speciosa* (Hall).

correction (Sokal and Rolfe, 1995). This indicates that species with larger geographic ranges during the *linguiformis* Zone were more successful during the crisis interval than those with smaller ranges. It is also important to note that geographic range prior to the crisis interval (i.e. in the *punctata* or Lower *varcus* Zones) and larger geographic range overall (as shown by the analysis of pooled species) did not confer a significant survival advantage during the biodiversity crisis interval.

## 4. Conclusions

The results presented above suggest that GARP models provide both robust and useful characterizations of species ranges. The general congruence between GARP predictions and the palaeobiological understanding of species ranges supports the accuracy of the predicted ranges. In addition, both the expansion and refinement of ranges available with GARP versus GIS polygon enclosures further supports the utility of this technique.

### *4.1. Implications for understanding the Late Devonian biodiversity crisis*

Both changes in species geographic ranges and the sizes of species ranges impact species survivorship through the Late Devonian biodiversity crisis interval. The timing of range changes is critical in conferring a survival advantage. Both a broad range during the terminal Frasnian *linguiformis* Zone and an increasing range entering that interval are related to survival into the Famennian. Based on this analysis, neither a large range nor expansion events prior to the *linguiformis* Zone, however, appear to have enhanced species survival. Species extant in the middle Frasnian *punctata* Zone that persisted into the Famennian, in fact, tended to have slightly smaller ranges (although not statistically significantly so) than their counterparts that did not survive the biodiversity crisis. Likewise, broad Lower *varcus* Zone species ranges conferred no advantage to species during the crisis. Moreover, differences in range size were not statistically associated with survival when species of all time periods were pooled, which further underscores the unique importance of large ranges during the *linguiformis* Zone for surviving the biodiversity crisis interval.

The Late Devonian biodiversity crisis has been attributed to a set of five pulses of extinction: one immediately prior to the *linguiformis* Zone in the Late *rhenana* Zone, three at the end of the *linguiformis* Zone, and one in the following Early *crepida* Zone (McGhee, 2001). Of the three time slices examined in this analysis, only the *linguiformis* Zone interval occurs within the biodiversity crisis window. The Lower *varcus* and *punctata* zones preceded the crisis interval by approximately 14.2 and 3.8 million years, respectively (Tucker et al., 1998). The Late Devonian extinction was a temporally protracted event (e.g. McGhee, 1989, 1996). The results of this study indicate that the effects of geographic range on species survival did not extend as far back as the middle Frasnian *punctata* Zone. An important component of the biodiversity crisis, however, was a decline in speciation rates (McGhee, 1989, 1996), and the affects of geographic range on speciation rate may have operated earlier in the Frasnian (Rode, 2004; Rode and Lieberman, 2003, 2004).

### *4.2. Further applicability*

The results presented herein indicate that ecological niche modeling methods such as GARP may be robust tools for predicting the geographic ranges of fossil taxa. The design of the algorithm, which permits *ad hoc* sampling readily available from museum collections, and the use of a relatively limited number of environmental coverages at a relatively coarse spatial scale (km versus meters), is ideally suited for handling fossil data in areas where the stratigraphy and sedimentology are well known and densely sampled. The success of modeling species ranges within this study suggests this method may be more broadly applicable to other palaeontological regions and time periods where extensive museum collections and fine-scale sedimentological data exist.

The level of analytical rigor of species range prediction achieved using GARP could be used as a tool to achieve a number of palaeontological or sedimentological goals. Range prediction maps could be consulted when creating search strategies for new field sites targeted for the collection of specific species. Combining range predictions for several species may allow the investigation of community patterns based on niche parameters, including the long-term stability of species associations, coordinated stasis, or Gleasonian versus Clementsian communities within specific environments or portions of time. In addition, comparison of species ranges within an evolutionary framework may permit the identification of speciation by vicariance, dispersal, or geodispersal (e.g. Wiley and Mayden, 1985; Lieberman, 2000). Resource partitioning or competitive exclusion could also potentially be examined (e.g. Anderson et al., 2002). Thus, this is potentially a new and valuable technique that can be applied to the study of the fossil record.

## Acknowledgments

We thank R. Goldstein, R.L. Kaesler, W.R. Van Schmus, and E.O. Wiley for comments on earlier versions of this manuscript. G.R. McGhee, Jr., D.J. Over, and C.W. Stock provided helpful reviews. This research was supported by the Department of Geology, University of Kansas, and NSF EPS 9874732, NSF EAR 0106885, the NASA Exobiology program, and a Self Faculty Fellowship to B.S.L., and a NSF Graduate Student Fellowship and a Self Graduate Fellowship to A.L.S.

## References

Adams, R.W., Anderson, W.A., Cohen, P., Fulreader, R.E., Moyer, P.T., Pefley, D.R., Sachs, K.N., Seaber, P.R., Stover, L.E., 1956. 28th Annual Meeting of the New York State Geologic Association Guidebook.

Anderson, R.P., Lew, D., Peterson, A.T., 2003. Evaluating predictive models of species' distributions: criteria for selecting optimal models. Ecol. Model. 162, 211–232.

Anderson, R.P., Peterson, A.T., Gómez-Laverde, M., 2002. Using niche-based GIS modeling to test geographic predictions of competitive exclusion and competitive release in South American pocket mice. Oikos 98, 3–16.

Applebaum, R.H.S., 1993. Dynamic Shelf Processes and Tempestite Sedimentology in the Upper Devonian Sonyea Group near Bainbridge, New York. Unpublished M.S. thesis, State University of New York at Oneota, p. 125.

Babcock, L.E., Wegweiser, M.D., 1998. Upper Devonian (Chautauquan) stratigraphy, southern Late Erie shoreline region. In: Harper, J.A. (Ed.), Geotectonic Environment of the Lake Erie Crustal Block. 63rd Annual Field Conference of Pennsylvania Geologists, pp. 4–14.

Batt, R.J., 1996. Faunal and lithological evidence for small-scale cyclicity in the Wanakah Shale (Middle Devonian) of western New York. Palaios 11, 230–243.

Batt, R.J., 1999. Fossil faunas and microstratigraphy of the Upper Ludlowville Formation: Small-scale sea level fluctuations and converging unconformities. In: Lash, G.G. (Ed.), 71st Annual Meeting, Field Trip Guidebook. New York State Geological Association, pp. Sun. C1–C37.

Bishuk, D., Jr., Applebaum, R., Ebert, J.R., 1991. Storm-dominated shelf and tidally-influenced foreshore sedimentation, Upper Devonian Sonyea Group, Bainbridge to Sidney Center, New York. In: Ebert, J.R. (Ed.), 63rd Annual Meeting, Field Trip Guidebook, New York State Geological Association, pp. 413–452.

Bowen, Z.P., Rhoads, D.C., McAlester, A.L., 1974. Marine benthic communities in the Upper Devonian of New York. Lethaia 7, 93–120.

Bowen, Z.P., Sutton, R.G., McAlester, A.L., Rhoads, D.C., 1970. Upper Devonian deltaic environments. In: Heaslip, W.G. (Ed.), Field Trip Guidebook, 42nd Annual Meeting. New York State Geological Association, pp. B1–B11.

Brenchley, P.J., Harper, D.A.T., 1998. Palaeoecology: Ecosystems, Environments and Evolution. Chapman and Hall, London.

Brett, C.E., Baird, G.C., 1994. Depositional sequences, cycles, and foreland basin dynamics in the Late Middle Devonian (Givetian) of the Genesee Valley and western Finger Lakes Region. In: Brett, C.E., Scatterday, J. (Eds), Field Trip Guidebook, 66th Annual Meeting. New York State Geological Association, pp. 505–586.

Brett, C.E., Speyer, S.E., Baird, G.C., 1986. Storm-generated sedimentary units: Tempestite proximality and event stratification in the Middle Devonian Hamilton Group of New York. Bulletin of the New York State Museum 457, 129–156.

Bridge, J.S., Dingman, L.E., 1981. Depositional environment of the Oneota Formation (Catskill Facies), near Unadilla, New York. In: Enos, P. (Ed.), Guidebook for Field Trips in South-Central New York, 53rd Annual Meeting. New York State Geological Association, pp. 255–278.

Brooks, D.R., McLennan, D.A., 1991. Phylogeny, Ecology, and Behavior. University of Chicago Press, Chicago.

Brooks, D.R., McLennan, D.A., 2002. The Nature of Diversity: An Evolutionary Voyage of Discovery. University of Chicago Press, Chicago.

Brown, J.H., Lomolino, M.V., 1998. Biogeography, 2nd ed., Sinauer Associates, Sunderland.

Buchan, L.A.J., Padilla, D.K., 2000. Predicting the likelihood of Eurasian watermilfoil presence in lakes, a macrophyte monitoring tool. Ecol. Appl. 10, 1442–1455.

Chong, G.W., Reich, R.M., Kalkhan, M.A., Stohlgren, T.J., 2001. New approaches for sampling and modeling native and exotic plant species richness. Western North American Naturalist 61, 328–335.

Colton, G.W., de Witt, W., Jr., 1958. Stratigraphy of the Sonyea Formation of Late Devonian age in western and west-central New York. USGS Oil and Gas Investigations, Chart OC-54.

Davis, J.C., 2002. Statistics and Data Analysis in Geology, 3rd ed., Wiley and Sons, New York.

Dennison, J.M., 1985. Catskill Delta shallow marine strata. In: Woodrow, D.L., Sevon, W.D. (Eds), The Catskill Delta. GSA Special Paper 201, pp. 91–106.

Dennison, J.M., Hasson, K.O., 1976. Stratigraphic cross section of Hamilton Group (Devonian) and adjacent strata along south border or Pennsylvania. AAPG Bulletin 60, 278–298.

Dennison, J.M., Schwietering, J.F., Avary, K.L., Hasson, K.O., 1979. Road log and stop descriptions. In: Avary, K.L. (Ed.), Devonian clastics in West Virginia and Maryland: Field Trip Guide. Eastern Section of AAPG Meeting, Morgantown, WV, pp. 1–64.

de Witt, W., Jr., 1960. Java Formation of Late Devonian age in western and central New York. AAPG Bulletin 44, 1933–1939.

de Witt, W., Jr., Roen, J.B., Wallace, L.G., 1993. Stratigraphy of Devonian black shales and associated rocks in the Appalachian Basin. In: Roen, J.B., Kepferle, R.C. (Eds), Petroleum Geology of the Devonian and Mississippian Black Shale of Eastern North America. USGS Bulletin 1909, pp. B1–B57.

Droser, M.L., Bottjer, D.J., Sheehan, P.M., McGhee, G.R., Jr., 2000. Decoupling the taxonomic and ecologic severity of Phanerozoic marine mass extinctions. Geology 28, 675–678.

Ehrets, J.R., 1981. The West Falls Group (Upper Devonian) Catskill Delta Complex: Stratigraphy, Sedimentary Environments and Aspects of Sedimentation. Unpublished M.S. thesis, University of Rochester, NY, p. 68.

Ellison, R.L., 1963. Faunas of the Mahantango Formation in South-Central Pennsylvania. Pennsylvanian Geological Survey, General Geology Report G39, 201–212.

Ellison, R.L., 1965. Stratigraphy and Paleontology of the Mahantango Formation in South-Central Pennsylvania. Pennsylvanian Geological Survey, General Geology Report 48, 1–298.

Environmental Systems Research Institute, Inc., 1999. ArcView GIS 3.2. Redlands, CA.

Epstein, J.B., 1986. The Valley and Ridge Province of eastern Pennsylvania – stratigraphic and sedimentologic contributions and problems. Geol. J. 21, 283–306.

Ettensohn, F.R., 1985. The Catskill Delta complex and the Acadian Orogeny: A model. In: Woodrow, D.L., Sevon, W.D. (Eds), The Catskill Delta: Denver, CO, Geological Society of America, Special Paper 201, pp. 39–49.

Faill, R.T., Wells, R.B., Nickelsen, R.P., Hoskins, D.M., 1973. Structure and Silurian-Devonian stratigraphy of the Valley and Ridge Province, central Pennsylvania, Guidebook for the 38th Annual Field Conference of Pennsylvania Geologists.

Ferguson, C.A., Bodenbender, B.E., Hones, J.L., Ahmed, K., 2001. Recording the fossil record: A GIS database of Middle Devonian fossils in the Michigan basin. Geological Society of America Annual Meeting, 2000, Abstracts with programs 109, A131.

Feria, T.P., Peterson, A.T., 2002. Prediction of bird community composition based on point-occurrence data and inferential algorithms: a valuable tool in biodiversity assessments. Diversity and Distributions, 8, 49–56.

Fletcher, F.W., 1962. Stratigraphy and structure of the "Catskill Group" in southeastern New York. In: Valentine, W.G. (Ed.), Guidebook to Field Trips, 34th Annual Meeting. New York State Geological Association, pp. D1–D20.

Frakes, L.A., 1963. Stratigraphy of the Nonred Upper Devonian across Pennsylvania. Pennsylvanian Geological Survey, 4th Series, Bulletin (General Geology Report), G39, 183–199.

Frakes, L.A., 1964. Paleogeography of the Trimmers Rock Member of the Fort Littleton Formation (Devonian) in Southern and Eastern Pennsylvania. Unpublished Ph.D. dissertation, University of California, Los Angeles, p. 396.

Goodwin, B.J., McAllister, A.J., Fahrig, L., 1998. Predicting invasiveness of plant species based on biological information. Conservation Biol. 13, 422–426.

Graham, R.W., 2000. FAUNMAP database: Filter effects from field to literature to database to analysis to interpretation. Geological Society of America, Abstracts with Programs 109, A131.

Grinnell, J., 1917. Field tests of theories concerning distributional control. American Naturalist 51, 115–128.

Haltuch, M.A., Berkman, P.A., Garton, D.W., 2000. Geographic information system (GIS) analysis of ecosystem invasion: Exotic mussels in Lake Erie. Limnology and Oceanography 45, 1778–1787.

Hasson, K.O., Dennison, J.M., 1979. Devonian shale stratigraphy between Perry Bay and the Fulton Lobe south-central Pennsylvania and Maryland. In: Dennison, J.M., Hasson, K.O., Hoskins, D.M., Jolley, R.M., Sevon, W.D. (Eds), Devonian Shales in South-Central Pennsylvania and Maryland. Guidebook for the 44th Annual Field Conference of Pennsylvania Geologists, pp. 1–17.

Hutchinson, G.E., 1957. Concluding remarks. Cold Spring Harbor Symposium on Quaternary Biology, 22, 415–427.

Jacobi, R.D., Smith, G., 1999. Structure and Upper Devonian stratigraphy in the Appalachian plateau of Allegany County, New York State, including the Clarendon-Linden fault system. In: Lash, G.G. (Ed.), 71st Annual Meeting, Field Trip Guidebook. New York State Geological Association, pp. Sat. C1–C44.

Juliusson, L., Graham, R., 1999. Geographic information systems and vertebrate paleontology. J. Vertebr. Paleontol. 19 (Suppl. to No. 3), 56.

Karman, A.A., 1968. Recognition of Basin, Slope, and Shelf Deposits in the Upper Devonian West Falls Group of New York State. Unpublished M.S. thesis, University of Rochester, NY, p. 58.

Kirchgasser, W.T., 1965. The Parrish Limestone (Upper Devonian) of West-Central New York. Unpublished M.S. thesis, Cornell University, Ithaca, NY, p. 177.

Kirchgasser, W.T., 1983. Lower Upper Devonian stratigraphy from the Batavia-Warsaw meridian to the Genesee Valley: Goniatite sequence and correlations. In: Hewitt, P.C. (Ed.), Guidebook to Field Trips, 45th Annual Meeting. New York State Geological Association, pp. C1–C27. .

Kirchgasser, W.T., Over, D.J., Woodrow, D.L., 1994. Frasnian (Upper Devonian) strata of the Genesee River Valley, western New York State. In: Brett, C.E., Scatterday, J. (Eds), Field Trip Guidebook, 66th Annual Meeting. New York State Geological Association, pp. 325–360.

Krajewski, S.A., Williams, E.G., 1971. Upper Devonian Sedimentation in Susquehanna County. Guidebook for the 36th Annual Field Conference of Pennsylvania Geologists, pp. 1–76.

Lafferty, A.G., Miller, A.I., Brett, C.E., 1994. Comparative spatial variability in faunal composition along two Middle Devonian paleoenvironmental gradients. Palaios 9, 224–236.

Leighton, L.R., 2000. Environmental distribution of spinose brachiopods from the Devonian of New York: Test of the soft-substrate hypothesis. Palaios 15, 184–193.

Lieberman, B.S., 2000. Paleobiogeography: Using Fossils to Study Global Change, Plate Tectonics, and Evolution. Kluwer Academic Press/Plenum Publishing, New York.

Linsley, D.M., 1994. Devonian paleontology of New York. Paleontological Research Institution, Special Publications 21.

Lundegard, P.D., Samuels, N.D., Pryor, W.P., 1985. Upper Devonian turbidite sequence, central and southern Appalachian basin: Contrasts with submarine fan deposits. In: Woodrow, D.L., Sevon, W.D. (Eds), The Catskill Delta. GSA Special Paper 201, pp. 107–121.
Mayer, S.M., 1994. Field studies of the Middle Devonian Ludlowville-Moscow sequence in the Genesee and Seneca Valleys, New York State. In: Brett, C.E., Scatterday, J. (Eds), Field Trip Guidebook, 66th Annual Meeting. New York State Geological Association, pp. 491–504.
Mayer, S.M., Baird, G.C., Brett, C.E., 1994. Correlation and facies divisions in the uppermost Ludlowville Formation (Givetian) across western and central New York State. Bulletin of the New York State Museum 481, 229–264.
McGhee, G.R., Jr., 1976. Late Devonian benthic marine communities of the central Appalachian Alleghany Front. Lethaia 9, 111–136.
McGhee, G.R., Jr., 1988. The Late Devonian extinction event: Evidence for abrupt ecosystem collapse. Paleobiology 14, 250–257.
McGhee, G.R., Jr., 1989. The Frasnian–Famennian extinction event. In: Donovan, S.K. (Ed.), Mass Extinctions: Processes and Evidence, Columbia University Press, New York, pp. 133–151.
McGhee, G.R., Jr., 1990. The Frasnian-Famennian extinction record in the eastern United States. In: Walliser, O.H., KauVman, E.G. (Eds), Extinction Events in Earth History. Springer, Berlin, pp. 161–168.
McGhee, G.R., Jr., 1996. The Late Devonian Mass Extinction: The Frasnian/Famennian Crisis. Columbia University Press, New York.
McGhee, G.R., Jr., 2001. The Late Devonian Mass Extinction. In: Briggs, D.E.G., Crowther, P.R. (Eds), Palaeobiology 2. Oxford, Blackwell Science, pp. 222–225.
McGhee, G.R., Jr., Sutton, R.G., 1981. Late Devonian marine ecology and zoogeography of the central Appalachian and New York. Lethaia 14, 27–43.
McGhee, G.R., Jr., Sutton, R.G., 1983. Evolution of late Frasnian (Late Devonian) marine environments in New York and the central Appalachians. Alcheringa 7, 9–21.
McGhee, G.R., Jr., Sutton, R.G., 1985. Late Devonian marine ecosystems of the lower West Falls Group in New York. In: Woodrow, D.L., Sevon, W.D. (Eds), The Catskill Delta. GSA Special Paper 201, pp. 199–210.
Metzger, W.J., Tesmer, I., Kirchgasser, W., 1974. Upper Devonian stratigraphy of Chautauqua County, New York. In: Peterson, D.N. (Ed.), Field Trip Guidebook, 46th Annual Meeting. New York Geological Association, B14–B23.
Miller, K.B., 1986. Depositional environments and sequences, "*Pleurodictyum* Zone", Ludlowville Formation of Western New York. Bulletin of the New York State Museum 457, 57–77.
Minitab Inc., 2003. Minitab Release 14. State College, Pennsylvania.
Oliver, W.A., Jr., 1976. Biogeography of the Devonian rugose corals. J. Paleontol. 50, 365–373.
Oliver, W.A., Jr., 1990. Extinctions and migrations of Devonian rugose corals in the Eastern Americas realm. Lethaia 23, 167–178.
Oliver, W.A., Klapper, G. (Eds), 1981. Devonian biostratigraphy of New York. International Union of Geological Sciences, Subcommission on Devonian Stratigraphy.
Oliver, W.A., Jr., Pedder, A.E.H., 1994. Crises in the Devonian history of rugose corals. Paleobiology 20, 178–190.
Over, D.J., 1997. Conodont biostratigraphy of the Java Formation (Upper Devonian) and the Frasnian-Famennian boundary in western New York State. In: Klapper, G., Murphy, M.A., Talent, J.A. (Eds), Paleozoic Sequence Stratigraphy, Biostratigraphy, and Biogeography: Studies in Honor of J. Granville ("Jess") Johnson. GSA Special Paper 321, pp. 161–177.
Over, D.J., Baird, G.C., Kirchgasser, W.T., 1999. Frasnian (Lower Upper Devonian) geology of western New York as seen along Eighteenmile Creek and Route 20A: Submarine discontinuities, gravity flows, and mass extinctions. In: Lash, G.G. (Ed.), Field Trip Guidebook, 71st Annual Meeting. New York State Geological Association, pp. Sun. B8–B21.
Patchen, D.G., Dugolinsky, B.K., 1979. Guidebook, Middle and Upper Devonian clastics, central and western New York State. West Virginian Geological and Economic Survey, p. 170.
Pepper, J.F., de Witt, W., Jr., 1950. Stratigraphy of the Upper Devonian Wiscoy sandstone and the equivalent Hanover shale in western and central New York. USGS Oil and Gas Investigation Chart, OC 45.
Peterson, A.T., 2001. Predicting species' geographic distributions based on ecological niche modeling. The Condor 103, 599–605.

Peterson, A.T., Ball, L.G., Cohoon, K.P., 2002a. Predicting distributions of Mexican birds using ecological niche modeling methods. Ibis 144, E27–E32.
Peterson, A.T., Cohoon, K.P., 1999. Sensitivity of distributional prediction algorithms to geographic data completeness. Ecol. Model. 117, 159–164.
Peterson, A.T., Ortega-Huerta, M.A., Bartley, J., Sánchez-Cordero, V., Soberón, J., Buddemeier, R.H., Stockwell, D.R.B., 2002b. Future predictions for Mexican faunas under global change scenarios. Nature 416, 626–629.
Peterson, A.T., Sánchez-Cordero, V., Beard, C.B., Ramsey, J.M., 2002c, Ecological niche modeling and potential reservoirs for Chagas disease, Mexico. Emerging Infectious Diseases 8, 662–667.
Peterson, A.T., Sánchez-Cordero, V., Soberón, J., Bartley, J., Buddemeier, R., Navarro-Sigüenza, A.G., 2001. Effects of global climate change on geographic distributions of Mexican Cracidae. Ecol. Model. 144, 21–30.
Peterson, A.T., Soberón, J., Sánchez-Cordero, V., 1999. Conservatism of ecological niches in evolutionary time. Science 285, 1265–1267.
Peterson, A.T., Vieglais, D.A., 2001. Predicting species invasions using ecological niche modeling: New approaches from bioinformatics attack a pressing problem. BioScience 51, 363–371.
Prave, A.R., Duke, W.L., 1991. Sedimentology and sequence stratigraphic framework of the Middle Devonian Mahantango Formation in Central Pennsylvania. In: Schultz, A., Compton-Gooding, E. (Eds), Geologic Evolution of the Eastern United States, Field Trip Guidebook, NE-SE GSA 1991, VMNH Guidebook 2. Geological Society of America, pp. 207–228.
Prave, A.R., Duke, W.L., Slattery, W., 1996. A depositional model for storm- and tide-influenced prograding siliciclastic shorelines from the Middle Devonian of the central Appalachian foreland basin, USA. Sedimentology 43, 611–629.
Rahmanian, V.D.K., 1979. Stratigraphy and Sedimentology of the Upper Devonian Catskill and Uppermost Trimmers Rock formations in Central Pennsylvania. Unpublished Ph.D. dissertation, Pennsylvania State University, p. 396.
Rickard, L.V., 1975. Correlation of the Silurian and Devonian rocks in New York State. New York State Museum and Science Service, Map and Chart Series, No. 24.
Rode, A.L., 2004. Phylogenetic revision of the Devonian bivalve, *Leptodesma (Leiopteria)*. Yale University Postilla, 229, 1–26.
Rode, A.L., Lieberman, B.S., 2002. Phylogenetic and biogeographic analysis of Devonian phyllocarid crustaceans. J. Paleontol. 76, 271–286.
Rode, A.L., Lieberman, B.S., 2003. GIS and phylogenetics, a combined approach to understanding biogeographic changes in the Late Devonian. Geological Society of America Annual Meeting, 2003, Abstracts with programs, 112, 157–158.
Rode, A.L., Lieberman, B.S., 2004. Using GIS to unlock the interactions between biogeography, environment, and evolution in middle and Late Devonian brachiopods and bivalves. Palaeogeography, Palaeoclimatology, Palaeogeography 211, 345–359.
Rode, A.L., Lieberman, B.S., 2005. Integrating biogeography and evolution using phylogenetics and PaleoGIS: A case study involving Devonian crustaceans. J. Paleontol. 79(2), 267–276.
Rodeheaver, J.S., 1992. Abundant and diverse fossils from the Middle Devonian Mahantango Formation, Mineral and Hampshire Counties, WV. The Compass of Sigma Gamma Epsilon 69, 181–198.
Roe, L.M., II, 1976. Sedimentary Environments of the Java Group (Upper Devonian): Three Dimensional Study. Unpublished Ph.D. dissertation, University of Rochester, NY, p. 153.
Sandberg, C.A., Ziegler, W., 1996. Devonian conodont biochronology in geologic time calibration. Senckenbergiana Lethaea, 76, 259–265.
Savarese, M., Gray, L.M., Brett, C.E., 1986. Faunal and lithologic cyclicity in the Centerfield Member (Middle Devonian, Hamilton Group) of Western New York: A reinterpretation of depositional history. Bulletin of the New York State Museum 457, 32–56.
Schultz, L.D., 1974. The Stratigraphy of the Trimmer Rock Formation in Northeastern Pennsylvania. Unpublished Ph.D. dissertation, Lehigh University, p. 132.
Sevon, W.D., 1985. Nonmarine facies of the Middle and Late Devonian Catskill coastal alluvial plain. In: Woodrow, D.L., Sevon, W.D. (Eds), The Catskill Delta. GSA Special Paper 201, pp. 79–90.
Smith, G.J., Jacobi, R.D., 2000. Tectonic and eustatic signals in the sequence stratigraphy of the Upper Devonian Canadaway Group, New York State. AAPG Bulletin 85, 325–357.
Sokal, R.R., Rohlf, F.J., 1995. Biometry, The Principles and Practice of Statistics in Biological Research. 3rd ed., W.H. Freeman and Company, New York.

Stockwell, D., Peters, D., 1999, The GARP modelling system: problems and solutions to automated spatial prediction. Int. J. Geogr. Inf. Sci. 13, 143–158.
Stockwell, D., Peterson, A.T., 2002. Effects of sample size on accuracy of species distribution models. Ecol. Model. 148, 1–13.
Sutton, R.G., 1963, Correlation of Upper Devonian Strata in South-Central New York. Pennsylvanian Geological Survey, 4th Series, Bulletin (General Geology Report), G39, 87–101.
Sutton, R.G., Bowen, Z.P., McAlester, A.L., 1970. Marine shelf environments of the Upper Devonian Sonyea Group of New York. GSA Bulletin 81, 2975–2992.
Sutton, R.G., McGhee, G.R., Jr., 1985. The evolution of Frasnian marine "community-types" of south-central New York. In: Woodrow, D.L., Sevon, W.D. (Eds), The Catskill Delta. GSA Special Paper 201, pp. 211–224.
Tesmer, I.H., 1966. Upper Devonian stratigraphy and paleontology of southwestern New York State (Erie, Chautauqua, and Cattaraugus Counties). In: Buehler, E.J. (Ed.), Geology of Western New York, Guidebook, 38th Annual Meeting. New York State Geological Association, pp. 47–52.
Tesmer, I.H., 1974. A brief description of Upper Devonian units to be observed on Chautauqua County field trip. In: Peterson, D.N. (Ed.), Field Trip Guidebook, 46th Annual Meeting. New York Geological Association, pp. B1–B13.
Tucker, R.D., Bradley, D.C., Ver Straeten, C.A., Harris, A.G., Ebert, J.R., McCutcheon, S.R., 1998. New U-Pb zircon ages and the duration and division of Devonian time. Earth Planetary Sci. Lett. 158, 175–786.
Ver Straeten, C.A., Brett, C.E., 1999. Lower and Middle Devonian foreland basin fill in the Catskill front: Stratigraphic synthesis, sequence stratigraphy, and the Acadian Orogeny. In: Garver, J.I., Smith, J.A. (Eds), Field Trip Guidebook for the 67th Annual Meeting. New York State Geological Association, pp. 313–356.
Walker, R.G., Sutton, R.G., 1967. Quantitative analysis of turbidites in the Upper Devonian Sonyea Group, New York. J. Sedimentary Petrology 37, 1012–1022.
Wiley, E.O., Mayden, R.L., 1985. Species and speciation in phylogenetic systematics, with examples from North American fish fauna. Annals of the Missouri Botanical Garden 72, 596–635.
Willard, B., 1934. Early Chemung shoreline in Pennsylvania. GSA Bulletin 45, 897–908.
Willard, B., 1935a. Hamilton Group of Central Pennsylvania. GSA Bulletin 46, 195–224.
Willard, B., 1935b. Portage Group in Pennsylvania. GSA Bulletin 46, 1195–1218.
Willard, B., 1935c. Hamilton Group along the Allegheny Front, Pennsylvania. GSA Bulletin 46, 1275–1290.
Williams, E.G., Slingerland, R., 1985. Field guide-Catskill sedimentation in central Pennsylvania. Field Conference of Pennsylvania Geologists 50, 45–62.
Woodrow, D.L., 1985. Paleogeography, paleoclimate, and sedimentary processes of the Late Devonian Catskill Delta. In: Woodrow, D.L., Sevon, W.D. (Eds), The Catskill Delta. GSA Special Paper 201, pp. 51–64.
Wygart, G.T., 1986. Deposition and early diagensis of a Middle Devonian marine shale: Ludlowville Formation, Western New York. Bulletin of the New York State Museum 457, 78–101.

Appendix 1.1. Raw environmental base data for the Lower *varcus* Zone

| Longitude | Latitude | Grain size/rock type | % Mud | % Silt | % Sand | % ls | Sedimentary structures | Bedding style /thickness | Substrate type | Water depth |
|---|---|---|---|---|---|---|---|---|---|---|
| –78.75 | 42.75 | Ludlowville: Wanakah Sh; med grey, soft, fossilif. Shale & shaly mudstone | 55 | 20 | 0 | 25 | ls concretions, fossils in tempestite layers | Thin | Muddy | Photic zone below SWB to above SWB for ls |
| –78.25 | 42.75 | Ludlowville: Wanakah Sh; dk grey sh, ls, calcareous grey sh, encrinte | 50 | 30 | 0 | 20 | Tempestite shell layers | Thin | Muddy | Below SWB to near normal WB |
| –77.75 | 42.75 | Ludlowville: Wanakah, grey fissile shale | 70 | 0 | 0 | 30 | Burried bottom ls assemblage | Thin | Muddy | SWB to normal WB for ls (20–25 m), below SWB for sh (100-150 m) |
| –77.25 | 42.75 | Ludlowville: Wanakah, black shale | 70 | 15 | 0 | 15 | Shell beds | Thin | Muddy | Below SWB |
| –76.75 | 42.75 | Ludlowville: Wanakah, black shale, micritic ls | 80 | 10 | 0 | 10 | Shell beds | Thin | Muddy | Below SWB |
| –76.25 | 42.75 | Ludlowville: Otisco Sh, Ivy Pt. Siltst.; siltstone, ls, concretions, mudst | 60 | 20 | 0 | 20 | Individual HCS, laminated mudstone | Thin to moderate | Muddy to sandy | Below SWB to lower shoreface |
| –75.75 | 42.75 | Ludlowville: upper, undif, micacous siltst, silty sh, sandy siltst | 30 | 60 | 0 | 10 | Laminated and ripple laminated | Thin to moderate | Silty | Below SWB to just above SWB |
| –75.25 | 42.75 | Ludlowville to Panther Mtn sst | 15 | 10 | 75 | 0 | | | | |

| | | | | | | | | | | |
|---|---|---|---|---|---|---|---|---|---|---|
| −74.75 | 42.75 | Panther Mtn sst, Ashokan Fm., med grain graywacke sst, olive brown mudst, dk shale | 20 | 0 | 80 | 0 | Fluvial & tidal influences; channel sands; x-beds to laminated | Thick | Sandy | Intertidal to subaerial |
| −74.25 | 42.75 | Panther to Plattekill Fm. coarse sst, red-grey sh, siltst, mudst | 20 | 10 | 70 | 0 | x-beds, channel fill | Thick | Sandy | Subaerially exposed |
| −73.75 | 42.75 | No outcrop | | | | | | | | |
| −80.25 | 42.25 | No outcrop | | | | | | | | |
| −79.75 | 42.25 | No outcrop | | | | | | | | |
| −79.25 | 42.25 | No outcrop | | | | | | | | |
| −78.75 | 42.25 | No outcrop | | | | | | | | |
| −78.25 | 42.25 | No outcrop | | | | | | | | |
| −77.75 | 42.25 | No outcrop | | | | | | | | |
| −77.25 | 42.25 | No outcrop | | | | | | | | |
| −76.75 | 42.25 | No outcrop | | | | | | | | |
| −76.25 | 42.25 | No outcrop | | | | | | | | |
| −75.75 | 42.25 | No outcrop | | | | | | | | |
| −75.25 | 42.25 | Plattekill Fm. coarse sst, red-grey sh, siltst, mudst | 20 | 10 | 70 | 0 | x-beds, channel fill | Thick | Sandy | Subaeriall exposed |
| −74.75 | 42.25 | No outcrop | | | | | | | | |
| −74.25 | 42.25 | No outcrop | | | | | | | | |
| −73.75 | 42.25 | No outcrop | | | | | | | | |
| −80.25 | 41.75 | No outcrop | | | | | | | | |
| −79.75 | 41.75 | No outcrop | | | | | | | | |
| −79.25 | 41.75 | No outcrop | | | | | | | | |
| −78.75 | 41.75 | No outcrop | | | | | | | | |
| −78.25 | 41.75 | No outcrop | | | | | | | | |

Appendix 1.1. (*Continued*)

| Longitude | Latitude | Grain size/rock type | % Mud | % Silt | % Sand | % ls | Sedimentary structures | Bedding style /thickness | Substrate type | Water depth |
|---|---|---|---|---|---|---|---|---|---|---|
| –77.75 | 41.75 | No outcrop | | | | | | | | |
| –77.25 | 41.75 | No outcrop | | | | | | | | |
| –76.75 | 41.75 | No outcrop | | | | | | | | |
| –76.25 | 41.75 | No outcrop | | | | | | | | |
| –75.75 | 41.75 | No outcrop | | | | | | | | |
| –75.25 | 41.75 | No outcrop | | | | | | | | |
| –74.75 | 41.75 | No outcrop | | | | | | | | |
| –74.25 | 41.75 | No outcrop | | | | | | | | |
| –80.25 | 41.25 | No outcrop | | | | | | | | |
| –79.75 | 41.25 | No outcrop | | | | | | | | |
| –79.25 | 41.25 | No outcrop | | | | | | | | |
| –78.75 | 41.25 | No outcrop | | | | | | | | |
| –78.25 | 41.25 | No outcrop | | | | | | | | |
| –77.75 | 41.25 | No outcrop | | | | | | | | |
| –77.25 | 41.25 | Millsboro Shale and Mahantango; black shale to siltstone | 90 | 10 | 0 | 0 | Laminated and storm-derived ripples | Thin to moderate | Muddy to silty | Below to just at SWB |
| –76.75 | 41.25 | Millsboro Shale and Mahantango; black shale to siltstone | 80 | 20 | 0 | 0 | Laminated and storm-derived ripples | Thin to moderate | Muddy to silty | Below to just at SWB |
| –76.25 | 41.25 | Millsboro Shale and Mahantango; black shale to siltstone to fine sst. | 63 | 35 | 2 | 0 | Laminated beds, ripple forms, a few HCS | Thin to moderate | Muddy to silty | Below SWB to above SWB |
| –75.75 | 41.25 | Mahantango; mudstone to siltstone a few fine sst | 43 | 47 | 10 | 0 | Laminated beds, a few HCS, storm facies | Thin (mainly) to moderate | Muddy to silty | Below to just above SWB |

| | | | | | | | | | | |
|---|---|---|---|---|---|---|---|---|---|---|
| –75.25 | 41.25 | Mahantango; mudstone to fine sst (more mudst) | 52 | 44 | 4 | 0 | Laminated beds, a few HCS, storm facies | Thin (mainly) to moderate | Muddy to sandy | Below SWB to lower shoreface |
| –74.75 | 41.25 | Mahantango; siltstone to fine sst (more sst) | 20 | 28 | 52 | 0 | HCS, amalgamated or single, sto-rm facies only | Moderate to thick | Silty to sandy | At SWB to upper Shoreface |
| –78.25 | 40.75 | Millsboro Shale and Mahantango; black shale to siltstone | 80 | 20 | 0 | 0 | Laminated and storm-derived ripples | Thin to moderate | Muddy to silty | Below to just at SWB |
| –77.75 | 40.75 | Millsboro Shale and Mahantango; black shale to siltstone to fine sst. | 57 | 40 | 3 | 0 | Laminated beds, ripple forms, a few HCS | Thin to moderate | Muddy to silty | Below SWB to above SWB |
| –77.25 | 40.75 | Mahantango; SWB mudstone to siltstone a few fine sst | 62 | 25 | 13 | 0 | Laminated beds, a few HCS, storm facies | Thin (mainly) to moderate | Muddy to silty | Below to just above |
| –76.75 | 40.75 | Mahantango; mudstone to fine sst (more mudst) | 27 | 40 | 33 | 0 | Laminated beds, a few HCS, storm facies | Thin (mainly) to moderate | Tuddy to sandy | Below SWB to lower shoreface |
| –76.25 | 40.75 | Mahantango; mudstone to fine sst (more mudst) | 21 | 47 | 32 | 0 | Laminated beds, a few HCS, storm facies | Thin (mainly) to moderate | Muddy to sandy | Below SWB to lower shoreface |
| –75.75 | 40.75 | Mahantango; mudstone to fine sst (more mudst) | 14 | 46 | 40 | 0 | Laminated beds, a few HCS, storm facies | Thin (mainly) to moderate | Muddy to sandy | Below SWB to lower shoreface |
| –75.25 | 40.75 | Mahantango; mudstone to | 10 | 53 | 37 | 0 | Laminated beds, a few HCS, | Thin (mainly) to moderate | Muddy to sandy | Below SWB to lower shoreface |

Appendix 1.1. (*Continued*)

| Longitude | Latitude | Grain size/rock type | % Mud | % Silt | % Sand | % ls | Sedimentary structures | Bedding style /thickness | Substrate type | Water depth |
|---|---|---|---|---|---|---|---|---|---|---|
| | | fine sst (more mudst) | | | | | storm facies | | | |
| –78.75 | 40.25 | Mahantango & Millsboro; Blk sh, mudst, few sltst. | 48 | 52 | 0 | 0 | Laminated beds, ripple forms, storm facies | Thin (mainly) to moderate | Muddy to silty | Below to just above SWB |
| –78.25 | 40.25 | Mahantango; mudstone to siltstone a few fine sst | 36 | 50 | 14 | 0 | Laminated beds, massive siltstone, a few HCS, storm facies | Thin (mainly) storm facies | Muddy to silty | Below to just above SWB |
| –77.75 | 40.25 | Mahantango; mudstone to siltstone a few fine sst | 28 | 50 | 22 | 0 | Laminated beds, massive siltstone, a few HCS, storm facies | Thin (mainly) to moderate | Muddy to fine sandy | Below SWB to lower shoreface |
| –77.25 | 40.25 | Mahantango; siltstone to coarse sst | 15 | 41 | 44 | 0 | Amalgamated HCS, TXB, channel sands, mud draped ripples | Moderate to thick | Sandy | At SWB, upper shoreface to intertidal |
| –76.75 | 40.25 | Mahantango; fine to coarse sst | 0 | 6 | 94 | 0 | Amalgamated HCS, TXB, channel sands, mud draped ripples | Moderate to thick | Sandy | Above SWB, upper shoreface to intertidal |
| –76.25 | 40.25 | Mahantango, none | | | | | None | None | Subaerially exposed | Subaerially exposed |

| | | | | | | | | | | |
|---|---|---|---|---|---|---|---|---|---|---|
| −75.75 | 40.25 | Mahantango, none | | | | | None | None | Subaerially exposed | Subaerially exposed |
| −78.75 | 39.75 | Mahantango; mudstone to siltstone a few fine sst | 33 | 50 | 17 | 0 | Laminated beds, massive siltstone, a few HCS, storm facies | Thin (mainly) to moderate | Muddy to silty | Below to just above SWB |
| −78.25 | 39.75 | Mahantango; mudstone to siltstone a few fine sst | 25 | 50 | 25 | 0 | Laminated beds, massive siltstone, a few HCS, storm facies | Thin (mainly) to moderate | Silty | Just above SWB |
| −77.75 | 39.75 | Mahantango; mudstone, fine to coarse sst | 15 | 43 | 42 | 0 | Amalgamated HCS, TXB, channel sands, mud draped ripples | Few thin, mostly moderate to thick | Muddy to sandy (mostly) | At SWB to upper shoreface to intertidal |
| −79.75 | 39.25 | Millsboro Shale; black shale | 100 | 0 | 0 | 0 | Laminated | Thin | Muddy | Well below SWB |
| −79.25 | 39.25 | Millsboro Shale and Mahantango; black shale to siltstone | 75 | 25 | 0 | 0 | Laminated and storm-derived ripples | Thin to moderate | Muddy to silty | Below to just at SWB |
| −78.75 | 39.25 | Mahantango; mudstone to siltstone a few fine sst | 31 | 50 | 19 | 0 | Laminated beds, a few HCS, storm facies | Thin (mainly) to moderate | Muddy to silty | Below to just above SWB |
| −78.25 | 39.25 | Mahantango; mudstone to fine sst (more mudst) | 25 | 50 | 25 | 0 | Laminated beds, a few HCS, storm facies | Thin (mainly) to moderate | Silty | Just above SWB |
| −77.75 | 39.25 | No outcrop | | | | | | | | |

Appendix 1.1. (*Continued*)

| Longitude | Latitude | Grain size/rock type | % Mud | % Silt | % Sand | % ls | Sedimentary structures | Bedding style /thickness | Substrate type | Water depth |
|---|---|---|---|---|---|---|---|---|---|---|
| –79.75 | 38.75 | Millsboro Shale; black shale | 100 | 0 | 0 | 0 | Laminated | Thin | Muddy | Well below SWB |
| –79.25 | 38.75 | Millsboro Shale and Mahantango; black shale to siltstone | 77 | 23 | 0 | 0 | Laminated and storm-derived ripples | Thin to moderate | Muddy to silty | Below to just at SWB |
| –78.75 | 38.75 | Mahantango; mudstone to siltstone a few fine sst | 35 | 50 | 15 | 0 | Laminated beds, a few HCS, storm facies | Thin (mainly) to moderate | Muddy to silty | Below to just above SWB |
| –78.25 | 38.75 | Mahantango; mudstone to fine sst (more mudst) | 21 | 50 | 29 | 0 | Laminated beds, ripple forms, single and amalgamated HCS | Thin (mainly) to moderate | Silty | Just above SWB |
| –80.25 | 38.25 | Millsboro Shale; black shale | 100 | 0 | 0 | 0 | Laminated | Thin | Muddy | Well below SWB |
| –79.75 | 38.25 | Millsboro Shale and Mahantango; black shale to siltstone | 87 | 13 | 0 | 0 | Laminated and storm-derived ripples | Thin to moderate | Muddy to silty | Below to just at SWB |
| –79.25 | 38.25 | Mahantango; mudstone to fine sst (more mudst) | 46 | 40 | 14 | 0 | Laminated beds, ripple forms, single and amalgamated HCS | Thin (mainly) to moderate | Muddy to sandy | Well below SWB to lower shoreface |
| –78.75 | 38.25 | No outcrop | | | | | | | | |
| –80.75 | 37.75 | No outcrop | | | | | | | | |

| | | | | | | | | | | |
|---|---|---|---|---|---|---|---|---|---|---|
| −80.25 | 37.75 | Millsboro Shale and Mahantango; black shale to siltstone | 66 | 32 | 2 | 0 | Laminated and storm-derived ripples | Thin to moderate | Muddy to silty | Below to just at SWB |
| −79.75 | 37.75 | Mahantango; mudstone to fine sst (more mudst) | 34 | 50 | 16 | 0 | Laminated beds, a few HCS, storm facies | Thin (mainly) to moderate | Muddy to sandy | Below SWB to lower shoreface |
| −79.25 | 37.75 | no outcrop | | | | | | | | |
| −81.25 | 37.25 | Millsboro Shale and Mahantango; black shale to siltstone | 75 | 25 | 0 | 0 | Laminated and storm-derived ripples | Thin to moderate | Muddy to silty | Below to just at SWB |
| −80.75 | 37.25 | Mahantango; mudstone with a few siltstone to v. fine sst | 38 | 62 | 0 | 0 | Laminated beds, ripple forms | Thin | Muddy | Below SWB |
| −80.25 | 37.25 | Mahantango; mudstone to siltstone a few fine sst | 39 | 50 | 11 | 0 | Laminated beds, a few HCS, storm facies | Thin (mainly) to moderate | Muddy to silty | Below to just above SWB |
| −79.75 | 37.25 | No outcrop | | | | | | | | |
| −81.75 | 36.75 | No outcrop | | | | | | | | |

Appendix 1.1. Raw environmental base data for the Lower *varcus* Zone.

| Longitude | Latitude | Depositional environment | Ichnofacies/bioturbation | Oxygenation | Biofacies | Reference |
|---|---|---|---|---|---|---|
| −78.75 | 42.75 | Outer to middle shelf | *Cruziana* and *Zoophycus* | Normal marine | *Ambocoelia, Athyris,* and normal marine | Oliver and Klapper, 1981; Miller, 1986; Wygart, 1996; Batt, 1999 |
| −78.25 | 42.75 | outer to middle shelf | | Normal marine | | Batt, 1999 |
| −77.75 | 42.75 | Deep to distal shelf | *Zoophycus*, lots of bioturbation | Normal marine | *Ambocoelia, Palaeoneilo, Chonetids* | Oliver and Klapper, 1981; Savarese et al., 1986; Batt, 1999 |
| −77.25 | 42.75 | Shelf to basin | *Zoophycus* | Dysaerobic | Dysaerobic | Batt, 1999 |
| −76.75 | 42.75 | Basin | *Zoophycus* | Dysaerobic | Dysaerobic | Batt, 1999 |
| −76.25 | 42.75 | Shelf | *Zoophycus*, lots of bioturbation | Normal marine | Corals | Brett et al., 1986; Brett and Baird, 1994; Mayer, 1994 |
| −75.75 | 42.75 | Middle shelf | | Normal marine | *Cypricardella, Ambocoelia, Tropidoleptus* | Oliver and Klapper, 1981 |
| −75.25 | 42.75 | | | | | |
| −74.75 | 42.75 | Tidal to estuarine | | Subaerial | Nonmarine | Ver Straeten and Brett, 1999 |
| −74.25 | 42.75 | Alluvial fan & coastal plain | | Subaerial | Nonmarine | Ver Straeten and Brett, 1999 |
| −73.75 | 42.75 | | | | | |
| −80.25 | 42.25 | | | | | |
| −79.75 | 42.25 | | | | | |
| −79.25 | 42.25 | | | | | |
| −78.75 | 42.25 | | | | | |
| −78.25 | 42.25 | | | | | |
| −77.75 | 42.25 | | | | | |
| −77.25 | 42.25 | | | | | |

| | | | | | | |
|---|---|---|---|---|---|---|
| −76.75 | 42.25 | | | | | |
| −76.25 | 42.25 | | | | | |
| −75.75 | 42.25 | | | | | |
| −75.25 | 42.25 | Alluvial fan & coastal plain | | Subaerial | Nonmarine | Ver Straeten and Brett, 1999 |
| −74.75 | 42.25 | | | | | |
| −74.25 | 42.25 | | | | | |
| −73.75 | 42.25 | | | | | |
| −80.25 | 41.75 | | | | | |
| −79.75 | 41.75 | | | | | |
| −79.25 | 41.75 | | | | | |
| −78.75 | 41.75 | | | | | |
| −78.25 | 41.75 | | | | | |
| −77.75 | 41.75 | | | | | |
| −77.25 | 41.75 | | | | | |
| −76.75 | 41.75 | | | | | |
| −76.25 | 41.75 | | | | | |
| −75.75 | 41.75 | | | | | |
| −75.25 | 41.75 | | | | | |
| −74.75 | 41.75 | | | | | |
| −74.25 | 41.75 | | | | | |
| −80.25 | 41.25 | | | | | |
| −79.75 | 41.25 | | | | | |
| −79.25 | 41.25 | | | | | |
| −78.75 | 41.25 | | | | | |
| −78.25 | 41.25 | | | | | |
| −77.75 | 41.25 | | | | | |
| −77.25 | 41.25 | Deep to outer shelf | *Cruziana*; slight to intense | Dysaerobic | Anoxic to open marine | Prave et al., 1996: inferred |
| −76.75 | 41.25 | Deep to outer shelf | *Cruziana*; slight to intense | Dysaerobic | Anoxic to open marine | Prave et al., 1996: inferred |
| −76.25 | 41.25 | Outer shelf to middle shelf | *Cruziana*; bioturbation common | Nn | Open marine: brachs, coral, crinoids; bryo and mollusks rare | Prave et al., 1996: inferred |

Appendix 1.1. (*Continued*)

| Longitude | Latitude | Depositional environment | Ichnofacies/bioturbation | Oxygenation | Biofacies | Reference |
|---|---|---|---|---|---|---|
| –75.75 | 41.25 | Outer shelf to middle shelf | *Cruziana*; slight to intense | Normal marine | Open marine: brachs, coral, crinoids; bryo and mollusks rare | Prave et al., 1996: inferred |
| –75.25 | 41.25 | Outer shelf to inner shelf/shoreface | *Cruziana* and some *Skolithos* | Normal marine | Open marine to abraded brach valves | Prave et al., 1996: inferred |
| –74.75 | 41.25 | Middle shelf to inner shelf/shoreface | *Cruziana* and *Skolithos* | Normal marine | Open marine to abraded brach valves | Prave et al., 1996: section |
| –78.25 | 40.75 | Deep to outer shelf | *Cruziana*; slight to intense | Normal marine | Anoxic to open marine | Prave et al., 1996: inferred |
| –77.75 | 40.75 | Outer shelf to middle shelf | *Cruziana*; bioturbation common | Normal marine | Open marine: brachs, coral, crinoids; bryo and mollusks rare | Prave et al., 1996: inferred |
| –77.25 | 40.75 | Outer shelf to middle shelf | *Cruziana*; slight to intense | Normal marine | Open marine: brachs, coral, crinoids; bryo and mollusks rare | Prave et al., 1996: inferred |
| –76.75 | 40.75 | Outer shelf to inner shelf/shoreface | *Cruziana* and some *Skolithos* | Normal marine | Open marine to abraded brach valves | Prave et al., 1996: inferred; Faill et al., 1973: section |
| –76.25 | 40.75 | Outer shelf to inner shelf/shoreface | *Cruziana* and some *Skolithos* | Normal marine | Open marine to abraded brach valves | Prave et al., 1996: inferred |
| –75.75 | 40.75 | Outer shelf to inner shelf/shoreface | *Cruziana* and some *Skolithos* | Normal marine | Open marine to abraded brach valves | Prave et al., 1996: section |
| –75.25 | 40.75 | Outer shelf to inner shelf/shoreface | *Cruziana* and some *Skolithos* | Normal marine | Open marine to abraded brach valves | Prave et al., 1996: inferred |
| –78.75 | 40.25 | Basin, outer shelf | *Cruziana* | Dysaerobic | Open marine & anoxic | Prave et al., 1996: inferred; Dennison and Hasson, 1976: section |

| | | | | | | |
|---|---|---|---|---|---|---|
| −78.25 | 40.25 | Outer shelf to middle shelf | *Cruziana*; slight to intense | Normal marine | Open marine: brachs, coral, crinoids; bryo and mollusks rare | Prave et al., 1996: inferred; Dennison and Hasson, 1976: section |
| −77.75 | 40.25 | Outer shelf to inner shelf | *Cruziana*; slight to intense | Normal marine | Open marine: brachs, coral, crinoids; bryo and mollusks rare | Prave et al., 1996: inferred; Dennison and Hasson, 1976: section |
| −77.25 | 40.25 | Middle platform to prograding tidal delta | *Skolithos* | Normal marine | Thick shelled brachs, mainly abraded | Prave et al., 1996: section |
| −76.75 | 40.25 | Inner platform to prograding tidal delta | *Skolithos* | Normal marine | Thick shelled brachs, mainly abraded | Prave et al., 1996: section |
| −76.25 | 40.25 | Coastal plain | subaerially exposed | Subaerial | Subaerially exposed | Prave et al., 1996: inferred |
| −75.75 | 40.25 | Coastal plain | subaerially exposed | Subaerial | Subaerially exposed | Prave et al., 1996: inferred |
| −78.75 | 39.75 | Outer shelf to middle shelf | *Cruziana*; slight to intense | Normal marine | Open marine: brachs, coral, crinoids; bryo and mollusks rare | Prave et al., 1996: inferred; Dennison and Hasson, 1976: section |
| −78.25 | 39.75 | Middle shelf | *Cruziana*; slight to intense | Normal marine | Open marine: brachs, coral, crinoids; bryo and mollusks rare | Prave et al., 1996: section; Dennsion et al., 1979: section |
| −77.75 | 39.75 | Middle platform to prograding tidal delta | *Cruziana* (some) and *Skolithos* | Normal marine | Open marine to thick shelled brachs, mainly abraded | Prave et al., 1996: inferred |
| −79.75 | 39.25 | Deep shelf | ? | Anaerobic | Anoxic | Dennison and Hasson, 1976: inferred |
| −79.25 | 39.25 | Deep to outer shelf | *Cruziana;* slight to intense | Dysaerobic | Anoxic to open marine | Dennison and Hasson, 1976: inferred |
| −78.75 | 39.25 | Outer shelf to middle shelf | *Cruziana;* slight to intense | Normal marine | Open marine: brachs, coral, crinoids; bryo and mollusks rare | Dennison and Hasson, 1976: inferred |
| −78.25 | 39.25 | Outer shelf to middle shelf | *Cruziana;* slight to intense | | Open marine: brachs, coral, crinoids; bryo and mollusks rare | Dennison and Hasson, 1976: inferred |

Appendix 1.1. (*Continued*)

| Longitude | Latitude | Depositional environment | Ichnofacies/bioturbation | Oxygenation | Biofacies | Reference |
|---|---|---|---|---|---|---|
| –77.75 | 39.25 | | | | | |
| –79.75 | 38.75 | Deep shelf | ? | | Anoxic | Dennison and Hasson, 1976, inferred |
| –79.25 | 38.75 | Deep to outer shelf | *Cruziana*; slight to intense | Anaerobic | anoxic to open marine | Dennison and Hasson, 1976, inferred |
| –78.75 | 38.75 | Middle shelf | *Cruziana*; slight to intense | Dysaerobic | Open marine: brachs, coral, crinoids; bryo and mollusks rare | Dennison and Hasson, 1976, inferred |
| –78.25 | 38.75 | Outer shelf to middle shelf | *Cruziana*; slight to intense | Normal marine | Open marine: brachs, coral, crinoids; bryo and mollusks rare | Dennison and Hasson, 1976: inferred |
| –80.25 | 38.25 | Deep shelf | ? | Anaerobic | Anoxic | Dennison and Hasson, 1976: inferred |
| –79.75 | 38.25 | Deep to outer shelf | *Cruziana*; slight to intense | Dysaerobic | Anoxic to open marine | Dennison and Hasson, 1976: inferred; Hasson and Dennison, 1979, text |
| –79.25 | 38.25 | Deep outer shelf to inner shelf/shoreface | *Cruziana* and some *Skolithos* | Normal marine | Open marine to abraded brach valves | Dennison and Hasson, 1976: inferred |

| | | | | | | |
|---|---|---|---|---|---|---|
| –78.75 | 38.25 | | | | | |
| –80.75 | 37.75 | | | | | |
| –80.25 | 37.75 | Deep to outer shelf | *Cruziana;* slight to intense | Dysaerobic | Anoxic to open marine | Dennison and Hasson, 1976: inferred |
| –79.75 | 37.75 | Outer shelf to inner shelf/shoreface | *Cruziana* and some *Skolithos* | Normal marine | Open marine to abraded brach valves | Dennison and Hasson, 1976: inferred |
| –79.25 | 37.75 | | | | | |
| –81.25 | 37.25 | Deep to outer shelf | *Cruziana*; slight to intense | Dysaerobic | Anoxic to open marine | Dennison and Hasson, 1976: inferred |
| –80.75 | 37.25 | Outer shelf | *Cruziana*; bioturbation common | Normal marine | Open marine: brachs, coral, crinoids; bryo and mollusks rare | Dennison and Hasson, 1976: inferred |
| –80.25 | 37.25 | Outer shelf to middle shelf | *Cruziana*; slight to intense | Normal marine | Open marine: brachs, coral, crinoids; bryo and mollusks rare | Dennison and Hasson, 1976: inferred |
| –79.75 | 37.25 | | | | | |
| –81.75 | 36.75 | | | | | |

Appendix 1.2. Raw environmental base data for the *punctata* Zone.

| Longitude | Latitude | Grain size/rock type | % Mud | % Silt | % Sand | % ls | Sedimentary structures | Bedding style / thickness | Substrate type | Water depth |
|---|---|---|---|---|---|---|---|---|---|---|
| −78.75 | 42.75 | Cashaqua Sh: light to dark gray shale w/concretions | 90 | 0 | 0 | 10 | Concretions horizons | Thin | Muddy | Below wave base |
| −78.25 | 42.75 | Cashaqua Sh: olive gray mudst 80% w/concretions 20% | 80 | 0 | 0 | 20 | Concretions horizons | Thin | Muddy | Below wave base |
| −77.75 | 42.75 | Cashaqua Sh: gray-green sh, mdst, concretions | 70 | 5 | 5 | 20 | Concretions | Thin | Muddy | Below wave base to shallow basin |
| −77.25 | 42.75 | Cashaqua (1/2) Rock Stream Fm (1/2); blue-gray calc siltstone and shale | 50 | 33 | 10 | 7 | Rare shallow current ripples; isolated turbidite flows; bioturbated | Thin to moderate | Muddy to silty | Above or near SWB |
| −76.75 | 42.75 | Cashaqua/ Rock Stream Fm. Olive grey shale, 40% silt/sand, 60% mud | 50 | 29 | 26 | 5 | Rare shallow current ripples | Thin to moderate | Silty | Above or near SWB |
| −76.25 | 42.75 | Glen Aubrey | 69 | 8 | 22 | 0 | Rare shallow current ripples, HCS, scours | Thin to moderate | Muddy and sandy | Above or near FWWB |
| −75.75 | 42.75 | Glen Aubrey | 69 | 8 | 22 | 0 | Rare shallow current ripples, HCS, scours | Thin to moderate | Muddy and sandy | Above or near FWWB |
| −75.25 | 42.75 | Glen Aubrey/Walton | 35 | 15 | 50 | 0 | rare shallow current ripples | Thin to moderate | Muddy and sandy | Above or near FWWB |

| | | | | | | | | | | |
|---|---|---|---|---|---|---|---|---|---|---|
| −74.75 | 42.75 | Walton | 15 | 10 | 75 | 0 | Bar complexes, fluvial sedimentation | Thick | Sandy | Above FWWB to subaerial |
| −74.25 | 42.75 | Walton | 10 | 10 | 80 | 0 | Bar complexes, fluvial sedimentation | Thick | Sandy | Above FWWB to subaerial |
| −73.75 | 42.75 | Eroded | | | | | | | | |
| −80.25 | 42.25 | Cashaqua Sh. Gray shale with lst concretions | 90 | 0 | 0 | 10 | Concretions horizons | Thin | Muddy | Below SWB |
| −79.75 | 42.25 | Cashaqua Sh. Gray shale with lst concretions | 90 | 0 | 0 | 10 | Concretions horizons | Thin | Muddy | Below SWB |
| −79.25 | 42.25 | Cashaqua Sh. Gray shale with lst concretions | 90 | 0 | 0 | 10 | Concretions horizons | Thin | Muddy | Below SWB |
| −78.75 | 42.25 | Cashaqua Sh: light to dark gray shale w/concretions | 90 | 0 | 0 | 10 | Concretions horizons | Thin | Muddy | Below SWB |
| −78.25 | 42.25 | Cashaqua Sh: olive gray mudst 80% w/concretions 20% | 80 | 0 | 0 | 20 | Concretions horizons | Thin | Muddy | Below SWB |
| −77.75 | 42.25 | Cashaqua Sh: gray-green sh, mdst, concretions | 70 | 5 | 5 | 20 | Concretions | Thin | Muddy | Below wave to near SWB basin |
| −77.25 | 42.25 | Cashaqua (1/2) Rock Stream Fm (1/2); blue-gray calc siltst and shale | 40 | 45 | 12 | 3 | x-beds, isolated turbidites, bioturbated | Thin to moderate | Silty | Above or near SWB |
| −76.75 | 42.25 | Cashaqua/Rock Stream olive gray sh, 40% silt/s and, 60% mud | 40 | 45 | 20 | 5 | Rare shallow current ripples, x-beds, concretions | Thin to moderate | Silty to sandy | Above SWB to near FWWB |

Appendix 1.2. (*Continued*)

| Longitude | Latitude | Grain size/rock type | % Mud | % Silt | % Sand | % ls | Sedimentary structures | Bedding style / thickness | Substrate type | Water depth |
|---|---|---|---|---|---|---|---|---|---|---|
| –76.25 | 42.25 | Glen Aubrey mudst interbed w/silst thinner sh or sst | 85 | 0 | 15 | 0 | Groove casts, cross lamination, cuspate ripples | Thin to moderate | Muddy to silty | At or below FWWB, 30 to 120’ of water |
| –75.75 | 42.25 | Glen Aubrey, green shale, silst, sst | 75 | 20 | 10 | 0 | HCS, scours, wave ripples | moderate | Muddy to silty | At FWWB to intertidal |
| –75.25 | 42.25 | Glen Aubrey/Walton | 35 | 15 | 50 | 0 | Rare shallow current ripples, HCS, scours | Thin to moderate | Muddy and sandy | Above or near FWWB |
| –74.75 | 42.25 | Walton Fm. red shale and coarse sst | 20 | 10 | 70 | 0 | Tidal indicators, x-beds, shallow channels | Thick | Sandy | Intertidal to subaerial |
| –74.25 | 42.25 | Walton Fm. red shale and sst; red and gray (Onteora sst) | 10 | 10 | 80 | 0 | Fluvial system, x-beds, slickensides | Thick | Sandy | Subaerial |
| –73.75 | 42.25 | Walton Fm. red shale and sst | 5 | 5 | 90 | 0 | Fluvial system | Thick | Sandy | Subaerial |
| –80.25 | 41.75 | Cashaqua Sh. Gray shale with lst concretions | 90 | 0 | 0 | 10 | Concretions horizons | Thin | Muddy | Below SWB |
| –79.75 | 41.75 | Cashaqua Sh. Gray shale with lst concretions | 90 | 0 | 0 | 10 | Concretions horizons | Thin | Muddy | Below SWB |
| –79.25 | 41.75 | Cashaqua Sh. Gray shale with lst concretions | 90 | 0 | 0 | 10 | Concretions horizons | Thin | Muddy | Below SWB |
| –78.75 | 41.75 | Cashaqua Sh: light to dark gray shale w/concretions | 90 | 0 | 0 | 10 | Concretions horizons | Thin | Muddy | Below SWB |

| | | | | | | | | | | |
|---|---|---|---|---|---|---|---|---|---|---|
| −78.25 | 41.75 | Cashaqua Sh: olive gray mudst 80% w/concretions 20% | 80 | 0 | 0 | 20 | Concretions horizons | Thin | Muddy | Below SWB |
| −77.75 | 41.75 | Cashaqua Sh: gray-green sh, mdst, concretions | 70 | 5 | 5 | 20 | Concretions | Thin | Muddy | Below wave to near SWB basin |
| −77.25 | 41.75 | Rock Stream; sst, mudst | 70 | 4 | 26 | 0 | Current and wave ripples; moderate ripples | Moderate | Muddy and sandy | above SWB but below FWWB |
| −76.75 | 41.75 | Rock Stream; sst, mudst | 70 | 4 | 26 | 0 | Current and wave ripples; moderate ripples | Moderate | Muddy and sandy | Above SWB but below FWWB |
| −76.25 | 41.75 | Glen Aubrey mudst interbed w/silst thinner sh or sst | 69 | 8 | 22 | 0 | Rare shallow current ripples, HCS, scours | Thin to moderate | Muddy to sandy | At or below FWWB, 30 to 120' of water |
| −75.75 | 41.75 | Glen Aubrey mudst interbed w/silst thinner sh or sst | 69 | 8 | 22 | 0 | Rare shallow current ripples, HCS, scours | Thin to moderate | Muddy to sandy | At or below FWWB, 30 to 120' of water |
| −75.25 | 41.75 | Walton Fm. red shale and sst | 20 | 10 | 70 | 0 | Tidal indicators | Thick | Sandy | Intertidal to subaerial |
| −74.75 | 41.75 | Walton Fm. red shale and sst | 10 | 10 | 80 | 0 | Fluvial system | Thick | Sandy | Subaerial |
| −74.25 | 41.75 | Walton Fm. red shale and sst | 5 | 5 | 90 | 0 | Fluvial system | Thick | Sandy | Subaerial |
| −80.25 | 41.25 | No outcrop | | | | | | | | |
| −79.75 | 41.25 | No outcrop | | | | | | | | |
| −79.25 | 41.25 | No outcrop | | | | | | | | |
| −78.75 | 41.25 | No outcrop | | | | | | | | |
| −78.25 | 41.25 | No outcrop | | | | | | | | |
| −77.75 | 41.25 | No outcrop | | | | | | | | |
| −77.25 | 41.25 | No outcrop | | | | | | | | |
| −76.75 | 41.25 | No outcrop | | | | | | | | |

Appendix 1.2. (*Continued*)

| Longitude | Latitude | Grain size/rock type | % Mud | % Silt | % Sand | % ls | Sedimentary structures | Bedding style / thickness | Substrate type | Water depth |
|---|---|---|---|---|---|---|---|---|---|---|
| –76.25 | 41.25 | Subaerial redbeds | | | | | | | | |
| –75.75 | 41.25 | Subaerial redbeds | | | | | | | | |
| –75.25 | 41.25 | Subaerial redbeds | | | | | | | | |
| –74.75 | 41.25 | Eroded | | | | | | | | |
| –78.25 | 40.75 | Marine sh, siltst, sst | | | | | | | | |
| –77.75 | 40.75 | Marine sh, siltst, sst | | | | | | | | |
| –77.25 | 40.75 | Subaerial redbeds | | | | | | | | |
| –76.75 | 40.75 | Subaerial redbeds | | | | | | | | |
| –76.25 | 40.75 | Subaerial redbeds | | | | | | | | |
| –75.75 | 40.75 | Subaerial redbeds | | | | | | | | |
| –75.25 | 40.75 | Subaerial redbeds | | | | | | | | |
| –78.75 | 40.25 | Marine sh, siltst, sst | | | | | | | | |
| –78.25 | 40.25 | Marine sh, siltst, sst | | | | | | | | |
| –77.75 | 40.25 | Subaerial redbeds | | | | | | | | |
| –77.25 | 40.25 | Trimmers Rock Fm., siltstone to silty shale within sst beds and Redbeds | 50 | 35 | 15 | 0 | Graded beds, flute casts, ball and pillow | Thin to moderate | Muddy | Moderate |
| –76.75 | 40.25 | Subaerial redbeds | | | | | | | | |
| –76.25 | 40.25 | Eroded | | | | | | | | |
| –75.75 | 40.25 | Eroded | | | | | | | | |
| –78.75 | 39.75 | Bralier Fm., gray silst and silty shale | 20 | 75 | 5 | 0 | Flute casts | Thin | Silty | Below FWWB and near SWB |

| | | | | | | | | | | |
|---|---|---|---|---|---|---|---|---|---|---|
| −78.25 | 39.75 | Trimmers Rock Fm. equivalent, siltstone to silty shale within sst beds | 50 | 35 | 15 | 0 | | | | |
| −77.75 | 39.75 | Catskill, non-marine red beds | | | | | | | | |
| −79.75 | 39.25 | Chatanooga, black shale | 85 | 10 | 0 | 5 | | | | |
| −79.25 | 39.25 | "Portage" or Brallier, grey silty shale and siltst | 65 | 30 | 5 | 0 | Thickly laminated shales w/siltstone interbeds | Thin | Muddy to silty | Below SWB |
| −78.75 | 39.25 | Brallier Fm., gray silst and silty shale | 57 | 38 | 5 | 0 | | | | |
| −78.25 | 39.25 | "Chemung" or Trimmers Rock equivalent, silstone with shale and sst | 50 | 35 | 15 | 0 | | | | |
| −77.75 | 39.25 | Catskill, non-marine red beds | | | | | | | | |
| −79.75 | 38.75 | "Portage" or Brallier, grey silty shale and siltst | 65 | 30 | 5 | 0 | | | | |
| −79.25 | 38.75 | Brallier Fm., gray silst and silty shale | 57 | 38 | 5 | 0 | | | | |
| −78.75 | 38.75 | "Chemung" or Trimmers Rock equivalent, silstone with shale and sst | 50 | 35 | 15 | 0 | Poorly sorted siltstone, thickly aminated | Thin (90%) to moderate (10%) | Muddy to silty | Below SWB |
| −78.25 | 38.75 | Catskill, non-marine red beds | | | | | | | | |

Appendix 1.2. (*Continued*)

| Longitude | Latitude | Grain size/rock type | % Mud | % Silt | % Sand | % ls | Sedimentary structures | Bedding style / thickness | Substrate type | Water depth |
|---|---|---|---|---|---|---|---|---|---|---|
| –80.25 | 38.25 | "Portage" or Brallier, grey silty shale and siltst | 65 | 30 | 5 | 0 | | | | |
| –79.75 | 38.25 | Brallier Fm., gray silst and silty shale | 57 | 38 | 5 | 0 | | Siltstone | | |
| –79.25 | 38.25 | "Chemung" or Trimmers Rock equivalent, silstone with shale and sst | 50 | 35 | 15 | 0 | | | | |
| –78.75 | 38.25 | Catskill, non-marine red beds | | | | | | | | |
| –80.75 | 37.75 | Chatanooga, black shale | 85 | 10 | 0 | 5 | | | | |
| –80.25 | 37.75 | Dark shale basinal of Brallier | 75 | 22 | 0 | 3 | | | | |
| –79.75 | 37.75 | Brallier Fm., gray silst and silty shale | 57 | 38 | 5 | 0 | | Moderate | | |
| –79.25 | 37.75 | "Chemung" or Trimmers Rock | 50 | 35 | 15 | 0 | | | | |

| | | | | | | | | | | |
|---|---|---|---|---|---|---|---|---|---|---|
| | | equivalent, silstone with shale and sst | | | | | | | | |
| −81.25 | 37.25 | Dark shale basinal of Brallier | 75 | 22 | 0 | 3 | | | | |
| −80.75 | 37.25 | Brallier Fm., gray silst and silty shale | 57 | 38 | 5 | 0 | | Moderate | | |
| −80.25 | 37.25 | Brallier Fm., gray silst and silty shale | 20 | 40 | 40 | 0 | x-beds, lenticular-irregular beds, coursening upward, TXB, Bouma seq | Mod to thick | Sandy and silty | Below FWWB |
| −79.75 | 37.25 | Brallier Fm., gray silst and silty shale | 20 | 40 | 40 | 0 | | | | |
| −81.75 | 36.75 | Brallier Fm., gray silst and silty shale | 57 | 38 | 5 | 0 | | | | |

Appendix 1.2. Raw environmental base data for the *punctata* Zone.

| Longitude | Latitude | Depositional environment | Ichnofacies/ bioturbation | Oxygenation | Biofacies | Reference |
|---|---|---|---|---|---|---|
| –78.75 | 42.75 | Distal slope | ? | | Ammonites, conodonts | Over et al., 1999; Oliver and Klapper, 1981; Sutton et al., 1970; Sutton, 1963; Sutton and McGhee, 1985 |
| –78.25 | 42.75 | Distal slope | Some | | Naples fauna | Kirchgasser, 1983; Sutton et al., 1970; Sutton, 1960 |
| –77.75 | 42.75 | Dstal slope | High in middle, lower on top and bottom | | Middle-bivalves & plants, other bivalves and gastropods | Kirchgasser et al., 1994; Sutton et al., 1970; Sutton, 1960 |
| –77.25 | 42.75 | Outer shelf, distal platform | ? | Good | ***Rhipidomella***; bivalves, gast, ceph, brachs, arths, fish | Sutton and McGhee, 1985 |
| –76.75 | 42.75 | Outer shelf, distal platform | ? | Good | ***Rhipidomella***; bivalves, gast, ceph, brachs, arths, fish | Adams et al., 1956; Sutton et al., 1970; Sutton and McGhee, 1985; Sutton, 1960 |
| –76.25 | 42.75 | Inner shelf | ? | | ***Cypricardella*** | Sutton and McGhee, 1985; Sutton et al., 1970; Sutton, 1960 |
| –75.75 | 42.75 | Inner shelf | ? | | ***Cypricardella*** | Sutton and McGhee, 1985; Sutton et al., 1970; Sutton, 1960 |
| –75.25 | 42.75 | Inner shelf | ? | | ***Cypricardella*** | Sutton and McGhee, 1985; Sutton et al., 1970 |
| –74.75 | 42.75 | Alluvial plain | ? | Subaerial | | Sutton et al., 1970; Woodrow, 1985 |
| –74.25 | 42.75 | Alluvial plain | ? | Subaerial | | Sutton et al., 1970; Woodrow, 1985 |
| –73.75 | 42.75 | | | | | Sutton et al., 1970 |

| | | | | | | |
|---|---|---|---|---|---|---|
| −80.25 | 42.25 | Distal slope | | Mod | Ammonites, conodonts | Sutton et al., 1970 |
| −79.75 | 42.25 | Distal slope | | Mod | Molluscan | Tesmer, 1966; Sutton et al., 1970 |
| −79.25 | 42.25 | Distal slope | | Mod | Molluscan | Tesmer, 1966; Sutton et al., 1970 |
| −78.75 | 42.25 | Distal slope | ? | | Ammonites, conodonts | Tesmer, 1966; Sutton et al., 1970 |
| −78.25 | 42.25 | Distal slope | Some | | Naples fauna | Sutton et al., 1970 |
| −77.75 | 42.25 | Distal slope | High in middle, lower on top and bottom | | Middle-bivalves & plants, other bivalves and gastropods | Sutton et al., 1970 |
| −77.25 | 42.25 | Outer shelf, distal platform | | Good | ***Rhipidomella***; bivalves, ceph, gast; rare brachs, arths, fish | Sutton et al., 1970; Sutton, 1960 |
| −76.75 | 42.25 | Outer shelf | ? | Good | ***Rhipidomella***; bivalves, ceph, gast; rare brachs, arths, fish | Sutton and McGhee, 1985; Sutton et al., 1970; Sutton, 1960 |
| −76.25 | 42.25 | Prodelta, distal platform, and open shelf | Tracks, trails, burrows | Good | ***Cypricardella***; Productella, *Ambocoelia, Chonetes, Leptodesma,* | Bowen et al., 1970; Sutton et al., 1970; Sutton and McGhee, 1985; Bishuk et al., 1991 |
| −75.75 | 42.25 | Tidal flat/marsh, prodelta, delta platform | | | ***Cypricardella*** | Krajewski and Williams, 1971; Sutton et al., 1970; Sutton and McGhee, 1985 |
| −75.25 | 42.25 | Inner shelf to subaerial | ? | | ***Cyrpricardella*** | Krajewski and Williams, 1971; Sutton et al., 1970; Sutton and McGhee, 1985 |
| −74.75 | 42.25 | Tidal flat to alluvial plain | | Subaerial | Plant roots and stems | Fletcher, 1962; Woodrow, 1985; Krajewski and Williams, 1971; Sutton et al., 1970; Bridge and Dingman, 1981 |

Appendix 1.2. (*Continued*)

| Longitude | Latitude | Depositional environment | Ichnofacies/ bioturbation | Oxygenation | Biofacies | Reference |
|---|---|---|---|---|---|---|
| –74.25 | 42.25 | Alluvial plain | | Subaerial | | Fletcher, 1962; Woodrow, 1985; Krajewski and Williams, 1971; Sutton et al., 1970 |
| –73.75 | 42.25 | Alluvial plain | | Subaerial | | Inferred |
| –80.25 | 41.75 | Distal slope | | Mod | Ammonites, conodonts | Inferred |
| –79.75 | 41.75 | Distal slope | | Mod | Molluscan | Inferred |
| –79.25 | 41.75 | Distal slope | | Mod | Molluscan | Inferred |
| –78.75 | 41.75 | Distal slope | ? | | Ammonites, conodonts | Inferred |
| –78.25 | 41.75 | Distal slope | Some | | Naples fauna | Inferred |
| –77.75 | 41.75 | Distal slope | High in middle, lower on top and bottom | | Middle-bivalves & plants, other bivalves and gastropods | Inferred |
| –77.25 | 41.75 | Outer shelf | | | ***Rhipidomella*** | Krajewski and Williams, 1971; Sutton et al., 1970; Sutton and McGhee, 1985 |
| –76.75 | 41.75 | Outer shelf | | | ***Rhipidomella*** | Krajewski and Williams, 1971; Sutton et al., 1970; Sutton and McGhee, 1985 |
| –76.25 | 41.75 | Inner shelf | Tracks, trails, burrows | | ***Cypricardella*** | Krajewski and Williams, 1971; Sutton et al., 1970; Sutton and McGhee, 1985 |
| –75.75 | 41.75 | Inner shelf | Tracks, trails, burrows | | ***Cypricardella*** | Krajewski and Williams, 1971; Sutton et al., 1970; Sutton and McGhee, 1985 |

| | | | |
|---|---|---|---|
| −75.25 | 41.75 | Tidal flat | Krajewski and Williams, 1971; Sutton et al., 1970; Woodrow, 1985 |
| −74.75 | 41.75 | Alluvial plain | Krajewski and Williams, 1971; Sutton et al., 1970; Woodrow, 1985 |
| −74.25 | 41.75 | Alluvial plain | Krajewski and Williams, 1971; Sutton et al., 1970; Woodrow, 1985 |
| −80.25 | 41.25 | | |
| −79.75 | 41.25 | | |
| −79.25 | 41.25 | | |
| −78.75 | 41.25 | | |
| −78.25 | 41.25 | | |
| −77.75 | 41.25 | | |
| −77.25 | 41.25 | | |
| −76.75 | 41.25 | | |
| −76.25 | 41.25 | | |
| −75.75 | 41.25 | | |
| −75.25 | 41.25 | | |
| −74.75 | 41.25 | | |
| −78.25 | 40.75 | | |
| −77.75 | 40.75 | | |
| −77.25 | 40.75 | | |
| −76.75 | 40.75 | | |
| −76.25 | 40.75 | | |
| −75.75 | 40.75 | | |
| −75.25 | 40.75 | | |
| −78.75 | 40.25 | | |

Appendix 1.2. (*Continued*)

| Longitude | Latitude | Depositional environment | Ichnofacies/ bioturbation | Oxygenation | Biofacies | Reference |
|---|---|---|---|---|---|---|
| −78.25 | 40.25 | | | | | |
| −77.75 | 40.25 | | | | | |
| −77.25 | 40.25 | Shelf, turbidite flows | Low overall, higher in intervals | | Crinoids, brachs, bivalves, gast in siltst. | |
| −76.75 | 40.25 | | | | | |
| −76.25 | 40.25 | | | | | |
| −75.75 | 40.25 | | | | | |
| −78.75 | 39.75 | Turbidite basin | | Good | | Dennison et al., 1979 |
| −78.25 | 39.75 | | | | | |
| −77.75 | 39.75 | | | | | Dennison, 1985 |
| −79.75 | 39.25 | | | | | Dennison, 1985 |
| −79.25 | 39.25 | Turbidite basin | | | | Dennison, 1985; Dennison et al., 1979 |
| −78.75 | 39.25 | | | | | Dennison, 1985 |
| −78.25 | 39.25 | | | | | Dennison, 1985 |
| −77.75 | 39.25 | | | | | Dennison, 1985 |
| −79.75 | 38.75 | Turbidite basin | | | | Dennison, 1985; Dennison et al., 1979 |
| −79.25 | 38.75 | Turbidite basin | | | | Dennison, 1985; Dennison et al., 1979 |

| | | | | | |
|---|---|---|---|---|---|
| −78.75 | 38.75 | Turbidite basin | | | Dennison, 1985 |
| −78.25 | 38.75 | | | | Dennison, 1985 |
| −80.25 | 38.25 | | | | Dennison, 1985 |
| −79.75 | 38.25 | Turbidite basin | | | Dennison, 1985; Dennison et al., 1979 |
| −79.25 | 38.25 | | | | Dennison, 1985; Dennison et al., 1979 |
| −78.75 | 38.25 | | | | Dennison, 1985; Dennison et al., 1979 |
| −80.75 | 37.75 | | | | Dennison, 1985 |
| −80.25 | 37.75 | | | | Dennison, 1985 |
| −79.75 | 37.75 | | | | Dennison, 1985; Lundegard et al., 1985 |
| −79.25 | 37.75 | | | | Dennison, 1985 |
| −81.25 | 37.25 | | | | |
| −80.75 | 37.25 | | | | Lundegard et al., 1985 |
| −80.25 | 37.25 | Delta front, turbidite slopes | Low, vertical burrows | Good | Lundegard et al., 1985 |
| −79.75 | 37.25 | | | | |
| −81.75 | 36.75 | | | | Lundegard et al., 1985 |

Appendix 1.3. Raw environmental base data for the *linguiformis* Zone.

| Longitude | Latitude | Grain size/rock type | % Mud | % Silt | % Sand | % ls | Sedimentary structures | Bedding style/ thickness | Substrate type | Water depth |
|---|---|---|---|---|---|---|---|---|---|---|
| −78.75 | 42.75 | Hanover Fm; 10% calc., grey sh 75%, siltst 15%, few sst | 73 | 13 | 4 | 10 | Distal turbidites, concretions | Thin 90% to moderate 10% | Muddy to silty | Below SWB |
| −78.25 | 42.75 | Wiscoy Fm & Hanover; grey sh 30%, silst 60%, sst 5%, lst 5% | 30 | 60 | 5 | 5 | Turbidites (more proximal), calc. concretions | Thin 50% to moderate 50% | Mainly silty | Near to above SWB |
| −77.75 | 42.75 | Wiscoy Fm; calc silst 20%, fine sst 60%, gray shale 20%, thin lst 5% | 25 | 60 | 20 | 5 | No HCS | Silst = massive; moderate 80% to thick 20% | Silty to sandy | Above SWB to near or above FWWB |
| −77.25 | 42.75 | Wiscoy; grey to green, sst 52%; mudst 24%, sh 14%, siltst 4% | 24 | 28 | 52 | 0 | x-beds, current ripples | Moderate | Muddy and sandy | Above SWB to below FWWB |
| −76.75 | 42.75 | Wiscoy; sst 43%; mudst 18%, sh 51%, siltst3% | 69 | 3 | 43 | 0 | Current ripples | Moderate | Muddy and sandy | Above SWB to below FWWB |
| −76.25 | 42.75 | Manfield Sh and sst; 40% sh, 60% sst | 40 | 0 | 60 | 0 | x-bedding, parting lineati-on, oscillation ri-pples, load casts | Thin to thick | Muddy and sandy | Above FWWB to subtidal |
| −75.75 | 42.75 | Slide Mountain; Catskill Facies, sst, redbeds | 15 | 10 | 75 | 0 | Thick x-beds, shallow channels | Thick | Sandy | Subaerial to fluvial |

| | | | | | | | | | | |
|---|---|---|---|---|---|---|---|---|---|---|
| −75.25 | 42.75 | Slide Mountain; Catskill Facies, sst, redbeds | 10 | 10 | 80 | 0 | Thick x-beds, shallow channels | Thick | Sandy | Subaerial to fluvial |
| −74.75 | 42.75 | Slide Mountain; Catskill Facies, sst, redbeds | 5 | 5 | 90 | 0 | Thick x-beds, shallow channels | Thick | Sandy | Subaerial to fluvial |
| −74.25 | 42.75 | Slide Mountain; Catskill Facies, sst, redbeds | 3 | 2 | 95 | 0 | Thick x-beds, shallow channels | Thick | Sandy | Subaerial to fluvial |
| −73.75 | 42.75 | eroded | | | | | | | | |
| −80.25 | 42.25 | Hanover Sh; grey sh w/ dk grey sh and silst and lst | 85 | 7 | 0 | 8 | Calcareous nodules | Thin | Muddy | Below SWB |
| −79.75 | 42.25 | Hanover Sh; grey sh w/ dk grey sh and silst and lst | 85 | 7 | 0 | 8 | Calcareous nodules | Thin | Muddy | Below SWB |
| −79.25 | 42.25 | Hanover Sh; grey sh w/ dk grey sh and silst and lst | 85 | 7 | 0 | 8 | Calcareous nodules | Thin | Muddy | Below SWB |
| −78.75 | 42.25 | Hanover Sh; grey sh w/ dk grey sh and silst and lst | 73 | 13 | 4 | 10 | Distal turbidities, concretions | Thin (90%), moderate (10%) | Muddy | Below SWB |
| −78.25 | 42.25 | Hanover (type section); med gray to med green-gr calc sh and mudrock | 70 | 10 | 5 | 15 | lst concretions | Thin 95% to moderate 5% | Muddy | Below SWB to above SWB |
| −77.75 | 42.25 | Wiscoy and Hanover, shale, siltst, sandst | 40 | 35 | 25 | 0 | Flute casts, cuspidate ripples, turbidite deposits | Thin to moderate | Muddy to silty | Below SWB to near FWWB |

Appendix 1.3. (*Continued*)

| Longitude | Latitude | Grain size/rock type | % Mud | % Silt | % Sand | % ls | Sedimentary structures | Bedding style/ thickness | Substrate type | Water depth |
|---|---|---|---|---|---|---|---|---|---|---|
| −77.25 | 42.25 | Wiscoy; med-dk gray and dk green-gray silst and fine sst | 20 | 40 | 40 | 0 | HCS, bed rolls | Moderate | Silty to sandy | Above SWB to below FWWB |
| −76.75 | 42.25 | Wiscoy | 40 | 5 | 55 | 0 | Groove and flute casts | Moderate | Muddy to sandy | Above SWB to below FWWB |
| −76.25 | 42.25 | Manfield Sh and sst; 40% sh, 60% sst | 40 | 0 | 60 | 0 | x-bedding, parting lineat-ion, oscillation ri-pples, load casts | Thin to thick | Muddy and sandy | Above FWWB to subtidal |
| −75.75 | 42.25 | Slide Mountain; Catskill Facies, sst, redbeds | 15 | 10 | 75 | 0 | Thick x-beds, shallow channels | Thick | Sandy | Subaerial to fluvial |
| −75.25 | 42.25 | eroded | | | | | | | | |
| −74.75 | 42.25 | eroded | | | | | | | | |
| −74.25 | 42.25 | eroded | | | | | | | | |
| −73.75 | 42.25 | eroded | | | | | | | | |
| −80.25 | 41.75 | Hanover Sh; grey sh w/ dk grey sh and silst and lst | 85 | 7 | 0 | 8 | Calcareous nodules | Thin | Muddy | Below SWB |
| −79.75 | 41.75 | Hanover Sh; grey sh w/ dk grey sh and silst and lst | 85 | 7 | 0 | 8 | Calcareous nodules | Thin | Muddy | Below SWB |
| −79.25 | 41.75 | Hanover Sh; grey sh w/ dk grey sh and silst and lst | 85 | 7 | 0 | 8 | Calcareous nodules | Thin | Muddy | Below SWB |

| | | | | | | | | | | |
|---|---|---|---|---|---|---|---|---|---|---|
| −78.75 | 41.75 | Hanover Sh; grey sh w/ dk grey sh and silst and lst | 73 | 13 | 4 | 10 | Distal turbidites, concretions | Thin (90%), moderate (10%) | Muddy | Below SWB |
| −78.25 | 41.75 | Hanover-Wiscoy; sst 15%; mudst 18%; sh 59%, silst 8% | 77 | 8 | 15 | 0 | | Thin to moderate | Muddy | Above SWB and below FWWB |
| −77.75 | 41.75 | Hanover-Wiscoy; sst 15%; mudst 18%; sh 59%, silst 8% | 77 | 8 | 15 | 0 | | Thin to moderate | Muddy | Above SWB and below FWWB |
| −77.25 | 41.75 | Wiscoy; sst 52%; mudst 24%, sh 14%, siltst 4% | 38 | 4 | 52 | 0 | Current ripples | Moderate | Muddy and sandy | Above SWB to below FWWB |
| −76.75 | 41.75 | Wiscoy; sst 43%; mudst 18%, sh 51%, siltst3% | 69 | 3 | 43 | 0 | Current ripples | Moderate | Muddy and sandy | Above SWB to below FWWB |
| −76.25 | 41.75 | Manfield Sh and sst; 40% sh, 60% sst | 40 | 0 | 60 | 0 | x-bedding, parting lineation, oscillation ripp-les, load casts | Thin to thick | Muddy and sandy | Above FWWB to subaerial |
| −75.75 | 41.75 | Slide Mountain, red siltstone 40, sst 60% | 5 | 20 | 75 | 0 | Trough cross beds, ripples | Thick | Sandy | Above FWWB to subaerial |
| −75.25 | 41.75 | Eroded | | | | | | | | |
| −74.75 | 41.75 | Eroded | | | | | | | | |
| −74.25 | 41.75 | Eroded | | | | | | | | |
| −80.25 | 41.25 | Eroded | | | | | | | | |
| −79.75 | 41.25 | Eroded | | | | | | | | |
| −79.25 | 41.25 | Eroded | | | | | | | | |
| −78.75 | 41.25 | Eroded | | | | | | | | |
| −78.25 | 41.25 | Eroded | | | | | | | | |
| −77.75 | 41.25 | Eroded | | | | | | | | |
| −77.25 | 41.25 | | | | | | | | | |
| −76.75 | 41.25 | | | | | | | | | |

Appendix 1.3. (*Continued*)

| Longitude | Latitude | Grain size/rock type | % Mud | % Silt | % Sand | % ls | Sedimentary structures | Bedding style/ thickness | Substrate type | Water depth |
|---|---|---|---|---|---|---|---|---|---|---|
| −76.25 | 41.25 | Trimmers Rock Fm, silty sh, sst, siltst, gray, olive-green, brown | 45 | 40 | 13 | 2 | x-beds, laminated, load structures | Thin | Silty | Below SWB |
| −75.75 | 41.25 | | | | | | | | | |
| −75.25 | 41.25 | Trimmers Rock, type I, coarse sst to fine; lt to dk gray | 15 | 82 | 3 | 0 | Load casts, eroded bases | Thin to moderate | Silty | Below SWB |
| −74.75 | 41.25 | Trimmers Rock, type I, coarse sst to fine; mod to light gray | 8 | 90 | 2 | 0 | Load casts, few x-beds, planar, graded beds to massive | Thin to moderate | Silty | Below SWB |
| −78.25 | 40.75 | Trimmers Rock, Type I, dk. Brown, red, lt. Grey silty shale facies | 30 | 45 | 25 | 2 | x-beds, load casts, pillows, HCS | Thin | Silty to sandy | Within SWB |
| −77.75 | 40.75 | Trimmers Rock, silty sh and sst | 35 | 50 | 15 | 0 | x-beds, flaser bedding, wavy beds, HCS | Thin to thick | Silty | Within SWB |
| −77.25 | 40.75 | Trimmers Rock Fm, Type I (80%), type II (20%), grey, olive, red, sst and mudst | 39 | 51 | 10 | 0 | Cross beds, load casts, some is | Thin to moderate | Muddy and sandy | Below SWB |

| | | | | | | | | | | |
|---|---|---|---|---|---|---|---|---|---|---|
| −76.75 | 40.75 | Trimmers Rock, Type I and II, green-olive | 36 | 50 | 14 | 0 | x-beds, load casts, scour bases | Thin | Muddy to silty | Below SWB |
| −76.25 | 40.75 | Trimmers Rk/Catskill, Type I sst and sh, lt. Gray, olive, & lt. Brown | 15 | 79 | 5 | 1 | Load structures, laminated | Thin | Muddy to silty | Below SWB |
| −75.75 | 40.75 | Trimmers Rock, silty sh and sst | 10 | 82 | 8 | 0 | Fining upward | Thin | Silty | Below SWB |
| −75.25 | 40.75 | Trimmer Rock, silty and silty sh, few sst | 8 | 82 | 10 | 0 | Bouma sequences, load structures, etc. | Thin to moderate | Silty | Below SWB |
| −78.75 | 40.25 | Trimmers Rock, siltst to shale and sst | 25 | 20 | 55 | 0 | | Thin | Silty to sandy | Below SWB |
| −78.25 | 40.25 | Trimmers Rock, Type I, green/red/ lt.gray shale and sst | 25 | 20 | 55 | 0 | x-beds, loads | Thin | Sandy and muddy | Below SWB |
| −77.75 | 40.25 | | | | | | | | | |
| −77.25 | 40.25 | Trimmers Rock Fm., dk to lt olive, brn, red,siltst to silty sh, top= green shale | 60 | 30 | 10 | 0 | Load casts, graded beds, flute casts | Thin to moderate | Muddy | Between SWB and FWWB |
| −76.75 | 40.25 | Trimmer Rock Fm, sst to shale, red, olive, lt brown, gray | 70 | 20 | 1 | 0 | Load casts, snowballs | Thin to moderate | Muddy | Below SWB |
| −76.25 | 40.25 | Trimmers Rock Fm, shales to type I sst | 39 | 56 | 5 | 0 | x-beds, load casts, eroded base, laminated beds | Thin to moderate | Muddy to silty | Below SWB |

Appendix 1.3. (*Continued*)

| Longitude | Latitude | Grain size/rock type | % Mud | % Silt | % Sand | % ls | Sedimentary structures | Bedding style/ thickness | Substrate type | Water depth |
|---|---|---|---|---|---|---|---|---|---|---|
| −75.75 | 40.25 | | | | | | | | | |
| −78.75 | 39.75 | Foreknobs Fm: Pound sst./Trimmer Rock, siltst-sh | 80 | 18 | 2 | 0 | HCS | Thin | Muddy to silty | Within SWB |
| −78.25 | 39.75 | Trimmers Rock, lt gray, brown, green sh and sst, Types I and II | 80 | 10 | 10 | 0 | Load casts, x-beds | Thin | Muddy | Within SWB |
| −77.75 | 39.75 | | | | | | | | | |
| −79.75 | 39.25 | | | | | | | | | |
| −79.25 | 39.25 | Foreknobs Fm: Pound sst.; yellow-gray sst | 5 | 25 | 70 | 0 | Cross beds | Thick | Sandy | Above SWB to intertidal zone |
| −78.75 | 39.25 | Foreknobs Fm: Pound sst.; yellow-gray sst | | | | | Cross beds | Thick | Sandy | Above SWB to intertidal zone |
| −78.25 | 39.25 | Foreknobs Fm: Pound sst.; yellow-gray sst | | | | | Cross beds | Thick | Sandy | Above SWB to intertidal zone |
| −77.75 | 39.25 | exposed | | | | | | | | |
| −79.75 | 38.75 | Chemung | | | | | | | | |

| | | | | | | |
|---|---|---|---|---|---|---|
| −79.25 | 38.75 | Foreknobs Fm: Pound sst.; fine to med yellow-gray sst | Cross beds | Thick | Sandy | Above SWB to intertidal zone |
| −78.75 | 38.75 | exposed | | | | |
| −78.25 | 38.75 | exposed | | | | |
| −80.25 | 38.25 | Brallier | | | | |
| −79.75 | 38.25 | Foreknobs Fm: Pound sst.; yellow-gray sst | Cross beds | Thick | Sandy | Above SWB to intertidal zone |
| −79.25 | 38.25 | Foreknobs Fm: Pound sst.; yellow-gray sst | Cross beds | Thick | Sandy | Above SWB to intertidal zone |
| −78.75 | 38.25 | exposed | | | | |
| −80.75 | 37.75 | "Portage" gray shale | | | | |
| −80.25 | 37.75 | Brallier silst | | | | |
| −79.75 | 37.75 | Chemung | | | | |
| −79.25 | 37.75 | Catstill | | | | |
| −81.25 | 37.25 | "Portage" gray shale | | | | |
| −80.75 | 37.25 | "Portage" gray shale | | | | |
| −80.25 | 37.25 | Brallier | | | | |
| −79.75 | 37.25 | Chemung | | | | |

Appendix 1.3. Raw environmental base data for the *linguiformis* Zone.

| Longitude | Latitude | Depositional environment | Ichnofacies/ bioturbation | Oxygenation | Biofacies | Reference |
|---|---|---|---|---|---|---|
| −78.75 | 42.75 | Proximal basin | Gutter casts, escape burrows | | Gast, carb. plants, crinoids, pterioids, ammonites, cephalopods | Jacobi and Smith, 1999; Over, 1997; Sutton and McGhee, 1985; deWitt, 1960; Smith and Jacobi, 2000; Pepper and deWitt, 1950 |
| −78.25 | 42.75 | Shelf to lower shoreface | *Skolithos* to offshore | | *Ambocoelia, Cariniferella* | Jacobi and Smith, 1999; Over, 1997 Sutton and McGhee, 1985 |
| −77.75 | 42.75 | Lower shoreface to lagoon | *Skolithos; Arenicolites/ Teichichnus* | | Fossils scarce, *Ambocoelia, Cariniferella* | Jacobi and Smith, 1999; Over, 1997; Sutton and McGhee, 1985 |
| −77.25 | 42.75 | Middle shelf | | | *Tylothyris-Schizophoria* | McGhee and Sutton, 1981 |
| −76.75 | 42.75 | Inner shelf | | | *Cyrtospirifer-Douvillina* | McGhee and Sutton, 1981 |
| −76.25 | 42.75 | Nearshore marine, estuary, lagoon, distributary mouth bar | Tracks and trails | | *Cyrtospirifer-Douvillina* | inferred |
| −75.75 | 42.75 | Alluvial fans | Root traces | Subaerial | | Woodrow, 1985 |
| −75.25 | 42.75 | Alluvial fans | Root traces | Subaerial | | Woodrow, 1985 |
| −74.75 | 42.75 | Alluvial fans | Root traces | Subaerial | | Woodrow, 1985 |
| −74.25 | 42.75 | Alluvial fans | Root traces | Subaerial | | Woodrow, 1985 |
| −73.75 | 42.75 | | | | | |
| −80.25 | 42.25 | Proximal basin | Low | Low | Conodonts and ammonites | inferred |
| −79.75 | 42.25 | Proximal basin | Low | Low | Conodonts and ammonites | inferred |
| −79.25 | 42.25 | Proximal basin | Low | Low | Conodonts and ammonites | Jacobi and Smith, 1999; Metzger et al., 1974; Tesmer, 1974; Leighton, 2000; McGhee and Sutton, 1981 |

| | | | | | | |
|---|---|---|---|---|---|---|
| −78.75 | 42.25 | Shelf to lower shelf | *Skolithos* to offshore | | *Ambocoelia, Cariniferella* | Jacobi and Smith, 1999; Metzger et al., 1974; Tesmer, 1974; Leighton, 2000; McGhee and Sutton, 1981 |
| −78.25 | 42.25 | Shelf, slope, and basin | *Cruziana* | | Ammonites, *Ambocoelia, Cariniferella* | Smith and Jacobi, 2000; Jacobi and Smith, 1999; Tesmer, 1974; Leighton, 2000; McGhee and Sutton, 1981 |
| −77.75 | 42.25 | Distal slope, open shelf, inner platform to lower shoreface | *Cruziana/Skolithos* | | Brachs, gast, bivalves, *Ambocoelia, Cariniferella* | deWitt, 1960; McGhee and Sutton, 1981; Leighton, 2000; Jacobi and Smith, 1999; Metzger et al., 1974 |
| −77.25 | 42.25 | Nearshore to inner platform; open shelf, prodelta, inner & outer platform | *Zoophycus/Skolithos* | | Rugose corals, *Tylothyris-Schizophoria* | deWitt, 1960; McGhee and Sutton, 1981; Leighton, 2000; Jacobi and Smith, 1999; Metzger et al., 1974 |
| −76.75 | 42.25 | Middle shelf | | | *Cyrtospirifer-Douvillina, Tylothyris-Schizophoria* | Woodrow, 1985 |
| −76.25 | 42.25 | Nearshore marine, estuary, lagoon, distributary mouth bar | Tracks and trails | | *Cyrtospirifer-Douvillina* | |
| −75.75 | 42.25 | Alluvial fans | Root traces | Subaerial | | Krajecski and Williams, 1971; Woodrow, 1985 |

Appendix 1.3. (*Continued*)

| Longitude | Latitude | Depositional environment | Ichnofacies/ bioturbation | Oxygenation | Biofacies | Reference |
|---|---|---|---|---|---|---|
| –75.25 | 42.25 | | | | | Krajewski and Williams, 1971 |
| –74.75 | 42.25 | | | | | Krajewski and Williams, 1971 |
| –74.25 | 42.25 | | | | | Krajewski and Williams, 1971 |
| –73.75 | 42.25 | | | | | Krajewski and Williams, 1971 |
| –80.25 | 41.75 | Proximal basin | Low | Low | Conodonts and ammonites | inferred |
| –79.75 | 41.75 | Proximal basin | Low | Low | Conodonts and ammonites | inferred |
| –79.25 | 41.75 | Proximal basin | Low | Low | Conodonts and ammonites | inferred |
| –78.75 | 41.75 | Shelf to lower shelf | *Skolithos*to offshore | | *Ambocoelia, Cariniferella* | inferred |
| –78.25 | 41.75 | Outer shelf | *Cruziana* | | *Ambocoelia-Cariniferella* | McGhee and Sutton, 1981 |
| –77.75 | 41.75 | Outer shelf | *Cruziana* | | *Ambocoelia-Cariniferella* | McGhee and Sutton, 1981 |
| –77.25 | 41.75 | Middle shelf | | | *Tylothyris-Schizophoria* | McGhee and Sutton, 1981 |
| –76.75 | 41.75 | Inner shelf | | | *Cyrtospirifer-Douvillina* | McGhee and Sutton, 1981 |
| –76.25 | 41.75 | Inner shelf, nearshore marine, estuary, lagoon, distributary mouth bar | Tracks and trails | | *Cyrtospirifer-Douvillina* | McGhee and Sutton, 1981 |
| –75.75 | 41.75 | Beach, tidal flat | Some burrows in red siltst | | None | Krajewski and Williams, 1971; Woodrow, 1985 |
| –75.25 | 41.75 | | | | | Krajewski and Williams, 1971 |
| –74.75 | 41.75 | | | | | Krajewski and Williams, 1971 |
| –74.25 | 41.75 | | | | | Krajewski and Williams, 1971 |
| –80.25 | 41.25 | | | | | Krajewski and Williams, 1971 |

| | | | | | | |
|---|---|---|---|---|---|---|
| −79.75 | 41.25 | | | | | Krajewski and Williams, 1971 |
| −79.25 | 41.25 | | | | | |
| −78.75 | 41.25 | | | | | |
| −78.25 | 41.25 | | | | | |
| −77.75 | 41.25 | | | | | |
| −77.25 | 41.25 | | | | | |
| −76.75 | 41.25 | | | | | |
| −76.25 | 41.25 | Distal slope to proximal basin floor | Some | | Isolated | Frakes, 1964 |
| −75.75 | 41.25 | | | | | Schultz, 1974 |
| −75.25 | 41.25 | Shelf slope | Common | Good | Isolated fossils | Frakes, 1964; Schultz, 1974 |
| −74.75 | 41.25 | Turbidites stacked, shelf slope | Common | Good | Isolated fossils | Frakes, 1964; Schultz, 1974 |
| −78.25 | 40.75 | Shallow marine, prodelta, subtidal | | Good | | Frakes, 1964; Rahmanian, 1979; Williams and Slingerland, 1985 |
| −77.75 | 40.75 | Shallow marine, subtidal, shelf | Common | | | Rahmanian, 1979; Williams and Slingerland, 1985 |
| −77.25 | 40.75 | Shelf slope to basin floor | Common | | Sparse fossils | Frakes, 1964 |
| −76.75 | 40.75 | Shelf slope to basin floor | Common | | Sparse fossils | Frakes, 1964 |
| −76.25 | 40.75 | Basinal or distal slope | Common | | Sparse fossils | Frakes, 1964; Schultz, 1974 |
| −75.75 | 40.75 | Slope | Common | | In lenses | Frakes, 1964; Schultz, 1974 |
| −75.25 | 40.75 | Slope | | | | Frakes, 1964; Schultz, 1974 |
| −78.75 | 40.25 | Slope to basin | | | | Rahmanian, 1979 |
| −78.25 | 40.25 | Distal slope and basinal floor | | | | Frakes, 1964; Rahmanian, 1979 |

Appendix 1.3. (*Continued*)

| Longitude | Latitude | Depositional environment | Ichnofacies/ bioturbation | Oxygenation | Biofacies | Reference |
|---|---|---|---|---|---|---|
| −77.75 | 40.25 | | | | | |
| −77.25 | 40.25 | Shelf, turbidite flows | Overall low, higher in intervals | High | Crinoid columnals, brachs & bivalves, gastropods | Frakes, 1964 |
| −76.75 | 40.25 | Platform slope | Common | | Sparse fossils | Frakes, 1964 |
| −76.25 | 40.25 | Platform slope | Common | | Sparse fossils | Frakes, 1964 |
| −75.75 | 40.25 | | | | | |
| −78.75 | 39.75 | Shallow marine to beach, prodelta, slope | | | | Rahmanian, 1979, Dennison, 1979 |
| −78.25 | 39.75 | Shallow marine to distal platfrom | | | | Frakes, 1964; Rahmanian, 1979 |
| −77.75 | 39.75 | | | | | |
| −79.75 | 39.25 | | | | | |
| −79.25 | 39.25 | Nearshore bar sands | *Skolithos* | High | *Cyrtospirifer-Camarotoechia*; crinoids, brach, plant stems | McGhee, 1976; McGhee and Sutton, 1981; Dennison et al., 1979 |
| −78.75 | 39.25 | Nearshore bar sands | *Skolithos* | High | *Cyrtospirifer-Camarotoechia* | McGhee, 1976; McGhee and Sutton, 1981; Dennison, et al., 1979 |
| −78.25 | 39.25 | Nearshore bar sands | *Skolithos* | High | *Cyrtospirifer-Camarotoechia* | McGhee, 1976; McGhee and Sutton, 1981 |
| −77.75 | 39.25 | | | | | |
| −79.75 | 38.75 | | | | | Dennison, 1985 |

| | | | | | | |
|---|---|---|---|---|---|---|
| −79.25 | 38.75 | Nearshore bar sands | *Skolithos* | High | *Cyrtospirifer-Camarotoechia, Schizophoria, Atrypa* | McGhee, 1976; McGhee and Sutton, 1981; Dennison, et al., 1979; Dennison, 1985 |
| −78.75 | 38.75 | | | | | Dennison, 1985 |
| −78.25 | 38.75 | | | | | Dennison, 1985 |
| −80.25 | 38.25 | | | | | Dennison, 1985 |
| −79.75 | 38.25 | Nearshore bar sands | *Skolithos* | High | *Cyrtospirifer-Camarotoechia* | McGhee, 1976; McGhee and Sutton, 1981; Dennison, et al., 1979; Dennison, 1985 |
| −79.25 | 38.25 | Nearshore bar sands | *Skolithos* | High | *Cyrtospirifer-Camarotoechia* | McGhee, 1976; McGhee and Sutton, 1981; Dennison, 1985 |
| −78.75 | 38.25 | | | | | Dennison, 1985 |
| −80.75 | 37.75 | | | | | Dennison, 1985 |
| −80.25 | 37.75 | | | | | Dennison, 1985 |
| −79.75 | 37.75 | | | | | Dennison, 1985 |
| −79.25 | 37.75 | | | | | Dennison, 1985 |
| −81.25 | 37.25 | | | | | Dennison, 1985 |
| −80.75 | 37.25 | | | | | Dennison, 1985 |
| −80.25 | 37.25 | | | | | Dennison, 1985 |
| −79.75 | 37.25 | | | | | Dennison, 1985 |

Appendix 2. Species occurrence data used in the GARP modeling analysis.

Appendix 2.1 Occurrence data points of species extant during the Lower *varcus* Zone.

| Species | Longitude | Latitude |
|---|---|---|
| *Athyris cora* | –76.42 | 42.71 |
| *Athyris cora* | –76.11 | 42.80 |
| *Athyris cora* | –75.92 | 42.82 |
| *Athyris cora* | –75.53 | 42.82 |
| *Athyris cora* | –78.78 | 42.65 |
| *Athyris cora* | –75.91 | 42.88 |
| *Athyris spiriferiodes* | –77.30 | 42.88 |
| *Athyris spiriferiodes* | –77.90 | 42.77 |
| *Athyris spiriferiodes* | –76.53 | 42.55 |
| *Athyris spiriferiodes* | –77.90 | 42.77 |
| *Athyris spiriferiodes* | –77.44 | 42.23 |
| *Athyris spiriferiodes* | –78.88 | 42.75 |
| *Athyris spiriferiodes* | –78.67 | 42.77 |
| *Athyris spiriferiodes* | –78.37 | 42.88 |
| *Athyris spiriferiodes* | –78.52 | 42.85 |
| *Athyris spiriferiodes* | –78.23 | 42.90 |
| *Athyris spiriferiodes* | –76.58 | 42.97 |
| *Athyris spiriferiodes* | –78.20 | 42.15 |
| *Athyris spiriferiodes* | –74.77 | 42.60 |
| *Athyris spiriferiodes* | –75.53 | 42.82 |
| *Athyris spiriferiodes* | –77.30 | 42.88 |
| *Athyris spiriferiodes* | –76.90 | 40.34 |
| *Cariniferella carinata* | –77.30 | 42.88 |
| *Cariniferella carinata* | –78.37 | 42.88 |
| *Cariniferella carinata* | –74.86 | 42.38 |
| *Cariniferella carinata* | –75.88 | 42.77 |
| *Cariniferella carinata* | –76.53 | 42.55 |
| *Cariniferella carinata* | –77.90 | 42.77 |
| *Cariniferella carinata* | –77.28 | 42.83 |
| *Cariniferella carinata* | –76.27 | 42.17 |
| *Cariniferella carinata* | –75.74 | 42.71 |
| *Cariniferella carinata* | –78.88 | 42.75 |
| *Cariniferella carinata* | –83.73 | 42.77 |
| *Cariniferella carinata* | –78.77 | 42.88 |
| *Cariniferella carinata* | –78.35 | 42.90 |
| *Cariniferella carinata* | –78.23 | 42.90 |
| *Cariniferella carinata* | –78.10 | 42.92 |
| *Cariniferella carinata* | –78.28 | 42.92 |
| *Cariniferella carinata* | –76.55 | 42.53 |
| *Cariniferella carinata* | –75.18 | 42.62 |
| *Cariniferella carinata* | –78.83 | 42.70 |
| *Cariniferella carinata* | –75.53 | 42.72 |
| *Cariniferella carinata* | –77.30 | 42.88 |

Appendix 2.1 (*Continued*)

| Species | Longitude | Latitude |
|---|---|---|
| *Cypricardella bellistriata* | −75.18 | 42.62 |
| *Cypricardella bellistriata* | −78.67 | 42.77 |
| *Cypricardella bellistriata* | −76.74 | 42.95 |
| *Cypricardella bellistriata* | −76.95 | 40.47 |
| *Leptodesma (Leiopteria) spinerigum* | −75.25 | 42.81 |
| *Leptodesma (Leiopteria) spinerigum* | −78.73 | 39.67 |
| *Leptodesma (Leiopteria) spinerigum* | −75.92 | 42.82 |
| *Leptodesma (Leiopteria) spinerigum* | −75.52 | 42.53 |
| *Leptodesma (Leiopteria) spinerigum* | −75.62 | 42.82 |
| *Mucrospirifer mucronatus* | −78.10 | 42.92 |
| *Mucrospirifer mucronatus* | −74.86 | 42.38 |
| *Mucrospirifer mucronatus* | −77.90 | 42.77 |
| *Mucrospirifer mucronatus* | −77.28 | 42.83 |
| *Mucrospirifer mucronatus* | −78.77 | 42.88 |
| *Mucrospirifer mucronatus* | −78.23 | 42.90 |
| *Mucrospirifer mucronatus* | −78.10 | 42.92 |
| *Mucrospirifer mucronatus* | −78.10 | 42.92 |
| *Mucrospirifer mucronatus* | −78.97 | 42.72 |
| *Palaeoneilo constricta* | −78.76 | 39.61 |
| *Palaeoneilo constricta* | −78.54 | 39.52 |
| *Palaeoneilo constricta* | −78.73 | 39.67 |
| *Palaeoneilo constricta* | −75.50 | 42.64 |
| *Palaeoneilo constricta* | −75.57 | 42.69 |
| *Palaeoneilo constricta* | −77.02 | 42.76 |
| *Palaeoneilo constricta* | −78.78 | 42.65 |
| *Palaeoneilo constricta* | −74.31 | 42.67 |
| *Palaeoneilo constricta* | −76.42 | 42.71 |
| *Palaeoneilo constricta* | −75.62 | 42.82 |
| *Palaeoneilo constricta* | −76.74 | 42.95 |
| *Palaeoneilo constricta* | −78.44 | 40.39 |
| *Palaeoneilo constricta* | −76.95 | 40.47 |
| *Paracyclas lirata* | −74.77 | 42.60 |
| *Paracyclas lirata* | −78.09 | 39.68 |
| *Paracyclas lirata* | −74.86 | 42.38 |
| *Paracyclas lirata* | −75.18 | 42.48 |
| *Paracyclas lirata* | −75.09 | 42.53 |
| *Paracyclas lirata* | −74.86 | 42.38 |
| *Paracyclas lirata* | −75.53 | 42.82 |
| *Paracyclas lirata* | −78.04 | 39.86 |
| *Paracyclas lirata* | −76.95 | 40.47 |
| *Spinatrypa spinosa* | −74.77 | 42.60 |
| *Spinatrypa spinosa* | −77.77 | 42.73 |
| *Spinatrypa spinosa* | −78.67 | 42.77 |
| *Spinatrypa spinosa* | −76.53 | 42.01 |
| *Spinatrypa spinosa* | −75.09 | 42.53 |

Appendix 2.1 (*Continued*)

| Species | Longitude | Latitude |
|---|---|---|
| *Spinatrypa spinosa* | –76.42 | 42.71 |
| *Spinatrypa spinosa* | –76.11 | 42.80 |
| *Spinatrypa spinosa* | –76.05 | 42.83 |
| *Spinatrypa spinosa* | –76.53 | 42.55 |
| *Spinatrypa spinosa* | –76.42 | 42.71 |
| *Spinatrypa spinosa* | –78.97 | 42.72 |
| *Spinatrypa spinosa* | –77.90 | 42.77 |
| *Spinatrypa spinosa* | –77.90 | 42.77 |
| *Spinatrypa spinosa* | –78.81 | 42.77 |
| *Spinatrypa spinosa* | –75.43 | 42.80 |
| *Spinatrypa spinosa* | –78.18 | 43.00 |
| *Spinocyrtia granulosa* | –78.78 | 42.65 |
| *Spinocyrtia granulosa* | –78.97 | 42.72 |
| *Spinocyrtia granulosa* | –77.03 | 42.75 |
| *Spinocyrtia granulosa* | –78.67 | 42.77 |
| *Spinocyrtia granulosa* | –78.98 | 42.81 |
| *Spinocyrtia granulosa* | –78.37 | 42.88 |
| *Spinocyrtia granulosa* | –78.78 | 42.65 |

Appendix 2.2 Occurrence data points of species extant during the *punctata* Zone.

| Species | Longitude | Latitude |
|---|---|---|
| *Cupularostrum exima* | −76.72 | 42.03 |
| *Cupularostrum exima* | −76.72 | 42.03 |
| *Cupularostrum exima* | −76.42 | 42.22 |
| *Cupularostrum exima* | −76.42 | 42.22 |
| *Cupularostrum exima* | −75.87 | 42.33 |
| *Cupularostrum exima* | −76.30 | 42.38 |
| *Eoschizodus chemungensis* | −76.57 | 42.02 |
| *Eoschizodus chemungensis* | −77.98 | 42.17 |
| *Eoschizodus chemungensis* | −75.54 | 42.31 |
| *Eoschizodus chemungensis* | −75.54 | 42.31 |
| *Eoschizodus chemungensis* | −75.31 | 42.33 |
| *Goniophora chemungensis* | −76.73 | 42.02 |
| *Goniophora chemungensis* | −76.57 | 42.02 |
| *Goniophora chemungensis* | −76.72 | 42.03 |
| *Goniophora chemungensis* | −76.72 | 42.03 |
| *Goniophora chemungensis* | −76.72 | 42.03 |
| *Goniophora chemungensis* | −76.72 | 42.03 |
| *Goniophora chemungensis* | −76.72 | 42.03 |
| *Goniophora chemungensis* | −76.72 | 42.03 |
| *Goniophora chemungensis* | −76.72 | 42.03 |
| *Goniophora chemungensis* | −76.72 | 42.03 |
| *Goniophora chemungensis* | −75.87 | 42.33 |
| *Goniophora chemungensis* | −76.05 | 42.79 |
| *Grammysia elliptica* | −76.73 | 42.03 |
| *Grammysia elliptica* | −76.72 | 42.03 |
| *Grammysia elliptica* | −76.72 | 42.03 |
| *Grammysia elliptica* | −76.72 | 42.03 |
| *Grammysia elliptica* | −76.72 | 42.03 |
| *Grammysia elliptica* | −76.72 | 42.03 |
| *Grammysia elliptica* | −78.05 | 42.27 |
| *Grammysia elliptica* | −75.57 | 42.30 |
| *Grammysia elliptica* | −75.54 | 42.31 |
| *Grammysia elliptica* | −75.87 | 42.33 |
| *Grammysia elliptica* | −75.50 | 42.36 |
| *Leptodesma (Leioptera) nitida* | −76.57 | 42.02 |
| *Leptodesma (Leioptera) nitida* | −76.57 | 42.02 |
| *Leptodesma (Leioptera) nitida* | −76.72 | 42.03 |
| *Leptodesma (Leioptera) nitida* | −76.72 | 42.03 |
| *Leptodesma (Leioptera) nitida* | −76.72 | 42.03 |
| *Leptodesma (Leioptera) nitida* | −76.72 | 42.03 |
| *Leptodesma (Leioptera) nitida* | −76.72 | 42.03 |
| *Leptodesma (Leioptera) nitida* | −77.98 | 42.17 |
| *Leptodesma (Leioptera) nitida* | −76.42 | 42.22 |
| *Palaeoneilo constricta* | −76.57 | 42.02 |

Appendix 2.2 (*Continued*)

| Species | Longitude | Latitude |
|---|---|---|
| *Palaeoneilo constricta* | –75.87 | 42.33 |
| *Palaeoneilo constricta* | –76.40 | 42.37 |
| *Palaeoneilo constricta* | –76.03 | 42.44 |
| *Palaeoneilo constricta* | –76.37 | 42.59 |
| *Praewaagenoconcha speciosa* | –76.57 | 42.02 |
| *Praewaagenoconcha speciosa* | –76.72 | 42.03 |
| *Praewaagenoconcha speciosa* | –76.72 | 42.03 |
| *Praewaagenoconcha speciosa* | –78.03 | 42.22 |
| *Praewaagenoconcha speciosa* | –75.57 | 42.30 |
| *Praewaagenoconcha speciosa* | –75.87 | 42.33 |
| *Praewaagenoconcha speciosa* | –75.87 | 42.33 |
| *Praewaagenoconcha speciosa* | –75.87 | 42.33 |
| *Ptychopteria chemungensis* | –76.57 | 42.02 |
| *Ptychopteria chemungensis* | –76.57 | 42.02 |
| *Ptychopteria chemungensis* | –76.57 | 42.02 |
| *Ptychopteria chemungensis* | –76.64 | 42.03 |
| *Ptychopteria chemungensis* | –76.72 | 42.03 |
| *Ptychopteria chemungensis* | –75.87 | 42.33 |
| *Tylothyris mesacostalis* | –76.73 | 42.03 |
| *Tylothyris mesacostalis* | –76.72 | 42.03 |
| *Tylothyris mesacostalis* | –75.54 | 42.31 |
| *Tylothyris mesacostalis* | –75.87 | 42.33 |
| *Tylothyris mesacostalis* | –75.87 | 42.33 |
| *Tylothyris mesacostalis* | –76.36 | 42.35 |
| *Tylothyris mesacostalis* | –76.18 | 42.60 |

Appendix 2.3 Occurrence data points of species extant during the *linguiformis* Zone.

| Species | Longitude | Latitude |
|---|---|---|
| *Ambocoelia gregaria* | –77.50 | 41.88 |
| *Ambocoelia gregaria* | –78.19 | 41.95 |
| *Ambocoelia gregaria* | –76.92 | 42.07 |
| *Ambocoelia gregaria* | –75.82 | 42.08 |
| *Ambocoelia gregaria* | –75.93 | 42.17 |
| *Ambocoelia gregaria* | –79.11 | 42.22 |
| *Ambocoelia gregaria* | –76.61 | 42.22 |
| *Ambocoelia gregaria* | –77.79 | 42.25 |
| *Ambocoelia gregaria* | –79.10 | 42.29 |
| *Ambocoelia umbonata* | –79.16 | 41.85 |
| *Ambocoelia umbonata* | –76.27 | 42.17 |
| *Ambocoelia umbonata* | –77.67 | 42.33 |
| *Ambocoelia umbonata* | –77.55 | 42.27 |
| *Athyris angelica* | –80.35 | 41.73 |
| *Athyris angelica* | –77.08 | 41.81 |
| *Athyris angelica* | –79.16 | 41.85 |
| *Athyris angelica* | –78.07 | 42.22 |
| *Athyris angelica* | –78.28 | 42.22 |
| *Athyris angelica* | –77.77 | 42.27 |
| *Athyris angelica* | –77.55 | 42.27 |
| *Athyris angelica* | –78.18 | 42.30 |
| *Athyris angelica* | –78.13 | 42.33 |
| *Athyris angelica* | –77.67 | 42.33 |
| *Athyris angelica* | –78.11 | 42.34 |
| *Cariniferella carinata* | –77.13 | 41.91 |
| *Cariniferella carinata* | –76.62 | 42.01 |
| *Cariniferella carinata* | –76.73 | 42.02 |
| *Cariniferella carinata* | –76.82 | 42.05 |
| *Cariniferella carinata* | –76.81 | 42.09 |
| *Cariniferella carinata* | –77.09 | 42.16 |
| *Cariniferella carinata* | –76.50 | 42.25 |
| *Cariniferella carinata* | –76.36 | 42.35 |
| *Cariniferella tioga* | –76.71 | 41.72 |
| *Cariniferella tioga* | –76.53 | 42.01 |
| *Cariniferella tioga* | –76.53 | 42.01 |
| *Cariniferella tioga* | –76.73 | 42.02 |
| *Cariniferella tioga* | –77.04 | 42.15 |
| *Cariniferella tioga* | –76.92 | 42.22 |
| *Cariniferella tioga* | –76.61 | 42.22 |
| *Cariniferella tioga* | –76.67 | 42.23 |
| *Cariniferella tioga* | –76.50 | 42.25 |
| *Cariniferella tioga* | –76.48 | 42.35 |
| *Cupularostrum contracta* | –77.08 | 41.81 |
| *Cupularostrum contracta* | –77.08 | 41.81 |

Appendix 2.3 (*Continued*)

| Species | Longitude | Latitude |
|---|---|---|
| *Cupularostrum contracta* | –77.50 | 41.88 |
| *Cupularostrum contracta* | –76.82 | 42.05 |
| *Cupularostrum contracta* | –79.48 | 42.08 |
| *Cupularostrum contracta* | –79.48 | 42.08 |
| *Cupularostrum contracta* | –76.81 | 42.09 |
| *Cupularostrum contracta* | –78.40 | 42.23 |
| *Cupularostrum contracta* | –77.79 | 42.25 |
| *Cupularostrum contracta* | –79.57 | 42.31 |
| *Cupularostrum contracta* | –77.67 | 42.33 |
| *Cupularostrum exima* | –76.71 | 41.68 |
| *Cupularostrum exima* | –80.06 | 41.80 |
| *Cupularostrum exima* | –79.20 | 42.02 |
| *Cupularostrum exima* | –79.10 | 42.29 |
| *Cupularostrum exima* | –75.97 | 42.33 |
| *Cyrtospirifer chemungensis* | –76.52 | 41.92 |
| *Cyrtospirifer chemungensis* | –77.12 | 42.02 |
| *Cyrtospirifer chemungensis* | –77.12 | 42.02 |
| *Cyrtospirifer chemungensis* | –77.13 | 42.03 |
| *Cyrtospirifer chemungensis* | –76.42 | 42.07 |
| *Cyrtospirifer chemungensis* | –76.25 | 42.10 |
| *Douvillina cayuta* | –76.71 | 41.68 |
| *Douvillina cayuta* | –77.08 | 41.81 |
| *Douvillina cayuta* | –77.13 | 41.91 |
| *Douvillina cayuta* | –77.11 | 41.96 |
| *Douvillina cayuta* | –77.13 | 42.00 |
| *Douvillina cayuta* | –76.62 | 42.01 |
| *Douvillina cayuta* | –76.46 | 42.01 |
| *Douvillina cayuta* | –76.37 | 42.02 |
| *Douvillina cayuta* | –77.14 | 42.03 |
| *Douvillina cayuta* | –76.72 | 42.03 |
| *Douvillina cayuta* | –76.87 | 42.08 |
| *Douvillina cayuta* | –76.81 | 42.09 |
| *Douvillina cayuta* | –76.82 | 42.17 |
| *Douvillina cayuta* | –76.61 | 42.22 |
| *Douvillina cayuta* | –76.50 | 42.25 |
| *Floweria chemungensis* | –79.19 | 41.83 |
| *Floweria chemungensis* | –77.12 | 42.02 |
| *Floweria chemungensis* | –77.13 | 42.03 |
| *Floweria chemungensis* | –76.05 | 42.17 |
| *Floweria chemungensis* | –75.93 | 42.17 |
| *Floweria chemungensis* | –78.40 | 42.23 |
| *Floweria chemungensis* | –77.79 | 42.25 |
| *Floweria parva* | –80.33 | 42.05 |
| *Floweria parva* | –78.07 | 42.22 |
| *Floweria parva* | –78.07 | 42.22 |

Appendix 2.3 (*Continued*)

| Species | Longitude | Latitude |
|---|---|---|
| *Floweria parva* | –77.55 | 42.27 |
| *Floweria parva* | –78.18 | 42.30 |
| *Leptodesma (Leptodesma) spinerigum* | –79.41 | 39.41 |
| *Leptodesma (Leptodesma) spinerigum* | –78.84 | 39.63 |
| *Leptodesma (Leptodesma) spinerigum* | –78.93 | 39.66 |
| *Leptodesma (Leptodesma) spinerigum* | –76.71 | 41.67 |
| *Leptodesma (Leptodesma) spinerigum* | –77.08 | 41.81 |
| *Leptodesma (Leptodesma) spinerigum* | –77.08 | 41.81 |
| *Leptodesma (Leptodesma) spinerigum* | –77.13 | 42.00 |
| *Leptodesma (Leptodesma) spinerigum* | –76.37 | 42.02 |
| *Leptodesma (Leptodesma) spinerigum* | –79.20 | 42.02 |
| *Leptodesma (Leptodesma) spinerigum* | –77.14 | 42.03 |
| *Leptodesma (Leptodesma) spinerigum* | –76.72 | 42.03 |
| *Leptodesma (Leptodesma) spinerigum* | –80.33 | 42.05 |
| *Leptodesma (Leptodesma) spinerigum* | –75.82 | 42.08 |
| *Leptodesma (Leptodesma) spinerigum* | –77.09 | 42.16 |
| *Nervostrophia nervosa* | –76.71 | 41.67 |
| *Nervostrophia nervosa* | –76.46 | 42.01 |
| *Nervostrophia nervosa* | –76.73 | 42.02 |
| *Nervostrophia nervosa* | –76.57 | 42.02 |
| *Nervostrophia nervosa* | –76.78 | 42.03 |
| *Nervostrophia nervosa* | –76.45 | 42.04 |
| *Nervostrophia nervosa* | –76.87 | 42.08 |
| *Nervostrophia nervosa* | –76.55 | 42.20 |
| *Nervostrophia nervosa* | –76.53 | 42.32 |
| *Praewaagenoconcha speciosa* | –80.15 | 41.64 |
| *Praewaagenoconcha speciosa* | –77.08 | 41.81 |
| *Praewaagenoconcha speciosa* | –77.08 | 41.81 |
| *Praewaagenoconcha speciosa* | –77.50 | 41.88 |
| *Praewaagenoconcha speciosa* | –77.50 | 41.88 |
| *Praewaagenoconcha speciosa* | –77.13 | 41.91 |
| *Praewaagenoconcha speciosa* | –79.48 | 42.07 |
| *Praewaagenoconcha speciosa* | –76.05 | 42.17 |
| *Praewaagenoconcha speciosa* | –78.40 | 42.23 |
| *Praewaagenoconcha speciosa* | –77.98 | 42.25 |
| *Praewaagenoconcha speciosa* | –77.67 | 42.33 |
| *Praewaagenoconcha speciosa* | –76.48 | 42.35 |
| *Productella rectispina* | –80.35 | 41.73 |
| *Productella rectispina* | –80.06 | 41.80 |
| *Productella rectispina* | –77.08 | 41.81 |
| *Productella rectispina* | –77.08 | 41.81 |
| *Productella rectispina* | –79.16 | 41.85 |
| *Productella rectispina* | –76.58 | 42.02 |
| *Productella rectispina* | –78.18 | 42.30 |
| *Productella rectispina* | –78.18 | 42.30 |

Appendix 2.3 (*Continued*)

| Species | Longitude | Latitude |
|---|---|---|
| *Pseudatrypa devoniana* | –76.71 | 41.72 |
| *Pseudatrypa devoniana* | –80.35 | 41.73 |
| *Pseudatrypa devoniana* | –80.06 | 41.80 |
| *Pseudatrypa devoniana* | –77.08 | 41.81 |
| *Pseudatrypa devoniana* | –77.08 | 41.81 |
| *Pseudatrypa devoniana* | –76.57 | 42.03 |
| *Pseudatrypa devoniana* | –80.33 | 42.05 |
| *Pseudatrypa devoniana* | –78.07 | 42.22 |
| *Pseudatrypa devoniana* | –77.77 | 42.27 |
| *Ptychopteria chemungensis* | –76.52 | 41.96 |
| *Ptychopteria chemungensis* | –76.46 | 42.01 |
| *Ptychopteria chemungensis* | –76.53 | 42.01 |
| *Ptychopteria chemungensis* | –76.37 | 42.02 |
| *Ptychopteria chemungensis* | –76.37 | 42.02 |
| *Ptychopteria chemungensis* | –76.40 | 42.07 |
| *Ptychopteria chemungensis* | –76.31 | 42.16 |
| *Ptychopteria chemungensis* | –76.31 | 42.16 |
| *Ptychopteria chemungensis* | –75.53 | 42.23 |
| *Ptychopteria chemungensis* | –76.50 | 42.25 |
| *Ptychopteria chemungensis* | –76.50 | 42.25 |
| *Schizophoria impressa* | –80.35 | 41.73 |
| *Schizophoria impressa* | –77.08 | 41.81 |
| *Schizophoria impressa* | –77.08 | 41.81 |
| *Schizophoria impressa* | –77.14 | 42.03 |
| *Schizophoria impressa* | –77.13 | 42.03 |
| *Schizophoria impressa* | –76.42 | 42.07 |
| *Schizophoria impressa* | –76.49 | 42.21 |
| *Schizophoria impressa* | –78.40 | 42.23 |
| *Schizophoria impressa* | –77.55 | 42.27 |
| *Schizophoria impressa* | –79.72 | 42.28 |
| *Schizophoria impressa* | –78.18 | 42.30 |
| *Schizophoria impressa* | –75.77 | 42.33 |
| *Schizophoria impressa* | –78.11 | 42.34 |
| *Schizophoria impressa* | –76.50 | 42.43 |
| *Schizophoria impressa* | –76.50 | 42.44 |
| *Spinatrypa spinosa* | –77.08 | 41.81 |
| *Spinatrypa spinosa* | –77.50 | 41.88 |
| *Spinatrypa spinosa* | –76.57 | 42.02 |
| *Spinatrypa spinosa* | –77.13 | 42.05 |
| *Spinatrypa spinosa* | –76.10 | 42.17 |
| *Spinatrypa spinosa* | –78.07 | 42.22 |
| *Spinatrypa spinosa* | –76.50 | 42.25 |
| *Spinatrypa spinosa* | –77.67 | 42.33 |
| *Strophonella hybrida* | –77.08 | 41.81 |
| *Strophonella hybrida* | –78.07 | 42.22 |

Appendix 2.3 (*Continued*)

| Species | Longitude | Latitude |
|---|---|---|
| *Strophonella hybrida* | −77.55 | 42.27 |
| *Strophonella hybrida* | −78.02 | 42.28 |
| *Strophonella hybrida* | −78.18 | 42.30 |
| *Strophonella hybrida* | −79.16 | 41.85 |
| *Tylothyris mesacostalis* | −76.71 | 41.72 |
| *Tylothyris mesacostalis* | −77.30 | 41.75 |
| *Tylothyris mesacostalis* | −77.08 | 41.81 |
| *Tylothyris mesacostalis* | −77.10 | 41.84 |
| *Tylothyris mesacostalis* | −77.50 | 41.88 |
| *Tylothyris mesacostalis* | −77.50 | 41.88 |
| *Tylothyris mesacostalis* | −77.13 | 41.91 |
| *Tylothyris mesacostalis* | −78.19 | 41.95 |
| *Tylothyris mesacostalis* | −76.62 | 42.01 |
| *Tylothyris mesacostalis* | −76.46 | 42.01 |
| *Tylothyris mesacostalis* | −76.53 | 42.01 |
| *Tylothyris mesacostalis* | −76.33 | 42.02 |
| *Tylothyris mesacostalis* | −76.64 | 42.03 |
| *Tylothyris mesacostalis* | −76.61 | 42.22 |
| *Tylothyris mesacostalis* | −75.53 | 42.23 |
| *Tylothyris mesacostalis* | −78.40 | 42.23 |
| *Tylothyris mesacostalis* | −76.50 | 42.25 |
| *Tylothyris mesacostalis* | −77.79 | 42.25 |
| *Tylothyris mesacostalis* | −78.16 | 42.31 |
| *Tylothyris mesacostalis* | −76.48 | 42.35 |
| *Tylothyris mesacostalis* | −76.10 | 42.17 |

Understanding Late Devonian and Permian-Triassic Biotic and Climatic Events:
Towards an Integrated Approach
edited by D.J. Over, J.R. Morrow and P.B. Wignall



## *Chapter 7*

# *Subdivision of the terminal Frasnian* linguiformis *conodont Zone, revision of the correlative interval of Montagne Noire Zone 13, and discussion of stratigraphically significant associated trilobites*

*Catherine Girard*[a,*], *Gilbert Klapper*[b] *and Raimund Feist*[c]

[a]*Paléoenvironnements et Paléobiosphère, Université Lyon I, UMR 5125 CNRS, F-69622 Villeurbanne Cedex, France*
[b]*1010 Eastwood Road, Glencoe, Illinois 60022-1125, USA*
[c]*Institut des Sciences de l'Evolution, Université Montpellier II, UMR 5554 CNRS, F-34095 Montpellier Cedex 05, France*

**Abstract**

The terminal zone of the Frasnian standard conodont zonation, the *linguiformis* Zone has been subdivided into a lower part defined by the total range of *Palmatolepis linguiformis* and an upper part defined by the absence of the nominal species together with the rare occurrences of *Pa. ultima (= Pa. praetriangularis)* and *Ancyrognathus (= Ancyroides) ubiquitus.* This subdivision has proved to be effective in sections on different continental blocks. Furthermore, the two subdivisions correlate exactly with subdivisions formally recognized herein of Montagne Noire Zone 13 (13b and 13c). The lower boundary of the *linguiformis* Zone lies closely below the base of the Upper Kellwasser event bed, the top of which coincides with the end-Frasnian extinction. Nonetheless, prominent gaps in the geographic distribution of *Pa. linguiformis* have led to some difficulty in the identification of the zone. These gaps have been partially closed by discoveries of the species in areas where it was unknown until now, notably the central Moroccan Meseta and especially the Frasnian/Famennian boundary-stratotype section at Coumiac Upper Quarry in the Montagne Noire, southern France. Nevertheless, anomalous areas of absence, such as the Anti-Atlas Mts. of southern Morocco and western Canada still exist. The trilobite *Palpebralia brecciae* is widespread in Europe, North Africa, and Australia, providing evidence for correlation of the lower part of the *linguiformis* Zone in sections where the nominal conodont species has not been found. Analysis of variation in the relative abundance of the principal conodont genera leads to the conclusion that no substantial difference exists in the conodont associations among representative sections in four of the continental blocks.

*Keywords:* Conodont distribution; *linguiformis* Zone; Late Frasnian; trilobite association

## 1. Introduction

The latest Frasnian *linguiformis* Zone was first established in Frasnian/Famennian boundary sections of the Rhenish Slate Mountains, Germany, with *Palmatolepis linguiformis* Müller, 1956, as the defining species for the lower boundary of the zone

*Corresponding author. Fax: +33-472-448382. *E-mail address:* Catherine.Girard@univ-lyon1.fr (C. Girard).

(Sandberg et al., 1988; Ziegler and Sandberg, 1990). This stratigraphic level is closely below the base of the Upper Kellwasser bed (UKW). In contrast to areas situated on the Avalonian margin of Laurussia, occurrences of *Pa. linguiformis* in Gondwana-related terranes with well-developed UKW were mostly restricted until now to the southeastern Harz Mountains and Thuringia, Germany. *Pa. linguiformis* is also known from the Carnic Alps on the Italian–Austrian border, either from extremely condensed sections (Joachimski et al., 1994) or from sequences without development of UKW facies (Spalletta et al., 2001). However, the species was previously thought to be absent from sections of the Moroccan Meseta and especially from the Montagne Noire, southern France, where strata equivalent to the *linguiformis* Zone had been recognized through graphic correlation of the Frasnian/Famennian boundary-stratotype section at Coumiac (Becker et al., 1989). The latter method of correlation was supported by the occurrence of other key conodont species and a remarkably widespread and short-ranging trilobite species, *Palpebralia brecciae*. Discoveries reported herein of *Pa. linguiformis* in the key sections of Coumiac and Mrirt in the central Moroccan Meseta, in addition to new data from sites in Germany and Western Australia, allow us to re-evaluate both the *linguiformis* Zone in terms of its palaeogeographical distribution, and the efficacy of the recently proposed subdivision of the zone by Sandberg et al. (2002). Numerical patterns such as the increase in conodont biomass and diversity as well as shifts in biofacies help to identify intrazonal events such as the significance of the level and fauna just below the UKW. Evolutionary trends in associated trilobite populations provide evidence for correlation of the pre-UKW part of the *linguiformis* Zone in sites where the index species is absent or extremely rare.

## 2. The *linguiformis* Zone and its subdivision

The *linguiformis* Zone was introduced by Sandberg et al. (1988, p. 270) as a formal replacement for the Uppermost *gigas* Zone of Ziegler (1971, p. 267) and it is the highest zone of the Frasnian standard zonation revised by Ziegler and Sandberg (1990). In all three papers, the lower boundary of the zone is defined by the first occurrence of the nominal species, *Pa. linguiformis*, and that definition remains unchanged.

Sandberg et al. (2002, p. 474, Fig. 1) proposed a subdivision of the *linguiformis* Zone into a lower (or early) part distinguished by the first occurrence of the nominal species and its total range, and an upper part without that species. The earliest occurrence of *Palmatolepis ultima (= Pa. praetriangularis*, synonymy in Klapper et al., 2004) and the associated *Ancyrognathus (=Ancyroides) ubiquitus* characterize the upper part of the zone. This subdivision is similar to an earlier proposal of Schülke (1995, p. 25), wherein the upper part was characterized by the first occurrence of *Pa. ultima*, but his generalized diagram (Schülke, 1995, Fig. 16) shows *Pa. linguiformis* ranging to the top of the zone. However, Schülke's (1995, Table 1) data for Steinbruch Schmidt and later for Sessacker Trench VI (Schülke, 1998, Table 1; Schindler et al., 1998, Table 1), both in the Rhenish Slate Mountains, Germany, show that the nominal species does not range to the top of the zone. Many sections now show that *Pa. linguiformis* does not range into the upper part of the zone and thus we follow the revision of the zone given by Sandberg et al. (2002).

The reference section for the *linguiformis* Zone is Steinbruch Schmidt near Bad Wildungen in the Kellerwald, eastern Rhenish Slate Mountains (Ziegler and Sandberg, 1990, p. 21). There, the lower part of the *linguiformis* Zone occurs through 0.77 m of limestone (Sandberg et al., 1988, Fig. 6 and Table 1, beds 11–13) directly below the Upper Kellwasser dark limestone and shale beds (UKW). The lower part of the zone continues higher in beds 14 and 15 of the UKW. The highest limestone bed (16; Sandberg et al., 1988, Table 1) of the UKW is in the upper part of the *linguiformis* Zone, above the highest occurrence of *Pa. linguiformis*, but with *Pa. ultima* and *Ancyrognathus ubiquitus*. Ziegler and Sandberg (1990) documented a number of other occurrences of the zone. However, a small specimen identified as *Pa. linguiformis* from the Perdrix Formation in the Alberta Rockies, western Canada (Pollock, 1968, Pl. 61, Fig. 30) probably refers instead to *Pa. ederi* Ziegler and Sandberg (1990). This similar but earlier species occurs in Zones 10–11 of the Montagne Noire zonation, and ranges into the *linguiformis* Zone in Germany and Nevada according to Ziegler and Sandberg (1990, p. 63).

The *linguiformis* Zone is widely distributed geographically on the basis of the nominal species as indicated in our summary of regional occurrences (see Section 3). However, there were significant gaps in the distribution of *Pa. linguiformis* in areas where the species was not found until now. New discoveries reported herein, including the Montagne Noire and the central Moroccan Meseta, have modified this view. Nonetheless, the Anti-Atlas Mountains of southern Morocco, including the well-studied Tafilalt area, is still a region where this species is anomalously absent, as is also the case in western Canada (Alberta Rockies and subsurface; Northwest Territories). Furthermore, where *Pa. linguiformis* has recently been found in some areas such as the Neuville section in southern Belgium (Bultynck and Martin, 1995; Bultynck et al., 1998) and the Montagne Noire (herein), there are also nearby sections where it has not been found as yet.

Where the nominal species of the *linguiformis* Zone is absent, it has been possible to correlate the equivalents of the lower part of the zone using the trilobite *Palpebralia brecciae*. This was demonstrated by Feist and Schindler (1994) at sections in the Rhenish Slate and Harz Mountains, Germany, and at Coumiac in the Montagne Noire.

Little, if any, controversy exists regarding the taxonomic identity of *Pa. linguiformis*, in contrast to some other *Palmatolepis* species integral to the two Frasnian conodont zonations (Klapper and Becker, 1999, pp. 342). Excellent illustrations of characteristic Pa elements of the species are given by Sandberg et al. (1994, Pl. 2, Figs. 1–4). Ziegler and Sandberg (1990, p. 59–60) provide a synonymy list for the species through that year, and we include more recent references in the summary of regional occurrences.

## 3. Occurrences of *Palmatolepis linguiformis*

### *3.1. Geographic distribution of* Pa. linguiformis

In the following summary, occurrences of *Pa. linguiformis* in different continental blocks are given (Fig. 1). Associated trilobites are listed where present.

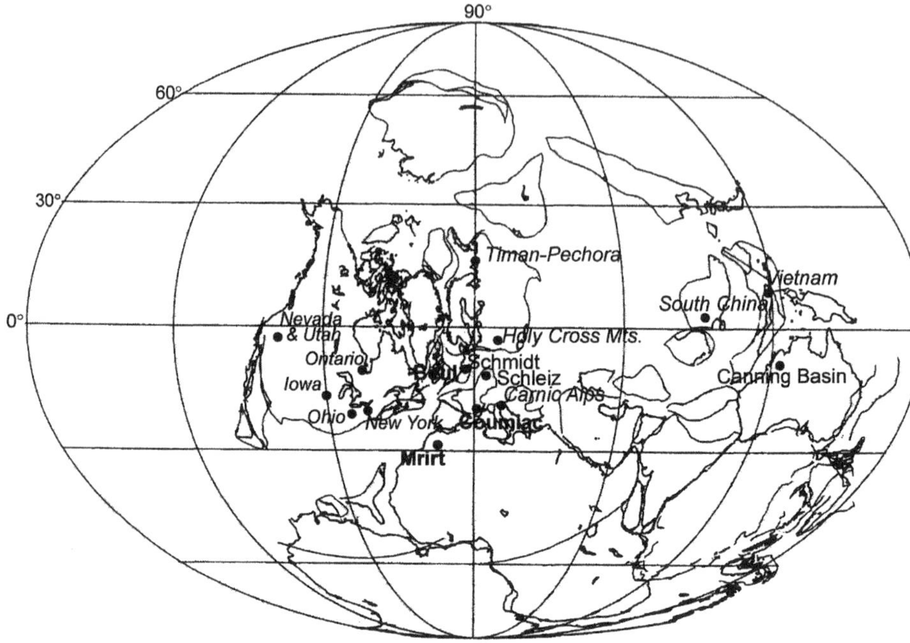


*Figure 1.* Palaeogeographic map (after Scotese and McKerrow, 1990) showing representative sites where *Palmatolepis linguiformis* was recovered. Localities in italics: occurrence of *Pa. linguiformis* without associated trilobites. In Roman: occurrence of *Pa. linguiformis* associated with *Palpebralia brecciae*. In boldface: new occurrences of *Pa. linguiformis* with *Palpebralia brecciae*.

### *3.1.1. Laurentia*

- Nevada, Devils Gate (Sandberg et al., 1988; Ziegler and Sandberg, 1990; Sandberg et al., 2003, Fig. 5); Northern Antelope Range (Johnson et al., 1980; Morrow, 2000; Sandberg et al., 2003); Badger Spring area, Adobe Range (Ziegler and Sandberg, 1990); Blue Basin quadrangle, Independence Mountains (Ziegler and Sandberg, 1990).

There are other Nevada localities where *Pa. linguiformis* has been recovered, even though it was not mentioned in the text of the original references. However, the identification of the zone was determined by the occurrence of the nominal species (C.A. Sandberg, 2004, pers. com.): March Spring, Toquima Range (Sandberg et al., 2003, p. 191); Ruby Crest North, South Ruby Mountains (Sandberg et al., 1989, Fig. 10), Indian Creek, Cherry Creek Range (Sandberg et al., 1989, Fig. 11), Tempiute Mountain, Timpahute Range (Morrow and Sandberg, 2003, Fig. 9); Northern Pancake Range, NPA and Black Point South sections (Morrow and Sandberg, 2003, Figs. 12 and 13), Red Hills, White Pine County (Sandberg et al., 1989, Fig. 10);

- Utah, Granite Mountain (Morrow, 2000; Morrow and Sandberg, 2003); (C.A. Sandberg, 2004, pers. com., same as for the Nevada localities); Coyote Knolls, Tule Valley (Sandberg et al., 1988; Sandberg et al., 1997); Big Jensen Pass, South Burbank Hills (Sandberg et al., 1989, Fig. 10);

- New York, upper Hanover Shale, Irish Gulf (Over, 1997, 2002);
- Ohio, upper Olentangy Shale, type section at Delaware (Over and Rhodes, 2000; Over, 2002); upper Olentangy at Lazarus Camp (Dzik, 2002); pyritic lag below Huron Shale, Schaeffer Farm (Over, 2002);
- Indiana, lower Blackiston Formation, southeast of Sellersberg (Clark and Becker, 1960; Ziegler, 1973) [= Selmier Member in present terminology]; top of Selmier Member, New Albany Shale, drill holes SDH 193 and INSC-3 (Sandberg et al., 1994); same stratigraphic level, Berry Materials Quarry, North Vernon (Over, 2002);
- Michigan, Antrim Shale, Lachine Member, Paxton Quarry, Alpena (Over, 2002);
- Wisconsin, "Kenwood" Shale, Milwaukee (Schumacher, 1971);
- Iowa, Sweetland Creek Shale, type section (Klapper and Furnish, 1963; Johnson and Klapper, 1992; Over, 2002); overlying Grassy Creek Shale at the same section (Over, 2002);
- Texas, Houy Formation, stratigraphic leak in Pillar Bluff area, Llano region (Seddon, 1970);
- Ontario, Moose River Basin, Onakawana B drill hole (Klapper et al., 2004).

*3.1.2. Baltica*

- Russia, Timan-Pechora Basin; Bagan well 3 (Klapper et al., 1996); Tebuk-Viss region, drill holes 881, 888 (Ovnatanova et al., 1999); Syv'yu River section (Savage and Yudina, 1999);
- Kostomłoty, Holy Cross Mts., Poland (Szulczewski, 1971);
- BK-70 drill hole, near Olkusz, Cracow area, southern Poland (Narkiewicz, 1978);
- Janczyce 1 drill hole, Holy Cross Mts. (Matyja and Narkiewicz, 1992);
- Kowala Quarry, Holy Cross Mts. (Racki et al., 2002; Dzik, 2002);
- Plucki Trench, Holy Cross Mts. (Racki et al., 2002; Dzik, 2002).

*3.1.3. Rhenohercynian (Avalonian) margin of Old Red Continent (Laurussia)*

- Steinbruch Schmidt, Braunau, near Bad Wildungen, Kellerwald, E Rhenish Slate Mts. (Müller, 1956, holotype of *Pa. linguiformis*; Ziegler, 1958, 1973; Sandberg et al., 1988; Schindler, 1990; Ziegler and Sandberg, 1990), associated trilobites: *Palpebralia brecciae, Harpes neogracilis, Otarion stigmatophthalmus* (Feist and Schindler, 1994);
- Sessacker Trench VI, near Oberscheld, Dill Syncline, S Rhenish Slate Mts. (Schülke, 1995, 1998; Schindler et al., 1998), associated trilobites: *Palpebralia brecciae, P. nodannulata, Acuticryphops acuticeps, Pteroparia aekensis* (Feist and Schindler, 1994);
- Beul, NE Rhenish Slate Mts. (new occurrence of *Pa. linguiformis*) (Pl. 1), associated trilobites: *Palpebralia brecciae* (Feist and Schindler, 1994);
- Martenberg, NE Rhenish Slate Mts. (Ziegler and Sandberg, 1990; Klapper and Becker, 1999);
- Burg Berg, NE Rhenish Slate Mts. (Ziegler and Sandberg, 1990);
- Borghausen, NE Rhenish Slate Mts. (Ziegler and Sandberg, 1990);
- Beringhausener Tunnel (BHT), NE Rhenish Slate Mts. (Ziegler, unpublished MS), new occurrence of associated trilobites: *Palpebralia brecciae, Acuticryphops acuticeps*;
- Benner Bicken, E Rhenish Slate Mts. (Müller, 1956; Ziegler and Sandberg, 1990);
- Weiherdamm iron mine, Oberscheld, S Rhenish Slate Mts. (Müller, 1956);

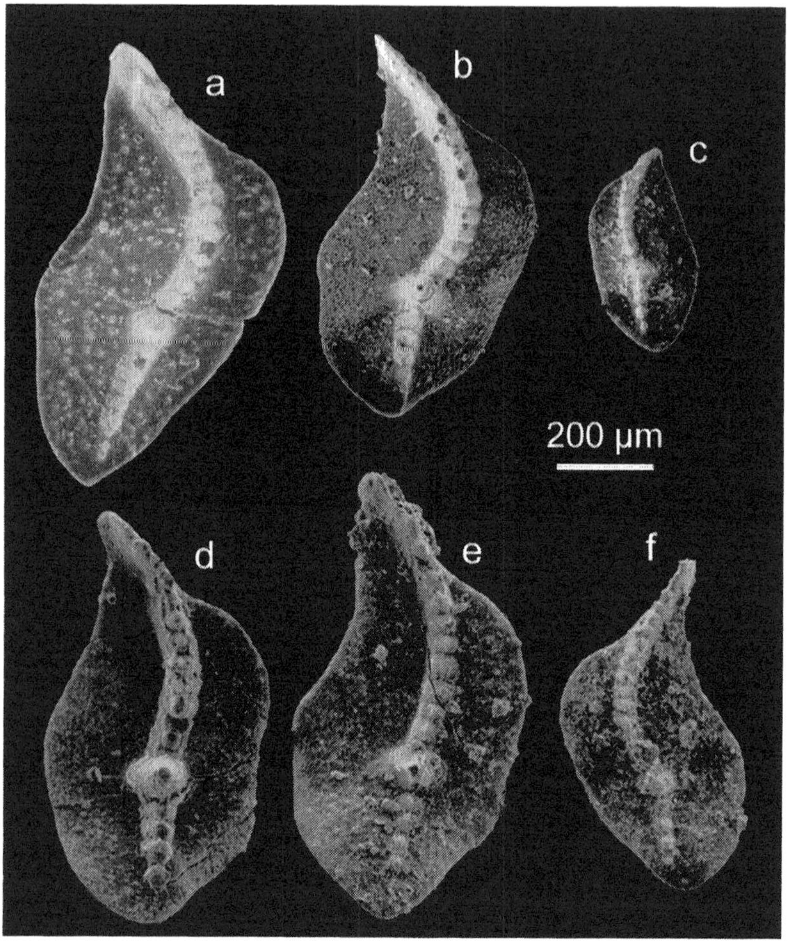


*Plate 1. Palmatolepis linguiformis* MÜLLER, 1956 (all are upper views of Pa elements): (a) specimen from Coumiac stratotype section, France (UL GFK01); (b) specimen from Wolayer Gletcher, Austria (UL GFK02); (c) small specimen from Beul, Germany (UL GFK03); (d–f) specimens from Mrirt, Morocco (UL GFK04, UL GFK05, UL GFK06).

- Aeke, Hühnertal, NW Harz Mts. (Schindler, 1990), associated trilobites: *Palpebralia brecciae, P.* cf. *nodannulata, Harpes neogracilis, Acuticryphops acuticeps, Pteroparia aekensis* (Feist and Schindler, 1994);
- Neuville, Dinant synclinorium, Belgium (Bultynck and Martin, 1995; Bultynck et al., 1998).

### *3.1.4. North Gondwana margin*

- Antoinettenweg, SE Harz Mts. (Hüneke, 1995);
- Schleiz, Thuringia, associated trilobites: *Palpebralia* cf. *brecciae, Harpes neogracilis, Acuticryphops acuticeps* (Weyer et al., 2003);

- Kahlleite, Thuringia (Bartzsch et al., 2001), new data on associated trilobites: *Palpebralia brecciae, Harpes neogracilis, Acuticryphops acuticeps, Bradocryphaeus feisti*;
- Wolayer Gletscher, Carnic Alps, Austria (Joachimski et al., 1994), associated trilobites: *Palpebralia* sp.;
- Pramosio A section, Carnic Alps, Italy (Spalletta and Perri, 1998; Spalletta et al., 2001);
- Coumiac, Upper Quarry, Montagne Noire (new occurrence of *Pa. linguiformis*) (Pl. 1), associated trilobites: *Palpebralia brecciae, P. palpebralis, Harpes neogracilis, Acuticryphops acuticeps, Bradocryphaeus feisti* (Feist and Schindler, 1994);
- Mrirt, central Moroccan Meseta (new occurrence of *Pa. linguiformis*) (Pl. 1), associated trilobites: *Palpebralia brecciae, Harpes neogracilis, Acuticryphops acuticeps, Otarion stigmatophthalmus, Gondwanaspis mrirtensis* (Feist, 2002).

#### *3.1.5. Northeast Gondwana margin*

- Xom Nha, central Vietnam (Ta Hoà Phuong, 1998);
- Maanshan section, Xiangzhou County, Guangxi, South China (Han, 1987; Jia et al., 1989, same illustration as preceding; Ji, 1989a);
- Yangti section near Guilin, Yangsuo County, Guangxi, South China (Ji, 1989b);
- Longmen and Dongcun sections, Guilin area, Yangsuo County, Guangxi, South China (Wang, 1994; Wang and Ziegler, 2002);
- Mostyn Vale Formation, type section, west of Keepit Dam, New South Wales, Australia (Wright et al., 1990);
- Calix Corner and Phacopid Gully (PG), Canning Basin, northern part of Western Australia (new occurrences of *Pa. linguiformis*), associated trilobites: *Palpebralia brecciae, Harpes neogracilis, Gondwanaspis* sp., *Acuticryphops* sp., *Otarion* sp., styginids;
- Horse Spring, Canning Basin (Becker et al., 1991; Klapper et al., 1995; Klapper, in prep.), associated trilobites: *Palpebralia brecciae*, *Palpebralia* n.sp.

### *3.2. New occurrences of* **Palmatolepis linguiformis**

#### *3.2.1. Montagne Noire (southern France)*

In the Coumiac Upper Quarry, latest Frasnian conodonts including *Pa. linguiformis* were recovered from Bed UQ31f, directly below the UKW. This bed is rich in conodonts (Table 1) and was previously regarded as in the upper part of the Montagne Noire Zone 13 owing to the presence of *Pa. boogaardi* (Klapper and Foster, 1993, Figs. 2, 14.7). The discovery of *Pa. linguiformis* allows us to identify the lower part of the *linguiformis* Zone precisely in this, the Frasnian/Famennian boundary-stratotype section. As elsewhere, the occurrence of *Pa. linguiformis* in Bed UQ31f is directly below the base of the hypoxic UKW (Bed UQ31g). Bed 31g correlates with the upper part of the *linguiformis* Zone because it has rare *Pa. ultima (= Pa. praetriangularis)* together with *Ancyrognathus ubiquitus* (cited in Becker et al., 1989, Fig. 3; Klapper et al., 1994, p. 435), and it lacks *Pa. linguiformis* despite repeated sampling by all three of us.

*Table 1.* Numerical analysis of the conodont communities from some of the different areas studied here. Raw numbers of conodonts found are given as well as real abundance per sample and as a percentage of abundance. Abbreviations: Po = *Polygnathus*, Pa = *Palmatolepis*, An = *Ancyro-group*, Ic = *Icriodus*. Wt = weight of dissolved rock-sample in grams. The label "Pa" corresponds to the total of *Palmatolepis* without the number of *Pa. linguiformis* specimens. The total of *Pa. linguiformis* recovered is indicated as "ling".

| | Section | Wt | Po | Pa | ling | An | Ic | Total | %Po | %Pa | %Paling | %An | %Ic |
|---|---|---|---|---|---|---|---|---|---|---|---|---|---|
| Morocco | Mrirt | 1350 | 5400 | 31,020 | 3 | 1269 | 270 | 39,039 | 14.22 | 81.71 | 0.01 | 3.34 | 0.71 |
| France | Coumiac | 2000 | 776 | 3327 | 1 | 120 | 256 | 6223 | 17.32 | 74.26 | 0.02 | 2.68 | 5.71 |
| Germany | Beul | 25 | 106 | 264 | 1 | 6 | 6 | 383 | 27.68 | 68.93 | 0.26 | 1.57 | 1.57 |
| | Beringhausener | 170 | 710 | 2586 | 1 | 46 | 0 | 3343 | 21.24 | 77.36 | 0.03 | 1.38 | 0.00 |
| Austria | Wolayer Glet. | 90 | 527 | 1057 | 39 | 8 | 27 | 1658 | 31.79 | 63.75 | 2.35 | 0.48 | 1.63 |
| Thuringia | Schleiz | | 280 | 910 | 10 | 106 | 60 | 1366 | 20.50 | 66.62 | 0.73 | 7.76 | 4.39 |
| Vietnam | Xom Nha | 360 | 83 | 165 | 24 | 0 | 0 | 272 | 30.51 | 60.66 | 8.82 | 0.00 | 0.00 |
| Australia | Calix Corner | 300 | 32 | 40 | 2 | 1 | 10 | 85 | 37.65 | 47.06 | 2.35 | 1.18 | 11.76 |
| | Phacopid Gully | 200 | 39 | 65 | 3 | 8 | 11 | 126 | 30.95 | 51.59 | 2.38 | 6.35 | 8.73 |
| | Horse Spring | 640 | 30 | 47 | 8 | 52 | 3 | 140 | 21.43 | 33.57 | 5.71 | 37.14 | 2.14 |

### *3.2.2. Central Moroccan Meseta*

A closely comparable situation to the Coumiac Upper Quarry occurs in the Mrirt section of Morocco, where *Pa. linguiformis* has been discovered by one of us (C.G.) in the bed directly below the UKW. This bed is extremely rich in conodonts (Table 1) and trilobites (Feist, 2002). Previously, it had been assigned by Lazreq (1992, 1999, Fig. 11, bed 43) to the Upper *rhenana* Zone (sensu Ziegler and Sandberg, 1990) based on the presence of *Pa. bogartensis (= Pa. rotunda)*, whereas the superjacent UKW with *Ancyrognathus ubiquitus* was regarded by Lazreq to represent the *linguiformis* Zone. Now we regard the bed directly below the UKW as in the lower part of the *linguiformis* Zone and the UKW as in the upper part of that zone.

The discovery of *Pa. linguiformis* both in the boundary-stratotype section at Coumiac and at Mrirt is important, as these occurrences are from the North-Gondwanan margin adjacent to cratonic Gondwana. Cratonic Gondwana, however, does not yet reveal any *Pa. linguiformis* in regions such the Tafilalt area of southern Morocco, where the highest Frasnian succession has been intensively studied (Walliser et al., 1989; Belka and Wendt, 1992; as well as many unpublished conodont collections made by Z. Belka and by M. R. House and R. T. Becker). In this region, the Late Frasnian succession is generally represented by oxygen-depleted, carbon-rich deposits of Kellwasser-type (Belka and Wendt, 1992), and trilobites have not been recovered from this hostile environment.

### *3.2.3. NE Rhenish Slate Mountains (Germany)*

Among numerous sections in the Sauerland that display the Frasnian–Famennian transition including a well-developed UKW, the artificial trench at the top of Beul hill (Feist and Schindler, 1994), and the BHT section (Ziegler, unpublished MS) are of special interest as *Pa. linguiformis* and *Palpebralia brecciae* were found together in the bed directly below the UKW.

## 4. Revision of Zone 13 of the Montagne Noire zonation

A 13-fold Frasnian zonation was proposed for the conodont sequence in the Montagne Noire, southern France (Klapper, 1989). Originally it was thought of as a local zonation applicable to about 15 Frasnian sections in the southeastern part of the Montagne Noire. However, the Montagne Noire zonation has since been widely replicated in other Devonian tropical–subtropical areas, including for example western New York State (Kirchgasser, 1994; Kralick, 1994), Sweetland Creek in Iowa (Johnson and Klapper, 1992), the Alberta Rockies and central Alberta subsurface in western Canada (McLean and Klapper, 1998), the Timan-Pechora region of the Russian Platform (Klapper et al., 1996), and the Canning Basin of Western Australia (Klapper, in prep.). The correlation of the Montagne Noire zonation with the widely used Frasnian standard zonation of Ziegler and Sandberg (1990) has been demonstrated through re-study of the Martenberg reference section in the Rhenish Slate Mountains (Klapper and Becker, 1999).

Zone 13 is the highest Frasnian zone in the Montagne Noire zonation. It first became clear in the study of Canning Basin sequences in the 1990s that three subdivisions of Zone 13 can be recognized (here named 13a, 13b, and 13c), at Horse Spring and other Canning Basin sections (Klapper, in prep.). Zone 13a contains *Pa. bogartensis*, *Pa. boogaardi*, *Pa. rhenana*, *Pa. eureka*,, and *Pa. juntianensis*. Zone 13b is characterized especially by *Pa. linguiformis*, as well as by the highest occurrences of *Pa. boogaardi*, *Pa. rhenana*, and *Pa. juntianensis*. Zone 13b is succeeded by the uppermost part of Zone 13 (13c), characterized by a low-diversity fauna consisting of *Pa. bogartensis* to the exclusion of almost all other species of *Palmatolepis* including *Pa. linguiformis*, and the presence of a few long-ranging Frasnian species of *Ancyrodella* and *Polygnathus*. Species of *Icriodus* are rare in Zones 13b and 13c at Horse Spring as well as some other sections in the Canning Basin conodont succession.

The three-fold subdivision of Zone 13 was described for the first time in the Onakawana B core in northern Ontario, Canada (Klapper et al., 2004, p. 375). The lower boundary of Zone 13b is defined by the lowest occurrence of *Pa. linguiformis*, as is the lower boundary of the *linguiformis* Zone in the Frasnian standard zonation of Ziegler and Sandberg (1990, p. 21). The succession of Zone 13b with *Pa. linguiformis*, overlain by the relatively low-diversity fauna of Zone 13c without *Pa. linguiformis*, is correlative with the two subdivisions of the *linguiformis* Zone proposed by Sandberg et al. (2002). Zone 13b occurs at the Coumiac Upper Quarry in Bed UQ31f, which has *Pa. linguiformis*; Zone 13c is in the UKW, UQ31g, which contains rare specimens of both *Ancyrognathus ubiquitus* and *Pa. ultima*. At Steinbruch Schmidt Zone 13b equates with beds 11–15 and Zone 13c with bed 16 (Sandberg et al., 1988, Table 1). The same subdivision is recognizable at Devils Gate, Nevada (Sandberg et al., 1988, Table 3) and at Kowala in the Holy Cross Mountains, Poland (Racki et al., 2002, Appendix A), among other sections.

Prior to the discovery of *Pa. linguiformis* in the Coumiac Upper Quarry Bed UQ31f, a correlation with the *linguiformis* Zone was recognized from other evidence. Graphic correlation of the Upper Coumiac section with a Frasnian composite standard (Klapper et al., 1995; Klapper, 1997) indicated that Beds 31f and 31g correlate with the *linguiformis* Zone (Becker et al., 1989, p. 262). Secondly, the occurrence of *Palpebralia brecciae* in Beds 31e and 31f (Feist and Schindler, 1994, Figs. 5 and 6) provided supporting evidence for correlation with the lower part of the *linguiformis* Zone, before the nominal species was discovered in Bed 31f.

## 5. Frasnian/Famennian boundary

The Frasnian/Famennian (F/F) stage boundary (GSSP) at the stratotype section, Coumiac Upper Quarry, was defined to coincide with the lower boundary of the Lower *triangularis* Zone, at the base of Bed UQ32a (Klapper et al., 1994; House et al., 2000). Bed 32a contains a relatively low-diversity conodont fauna, with the *Palmatolepis* component dominated by *Pa. ultima (= Pa. praetriangularis). Pa. triangularis* in the strict sense as discussed by Klapper et al. (2004) does not occur until higher in Bed 32c of the Upper Coumiac section. Thus, the base of the Lower *triangularis* Zone must be identified by criteria other than the first occurrence of the nominal species. This can be accomplished by using the very abundant occurrence of *Pa. ultima* together with the first occurrence of *Pa. subperlobata* to identify the base of the Famennian elsewhere. Since the definition and identification of the F/F boundary are not within the main focus of the present paper, the reader should see the fuller discussion in Klapper et al. (2004).

## 6. Numerical patterns in conodont diversity and biofacies

The variation of relative abundances of the different conodont genera was investigated according to the method first indicated by Seddon and Sweet (1971) that was later extensively developed by Sandberg (1976), Sandberg et al. (1988), and Girard and Feist (1997). In this methodology, the relative abundances of the principal groups (*Palmatolepis*, *Icriodus*, *Polygnathus*, and the "Ancyro-group", i.e. *Ancyrodella* and *Ancyrognathus*) are represented as a percentage of the total conodont fauna.

Sandberg et al. (1988, Fig. 15) give conodont biofacies assignments for seven environmental settings ranging from deep basin to inner shelf in the *linguiformis* Zone and the succeeding Lower *triangularis* Zone (Lower Famennian). In order for us to compare geographically different conodont associations, all located on outer-shelf platforms, representative sections have been taken into consideration (Table 1). For each section, the highest level directly below the UKW or its correlative equivalent has been sampled for numerical comparisons. The purpose is to test whether or not there is any palaeogeographic differentiation within outer-shelf environmental settings of the lower (early) *linguiformis* Zone.

No substantial difference can be observed among North Gondwana (Coumiac and Mrirt), Rhenish Slate Mountains (BHT and Beul), Vietnam (Xom Nha), and Canning Basin associations (Fig. 2), corroborating previous testing done for German, Belgian, and Nevada-Utah sections (Sandberg et al., 1988). Comparisons of the taxonomic diversity and conodont biofacies indicate a clear predominance of *Palmatolepis*. However, the percentage of *Pa. linguiformis* specimens within the genus *Palmatolepis* does not exceed 10% for

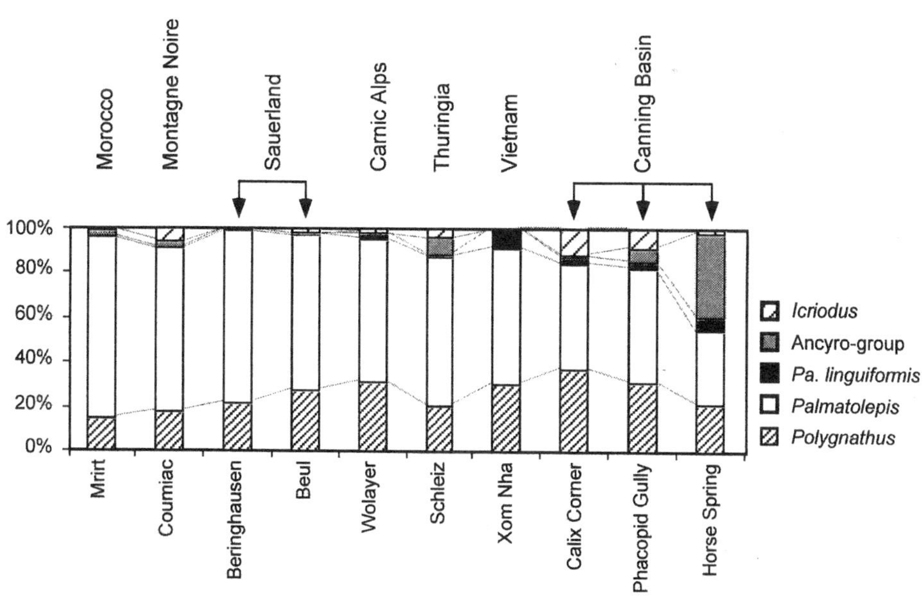


*Figure 2.* Variation in relative proportion of platform (Pa) elements of the conodont genera composing the total assemblage of the studied levels. Percentage of *Palmatolepis linguiformis* has been separated from other species of *Palmatolepis*.

the sections presented here (Table 1 and Fig. 2), and values are often lower than 1% of the total *Palmatolepis* fauna.

This predominance of *Palmatolepis* is seen in other offshore sections including, for example Steinbruch Schmidt (Sandberg et al., 1988); Antoinettenweg, SE Harz Mountains (Hüneke, 1995), Coumiac, Montagne Noire (Girard and Feist, 1997), South China (Wang, 1994; Wang and Ziegler, 2002) and may correspond to the well-known sea-level rise prior to the Upper Kellwasser event (Sandberg et al., 1988, 2002). This eustatic rise has been discussed extensively by Sandberg et al. (2002, p. 479) and by Morrow and Sandberg (2003, pp. 354–356). The sea-level rise is interpreted to have caused stratification of the water column and a rise of the oxygen-minimum zone, thus producing the black shales of the UKW (Sandberg et al., 1988, p. 294, Fig. 15; Sandberg et al., 2002, Fig. 5, Table 1, point no. 6). The interpretation of global deepening in the early part of the *linguiformis* Zone is further supported by the trends toward eye-reduction in most of the terminal Frasnian evolutionary lineages of trilobites (see next section).

## 7. Trilobites in the *linguiformis* Zone

Following the severe reduction in diversity and extinctions that trilobites suffered during the late Givetian Taghanic onlap, surviving lines of the Styginidae, Harpetidae, Odontopleuridae, Tropidocoryphidae, Dalmanitidae, Aulacopleuridae, Proetidae, and Phacopidae ranged to the base of the UKW though impoverished in lower rank taxonomy. At this point, trilobites simultaneously disappeared almost completely; only proetids and phacopids survived and radiated in post-Kellwasser times with new lineages of cryptogenic origin (Feist, 1992). Precision on trilobite occurrences in the lower, pre-UKW part of the *linguiformis* Zone was documented from the European margin of Laurussia and from Gondwana related terranes in southern Europe and North Africa (Feist and Schindler, 1994; Feist, 2002; Weyer et al., 2003). These occurrences are updated herein (see list of *linguiformis* sites above).

According to previously available data, *linguiformis* Zone trilobite populations of different, palaeogeographically independent sites are surprisingly homogeneous in their generic composition, and they share several taxa at the specific level. However, these belong to the same widespread outer-shelf lime-mud cephalopod biofacies, whereas contemporaneous nearshore trilobite remains from siliciclastic deposits and shallow carbonate-platform areas remain poorly known. An attempt to relate deeper ramp faunas to others from reef habitats is currently being made in the investigation of terminal Frasnian perireefal communities from the Canning Basin (Feist and McNamara, in prep.). Though far distant from European–North African faunas, these Australian communities present a comparatively high similarity factor at the generic level with only a few endemic elements. A species in common is *Palpebralia brecciae*. By contrast, no *linguiformis* Zone trilobites have not yet been discovered from either the Americas or Asia. This apparent absence might result in part from the poor availability of suitable facies for preservation, but more probably from sampling biases and insufficiently detailed investigations.

Trilobites from eight orders are known from the interval between the two Kellwasser horizons and five species are restricted to the lower part of the *linguiformis* Zone (Fig. 3). The most frequent and widely distributed are proetids of the genus *Palpebralia*. Among these, *Palpebralia brecciae*, which descended from *P. palpebralis*, is present in most

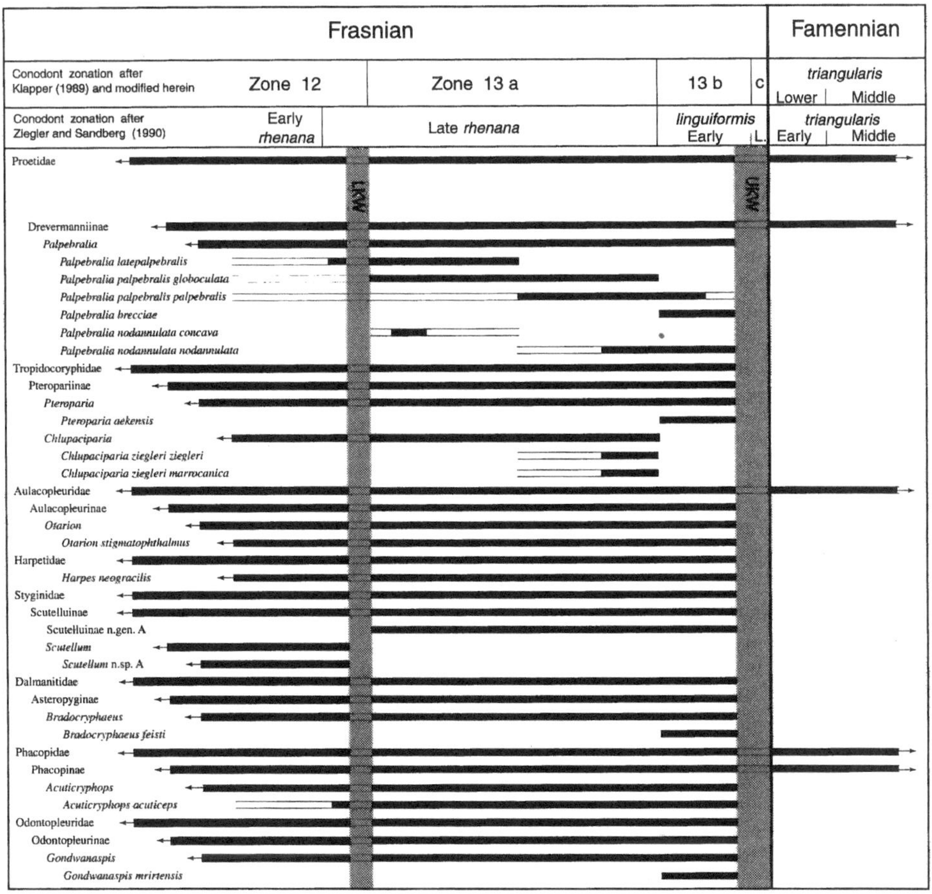


*Figure 3.* Range chart of trilobites between the Lower (LKW) and Upper Kellwasser (UKW) beds (gray shading) including the lower part of the *linguiformis* Zone (black bars: observed occurrences, open bars: inferred ranges) (after Feist and Schindler, 1994, modified). The boundary between the lower (early) and upper (late) parts of the *linguiformis* Zone coincides with the boundary between Zones 13b and 13c. Stepwise extinction of trilobite species at the top of Zone 13a and within Zone 13b is evident.

European–North African and NW Australian sites between the base of the *linguiformis* Zone and the base of the UKW. It represents therefore a marker species for correlation of the lower part of this zone, especially where the nominal conodont index is absent or extremely rare. This situation occurs both in the Coumiac Upper Quarry Frasnian/Famennian boundary-stratotype section (one single specimen of *Pa. linguiformis* among 3327 specimens of *Palmatolepis* in the collections of C.G. and no specimens of this species in collections made by the other two authors) and at Mrirt (three specimens of *Pa. linguiformis* among 31,020 *Palmatolepis*). Other trilobite taxa restricted to the *linguiformis* Zone such as *Pteroparia aekensis*, *Bradocryphaeus feisti*, and *Gondwanaspis mrirtensis* are known as yet, each from only two different sites (Sessacker/Aeke, Coumiac/Kahlleite, Mrirt/Coumiac, respectively). *Palpebralia palpebralis*, *P. nodannulata*, *Otarion stigmatophthalmus*, *Harpes neogracilis*, *Acuticryphops acuticeps*, and *Gondwanaspis* sp. are widely distributed in strata between the LKW and UKW, all crossing the lower boundary of the *linguiformis* Zone. Endemic taxa of the Canning Basin restricted to the *linguiformis* Zone in that area comprise new species of *Palpebralia* and a new styginid with reduced eyes.

Most if not all of the terminal Frasnian evolutionary lineages exhibit paedomorphic trends toward eye-reduction. The last offshore communities, immediately before their sudden demise related to the Upper Kellwasser Event, are all composed of reduced-eyed or even blind forms. This clearly reflects deepening trends of global sea level.

One such lineage is comprised of different morphs in succeeding *Acuticryphops acuticeps* populations with decreasing mean numbers of ocular lenses displayed on the visual field of the eyes. In the Coumiac section, from an initial mean of 9.59 lenses above the LKW this number drops to 4.42 below the base of the UKW. Morphs with three or fewer lenses occur exclusively in the *linguiformis* Zone (Feist, 1995). The presence of these morphs might eventually contribute to the identification of the lower part of this zone.

## Acknowledgments

C. G. thanks Sabrina Renaud for numerous discussions. G. K. thanks P.E. Playford, R.S. Nicoll, R.W. Brown, M.R. House, R.T. Becker, and W.T. Kirchgasser for field assistance at various times during five expeditions to the Canning Basin from 1988–1995. R. F. is grateful to D. Korn, K. Bartzsch, D. Weyer, and E. Schindler for their help in collecting trilobite samples and providing field data in the Rhenish and Thuringian Slate Mountains. We are grateful to C. Sandberg for a thorough review of an earlier version of the manuscript, as well as I. Schülke for his review; however, we are responsible for any remaining errors of interpretation.

## References

Bartzsch, K., Blumenstengel, H., Gereke, M., Schindler, E., Weyer, D., Ziegler, W., 2001. Post-conference field trip (N, part 1): Thüringer Schiefergebirge. In: Jansen, U., Königshof, P., Plodowski, G., Schindler, E. (Eds), Field Trips Guidebook. 15th International Senckenberg Conference May 11–21, pp. 87–113.

Becker, R.T., Feist, R., Flajs, G., House, M.R., Klapper, G., 1989. Frasnian–Famennian extinction events in the Devonian at Coumiac, southern France. C.R. Acad., Paris, 309, Sér. II, 259–266.

Becker, R.T., House, M.R., Kirchgasser, W.T., Playford, P.E., 1991. Sedimentary and faunal changes across the Frasnian/Famennian boundary in the Canning Basin of Western Australia. Hist. Biol. 5, 183–196.

Belka, Z., Wendt, J., 1992. Conodont biofacies patterns in the Kellwasser Facies (upper Frasnian/lower Famennian) of the eastern Anti-Atlas, Morocco. Palaeogeogr. Palaeoclimatol. Palaeoecol. 91, 143–173.

Bultynck, P., Helsen, S., Hayduckiewich, J., 1998. Conodont succession and biofacies in upper Frasnian formations (Devonian) from the southern and central parts of the Dinant Synclinorium (Belgium) – (Timing of facies shifting and correlation with late Frasnian events). Bull. Inst. r. Sci. nat. Belg. 68, 25–75.

Bultynck, P., Martin, F., 1995. Assessment of an old stratotype: the Frasnian/Famennian boundary at Senzeilles, southern Belgium. Bull. Inst. r. Sci. nat. Belg. 65, 5–34.

Clark, D.L., Becker, J.H., 1960. Upper Devonian correlations in western Utah and eastern Nevada. Geol. Soc. Am. Bull. 71, 1661–1674.

Dzik, J., 2002. Emergence and collapse of the Frasnian conodont and ammonoid communities in the Holy Cross Mountains, Poland. Acta Palaeontol. Pol. 47, 565–650.

Feist, R., 1992. The Late Devonian trilobite crises. Hist. Biol. 5, 197–214.

Feist, R., 1995. Effect of paedomorphosis in eye-reduction on patterns of evolution and extinction in trilobites. In: McNamara, K.J. (Ed.), Evolutionary Change and Heterochrony. Wiley, New York, pp. 225–244.

Feist, R., 2002. Trilobites from the latest Frasnian Kellwasser crisis in north Africa (Mrirt, central Moroccan Meseta). Acta Palaeontol. Pol. 47, 203–210.

Feist, R., Schindler, E., 1994. Trilobites during the Frasnian Kellwasser Crisis in European Late Devonian cephalopod limestones. In: Königshof, P., Werner, R. (Eds), Willi Ziegler-Festschrift II. Courier Forschungsinstitut Senckenberg 169, 195–223.

Girard, C., Feist, R., 1997. Eustatic trends in conodont diversity across the Frasnian–Famennian boundary in the stratotype area, Montagne Noire, southern France. Lethaia 29, 329–337.

Han, Y., 1987. Study on Upper Devonian Frasnian/Famennian boundary in Ma-anshan, Zhongping, Xiangzhou, Guangxi. Chinese Acad. Sci. Bull. 17, 171–194.

House, M.R., Becker, R.T., Feist, R., Flajs, G., Girard, C., Klapper, G., 2000. The Frasnian/Famennian GSSP at Coumiac, southern France. In: Bultynck, P. (Ed.), Subcommission on Devonian Stratigraphy: Recognition of Devonian Series and Stage Boundaries in Geological Areas. Courier Forschungsinstitut Senckenberg 225, 59–75.

Hüneke, H. 1995. Early Devonian (Emsian) to Late Devonian (Famennian) stratigraphy and conodonts of the Antoinettenweg section in the Lower Harz Mountains (Germany). Courier Forschungsinstitut Senckenberg 188, 99–131.

Ji, Q., 1989a. On the Frasnian–Famennian mass extinction event in South China. In: Ziegler, W. (Ed.), Papers on Ordovician to Triassic Conodonts. Courier Forschungsinstitut Senckenberg 117, 275–301.

Ji, Q., 1989b. On the Frasnian conodont biostratigraphy in the Guilin area of Guangxi, South China. In: Ziegler, W. (Ed.), Papers on Ordovician to Triassic Conodonts. Courier Forschungsinstitut Senckenberg 117, 303–319.

Jia, H.C., Xian, S.Y., Yang, D.L., Zhou, H.L., Han, Y.J., Chen, Z.H., Wang, J.X., Wang, R.G., Wang, S.T., Zhang, Z.X., Wei, M., 1989. An ideal Frasnian/Famennian boundary in Ma-anshan, Zhongping, Xiangzhou, Guangxi, south China. In: McMillan, N.J., Embry, A.F., Glass, D.J. (Eds), Devonian of the World. Paleontology, Paleoecology, Biostratigraphy. Can. Soc. Petrol. Geol. Mem. 14 (3), 79–92 [imprint, 1988].

Joachimski, M.M., Buggisch, W., Anders, T., 1994. Mikrofacies, Conodontenstratigraphie und Isotopengeochimie des Frasne/Famenne-Grenzprofils Wolayer Gletscher (Karnische Alpen). Abh. Geol. B.-A. 50, 183–195.

Johnson, J.G., Klapper, G., 1992. North American Midcontinent Devonian T-R cycles, Norman, Oklahoma. In: Chaplin, J.R., Barrick, J.E. (Eds), Special Papers in Paleontology and Stratigraphy: A Tribute to Thomas W. Amsden. Oklahoma Geol. Surv. Bull. 145, 127–135.

Johnson, J.G., Klapper, G., Trojan, W.R., 1980. Brachiopod and conodont successions in the Devonian of the northern Antelope Range, Marburg, Germany, central Nevada. Geol. Palaeontol. 14, 77–116.

Kirchgasser, W.T., 1994. Early morphotypes of *Ancyrodella rotundiloba* at the Middle-Upper Devonian boundary, Genesee Formation, west-central New York. In: Landing, E. (Ed.), Studies in Stratigraphy and Paleontology in Honor of Donald W. Fisher. NY State Museum Bull. 481, 117–134.

Klapper, G., 1989. The Montagne Noire Frasnian (Upper Devonian) conodont succession. In: McMillan, N.J., Embry, A.F., Glass, D.J. (Eds), Devonian of the World. Paleontology, Paleoecology, Biostratigraphy. Can. Soc. Petrol. Geol. Mem. 14 (3), 449–468 [imprint, 1988].

Klapper, G., 1997. Graphic correlation of Frasnian (Upper Devonian) sequences in Montagne Noire, France, and western Canada. In: Klapper, G., Murphy, M.A., Talent, J.A. (Eds), Paleozoic Sequence Stratigraphy, Biostratigraphy, and Biogeography: Studies in Honor of J. Granville ("Jess") Johnson. Geol. Soc. Am. Spec. Pap. 321, 113–129.

Klapper, G., Becker, T.R., 1999. Comparison of Frasnian (Upper Devonian) conodont zonations. In: Serpagli, E., Coradini, C. (Eds), Studies on Conodonts: Proceedings of the Seventh European Conodont Symposium, Bologna-Modena 23–25 June, 1998. Boll. Soc. Paleont. Ital. 37(2–3), 339–348.

Klapper, G., Feist, R., Becker, R.T., House, M.R., 1994. Definition of the Frasnian/Famennian Stage boundary. Episodes 16, 433–441 [imprint, 1993].

Klapper, G., Foster, C.T., Jr., 1993. Shape analysis of Frasnian species of the Late Devonian conodont genus *Palmatolepis*. The Paleontol. Soc. Mem. 32 (J. Paleontol. 67, Suppl. 4), 1–35.

Klapper, G., Furnish, W.M., 1963. Conodont zonation of the Early Upper Devonian in eastern Iowa. Iowa Acad. Sci. Proc. 69, 400–410.

Klapper, G., Kirchgasser, W.T., Baesemann, J.F., 1995. Graphic corelation of a Frasnian (Upper Devonian) composite standard. In: Mann, K.O., Lane, H.R. (Eds), Graphic Correlation. SEPM Society for Sedimentary Geology Special Publication 53, 177–184.
Klapper, G., Kuz'min, A.V., Ovnatanova, N.S., 1996. Upper Devonian conodonts from the Timan-Pechora region, Russia, and correlation with a Frasnian composite standard. J. Paleontol. 70, 131–152.
Klapper, G., Uyeno, T.T., Armstrong, D.K., Telford, P.G., 2004. Conodonts of the Williams Island and Long Rapids Formations (Upper Devonian, Frasnian–Famennian) of the Onakawana B drillhole, Moose River Basin, northern Ontario, with a revision of lower Famennian species. J. Paleontol. 78, 371–387.
Kralick, J.A., 1994. The conodont genus *Ancyrodella* in the middle Genesee Formation (lower Upper Devonian, Frasnian), western New York. J. Paleontol. 68, 1384–1395.
Lazreq, N., 1992. The Upper Devonian of M'rirt (Morocco). In: Ziegler, W. (Ed.), Papers on Paleozoic Conodonts from Eurasia and North Africa. Courier Forschungsinstitut Senckenberg 154, 107–123.
Lazreq, N., 1999. Biostratigraphie des conodontes du Givétien au Famennien du Maroc central – Biofaciès et événement Kellwasser. Courier Forschungsinstitut Senckenberg 214, 1–111.
Matyja, H., Narkiewicz, M., 1992. Conodont biofacies succession near the Frasnian/Famennian boundary: some Polish examples. In: Ziegler, W. (Ed.), Papers on Paleozoic Conodonts from Eurasia and North Africa. Courier Forschungsinstitut Senckenberg 154, 125–147.
McLean, R.A., Klapper, G., 1998. Biostratigraphy of Frasnian (Upper Devonian) strata in western Canada, based on conodonts and rugose corals. Bull. Can. Petrol. Geol. 46, 515–563.
Morrow, J.R., 2000. Shelf-to-basin lithofacies and conodont paleoecology across Frasnian–Famennian (F-F, mid-Late Devonian) boundary, central Great Basin (Western U.S.A.). Courier Forschungsinstitut Senckenberg 219, 1–57.
Morrow, J.R., Sandberg, C.A., 2003. Late Devonian sequence and event stratigraphy across the Frasnian–Famennian (F-F) boundary, Utah and Nevada, Chapter 10. In: Harries, P.J. (Ed.), Approaches in High-Resolution Stratigraphic Paleontology, Topics in Geobiology Series 21. Kluwer Academic Publishers, Dordrecht, The Netherlands, pp. 351–419.
Müller, K.J., 1956. Die Gattung *Palmatolepis*. Abh. Senckenberg naturforsch. Gesellsch. 494, 1–70.
Narkiewicz, M., 1978. Stratigraphy and facies development of the Upper Devonian in the Olkusz-Zawiercie area, southern Poland. Acta Geol. Pol. 28, 415–470 [in Polish, with English summary].
Over, D.J., 1997. Conodont biostratigraphy of the Java Formation (Upper Devonian) and the Frasnian–Famennian boundary in western New York State. In: Klapper, G., Murphy, M. A., Talent, J.A. (Eds), Paleozoic Sequence Stratigraphy, Biostratigraphy, and Biogeography: Studies in Honor of J. Granville ("Jess") Johnson. Geol. Soc. Am. Spec. Pap. 321, 161–177.
Over, D.J., 2002. The Frasnian/Famennian boundary in central and eastern United States (Amsterdam). In: Racki, G., House, M.R. (Eds), Late Devonian Biotic Crisis: Ecological, Depositional and Geochemical Records. Palaeogeogr. Palaeoclimatol. Palaeoecol. 181, 153–169.
Over, D.J., Rhodes, M.K., 2000. Conodonts from the upper Olentangy Shale (Upper Devonian, central Ohio) and stratigraphy across the Frasnian–Famennian boundary. J. Paleontol. 74, 101–112.
Ovnatanova, N.S., Kuzmin, A.V., Menner, V.V., 1999. The succession of Frasnian conodont assemblages in the type sections of the southern Timan-Pechora Province (Russia). In: Serpagli, E., Coradini, C. (Eds), Studies on Conodonts: Proceedings of the Seventh European Conodont Symposium, Bologna-Modena 23–25 June, 1998. Boll. Soc. Paleont. Ital. 37(2–3), 349–360.
Pollock, C.A., 1968. Lower Upper Devonian conodonts from Alberta, Canada. J. Paleontol. 42, 415–443.
Racki, G., Racka, M., Matyja, H., Devleeschouwer, X., 2002. The Frasnian/Famennian boundary interval in the South Polish-Moravian shelf basins: integrated event-stratigraphical approach (Amsterdam). In: Racki, G., House, M.R. (Eds), Late Devonian Biotic Crisis: Ecological, Depositional and Geochemical Records. Palaeogeogr., Palaeoclimatol., Palaeoecol. 181, 251–297.
Sandberg, C.A., 1976. Conodont biofacies of Late Devonian *Polygnathus styriacus* Zone in Western United States. In: Barnes, C.R. (Ed.), Conodont Paleoecology. Geol. Assoc. Can. Spec. Pap. 15, 171–186.
Sandberg, C.A., Hasenmueller, N.R., Rexroad, C.B., 1994. Conodont biochronology, biostratigraphy, and biofacies of Upper Devonian part of New Albany Shale, Indiana. In: Königshof, P., Werner, R. (Eds), Willi Ziegler-Festschrift I. Courier Forschungsinstitut Senckenberg 168, 227–253.
Sandberg, C.A., Morrow, J.R., Poole, F.G., Ziegler, W., 2003. Middle Devonian to Early Carboniferous event stratigraphy of Devils Gate and Northern Antelope Range sections, Nevada, U.S.A. Courier Forschungsinstitut Senckenberg 242, 187–207.

Sandberg, C. A., Morrow, J. R., Warme, J. E., 1997. Late Devonian Alamo Impact Event, global Kellwasser Events, and major eustatic events, eastern Great Basin, Nevada and Utah. Brigham Young Univ. Geol. Stud. 42(1), 129–160.

Sandberg, C.A., Morrow, J.R., Ziegler, W., 2002. Late Devonian sea-level changes, catastrophic events, and mass extinctions. In: Koeberl, C., MacLeod, K.G. (Eds), Catastrophic Events and Mass Extinctions: Impacts and Beyond. Geol. Soc. Am. Spec. Pap. 356, 473–487.

Sandberg, C.A., Poole, F.G., Johnson, J.G., 1989. Upper Devonian of Western United States. In: McMillan, N.J., Embry, A.F., Glass, D.J. (Eds), Devonian of the World. Regional Syntheses. Can. Soc. Petrol. Geol. Mem. 14 (1), 183–220 [imprint, 1988].

Sandberg, C.A., Ziegler, W., Dreesen, R., Butler, J.L., 1988. Late Frasnian mass extinction: conodont event stratigraphy, global changes, and possible causes. Courier Forschungsinstitut Senckenberg 102, 263–307.

Savage, N.M., Yudina, A.B., 1999. Late Devonian Syv'yu River section, Timan-Pechora Basin, northwestern Russia. In: Serpagli, E., Coradini, C. (Eds), Studies on Conodonts: Proceedings of the Seventh European Conodont Symposium, Bologna-Modena 23–25 June, 1998. Boll. Soc. Paleont. Ital. 37(2–3), 361–373.

Schindler, E., 1990. Die Kellwasser-Krise (hohe Frasne-Stufe, Ober-Devon). Göttinger Arb. Geol. Paläont. 46, 1–115.

Schindler, E., Schülke, I., Ziegler, W. 1998. The Frasnian/Famennian boundary at the Sessacker Trench section near Oberscheld (Dill Syncline, Rheinisches Schiefergebirge, Germany). Senckenberg. lethaea 77(1/2), 243–261.

Schülke, I., 1995. Evolutive Prozesse bei *Palmatolepis* in der frühen Famenne-Stufe (Conodonta, Ober-Devon). Göttinger Arb. Geol. Paläont. 67, 1–108.

Schülke, I., 1998. Conodont community structure around the "Kellwasser mass extinction event" (Frasnian/ Famennian boundary interval). Senckenberg. lethaea 77, 87–99.

Schumacher, D., 1971. Conodonts and biostratigraphy of the "Kenwood" Shale (Upper Devonian). In: Clark, D.L. (Ed.), Conodonts and Biostratigraphy of the Wisconsin Paleozoic. Wisconsin Geological and Natural History Survey, Information Circular 19, 68–77, 100–107.

Scotese, C.R., McKerrow, W.S., 1990. Revised world maps and introduction. In: McKerrow, W.S., Scotese, C.R. (Eds), Palaeozoic Palaeogeography and Biogeography. Geol. Soc. Lon. Mem. 12, 1–24.

Seddon, G., 1970. Pre-Chappel conodonts of the Llano region, Texas. Bureau Econom. Geol., Rep. Invest. 68, 1–130.

Seddon, G., Sweet, W.C., 1971. An ecologic model for conodonts. J. Paleontol. 45, 869–880.

Spalletta, C., Farabegoli, E., Klapper, G., Perri, M.C., 2001. Frasnian–Famennian boundary at Pramosio, Carnic Alps, Italy. 15th International Senckenberg Conference, Frankfurt am Main, Joint Meeting IGCP 421/SDS, May 11–21, Abstracts, 89.

Spalletta, C., Perri, M.C., 1998. Stop 2.2 – The Frasnian–Famennian boundary at the Pramosio A section (Carnic Alps, Italy). In: Perri, M.C., Spalletta, C. (Eds), Southern Alps Field Trip Guidebook, June 27–July 2, ECOS VII, Vol. 60, Special Issue. Giornale di Geologia, Serie 3ª, pp. 198–205.

Szulczewski, M., 1971. Upper Devonian conodonts, stratigraphy and facial development in the Holy Cross Mts. Acta Geol. Pol. 21, 1–129.

Ta Hoà Phuong, 1998. Upper Devonian conodont biostratigraphy in Viêt Nam. J. Geol. Geol. Surv. Viêt Nam Ser. B 11–12, 76–84.

Walliser, O.H., Groos-Uffenorde, H., Schindler, E., Ziegler, W., 1989. On the Upper Kellwasser Horizon (Boundary Frasnian/Famennian). In: Walliser, O.H., Ziegler, W. (Eds), Contributions to Devonian Palaeontology and Stratigraphy. Courier Forschungsinstitut Senckenberg 110, 247–256.

Wang, C.-Y., 1994. Application of the Frasnian standard conodont zonation in South China. In: Königshof, P., Werner, R. (Eds), Willi Ziegler-Festschrift I. Courier Forschungsinstitut Senckenberg 168, 83–129.

Wang, C.-Y., Ziegler, W., 2002. The Frasnian–Famennian conodont mass extinction and recovery in South China Frankfurt-am-Main, Germany. In: Weddige, K., Ziegler, W. (Eds), Advances in Conodont, Devonian and Carboniferous Research. Senckenberg. lethaea 82(2), 463–493.

Weyer, D., Feist, R., Girard, C., 2003. Conodonta, Trilobita, and Anthozoa near the Late Frasnian Upper Kellwasser Event of the Geipel Quarry section in Schleiz, Thuringian Mountains (Germany). Mitt. Mus. Nat. Kd. Berl., Geowiss. Reihe 6, 71–78.

Wright, A.J., Pickett, J.W., Sewell, D., Roberts, J., Jenkins, T.B.H., 1990. Corals and conodonts from the Late Devonian Mostyn Vale Formation, Keepit, New South Wales. In: Jell, P.A. (Ed.), Devonian and Carboniferous Coral Studies. Assoc. Australasian Palaeontol. Mem. 10, 211–254.

Ziegler, W., 1958. Conodontenfeinstratigraphische Untersuchungen an der Grenze Mitteldevon/Oberdevon und in der Adorfstufe. Notizbl. Hess. Landesamt Bodenforsch. 87, 7–77.
Ziegler, W., 1971. Conodont stratigraphy of the European Devonian. In: Sweet, W.C., Bergström, S.M. (Eds), Symposium on Conodont Biostratigraphy. Geol. Soc. Am. Mem. 127, 227–284.
Ziegler, W., 1973. *Palmatolepis linguiformis* Müller, 1956. In: Ziegler (Ed.), Catalogue of Conodonts. E. Schweizerbart'sche Verlagsbuchhandlung, Stuttgart, I, 283–285.
Ziegler, W., Sandberg, C.A., 1990. The Late Devonian standard conodont zonation. Courier Forschungsinstitut Senckenberg 121, 1–115.

Understanding Late Devonian and Permian-Triassic Biotic and Climatic Events:
Towards an Integrated Approach
edited by D.J. Over, J.R. Morrow and P.B. Wignall



## *Chapter 8*

# *Productivity and bottom water redox conditions at the Frasnian–Famennian boundary on both sides of the Eovariscan Belt: constraints from trace-element geochemistry*

*Laurent Riquier[a,]*, Nicolas Tribovillard[a], Olivier Averbuch[a], Michael M. Joachimski[b], Grzegorz Racki[c], Xavier Devleeschouwer[d], Abderrazzak El Albani[e] and Armelle Riboulleau[a]*

[a]*Laboratoire des Processus et Bilans des Domaines Sédimentaires, CNRS UMR 8110 & FR 1818, Université Lille 1, bâtiment SN5, F-59655 Villeneuve d'Ascq cedex, France*
[b]*Institute of Geology and Mineralogy, University of Erlangen-Nürnberg, Schlossgarten 5, D-91054 Erlangen, Germany*
[c]*Department of Palaeontology and Stratigraphy, University of Silesia, Bedzinska Str. 60, PL 41-200 Sosnowiec, Poland*
[d]*Geological Survey of Belgium, Rue Jenner 13, B-1000 Brussels, Belgium*
[e]*Laboratoire Hydrogéologie, Argiles, Sols et Altérations., CNRS UMR 6532, Université Poitiers, 40, Avenue du Recteur Pineau, F-86022 Poitiers cedex, France*

### Abstract

The Late Devonian Frasnian–Famennian (F–F) mass extinction event coincides in many places with the deposition of $C_{org}$-rich "Kellwasser" facies. Four F–F boundary sections representative of platform and basin environments from widely separated locations (Morocco, Germany, and France) were analysed for inorganic geochemistry, especially trace elements (redox and productivity proxies), in order to describe paleodepositionnal environments for the Kellwasser horizons. Ni/Co, V/Cr, U/Th, and V/(V+Ni) ratios, as well as redox trace metal concentrations indicate that oxygen-depleted conditions existed during the times of Kellwasser facies deposition. In platform settings, dysoxic conditions seem to be limited to the Late Frasnian. In basinal settings, oxygen depletion was stronger and persisted into the Early Famennian. Enrichments of Ba, Cu, Ni, that are limited to the Late Frasnian, show that surface productivity was relatively high and organic matter could accumulate, especially in the deeper environments. The stratigraphical distribution of several geochemical markers are linked with two positive excursions of the $\delta^{13}C_{carb}$ signal that result from enhanced organic matter burial. Reducing conditions likely resulted from high productivity of Late Devonian marine ecosystems. Intense nutrient supply resulted probably from the biogeochemical recycling of nutrients, and/or runoff from emerged lands. Coupled with other factors, such as rapid sea-level fluctuations and climatic changes, oxygen-depleted conditions and eutrophication would have modified Late Devonian environments and could be possible factors in the F–F mass mortality.

*Keywords:* Late Frasnian; inorganic geochemistry; anoxia; primary productivity; eutrophication

## 1. Introduction

The Late Devonian Frasnian–Famennian (F–F) boundary (374.5 Ma; Kaufmann et al., 2004) represents one of the five most severe biological crises of the Phanerozoic

*Corresponding author. Fax: +33-320-43-49-10. *E-mail address*: Laurent.Riquier@ed.univ-lille1.fr (L. Riquier).

(Sepkoski, 1986; Copper, 1986; McGhee, 1989, 1996; Hallam and Wignall, 1997; House, 2002). The biodiversity decrease particularly affected marine shallow-water tropical faunas of epicontinental reefal carbonate platforms. In outer shelfal (ca. 100–200 m depth) and epicontinental basin settings, the Upper Frasnian extinction interval is characterized by the occurrence of two distinctive black-coloured units, i.e., the so-called "Kellwasser" horizons (Schindler, 1990). The top of the Upper Kellwasser horizon usually marks the F–F boundary. These two beds have been recognized in various places, located on the borders of Laurussian (N. America, N. Europe) (Sandberg et al., 1989; Schindler, 1990, 1993; Buggisch, 1991) and Gondwanian (S. Europe, S. America, Africa) blocks (Feist, 1990; Wendt and Belka, 1991). They have generally been attributed to episodes of enhanced organic carbon preservation under pervasive anoxic bottom conditions (Joachimski and Buggisch, 1993; Joachimski et al., 2001) or increased carbon flux to the sea floor driven by high surface water productivity (May, 1995; Caplan and Bustin, 1998), or combinations of these processes (Murphy et al., 2000b). This paper presents the inorganic geochemistry record for four F–F boundary sequences deposited in different paleogeographic units on both sides of the incipient Eovariscan orogenic belt. The sections studied are distributed from platform to basin settings. Geochemical signatures of Gondwanian (Morocco and France) and Laurussian (Germany) sections were analysed, specially focussing on redox and productivity indicators. The purpose of this study is to provide insight to the paleoenvironmental conditions existing in marine setting and leading to organic matter (OM) accumulation. The results are compared with those obtained for time-equivalent deposits in the detailed studies of Joachimski et al. (2001, 2002), Racki et al. (2002), Yudina et al. (2002) and Bond et al. (2004) to establish a wider paleoenvironmental reconstruction.

## 2. Investigated sections

The four sections investigated, Bou-Ounebdou, Anajdam, La Serre, and Aeketal, are well-known F–F boundary localities (Fig. 1). During the last decade, detailed investigations of sedimentary sequences and faunas have been performed, particularly within the IGCP projects 216 on "Global biological events in Earth History" (Walliser, 1996) and 421 on "North Gondwanian Mid-Paleozoic bioevent/biogeography patterns in relation to crustal dynamics" (El Hassani and Tahiri, 1999) and by the Subcommission on Devonian Stratigraphy in order to supplement the F–F stratotype section at Coumiac. These studies allowed paleoenvironments assessment and precise dating using conodont biostratigraphy.

### 2.1. *Bou-Ounebdou and Anajdam (Western Meseta, Morocco)*

The Bou-Ounebdou and Anajdam sections are located in the M'rirt area, in the eastern part of central Morocco (western Meseta). Both sections are in the Azrou-Kenifra Basin, located at subtropical palaeolatitudes (15–30°S) on the northern margin of Gondwana during the Late Devonian (Fig. 2). The Anajdam and Bou-Ounebdou sections were described in detail by Lazreq (1992, 1999), Becker and House (2000), Walliser et al. (2000), and Chakiri (2002). The interval considered in this study records condensed sedimentation, spanning from the Late Frasnian (*rhenana* Zone) to Early Famennian (*triangularis* Zone). The Early

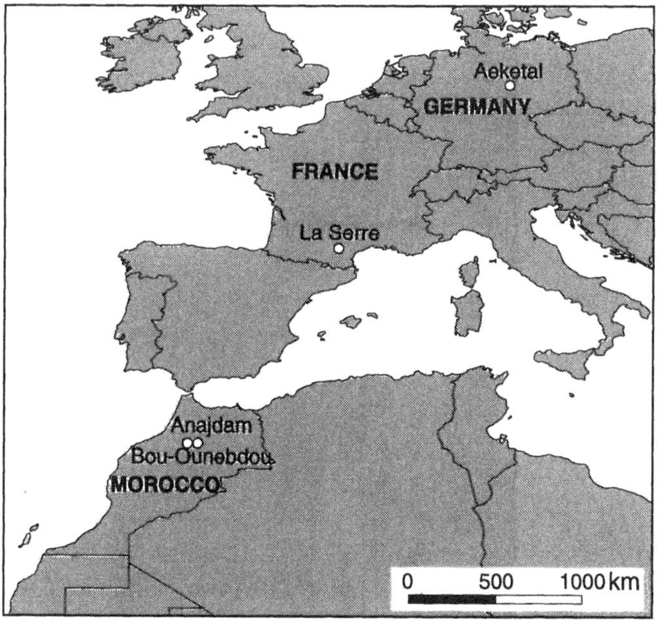


*Figure 1.* Index map of Western Europe and North Africa, showing several locations of studied sections in Germany, France, and Morocco.

*triangularis* Zone may be absent in the Bou-Ounebdou section (Lazreq, 1992; Becker, pers. comm.). The sequences mainly consist of a succession of cephalopod limestone beds (wackestone) with intercalated calcareous shale levels. In the upper Frasnian, two dark-coloured (supposedly organic C-rich) horizons were biostratigraphically identified as the Lower and Upper Kellwasser horizons, respectively (Lazreq, 1992). The Lower and Upper dark-coloured Kellwasser horizons are about 25 and 40 cm thick, respectively (Fig. 3). The faunal associations of conodonts, tentaculitids, ammonoids, trilobites, bivalves, and ostracods characterize deep, off shore platform settings where the majority of Kellwasser fossils are pelagic, such as *Manticoceras* ammonoids, orthocone nautiloids, bivalves of the *Buchiola*-group, conodonts, and tentaculitids (Lazreq, 1999; Chakiri, 2002).

## 2.2. *Aeketal (Harz Mountains, Germany)*

The Aeketal section is located in the north-western part of the Harz Mountains, included in the Rhenohercynian zone of the German Variscides. The Harz area is believed to have occupied a subtropical position (0–15°S) on the southern margin of Laurussia during deposition of Late Devonian sediments (Fig. 2). The Aeketal section, studied in details by Schindler (1990, 1993) and Feist and Schindler (1994), records condensed sedimentation spanning from Late Frasnian (*rhenana* Zone) to Early Famennian (*triangularis* Zone). The section mainly exposes light-grey limestones (calcilutites) and two black horizons of limestones and marls. The Upper Kellwasser horizon is 90 cm thick (Fig. 3). Diversified faunas, such as trilobites, ostracods, echinoderms, and ammonoids, have been found and both

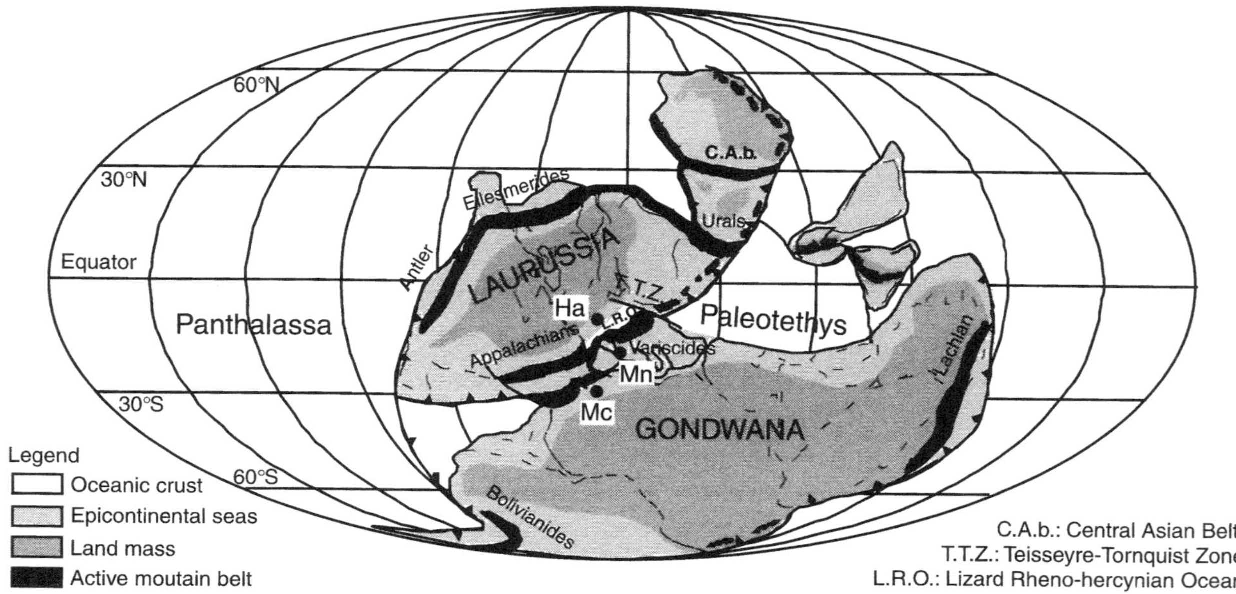


*Figure 2.* Schematic paleogeographic reconstruction for the Frasnian–Famennian boundary times showing main geological structures and the approximate location of studied areas. Mc: Maroc central (Central Morocco); Mn: Montagne Noire (France); Hz: Harz Mountains (Germany) (Modified from Averbuch et al., 2005).

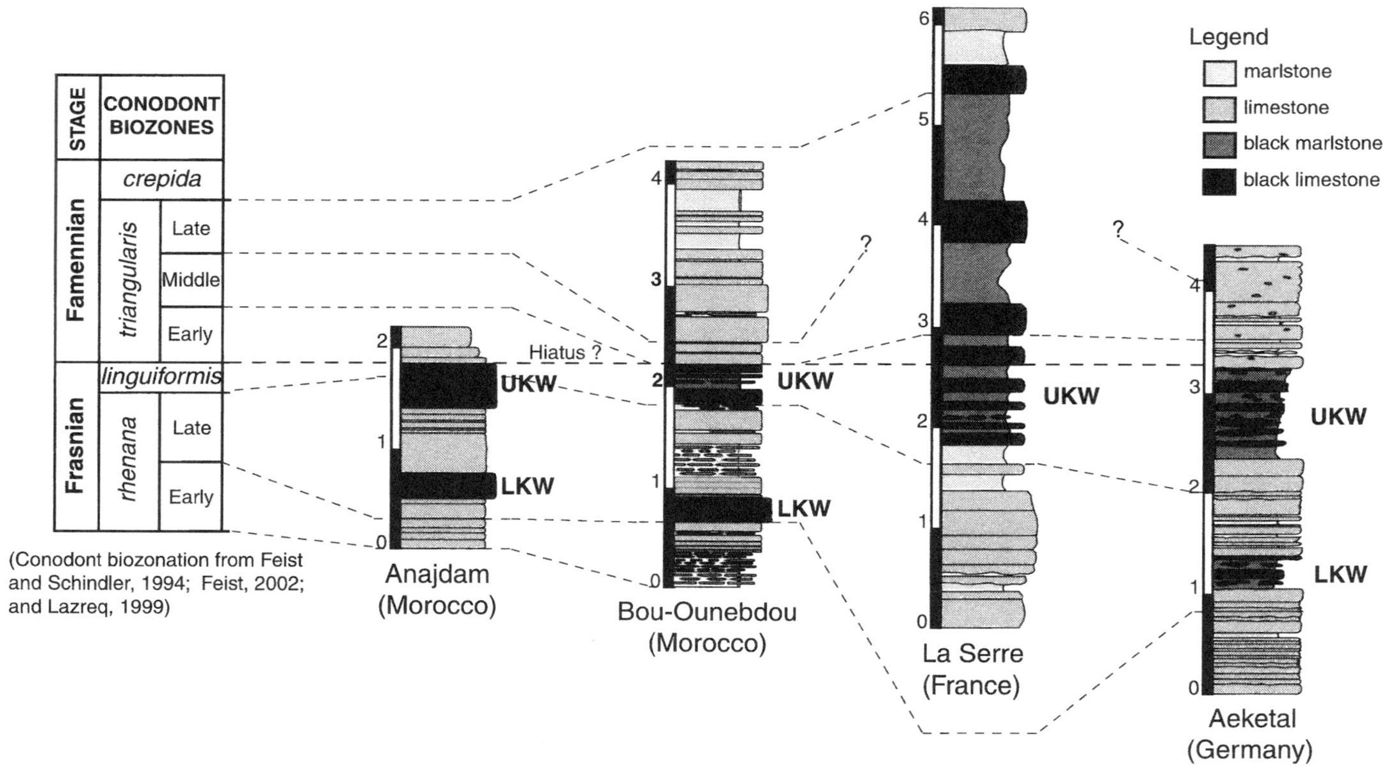


*Figure 3.* Simplified stratigraphic columns of the Frasnian–Famennian boundary sequence of the sections studied and biocorrelations. LKW: Lower Kellwasser, UKW: Upper Kellwasser, scale in meters.

Kellwasser horizons are dominated by pelagic fossils, such as conodonts, homoctenid cricoconarids, orthocone nautilids, and bivalves of the *Buchiola*-group. The Aeketal section is thought to correspond to a submarine-rise setting in a pelagic realm (Schindler, 1990).

### 2.3. *La Serre (Montagne Noire, France)*

The La Serre (Trench C) section is located on the southern slope of the La Serre hill, 2.4 km from Cabrières (Montagne Noire). Paleogeographically, the Montagne Noire area was located at paleolatitudes between 15 and 30°S (Fig. 2). This section displays a complete succession without sedimentary gaps. The interval considered in this study starts with light-grey pyritic calcilutites rich in pelagic and benthic biota that are abruptly overlain by an alternating sequence of dark brownish and black, fissile marls, and platy thin-bedded bituminous marly limestones. The levels at the base of this alternation are considered to be equivalent to the Upper Kellwasser horizon (Schindler, 1990; Girard and Albarède, 1996; Devleeschouwer, 1999). The $C_{org}$-rich Kellwasser facies prevailed for longer time periods than the Kellwasser horizon duration, and ended in the early Famennian *crepida* Zone (Fig. 3). At La Serre, the fine-grained, laminated sediments contain exclusively pelagic, nekto-planktonic biota of small size, characterized by homoctenids, conodonts, entomozoan ostracods, and buchiolid bivalves, but no cephalopods, in contrast to the normally much coarser-grained, cephalopod-rich Kellwasser limestone of the western Meseta and Aeketal sections. Based on palaeoecological criteria, the depositional environment of the La Serre section is interpreted as corresponding to an offshore basin (Feist, 1985; Paris et al., 1996; Lethiers et al., 1998).

## 3. Methods

For each section, about 20 bulk-rock samples of limestones to marls were analysed for major (Al, Fe, K, Mn, Ti, Si) and trace elements (Ba, Co, Cr, Cu, Mo, Ni, Pb, U, V, Zn). Elemental analyses were performed by ICP-AES (major or minor elements) and ICP-MS (trace elements), at the spectrochemical laboratory of the Service d'Analyse des Roches et des Minéraux of the Centre National de la Recherche Scientifique (Vandœ uvre-les-Nancy, France). The samples were prepared by fusion with $LiBO_2$ and $HNO_3$ dissolution. Precision and accuracy were both found to be better than 1% (mean 0.5%) for major–minor elements, 5% for Cr, Pb, U, V, and Zn, and 10% for Ba, Co, Cu, Mo, and Ni as checked by international standards (Carignan et al., 2001) and analysis of replicate samples, respectively. Calcium carbonate ($CaCO_3$) content was determined using a Bernard calcimeter (acid digestion). Total carbon was measured with a Leco CS-125 analyser (combustion at 1200°C) and total organic carbon (TOC) was calculated as the difference between total carbon and inorganic carbon, deduced from calcium carbonate content. A few tests of Rock Eval pyrolysis were also performed on the Morocco and Germany samples to compare the TOC values, using the Delsi Instrument 'Oil Show Analyser' of the Earth Sciences Department of Pierre et Marie Curie – Paris VI University.

The trace metal data are reported here as metal/aluminium ratios in order to eliminate the dilution effect of the biogenic phase (calcium carbonate and OM) and to compare different environmental settings with regard to variations in trace-element contents

(Brumsack, 1989; Arthur et al., 1990; Calvert and Pedersen, 1993). Higher element/Al ratios will indicate synsedimentary additions, if a constant composition of the background is assumed. Variations of the Al content in relation with diagenetic formation of clay minerals is unlikely, because in the present study Al is positively correlated with Ti ($R^2$= 0.99 for each section) that is not incorporated during authigenic clay–mineral formation (Pratt et al., 1986), an indication that Al is a reliable detrital proxy. All trace elements concentrations are reported as Al-normalized values, instead of Ti-normalized because Al-normalization is more usual in the literature, and facilitates comparisons.

## 4. Results

### *4.1. Carbon concentrations*

The four sections studied are composed of pure to shaly limestones. Carbonate contents mostly vary from 50 to 99% $CaCO_3$ (Figs. 4–7). At Anajdam, both Upper and Lower Kellwasser horizons are made of massive limestone; at Bou-Ounebdou, only the Lower Kellwasser horizon is made of massive limestone. In the other sections studied, the Kellwasser horizons correspond to marl–limestone alternations (Figs. 4–7). In Moroccan platform sections, no enrichment of TOC is recorded in the Kellwasser horizons. At Aeketal, samples from the grey limestone sequence are devoid of organic carbon, whereas the TOC content can reach 1.5 wt% in the Kellwasser horizons. At La Serre, the TOC content varies from 0 to 3.7 wt%, maximal values are recorded in the Early and Middle *triangularis* zones. Except for the La Serre section, all the sections studied belong to heavily tectonised and/or buried areas, and the OM endured severe loss and transformations. Consequently, the major part of the residual organic carbon is 'graphitic' carbon, and the TOC values have little significance (see Section 5). Conversely, in the case of the La Serre section and the Kowala section used for comparison (Joachimski et al., 2001, 2002), TOC values can be used directly because of (very) low thermal alteration (Devleeschouwer, 1999).

### *4.2. Trace metal concentrations*

Enrichment factors (EF) for selected trace elements were determined by comparing Al-normalized trace element ratio to those of average shale (Wedepohl, 1971, 1991). The enrichment factor for any element (e), hereafter $EF_{(e)}$, is equal to $(e/Al)_{sample}/(e/Al)_{shale}$. In plots of stratigraphical variations of normalized elements concentration, the ratio calculated for the average shale (Wedepohl, 1971) is indicated by a dashed line to provide the baseline for which $EF_{(e)} = 1$.

In the outer platform environments of Anajdam, Bou-Ounebdou, and Aeketal, the concentrations of redox-sensitive metals (Mo, U, V, Cr, Cu, Ni, Zn, and Pb), as well as markers of productivity (Ba) and OM-related trace elements (Cu, Ni, Zn), are generally elevated throughout the black-coloured Kellwasser horizons relative to overlying and underlying strata. Most of the trace metal elements, particularly U and Cu, show a relatively monotonous distribution in strata overlying and underlying the Kellwasser horizons. $EF_{(U)}$ and $EF_{(Cu)}$ have average values of 2.4 and 1.5, respectively. The Lower Kellwasser horizon is

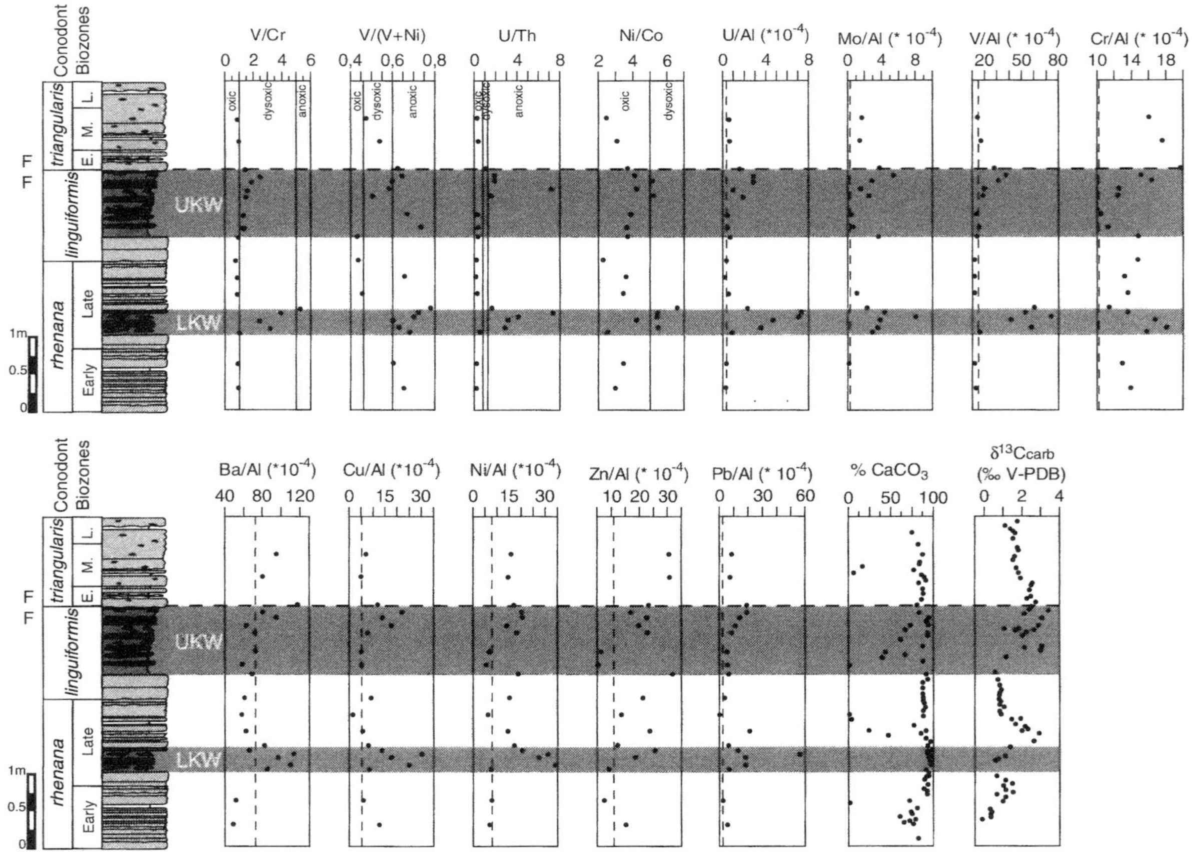


*Figure 4.* Stratigraphic distribution of carbonate content, $\delta^{13}C_{carb}$, Al-normalised trace-element concentrations and geochemical indices for the Aeketal section. Dashed lines indicate the metal/aluminium concentration ratios for average shales (Wedepohl, 1971, 1991) and the boundary of redox zones for V/Cr, Ni/Co, U/Th (from Jones and Manning, 1994) and V/(V+Ni) (from Hatch and Leventhal, 1992). Dark grey background corresponds to the black shale and limestone facies.

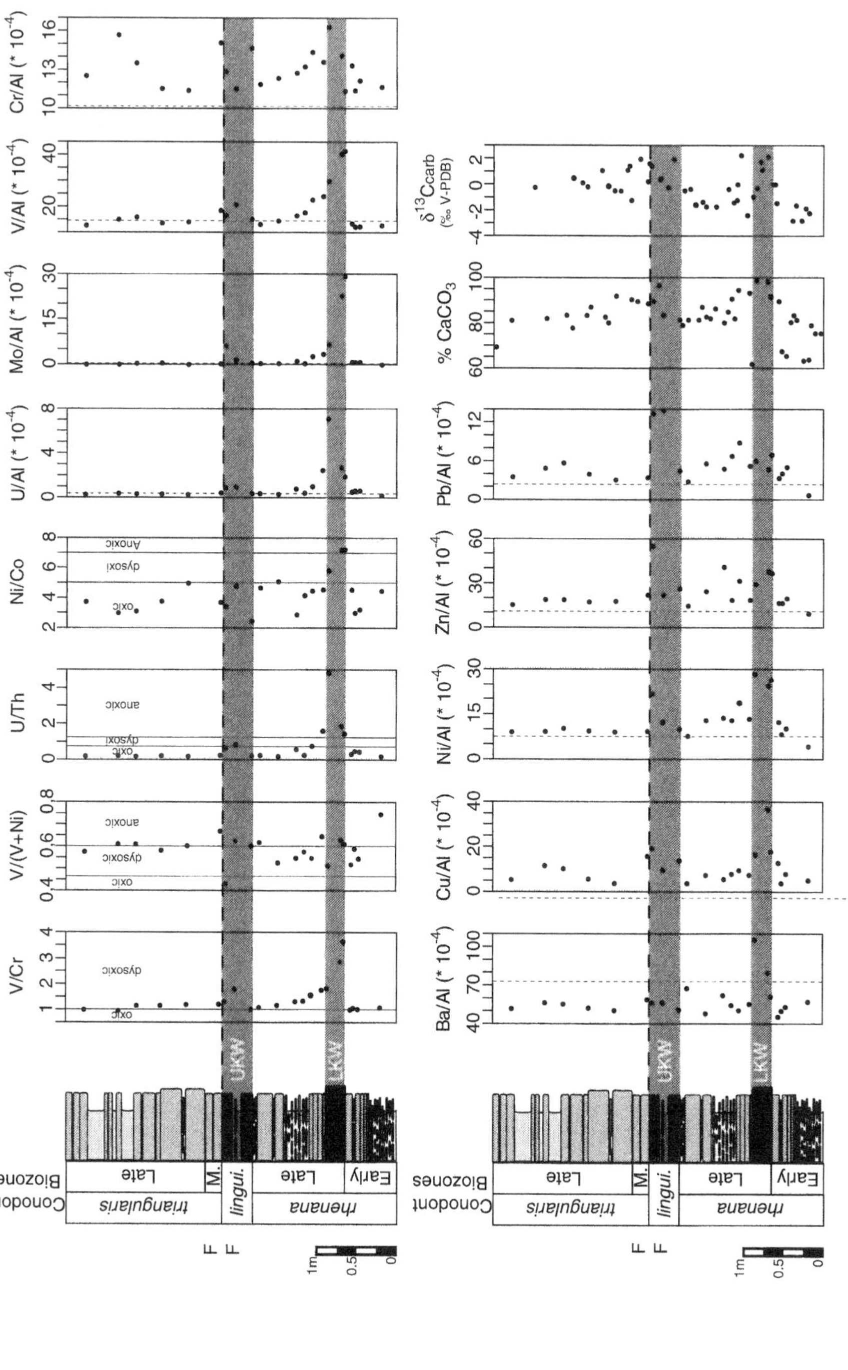


*Figure 5.* Stratigraphic distribution of carbonate content, $\delta^{13}C_{carb}$, Al-normalised trace-element concentrations and geochemical indices for the Bou-Ounebdou section. Dashed lines indicate the metal/aluminium concentration ratios for average shales (Wedepohl, 1971, 1991) and the boundary of redox zones for V/Cr, Ni/Co, U/Th (from Jones and Manning, 1994) and V/(V+Ni) (from Hatch and Leventhal, 1992). Dark grey background corresponds to the black shale and limestone facies.

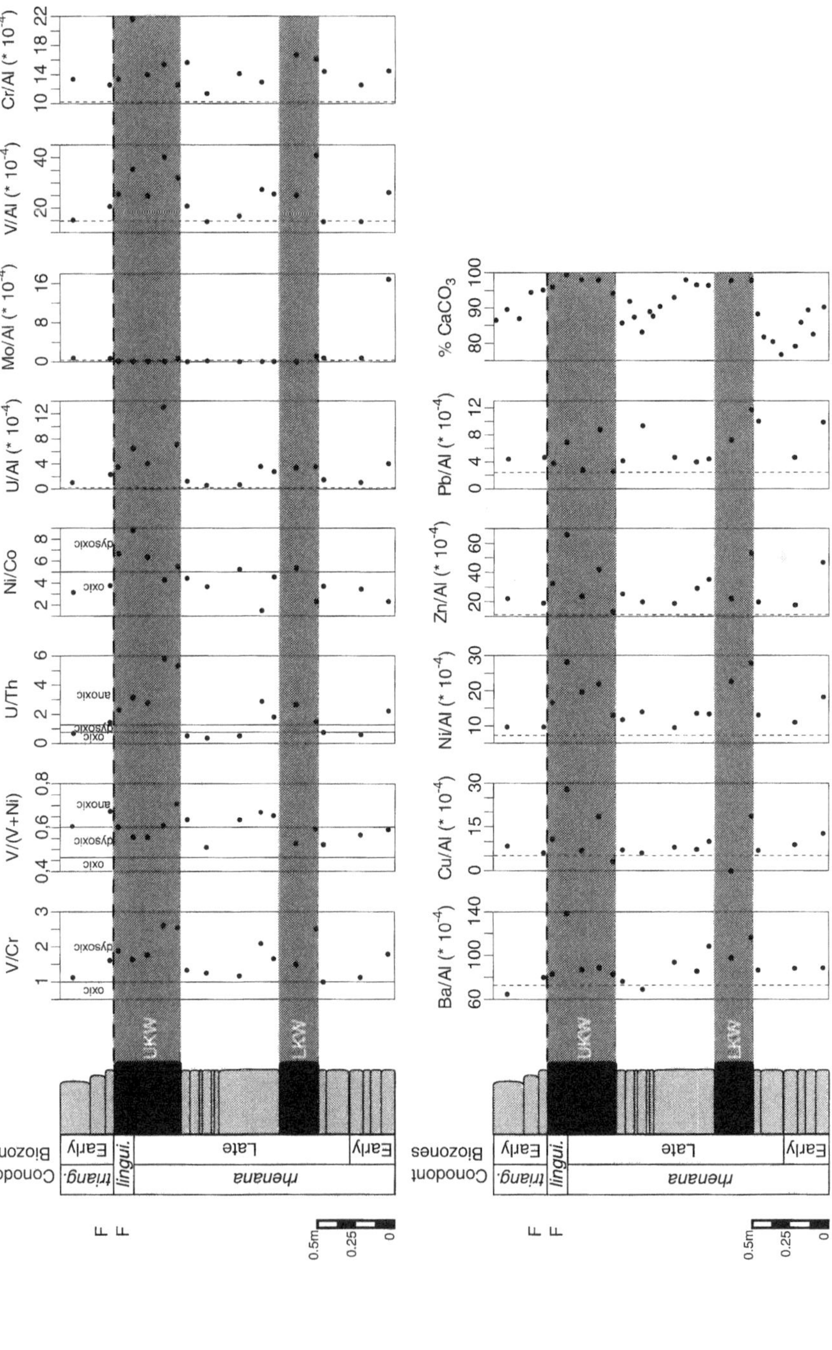


*Figure 6.* Stratigraphic distribution of carbonate content, Al-normalised trace-element concentrations and geochemical indices for the Anajdam section. Dashed lines indicate the metal/aluminium concentration ratios for average shales (Wedepohl, 1971, 1991) and the boundary of redox zones for V/Cr, Ni/Co, U/Th (from Jones and Manning, 1994) and V/(V+Ni) (from Hatch and Leventhal, 1992). Dark grey background corresponds to the black shale and limestone facies.

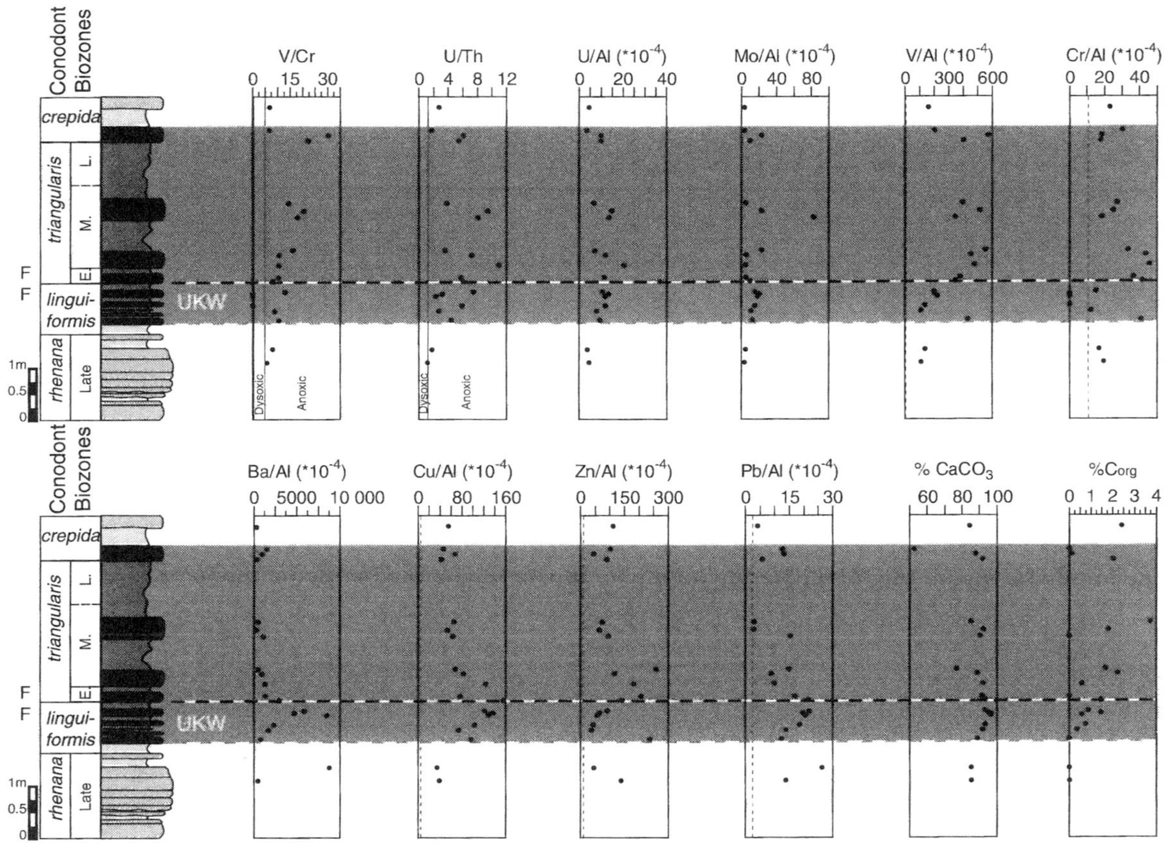


*Figure 7.* Stratigraphic distribution of carbonate content, Al-normalised trace-element concentrations and geochemical indices for the La Serre section. Dashed lines indicate the metal/aluminium concentration ratios for average shales (Wedepohl, 1971, 1991) and the boundary of redox zones for V/Cr, Ni/Co, U/Th (from Jones and Manning, 1994) and V/(V+Ni) (from Hatch and Leventhal, 1992). Dark grey background correspond to the black shale and limestone facies and white dashed lines indicate the time-equivalent upper Kellwaser horizon.

marked by a short positive peak of these ratios. The maximum $EF_{(U)}$ of the peaks varies between 10 at Anajdam and 23 at Bou-Ounebdou. $EF_{(Cu)}$ ranges from 3 to 7. The Upper Kellwasser horizon records a gradual and moderate rise of the ratios. $EF_{(Cu)}$ is about 3 in the three sections. At Aeketal and Bou-Ounebdou, samples from the *linguiformis* Zone are weakly to moderately enriched in U ($3 < EF_{(U)} < 8$), whereas at Anajdam $EF_{(U)}$ can reach 38. For the other trace elements, comparable trends are observed (Figs 4–6).

In the basinal environments at La Serre, trace metal elements are always highly enriched with respect to average shales, in contrast to the outer platform sections that exhibit relatively low EF. This concentration contrast is particularly exhibited with barium. Due to analytical problems and low sample quantities, Ni data are not available for this section. In Bou-Ounebdou, Anajdam, and Aeketal sections, the Kellwasser horizons are marked by a slight increase of barium ($EF_{(Ba)} < 2$), whereas in the La Serre section, $EF_{(Ba)}$ shows a remarkable increase from the base of the Upper Kellwasser horizon to the F–F boundary. $EF_{(Ba)}$ values rise up to 120, a few cm below the F–F boundary in the *linguiformis* Zone (Fig. 7). At La Serre, the concentrations of the other trace metals (Mo, U, V, Cr, Cu, and Zn) fluctuate widely through the F–F sequence. Peak values are observed immediately above the F–F boundary, in the Early to Middle *triangularis* Zone, $EF_{(U)}$ and $EF_{(Cu)}$ reach 109 and 30, respectively.

In the recent literature (Hatch and Leventhal, 1992; Jones and Manning, 1994), V/Cr, V/(V+Ni), Ni/Co, and U/Th indices have been used to derive information on paleo-oxygenation in the depositional environment. The four ratios show similar trends of higher values in the dark-coloured Kellwasser samples relative to overlying and underlying strata (Figs. 4–7); the highest values were recorded in the basinal La Serre section. In addition, the distribution of the four redox indices seems to confirm the diachronism of positive peaks between platform and basin settings observed with the trace metal concentrations. In platform environments, maximal values are recorded in the Kellwasser horizons during Late Frasnian. In the distal La Serre section, the samples of the time-equivalent Upper Kellwasser horizon (*linguiformis* Zone) do not show peak values; the highest values are found within the $C_{org}$-rich facies in the Early Famennian Early *triangularis* Zone.

### 4.3. *Summary*

The stratigraphical variations of the redox indices U/Th, V/Cr, Ni/Co, and V/(V+Ni) as well as the trace element concentrations show consistent fluctuations in space and time. Firstly, in the basinal environment at La Serre, the EF of trace elements are always higher than those recorded in outer platforms. Thus, the magnitude of trace metal enrichments varies with respects to the depositional setting. Secondly, the redox indices and the trace element concentrations are the highest within the black Kellwasser intervals. Moreover, it is noteworthy that in each platform setting, where the two Kellwasser horizons are found, the Lower Kellwasser records a rapid shift toward the highest values for most geochemical indices, particularly in central Morocco, whereas the Upper Kellwasser horizon is marked by a minor but more durable enrichment. In basinal setting, where only one prolonged interval of Kellwasser facies deposition is recognized, the increases of the redox indices are not limited to Late Frasnian times and extends into the Famennian strata. In contrast to other trace elements, Ba enrichments seem to be limited to Late Frasnian in all studied sections, both in proximal and distal settings. The element abundance lowers immediately above the F–F boundary.

## 5. Interpretation

Illite Crystallinity Index Standard (CIS) obtained from clay mineral analyses (unpublished data), indicate that the investigated sections have been affected by a pronounced diagenesis, especially in Germany. Illite CIS ranges from 0.30 to 0.41, which correspond to a maximal temperature of 300°C (Han et al., 2000). Nevertheless, the correlative geochemical trends observed in different lithologies and distant sites suggest that primary environmental signals were not significantly altered by diagenetic overprint, except for the organic-carbon content. In addition, Mongenot et al. (1996) have shown that strong thermal evolution of an OM-rich formation does not alter the trace-metal distribution in a detectable way, even in the case of severe OM destruction.

### *5.1. Paleo-oxygenation*

For each section, it is clear that the Kellwasser horizons are levels enriched in Mo, U, V, Cu, Cr, and Ni. These elements can be fixed in high amounts in sediments under reducing conditions (Brumsack, 1989; Hatch and Leventhal, 1992; Calvert and Pedersen, 1993). They may either be precipitated as autonomous sulphides, coprecipitated with iron sulphides, or bound to organic matter. Among the studied trace metals, V, U, and Mo are reputed as sensitive redox markers (a very large literature is devoted to the trace metals used as redox proxies, see Crusius et al., 1996; Helz et al., 1996; Dean et al., 1997; Zheng et al., 2000; Adelson et al., 2001; Lyons et al., 2003; Rimmer, 2004; Rimmer et al., 2004; Tribovillard et al., 2004a,b among recent papers with thorough presentations of these elements). At La Serre (distal setting), the enrichment in these elements is very strong, indicating the rapid development of reducing conditions in the sediments. Anoxia could have affected the water column itself as indicated by the ostracod fauna (Lethiers et al., 1998; Tribovillard et al., 2004a), which is in agreement with the high trace-element concentrations at La Serre. In outer platforms, where the increase of Al-normalized trace elements is obviously lower than in distal settings, the Kellwasser events seem to coincide with a period of oxygen decreasing, probably corresponding to dysoxic conditions only, and not to anoxic conditions. For instance, considering molybdenum abundance, the La Serre section shows much higher enrichments than the other sections studied. Following the papers by Helz et al. (1996), Zheng et al. (2000), Adelson et al. (2001), Vorlicek et al. (2002, 2004), and Lyons et al. (2003), it is currently admitted that Mo enrichment is tied to sulphidic conditions. It is inferred that, if sulphidic conditions unambiguously developed at La Serre, the depositional conditions for the other sections studied were probably no more severe than dysoxic.

Depositional conditions are assessed by cross-ploting of redox parameters V/Cr, V/(V+Ni), Ni/Co, and U/Th (Fig. 8). In this study, we use threshold values proposed by Hatch and Leventhal (1992) for V/(V+Ni) and by Jones and Manning (1994) for V/Cr, Ni/Co, and U/Th (Table 1). However, it must be kept in mind that these thresholds must be cautiously used because they were not specifically determined for Devonian environments, and thus the relative variations are of safer use. For each of the indices used, the Kellwasser samples show higher values and the data plot within the dysoxic and anoxic fields for proximal sections. In platform settings U/Th values indicate predominantly anoxic conditions for the dark-coloured horizons and oxic conditions below, between, and above the

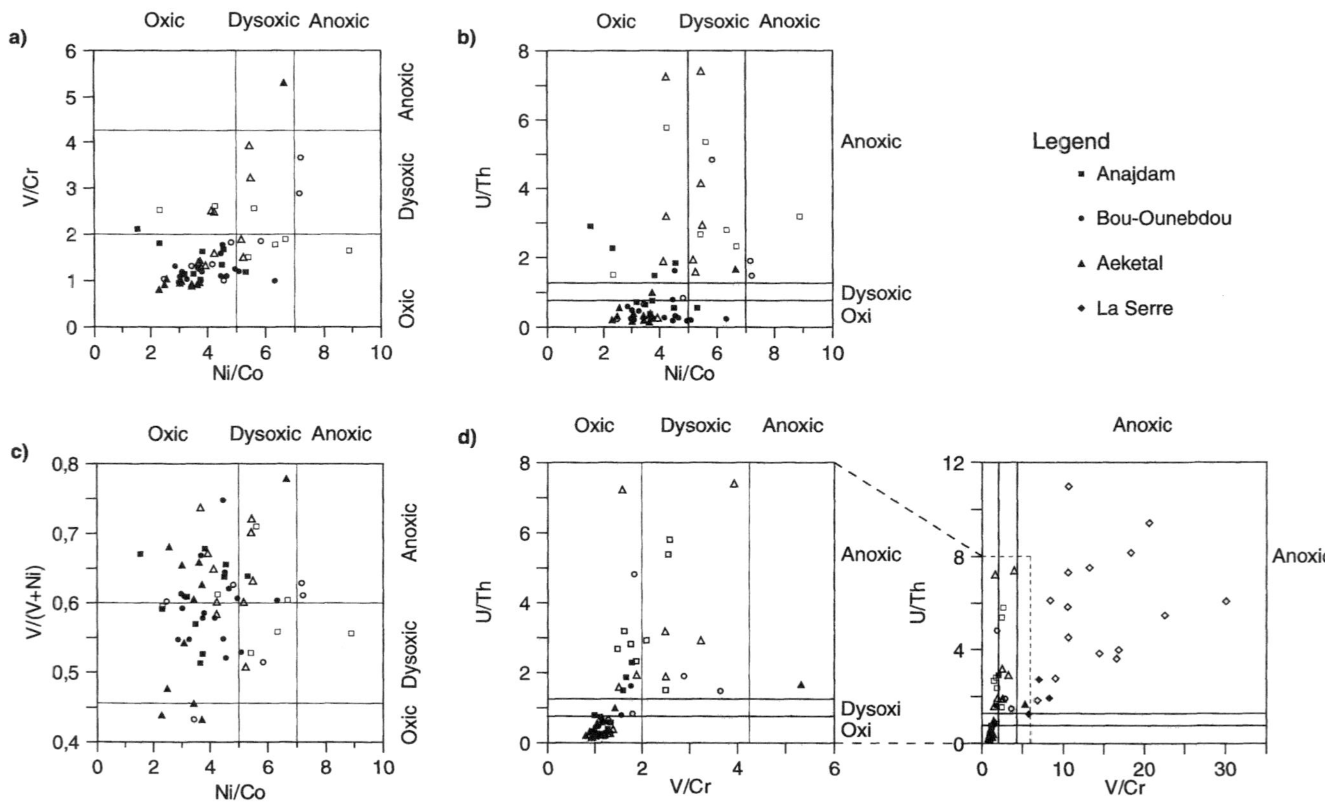


*Figure 8.* Crossplots of trace-elements ratios used as paleoredox proxies. (a) V/Cr vs. Ni/Co, (b) U/Th vs. Ni/Co, (c) V/(V+Ni) vs. Ni/Co, and (d) U/Th vs. V/Cr. Ranges for V/Cr, U/Th, and Ni/Co are from Jones and Manning (1994); ranges for V/(V+Ni) are from Hatch and Leventhal (1992). Ni concentrations not available for the La Serre section. Open symbols: Kellwasser samples.

*Table 1.* Summary of correlations between trace element ratio values and redox zones.

| | Oxic zone (8.0–2.0 $mlO_2/l$) | Dysoxic zone (2.0–0.2 $mlO_2/l$) | Anoxic zone (0.0 $mlO_2/l$) |
|---|---|---|---|
| U/Th[a] | <0.75 | 0.75–1.25 | >1.25 |
| V/Cr[a] | <2.00 | 2.00–4.25 | >4.25 |
| Ni/Co[a] | <5.00 | 5.00–7.00 | >7.00 |
| V/V+Ni[b] | <0.46 | 0.46–0.60 | 0.54–0.82 |

[a] Jones and Manning (1994).
[b] Hatch and Leventhal (1992).
Note : Bottom water oxygen levels from Tyson and Pearson, (1991).

Kellwasser levels. The V/Cr and Ni/Co ratio indicate oxic to dysoxic conditions for the Kellwasser interval and oxic conditions elsewhere. V/(V+Ni) ratio does not show clear distinctions between the two Kellwasser horizons and the other beds. Dysoxic to anoxic conditions for all sequences are suggested by V/(V+Ni) ratio. The chemical data indicate two pulses of bottom water dysoxia, coeval with the Kellwasser horizons in platform environments. Conversely, in the basinal La Serre section, the highest values of V/Cr and U/Th indicate that reducing conditions were more pronounced in deeper settings of the Gondwanan margins and anoxia was not limited to the Late Frasnian, but extended into the Early Famennian (*crepida* Zone; see also Tribovillard et al., 2004a). Thus, it clearly appears that the intensity and duration of bottom water oxygen depletion varied according to the paleodepth of deposition. Our data are in agreement with published ones concerning other F–F boundary sequences. In subpolar Ural area, Yudina et al. (2002) documented dysoxic conditions from the Early *rhenana* Zone to the Late *triangularis* Zone with a euxinic pulse in the Late *linguiformis* Zone. In Germany, the Steinbruch Benner section records two anoxic episodes in an otherwise well-oxygenated sequence (Bond et al., 2004), whereas in Poland, the basinal Kowala section records oxygen-depleted condition across the F–F boundary (Joachimski et al., 2001; Racki et al., 2002; Bond et al., 2004). In addition, Marynowski et al. (2000) and Joachimski et al. (2001) observed the presence of chlorobiacean pigment molecular markers at Kowala, an indication that anoxia rose into the photic zone.

### *5.2. Paleoproductivity*

Ba/Al, as well as Ni/Al, Cu/Al, and Zn/Al, shows noticeable variations from the Late Frasnian to Early Famennian. The distribution of dissolved Ba in the ocean appears to be related to biological production in surface water as Ba is incorporated in photosynthetic organisms (Dymond et al., 1992; MacManus et al., 1998). During decay, Ba can be partly released and recycled. It is supplied to the sediment and preserved primarily in the form of barite ($BaSO_4$) (Dymond et al., 1992). Ba abundance is commonly used as a paleoproductivity markers even if the interpretation is not always straight forward because barite can be dissolved in case of intense sulphate-reduction (McManus et al., 1998; Jacot des Combes et al., 1999). In that case, Ba can be released to pore waters, migrate with them, and re-precipitate when less reducing conditions are met. Thus barite migration fronts are formed (e.g. McManus et al., 1998). The distribution of dissolved Ni, Cu, and Zn in the ocean appears to be related to settling organic debris, since Ni, Cu, and Zn behave as micronutrients (Calvert and Pedersen, 1993; Algeo and Maynard, 2004). These elements are readily adsorbed onto organic particles settling through the water column. So, decaying OM easily forms organo-metal complexes that can be incorporated to the sediment (Algeo and Maynard, 2004; Tribovillard et al., 2004b). Thus, these elements are usually trapped with organic particles and are incorporated into iron sulphides during diagenesis. Consequently, they are preserved in the sediment even in case of complete OM remineralisation as well as in the case of marked thermal evolution (Mongenot et al., 1996). Thus, Ni, Cu, and Zn serve as indicators of organic inputs (Tribovillard et al., 2000, 2004b), and, in the present case, Cu abundance is a better proxy to OM inputs than TOC.

The highest values of paleoproductivity (Ba) and OM input (Cu) tracers are recorded in the basinal sequence at La Serre. As said above, reducing conditions developed at La Serre. Consequently, Ba diagenetical release and migration can be suspected. It implies that the strong Ba enrichments can be interpreted as an indication that barite was brought to the sediment by increased export production, but Ba can have been moved and concentrated after deposition. The level the richest in Ba are not necessarily precisely those recording the highest productivity. However, the Ba and Cu enrichments are observed in the same strata. As Cu is not known to be moved during early diagenesis, the good Ba–Cu covariations are an indication that Ba content was probably not deeply affected by post-deposition migrations. An independent indication of marked productivity at La Serre is reported by Paris et al. (1996) who based their interpretation on chitinozoan overabundance. The enrichment of these productivity tracers is outstanding compared to platform environments. However, in the La Serre section and in less distal settings, the increased abundance of Ba, and Cu is limited to the Late Frasnian. Consequently, it may be thought that the Late Frasnian, more particularly the Kellwasser intervals, are characterized by a rise in productivity and by enhanced OM accumulation. These conclusions agree with results obtained from carbon isotopic signal by Joachimski et al. (1994, 2002) on the sections studied here. These data are illustrated with Figures 4 and 5. In the Bou-Ounebdou and Aeketal sections, as well as in other F–F boundary sections in many places (Australia, Austria, Canada, China, France, Germany, Poland, USA; Joachimski et al., 2002), two positive excursions of $\delta^{13}C_{carb}$ with amplitude of around 3.5% have been recorded (Figs. 4 and 5). In the two sections considered here, a first excursion starts just above the base of the Late *rhenana* Zone and coincides with the deposition of the Lower Kellwasser horizon. Between the two Kellwasser horizons, $\delta^{13}C_{carb}$ values decrease toward the pre-excursion background level. The second excursion occurs below the base of the Upper Kellwasser horizon. This increase in $\delta^{13}C_{carb}$ starts in the uppermost Late *rhenana* Zone at Bou-Ounebdou and at the base of the *linguiformis* Zone at Aeketal. In both sections, the maximum $\delta^{13}C_{carb}$ values are recorded several centimetres above the top of the Kellwasser horizons. Then, a gradual decrease toward lighter values is recorded in the *triangularis* Zone. The two Late Frasnian positive shifts in $\delta^{13}C$ are explained by an increase in the organic carbon burial rate (Joachimski et al., 2002). This phenomenon may be due to either better preservation of organic matter, as a consequence of reducing conditions, and/or high primary productivity. A productivity increase results from higher concentration of $PO_4^{3-}$ and $NO_3^-$ that are biolimiting nutrients, and possibly of iron (Martin, 1990; Falkowski et al., 1998). The nutrient supply can originate from recycling processes such as upwelling, basinal turnover, P regeneration from anoxic sediments (Ingall and Van Cappellen, 1990; Murphy et al., 2000a) or continental runoff (Peterhänsel and Pratt, 2001). Reactive iron can be also significantly brought by winds, e.g. as clay-mineral coatings.

The contrasting record of paleoproductivity between deep and shallower water settings may be explained by environmental conditions. Generally speaking, organic productivity on platforms may be high as a consequence of an enhanced nutrient flux from continental weathering and, possibly, upwelling, which may favour the onset of dysoxic conditions and OM burial (Tyson and Pearson, 1991). Nevertheless, higher oxygenation levels and sensitive sea-level fluctuations in proximal environments generally do not allow good accumulation and preservation of OM. Platform settings do not allow favourable conditions for OM burial, particularly during sea-level lowstands (see discussion about this aspect in Tyson and Pearson, 1991). In basins, where terrestrial inputs are lower, primary productivity was seemingly high during the Kellwasser interval. OM preservation was high due to a marked oxygen deficiency, in addition to high productivity and relatively reduced sedimentation rate.

To conclude, it seems that the record of primary production was potentially related to the depositional setting, with deep-water environments showing the highest record for water mass fertility (geochemical data and independent paleoecological data). The sediments have recorded productivity indications in conjunction with favourable, oxygen-deficient paleoenvironmental conditions affecting basinal settings. In all studied sections, pulses of recorded productivity seem to be limited to the Late Frasnian. Similar trends have been observed for the Polish and Ural sections (Racki et al., 2002; Yudina et al., 2002). There, the highest concentrations of productivity proxies are recorded below and within the F–F boundary strata.

## 6. Discussion

Numerous authors have postulated that the usually organic C-rich Kellwasser horizons resulted from enhanced preservation of OM (Buggisch, 1991; Joachimski et al., 1994, 2002), under $O_2$-depleted conditions and high primary productivity (this study; Paris et al., 1996; Racki, 1998, 1999; Racki et al., 2002; Filipiak, 2002; Giles et al., 2002; Murphy et al., 2000a), presumably linked to short-term transgressive–regressive pulses (Buggisch, 1991; Sandberg et al., 2002) and an incipient climate cooling (Copper, 1986). Primary productivity and thus levels of dissolved oxygen in marine domains are controlled by nutrient availability in surface waters. The nutrients can be delivered mainly by vertical convection currents during oceanic overturn and fluvial runoff, together with submarine hydrothermalism and volcanism. The respective contributions of these various environmental nutrient sources, which associate several transport processes (fluvial, eolian, marine currents), depend on geographic context and consequently may vary locally (Mutti and Hallock, 2003).

The onset of eutrophic (namely, rich in nutrients) conditions triggers enhanced marine production. A high rate of primary production increases the accumulation of organic matter and accelerates consumption of oxygen within the bottom water. On platforms, anoxia develops in the bottom water when the respiratory demands of heterotrophic activity, i.e., OM degradation mainly carried out by aerobic bacteria, exceeds the supply of dissolved oxygen derived from photosynthetic activity and exchanges with the atmosphere (Brasier, 1995). Other processes, such as thermohaline stratification and restricted lateral circulation in the deep-water column, are generally invoked to explain the development of anoxic conditions (Cruse and Lyons, 2004). From these observations, several hypotheses and models have been developed to explain the onset of eutrophication in surface water and the development of bottom water anoxia during the time slice encompassing the F–F boundary.

### *6.1. Anoxia*

Late Devonian anaerobic environments are thought to have developed because of water stratification in basinal settings, that may have resulted from the formation of a halocline associated with repeated spills of warm, saline surface water generated by evaporation on sub-tropical platforms (Joachimski and Buggisch, 1993; Hallam and Wignall, 1997; Claeys et al., 1996; Marynowski et al., 2000; Joachimski et al., 2001). High evaporation rates caused an increase of water salinity, producing denser saline waters that subsequently sunk down and were carried to bottom waters in distal settings. Water stratification induced in turn bottom water anoxia. In modern shallow environments, water is usually mixed down to a depth of about 150–200 m by wind energy and diffusion, preventing the onset of water stratification and anoxia formation. In such conditions, the redox boundary or chemocline is situated at the sediment surface or, more frequently, within the sediment (the reader is referred to the articles of the volume edited by Tyson and Pearson, 1991). Consequently, most of OM is not accumulated in sediment because of aerobic degradation within the water column and at the sediment–water interface. Water stratification may take place in relatively shallow environments during sea-level rise, as it possibly occurred for the two Kellwasser events (e.g., Wignall and Newton, 2001). During transgression, former shallow continental areas were probably flooded, which increased the formation of warm, saline, and dense water masses by evaporation on sub-tropical platforms (Hallam and Wignall, 1997). The oxic–anoxic interface rose in the water column and previously well-oxygenated platform environments were covered by anoxic waters (Joachimski and Buggisch, 1993; Walliser, 1996). The lateral migration of oxygen-depleted water onto shelves in response to eustatic transgressions certainly led to reduced availability of ecological niches, potentially causing the death of nektobenthic organisms. In deeper, quiet water environments, water stagnation, density contrast, and thus anoxia, were relatively stable. Thus, reducing conditions could prevail in the basin depositional environments during F–F transition and could last until early Famennian (this study, Lethiers et al., 1998; Joachimski et al., 2001; Racki, 1998; Tribovillard et al., 2004a), whereas in platform environments, oxygen-depletion could develop only during the two pulses of sea-level rise corresponding to the Kellwasser horizons.

### *6.2. Nutrient recycling*

Water stratification and anoxia were perhaps not permanent because of episodic oceanic turnover or stratification disruption (Wilde and Berry, 1984; Halas et al., 1992; Becker and House, 1994; Racki et al., 2002). Long-term water stratification and bottom water anoxia development could facilitate the preferential release of P and N from OM during carbon oxidation (Ingall and Jahnke, 1997; Murphy et al., 2000a). Another source of nutrients in bottom water may be hydrothermal megaplumes (Vogt, 1989). According to some authors (Racki, 1998, 1999; Racki et al., 2002; Yudina et al., 2002), Variscan submarine volcanic eruptions may have episodically released Fe, P, and other trace elements, such as V and Cr. Upon release, episodic water mixing would have brought nutrient-rich bottom waters up into the photic zone, establishing eutrophic conditions and stimulating temporarily phytoplankton productivity. The transfer of dissolved P from anoxic bottom water to ocean surface creates a positive feedback loop between high surface productivity, anoxia and P regeneration in bottom water (Ingall and Jahnke, 1997). Intensification of upwelling can be caused by changes in water circulation and by wind-stress, in response to climatic or sea-level variations. The incipient climatic cooling, which destabilized the greenhouse climate during the Late Devonian (Copper, 1986; Buggisch, 1991), favoured seasonality contrasts and the effectiveness and the frequency of water mixing increased (Sageman et al., 2003). During the Kellwasser events, deepening phases probably prevented mixing in bottom water and enhanced water stratification whereas the shallowing phases, recorded at the end of the dark-coloured horizons, would have increased stratification disruptions.

### *6.3. Terrestrial nutrient inputs*

Based on isotopic C composition, Joachimski et al. (2001, 2002) suggest that upwelled nutrient-rich deep waters were not the only cause for eutrophication and that during the Kellwasser events, the productivity must have been stimulated by an enhanced input of continent-derived nutrient under greenhouse conditions (Algeo et al., 1995; Algeo and Scheckler, 1998; Joachimski et al., 2001; Girard and Lecuyer, 2002; Chen et al., 2002; Martin, 2003). Evidence of large-scale high continental input into the ocean is suggested by the strontium $^{87}Sr/^{86}Sr$ isotope signal (Veizer et al., 1997; Goddéris and Veizer, 2000), by the rare earth element distribution (Girard and Lécuyer, 2002), and by the distribution of the magnetic parameters of the formations that were deposited close to the paleo-reliefs (Devleeschouwer, 1999; Crick et al., 2002). Weathering of aluminosilicate rocks supplies nutrients and high continental weathering rates are controlled by enhanced humidity conditions and sea level falls (François et al., 1993; Royer et al., 2001; Berner and Kothavala, 2001). In addition, mountain uplift automatically enhances the erosion/weathering intensity and is thus considered to increase the nutrient flux to the seas independent of the sea-level fluctuations: Racki (1998), Gibbs et al. (1999), Peterhänsel and Pratt (2001), Tribovillard et al. (2004a), and Averbuch et al. (2005) suggest that tectonic activity must have been a factor enhancing long-term nutrient flux to the oceans at the Frasnian/Famennian boundary. Increases in seafloor spreading rates enhance volcanism and mountain building, and potentially control sea-level fluctuations (Racki, 1998). This mechanism would result in a higher supply of nutrients to shallow water environments. Volcanic activity is indicated by some bentonite layers, found between the two Kellwasser horizons at Steinbruch Schmidt (Germany) and in other localities (USA) (Tucker et al., 1998; Kaufmann et al., 2004). Volcanic ashes have a high fertilization potential. According to Racki et al. (2002) and Yudina et al. (2002), nutrients, supplied in Urals and Polish areas, resulted from replenishing by volcanism and hydrothermal activity. Variscan volcanism influence may not be excluded, particularly in the Laurussian margin as a nutrient source. However, the traces of volcanism remain relatively scarce. The onset and the acceleration of Ellesmerian–Eovariscan orogenic

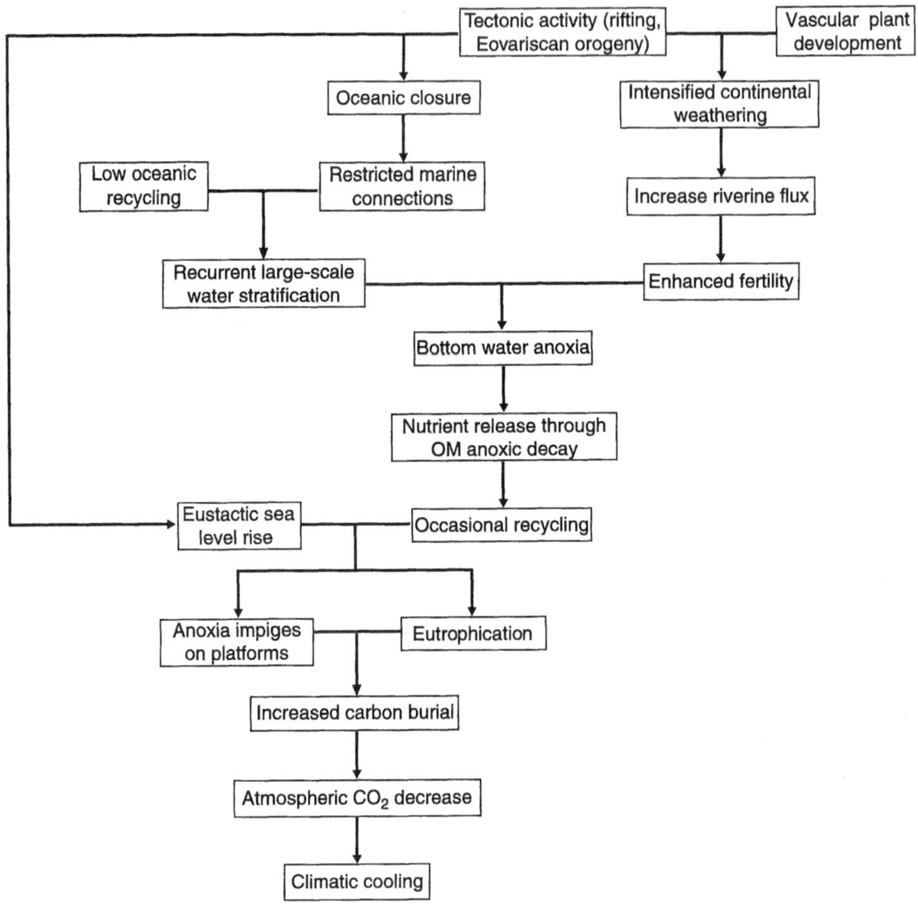


*Figure 9.* Flowchart illustrating possible links between several environmental processes during Late Devonian.

activities probably stimulated weathering on emerged landmasses. In addition, mountain building might modify atmospheric circulations and collision of Laurussia and northwestern Gondwana may cause changes in oceanic water circulations (Copper, 1986). Reduced circulation slows the ventilation of deep water.

Algeo et al. (1995) have proposed to link the enhanced continental weathering with the development of vascular land plants. Palaeobotanical innovations, i.e. arborescence, advanced roots systems as well as the seed habit, are believed to have led to an intensification of pedogenesis and enhanced chemical weathering by humic acid leaching (Retallack, 1997). The terrestrial nutrient hypothesis is interesting for long-term supply of continent-derived nutrients. Rejuvenation of continental rocks via repeated uplift associated with vascular plant development would ensure continued supply of nutrients in the Late Devonian.

Accompanying this long-term trend that can explain the long-term OM enrichment in basinal settings, the short-term Kellwasser events can be explained by two additional pulses of sea-level variation that would have allowed increased nutrient-cycling and thus anoxia to impinge on platforms and end with the deposition of the Kellwasser horizons (Fig. 9).

### *6.4. Climatic consequences*

Silicate alteration, enhanced by large uplifting starting during Frasnian, and the burial of large quantities of OM by the end of Frasnian, resulting from high primary production and reducing conditions, stimulated $CO_2$-pumping. The decreased concentration of this greenhouse gas could have resulted in climatic cooling (Berner, 1997; Berner and Kothavala, 2001; Royer et al., 2001). Basing on conodont apatite $\delta^{18}O$ signatures, Joachimski and Buggisch (2002) proposed a temperature decrease of 5–7°C for tropical sea-surface during the Kellwasser horizons. The cooling of sea water, which could not be tolerated by many organisms has been postulated as a factor involved in the Kellwasser crisis (Copper, 1986). The extinction of the Late Frasnian carbonate-producing reef ecosystem may have reduced $CO_2$ concentration in surface water (Copper, 2002). The climatic cooling may have caused a sea-level fall, recorded at the end of the Kellwasser deposition that led to emergence of platform and reduction of ecological niches.

## 7. Conclusions

The geochemical study of four F–F boundary sections located on both sides of the Eovariscan belt shows that the depositional conditions of the Kellwasser horizons were strongly influenced by productivity and anoxia. The geographical distribution of the studied sections indicates that it was not just local conditions. In basinal environments of La Serre and Kowala, reducing conditions developed in a dominant, but probably not permanent manner, and these conditions extended into the early Famennian, at least to the *crepida* Zone. In platform settings (Aeketal, Anajdam, and Bou-Ounebdou), dysoxic–anoxic conditions developed only twice (Late *rhenana* and *linguiformis* zones). The oxygen-depleted conditions probably developed in response to sustained productivity, coupled to recurrent water stratification and possible reduction of sedimentation rates during sea-level rises. The pulse of productivity, recorded in the Late Frasnian, is interpreted as sustained primarily mostly by an increased land-derived nutrient flux, enhanced by the weathering of the incipient Eovariscan belt uplift. This primary nutrient input is probably reinforced by episodes of marine nutrient recycling. The 'chain of reactions' is illustrated with Figure 9. Lastly, it can be hypothesized that OM storage and atmospheric $pCO_2$ drawdown of the Late Devonian induced a global climate cooling culminating with Devonian glaciations.

## Acknowledgements

We thank Nezha Lazreq, Catherine Crônier, Ivan Berra, Said Chakiri, and Alain Herbosh for their assistance during field work, Jacques Morel for the ICP-MS measurements at CRPG-CNRS (Vandœuvre-les-Nancy, France), and François Baudin for the Rock Eval pyrolysis. We thank Walter Dean and an anonymous reviewer who helped to improve this paper considerably. We thank D. Jeffrey Over for his comments and corrections. This is a contribution to the 'Eclipse' Program of the CNRS-INSU.

## References


Adelson, J.M., Helz, G.R., Miller, C.V., 2001. Reconstructing the rise of recent coastal anoxia; molybdenum in Chesapeake Bay sediments. Geochim. Cosmochim. Acta 65, 237–252.

Algeo, T.J., Berner, R.A., Maynard, J.B., Scheckler, S.E., 1995. Late Devonian oceanic anoxic events and biotic crises: rooted in the evolution of vascular land plants? GSA Today 5, 64–66.

Algeo, T.J., Scheckler, S.E., 1998. Terrestrial-marine teleconnections in the Devonian: links between the evolution of land plants, weathering processes, and marine anoxic events. Philos. Trans. R. Soc. London, B 353, 113–130.

Algeo, T.J., Maynard, J.B., 2004. Trace-element behavior and redox facies in core shales of Upper Pennsylvanian Kansas-type cyclothems. Chem. Geol. 206, 289–318.

Arthur, M.A., Jenkyns, H.C., Brumsack, H.J., Schlanger, S.O., 1990. Stratigraphy, geochemistry, and paleoceanography of organic carbon-rich Cretaceous sequences. In: Ginsburg, R.N., Beaudoin, B. (Eds), Cretaceous Resources, Events and Rhythms. Kluwer Academic Publishing, Amsterdam, pp. 75–119.

Averbuch, O., Tribovillard, N., Devleeschouwer, X., Riquier, L., Mistiaen, B., van Vliet-Lanoë, B., 2005. Mountain building-enhanced continental weathering and organic carbon burial as major causes for climatic cooling at the Frasnian–Famennian boundary (ca 376 Ma BP). Terra Nova 17(1), pp. 33–42.

Becker, R.T., House, M.R., 1994. Kellwasser events and goniatite successions in the Devonian of Montagne Noire with comments on possible causations. Cour. Forsch. Inst. Senckenb. 169, 45–77.

Becker, R.T., House, M.R., 2000. Sedimentary and faunal succession of the allochtonous Upper Devonian at Gara d'Mrirt (Eastern Moroccan Meseta). Notes Mém. Serv. Géol. Maroc 399, 109–114.

Berner, R.A, 1997. The rise of plants and their effects on weathering and atmospheric $CO_2$. Science 276, 544–546.

Berner, R.A., Kothavala, Z., 2001. Geocarb III: a revised model of atmospheric $CO_2$ over Phanerozoic times. Am. J. Sci. 301, 182–204.

Bond, D., Wignall, P.B., Racki, G., 2004. Extent and duration of marine anoxia during the Frasnian–Famnennian (Late Devonian) mass extinction in Poland, Germany, Austria and France. Geol. Mag. 141, 173–193.

Brasier, M.D., 1995. Fossil indicators of nutrient levels. 1: Eutrophication and climate change. In: Bosence, D.W.J., Allison, P.A. (Eds), Marine palaeoenvironmental analysis from fossils. Geol. Soc. London Spec. Publ. 83, pp. 113–132.

Brumsack, H.J., 1989. Geochemistry of recent TOC-rich sediments from the Gulf of California and the Black Sea. Geol. Rundsch. 78, 851–882.

Buggisch, W., 1991. The global Frasnian–Famennian "Kellwasser Event". Geol. Rundsch. 80, 49–72.

Calvert, S.E., Pedersen, T.F., 1993. Geochemistry of recent oxic and anoxic marine sediments: implications for the geological record. Mar. Geol. 113, 67–88.

Caplan, M.L., Bustin, R.M., 1998. Paleoceanographic controls on geochemical characteristic of organic-rich Exshaw mudrocks: role of enhanced primary productivity. Org. Geochem. 30, 161–188.

Caplan, M.L., Bustin, R.M., 1999. Devonian–Carboniferous Hangenberg mass extinction event, widespread organic-rich mudrocks and anoxia: causes and consequences. Palaeogeogr. Palaeoclimatol. Palaeoecol. 149, 187–207.

Carignan, J., Hild, P., Morel, J., Yeghicheyan, D., 2001. Routine analyses of trace elements in geochemical samples using flow injection and low pressure on-line liquid chromatography coupled to ICP-MS: a study of geochemical reference materials BR, DR-N, UB-N, AN-G and GH. Geostand Newslett. 25, 187–198.

Chakiri, S., 2002. Sédimentologie et géodynamique du Maroc Central Hercynien pendant le Dévonien. Unpublished PhD thesis, Ibn Tofail University, Kenitra.

Chen, D., Tucker, M.E., Shen, Y., Yans, J., Préat, A., 2002. Carbon isotope excursions and sea-level change: implication for the Frasnian–Famennian biotic crisis. J. Geol. Soc. 159, 623–626.

Claeys, P., Kyte, F.T., Herbosch, A., Casier, J.-G., 1996. Geochemistry of the Frasnian–Famennian boundary in Belgium: mass extinction, anoxic oceans, and microtektite layer, but no much iridium? Geol. Soc. Am. Spec. Pap. 307, 491–504.

Copper, P., 1986. Frasnian/Famennian mass extinction and cold-water oceans. Geology 14, 835–839.

Copper, P., 2002. Reef development at the Frasnian–Famennian mass extinction boundary. Palaeogeogr. Palaeoclimatol. Palaeoecol. 181, 27–65.

Crick, R.E., Ellwood, B.B., Feist, R., El Hassani, A., Schindler, E., Dreesen, R., Over, D.J., Girard, C., 2002. Magnetostratigraphy susceptibility of the Frasnian/Famennian boundary. Palaeogeography, Palaeoclimatology, Palaeoecology, 181, 67–90.

Cruse, A.M., Lyons, T.W., 2004. Trace metal records of regional paleoenvironmental variability in Pennsylvanian (Upper Carboniferous) black shales. Chem. Geol. 206, 319–345.

Crusius, J., Calvert, S., Pedersen, T., Sage, D., 1996. Rhenium and molybdenium enrichments in sediments as indicators of oxic, suboxic, and sulfidic conditions of deposition. Earth Planet. Sci. Let. 145, 65–78.

Dean, W.E., Gardner, J.V., Piper, D.Z., 1997. Inorganic geochemical indicators of glacial-interglacial changes in productivity and anoxia on the California continental margin. Geochim. Cosmochim. Acta 61, 4507–4518.

Devleeschouwer, X., 1999. La transition Frasnien/Famennien (Dévonien sup.) en Europe : sédimentologie, stratigraphie séquentielle et susceptibilité magnétique. Unpublished PhD thesis, FreeUniversity, Brussels.

Dymond, J., Suess, E., Lyle, M., 1992. Baryum in deep-sea sediments: a geochemical proxy for paleoproductivity. Paleoceanography 7, 163–181.

El Hassani, A., Tahiri, A., 1999. Excursion guide book. Sub-commission Devonain Stratigraphy- IGCP 421 Morrocco Meeting. April-Mai – Rabat-Errachidia, 160 pp.

Falkowski, P.G., *Barber*, R.T., Smetacek, V., 1998, Biogeochemical controls and feedbacks on ocean primary production. *Science* 281, 200–206.

Feist, R., 1985. Devonian stratigraphy of the South-eastern Montagne Noire (France). Cour. Forsch. Inst. Senckenb. 75, 331–352.

Feist, R., 1990. The Frasnian–Famennian boundary and adjacent strata of the eastern Montagne noire, France, IUGS Subcommission on Devonian Stratigraphy, Field Meeting 1990, Guide Book. Montpellier, 69 pp.

Feist, R., 2002. Montagne Noire 25–28 June 2002.–IGCP 421: North Gondwana mid-Paleozoic biodynamics and ECOS VIII: 8th European Conodont Symposium—Guidebook of the Field Excursion. Institut des Sciences de l'Evolution, University of Montpellier II, Montpellier, pp. 1–85.

Feist, R., Schindler, E., 1994. Trilobites during the Frasnian Kellwasser Crisis in European Late Devonian cephalopod limestones. Cour. Forschung. Inst. Senckenb. 169, 195–223.

Filipiak, P., 2002. Palynofacies around the Frasnian–Famennian boundary in the Holy Cross Mountains, southern Poland. Palaeogeogr,. Palaeoclimatol., Palaeoecol. 181, 313–324.

François, L., Walker, J.C.G., Opdyke, B.N., 1993. The history of global weathering and the chemical evolution of the ocean-atmosphere system. Geophys. Monogr. 74, vol.14, 143–159.

Gibbs, M.T., Bluth, G.J., Fawcett, P.J., Kump, L.R., 1999. Global chemical erosion over the last 250 My: variations due to changes in paleogeography, paleoclimate and paleogeology. Am. J. Sci. 299, 611–651.

Giles, K.A., McMillan, N.J., McCarson, B.L., 2002. Geochemical analysis and paleoecological implications of phosphatic microspherules (otoliths?) from Frasnian–Famennian boundary strata in the Great Basin, USA. Palaeogeogr. Palaeoclimatol. Palaeoecol. 181, 111–125.

Girard, C., Albarède, F., 1996. Trace elements in conodont phosphate from the Frasnian/Famennian boundary. Palaeogeogr. Palaeoclimatol. Palaeoecol. 126, 195–209.

Girard, C., Lécuyer, C., 2002. Variations in Ce anomalies of conodonts through the Frasnian–Famennian boundary of Poland (Kowala – Holy Cross Mountains): implications for the redox state of seawater and biodiversity. Palaeogeogr. Palaeoclimatol. Palaeoecol. 181, 299–311.

Goddéris, Y., Veizer, J., 2000. Tectonic control of chemical and isotopic composition of ancient oceans: the impact of continental growth. American Journal of Science, 200, 434–461.

Godderis and Veizer, as well as Crick et al, are not in the reference list.

Here are the complete references:

Goddéris Y., Veizer J., 2000. Tectonic control of chemical and isotopic composition of ancient oceans: the impact of continental growth. American Journal of Science, 200, 434-461.

Halas, S., Balinski, A., Gruszczynski, M., Hoffman, A., Malkowski, K., Narkiewicz, M., 1992. Stable isotope record at the Frasnian/Famennian boundary in southern Poland. N. Jb. Geol. Paläontol. Mh. 3, 129–138.

Hallam, A., Wignall, P.B., 1997. Mass extinctions and sea-level changes. Earth. Sci. Rev. 48, 217–250.

Han, G., Preat, A., Chamley, H, Deconinck, J.-F., Mansy, J.-L., 2000. Palaeozoic clay mineral sedimentation and diagenesis in the Dinant and Avesnes Basins (Belgium, France): relationships with Variscan tectonism. Sed. Geol. 136, 217–238.

Hatch, J.R., Leventhal, J.S., 1992. Relationship between inferred redox potential of the depositional environment and geochemistry of the Upper Pennsylvanian (Missourian) Stark Shale Member of the Dennis Limestone, Wabaunsee County, Kansas, U.S.A. Chem. Geol. 99, 65–82.

Helz, G.R., Miller, C.V., Charnock, J.M., Mosselmans, J.L.W., Pattrick, R.A.D., Garner, C.D., Vaughan, D.J., 1996. Mechanisms of molybdenum removal from the sea and its concentration in black shales: EXAFS evidences. Geochim. Cosmochim. Acta 60, 3631–3642.

House, M.R., 2002. Strength, timing, setting and cause of mid-Palaeozoic extinctions. Palaeogeogr. Palaeoclimatol. Palaeoecol. 181, 5–25.

Ingall, E.D., Van Cappellen, P., 1990. Relation between sedimentation rate and burial of organic phosphorus and organic carbon in marine sediments. Geochim. Cosmochim. Acta 54, 1219–1234.

Ingall, E., Jahnke, R., 1997. Influence of water-column anoxia on the elemental fractionation of carbon and phosphorus during diagenesis. Mar. Geol. 139, 219–229.

Jacot des Combes, H., Caulet, J. – P., Tribovillard N., 1999. Pelagic productivity changes in the equatorial area of the NW Indian Ocean during the last 400 kyr. Mar. Geol. 158, 27–55.

Joachimski, M.M., Buggisch, W., 1993. Anoxic events in the late Frasnian – Causes of the Frasnian–Famennian faunal crisis? Geology 21, 675–678.

Joachimski, M.M., Buggisch, W., 2002. Conodont apatite $\delta^{18}O$ signatures indicate climatic cooling as a trigger of the Late Devonian mass extinction. Geology 30, 711–714.

Joachimski, M.M., Buggisch, W., Anders, T., 1994. Mikrofazies, Conodontenstratigraphie und Isotopengeochimie des Frasne/Famenne Grenzprofils Wolayer Gletscher (Karnische Alpen). Abh. Geol. Bundesanst. Wien 50, 183–195.

Joachimski, M.M., Ostertag-Henning, C., Pancost, R.D., Strauss H., Freeman, K.H., Littke, R., Sinninghe Damsté, J.S., Racki, G., 2001. Water column anoxia, enhanced productivity and concomitant changes in $^{13}C$ and $^{34}S$ across the Frasnian–Famennian boundary (Kowala Holy Cross Mountains/Poland). Chem. Geol. 175, 109–131.

Joachimski, M.M., Pancost, R.D., Freeman, K.H., Ostertag-Henning, C., Buggisch, W., 2002. Carbon isotope geochemistry of the Frasnian–Famennian transition. Palaeogeogr. Palaeoclimatol. Palaeoecol. 181, 91–109.

Jones, B., Manning, D.A.C., 1994. Comparison of geochemical indices used for the interpretation of palaeoredox conditions in ancient mudstones. Chem. Geol. 111, 111–129.

Kaufmann, B., Trapp, E., Mezger, K., 2004. The numerical age of the Late Frasnian (Late Devonian) Kellwasser horizons – a new U-Pb zircon date from Steinbruch Schmidt (Kellerwald, Germany). J. Geol. 112, 495–501.

Lazreq, N., 1992. The upper Devonian of M'rirt (Morocco). Cour. Forsch. Inst. Senckenb. 154, 107–123.

Lazreq, N., 1999. Biostratigraphie des conodontes du Givétien au Famennien du Maroc Central. Biofaciès et événement Kellwasser. Cour. Forsch. Inst. Senckenb. 214, 1–111.

Lethiers, F., Baudin, F., Casier, J.-G., 1998. Ostracodes de la limite Frasnien–Famennien en environnement anoxique (La Serre, Montagne Noire, France). Rev. Micropal. 41, 321–336.

Lyons, T.W., Werne, J.P., Hollander, D.J., Murray, R.W., 2003. Contrasting sulfur geochemistry and Fe/Al and Mo/Al ratios across the last oxic-to-anoxic transition in the Cariaco Basin, Venezuela. Chem. Geol. 195, 131–157.

McGhee, G.R., 1989. The Frasnian–Famennian event. In: Donovan, S.K. (Eds), Mass Extinction: Processes and Evidences. Belhaven Press, London, pp. 133–151.

McGhee, G.R., 1996. The Late Devonian Mass Extinction. The Frasnian–Famennian Crisis. Columbia University Press, New York.

McManus, J., Berelson, W.M., Klinkhammer, G.P., Johnson, K.S., Coale, K.H., Anderson, R.F., Kumar, N., Burdige, D.J., Hammond, D.E., Brumsack, H.-J., McCorkle, D.C., Rushdi, A., 1998. Geochemistry of barium in marine sediments: implications for its use as a paleoproxy. Geochim. Cosmochim. Acta 62, 3453–3473.

Martin, J.H., 1990. Glacial–interglacial $CO_2$ change: the iron hypothesis. Paleoceanography 5, 1–13.

Martin, R.E., 2003. The fossil record of biodiversity: nutrients, productivity, habitat area and differential preservation. Lethaia 36, 179–194.

Marynowski, L., Narkiewicz, M., Grelowski, C., 2000. Biomarkers as environmental indicators in a carbonate complex, example from the Middle to Upper Devonian, Holy Cross Mountains, Poland. Sediment. Geol. 137, 187–212.

May, A., 1995. Relationship among sea-level fluctuation, biogeography and bioevents of the Devonian: an attempt to approach a powerful, but simple model for complex long-range control of biotic crises. Geolines 3, 38–49.

Mongenot, T., Tribovillard, N.-P., Desprairies, A., Lallier-Vergès, E., Laggoun-Defarge, F., 1996. Trace elements as palaeoenvironmental markers in strongly mature hydrocarbon source rocks: the Cretaceous La Luna Formation of Venezuela. Sediment. Geol. 103, 23–37.

Murphy, A.E., Sageman, B.B., Hollander, D.J., 2000a. Eutrophication by decoupling of the marine biogeochemical cycles of C, N and P: a mechanism for the Late Devonian mass extinction. Geology 28, 427–430.

Murphy, A.E., Sageman, B.B., Hollander, D.J., Lyons, T.L., Brett, C.E., 2000b. Black shale deposition and faunal overturn in the Devonian Appalachian Basin: clastic starvation, seasonal water-column mixing, and efficient biolimiting nutrient recycling. Paleoceanography 15, 280–291.

Mutti, M., Hallock, P., 2003. Carbonate systems along nutrient and temperature gradients: some sedimentological and geochemical constraints. Int. J. Earth Sci. 92, 465–475.

Paris, F., Girard, C., Feist, R., Winchester-Seeto, T., 1996. Chitinozoan bio-event in the Frasnian–Famennian boundary beds at La Serre (Montagne Noire, Southern France). Palaeogeogr. Palaeoclimatol. Palaeoecol. 121, 131–145.

Peterhansel, A., Pratt, B.R., 2001. Nutrient-triggered bioerosion on a giant carbonate platform masking the post-extinction Famennian benthic community. Geology 29, 1079–1082.

Pratt, L.M., Claypool, G.E., King, J.D., 1986. Geochemical imprint of depositional conditions on organic matter in laminated-bioturbated interbeds from fine-grained marine sequences. Mar. Geol. 70, 67–84.

Racki, G., 1998. Frasnian–Famennian biotic crisis: undervalued tectonic control? Palaeogeogr. Palaeoclimatol. Palaeoecol. 141, 177–198.

Racki, G., 1999. Silica-secreting biota and mass extinctions: survival patterns and processes. Palaeogeogr. Palaeoclimatol. Palaeoecol. 154, 107–132.

Racki, G., Racka, M, Matyja, H., Devleeschouwer, X., 2002. The Frasnian/Famennian boundary interval in the South Polish-Moravian shelf basins: integrated event-stratigraphical approach. Palaeogeogr. Palaeoclimatol. Palaeoecol. 181, 251–297.

Retallack, C.J., 1997. Early forest soils and their role in Devonian global change. Science 276, 583–585.

Rimmer, S.M., Thompson, J.A., Goodnight, S.A., Robl, T.L., 2004. Multiple controls on the preservation of organic matter in Devonian–Mississippian marine black shales: geochemical and petrographic evidence. Palaeogeogr., Palaeoclimatol., Palaeoecol. 215, 125–154

Rimmer, S.M., 2004. Geochemical paleoredox indicators in Devonian–Mississippian black shales, Central Appalachian Basin (USA). Chem. Geol. 206, 373–391.

Royer, D.L., Berner, R.A., Beerling, D.J., 2001. Phanerozoic atmospheric $CO_2$ change: evaluating geochemical and paleobiological approaches. Earth-Sci. Rev. 54, 349–392.

Sageman, B.B., Murphy, A.E., Werne, J.P., Ver Straeten, C.A., Hollander, D.J., Lyons, T.W., 2003. A tale of shales: the relative roles of production, decomposition, and dilution in the accumulation of organic-rich strata, Middle-Upper Devonian, Appalachian basin. Chem. Geol. 195, 229–273.

Sandberg, C.A., Poole, F., Johnson, J.G., 1989. Upper Devonian of the western United States. In: McMillan, N.J., Embry, A.F., Glass, D.J. (Eds), Devonian of the World, Vol. I. Regional Syntheses, Proceedings of the 2nd International Symposium on the Devonian System, Calcary, Canada. Can. Soc Petrol. Geol. pp. 183–220 (imprinted 1988).

Sandberg, C.A., Morrow, J.R., Ziegler, W., 2002. Late Devonian sea level changes, catastrophic events and mass extinctions. Geol. Soc. Amer. Spec. Pap. 356, 473–487.

Schindler, E., 1990. Die Kellwasser Krise (hohe Frasne-Stufe, Ober Devon). Gött. Arb. Geol. Paläeontol. 46 1–125.

Schindler, E., 1993. Event-stratigraphic markers within the Kellwasser Crisis near the Frasnian/Famennian boundary (Upper Devonian) in Germany. Palaeogeogr. Palaeoclimatol. Palaeoecol. 104, 115–125.

Sepkoski, J.J., 1986. Phanerozoic overview of mass extinctions. In: Raup, D.M., Jablonski, D. (Eds). Patterns and Processes in the History of Life. Springer, Berlin, pp. 277–295.

Streel, M., Caputo, M.V., Loboziak, S., Melo, J.H.G., 2000. Late Frasnian–Famennian climates based on palynomorph analyses and the question of the Late Devonian glaciations. Earth-Sci. Rev. 52, 121–173.

Tribovillard, N., Desprairies, A., Lallier-Vergès, E., Bertrand, P., Moureau, N., Ramdani, A., Ramanampisoa, L., 1994. Geochemical study of organic-matter rich cycles from the Kimmeridge clay formation of Yorkshire (UK): productivity versus anoxia. Palaeogeogr. Palaeoclimatol. Palaeoecol. 108, 165–181.

Tribovillard, N., Dupuis, C., Robin, E., 2000. Sedimentological and diagenetical conditions of the impact level of the Cretaceous/Tertiary boundary in Tunisia: no anoxia required. Bull. Soc. Géol. Fr. 171, 629–636.

Tribovillard, N., Averbuch, O., Devleeschouwer, X., Racki, G., Riboulleau, A., 2004a. Deepwater anoxia at the Frasnian–Famennian boundary (La Serre, France): a tectonically-induced Late Devonian oceanic anoxic event? Terra Nova, 16, 288–295.

Tribovillard, N., Ramdani, A., Trentesaux, A., 2004b. Controls on organic accumulation in Late Jurassic shales of Northwestern Europe as inferred from trace-metal geochemistry. In: Harris, N. (ed.), The Deposition of Organic-Carbon-Rich Sediments: Models, Mechanisms, and Consequences. SEPM Spec. Public., 82, 145–164.

Tucker, R.D., Bradley, D.C., Ver Straeten, C.A., Harris, A.G., Ebert, J.R., Mc Cutcheon, S.R., 1998. New U-Pb zircon ages and the duration and division of Devonian time. Earth Planet. Sc. Lett. 158, 175–186.

Tyson, R.V., Pearson, T.H., 1991. Modern and ancient continental shelf anoxia: an overview. In: Tyson, R.V., Pearson, T.H. (Eds), Modern and Ancient Continental Shelf Anoxia. Spec. Publ. Geol. Soc. 58, pp. 1–24.

Tyson, R.V., Pearson, T.H., 1991. Modern and ancient continental shelf anoxia. Geol. Soc. Spec. Publ.58, p. 470.

Veizer, J., Buhl, D., Diener, A., Ebneth, S., Podlaha, O.G., Bruckschen, P., Jasper, T., Korte, C., Schaaf, M., Ala, D., Azmy, K., 1997. Strontium isotope stratigraphy: potential resolution and event correlation. Palaeogeogr. Palaeoclimatol. Palaeoecol. 132, 65–77.

Vogt, P.R.,1989.Volcanogenic upwelling of anoxic, nutrient-rich water: a possible factor in carbonate-bank/reef demise and benthic faunal extinction? Geol. Soc. Am. Bull. 101, 1225–1245.

Vorlicek, T.P., Helz, G.R., 2002. Catalysis by mineral surfaces: implications for Mo geochemistry in anoxic environments. Geochim. Cosmochim. Acta 66, 3679–3692.

Vorlicek, T.P., Kahn, M.D., Kasuza, Y., Helz, G.R., 2004. Capture of molybdenum in pyrite-forming sediments: role of ligand-induced reduction by polysulfides. Geochim. Cosmochim. Acta 68, 547–556.

Walliser, O.H., 1996. Global events and event stratigraphy in the Phanerozoic: results of international interdisciplinary cooperation in the IGCP Project 216 "Global Biological Events in Earth History": Springer, Heidelberg, p. 333.

Walliser, O.H., El Hassani, H., Tahiri, A., 2000. Mrirt: a key area for the Variscan Meseta of Morocco. Notes Mém. Serv. Géol. Maroc 399, 93–108.

Wedepohl, K.H., 1971. Environmental influences on the chemical composition of shales and clays. In: Ahrens, L.H., Press, F., Runcorn, S.K., Urey, H.C. (Eds), Physics and chemistry of the Earth, Pergamon, Oxford, pp. 305–333.

Wedepohl, K.H., 1991. The composition of the upper Earth's crust and the natural cycles of selected metals. In: Merian, E. (Ed.), Metals and their Compounds in the Environment, VCH-Verlagsgesellschaft, Weinheim, pp. 3–17.

Wendt, J., Belka, Z., 1991. Age and depositional environment of Upper Devonian (Early Frasnian to Early Famennian) black shales and limestones (Kellwasser Facies) in the eastern Anti-Atlas, Morocco. Facies, 25, 51–90.

Wignall, P.B., Newton, R., 2001. Black shales on the basin margin: a model based on examples from the Upper Jurassic of the Boulonnais, northern France. Sedim. Geol. 144, 335–356.

Wilde, P., Berry, W.B.N., 1984. Destabilization of the oceanic density structure and its significance to marine "extinction" events. Palaeogeogr. Palaeoclimatol. Palaeoecol. 48, 143–162.

Yudina, A.B., Racki, G., Savage, N.S., Racka, M., Malkowski, K., 2002. The Frasnian–Famennian events in deep-shelf succession, Subpolar Urals: biotic, depositional and geochemical records. Acta Palaeontol. Pol. 47, 355–372.

Zheng, Y., Anderson, R.F., van Geen, A., Kuwabara, J., 2000. Authigenic molybdenium formation in marine sediments: a link to pore water sulfide in the Santa Barbara Basin. Geochim. Cosmochim. Acta 64, 4165–4178.

Understanding Late Devonian and Permian-Triassic Biotic and Climatic Events:
Towards an Integrated Approach
edited by D.J. Over, J.R. Morrow and P.B. Wignall



## Chapter 9

# Evidence for Late Devonian (Kellwasser) anoxic events in the Great Basin, western United States

*David Bond* and Paul B. Wignall*

*School of Earth Sciences, University of Leeds, Leeds LS2 9JT, UK*

**Abstract**

The Frasnian-Famennian (Late Devonian) mass extinction has often been related to the development of the Kellwasser anoxic events in Europe and North Africa but the synchronous development of the anoxia has not been reported from the Great Basin of the western United States. An integrated sedimentological, palaeoecological, and pyrite petrographic study has been undertaken on a range of F-F boundary sections from Nevada and Utah spanning a spectrum of carbonate and clastic depositional environments from distal basin, base-of-slope, mid-slope, and intra-shelf basin settings. These reveal a range of facies from oxic strata, fully bioturbated and lacking any pyrite, to euxinic strata characterised by fine lamination and pyrite framboid populations of small size and narrow size range. Oxygen-restricted deposition is seen in all sections at various times, but the only interval characterised by basin-wide euxinicity occurs at the end of the Frasnian Stage late in the *linguiformis* Zone. This is the peak of the F-F mass extinction and it is also contemporaneous with the Upper Kellwasser Horizon of Europe. The study therefore reinforces the claim that the mass extinction coincides with the global development of marine anoxia. Shallow-water sections were not studied but slope and base-of-slope sections record many sediment-gravity flows that transported an allochthonous fauna into deeper-water settings. This shallow-water fauna temporarily disappears late in the *linguiformis* Zone perhaps indicating the development of oxygen-restriction in shallow-water settings. Intriguingly the condensed, deepest water sections from the Woodruff basin record somewhat higher oxygenation levels than the contemporaneous slope sections. The most oxygen-restricted conditions may therefore have occurred in a mid-water oxygen-minimum zone that expanded its vertical range both upwards and downwards and became sulphidic late in the *linguiformis* Zone.

## 1. Introduction

The extinction losses near the Frasnian-Famennian (F-F, Late Devonian) boundary constitute one of the "big 5" crises of the fossil record (e.g. Raup and Sepkoski, 1982; Sepkoski, 1996). Several causal mechanisms have been proposed, and marine anoxia is widely regarded to have played a role in many of them. The extinctions are associated with two discrete anoxic events (House, 1985; Casier, 1987; Walliser et al., 1989; Goodfellow et al., 1989; Buggisch, 1991; Becker, 1993). These events are manifest in Germany as bituminous limestones known as the Lower and Upper Kellwasser horizons that developed in the Late *rhenana* and late in the *linguiformis* conodont zones, respectively (Buggisch, 1972; Schindler, 1990; Walliser, 1996). Elsewhere in Europe, the intensity and duration of anoxic

*Corresponding author. Fax: +44-113-343-5259. *E-mail address:* d.bond@earth.leeds.ac.uk (D. Bond).

facies varies according to water depth. For instance in Poland and France, anoxia is postulated to have developed during the Frasnian on basin slopes, before developing in shelf and shallow-water environments in the latest *linguiformis* Zone (Bond et al., 2004). Implicit in the anoxia kill hypothesis is the need for globally synchronous and widespread anoxia (e.g. House, 1985). Crucially therefore, the fact that the latest Frasnian event (corresponding to the Upper Kellwasser Horizon of Germany) was widespread and simultaneous in its development, increases the likelihood that it was a direct cause of the marine mass extinction. Age-equivalent anoxic facies are also known over a wide area from Morocco to the subpolar Urals (Wendt and Belka, 1991; Racki et al., 2002; Yudina et al., 2002).

Although the Kellwasser facies in Europe present clear evidence for widespread anoxia, previous work in the Great Basin of the western United States has suggested that the anoxic interval significantly pre-dated the F-F boundary. Furthermore, the F-F boundary interval may in fact record oxic deposition, perhaps as a result of relative sea-level fall (Bratton et al., 1999). Thus, the latest Frasnian anoxic event may have been of regional (i.e. European and North Africa) rather than global extent. This seriously undermines the role of anoxia in the mass extinction.

The oxygenation history proposed by Bratton et al. (1999) was based on a range of geochemical redox indicators, principally trace metal indices, from outcrop sections in the deserts of Nevada and Utah. A new approach for analysing palaeo-oxygenation levels has been applied to the sections originally used in the Bratton et al. (1999) study (Coyote Knolls and Whiterock Canyon), and additional sections (Devils Gate, Northern Antelope Range, Tempiute Mountain, and Warm Springs). In this study, we document the late Frasnian–early Famennian oxygenation history of the Great Basin, using integrated techniques of pyrite framboid and sedimentary fabric analysis. Unlike many geochemical indices, both these rock attributes are preserved even in heavily–weathered strata (e.g. Lüning et al., 2003). It is hoped to clarify the F-F redox record of the United States, and thus to determine how widespread the anoxic facies were. Comparisons with the timing of anoxia in Europe are made, in order to address the question of whether anoxia was implicated in the Late Devonian mass extinction.

## 2. Regional geology of the Great Basin

The Great Basin Devonian outcrops of Nevada and Utah record deposition adjacent to a foreland basin (Fig. 1). This formed the central part of a large continental carbonate shelf extending northwards from Mexico to western Canada, along the western edge of Laurasia (Sandberg and Poole, 1977; Sandberg et al., 1989; Morrow, 2000), located at 5–10° N (Scotese and McKerrow, 1990). The sections record deposition in two basins, the Pilot and the Woodruff basins, separated by the proto-Antler forebulge (Sandberg et al., 2003) (Fig. 2). To the east of the forebulge, the Pilot Basin was a rapidly subsiding, deep-water intra-shelf Basin with an infill dominated by siltstones, thin micrites, and sandy turbidites (Sandberg et al., 1989). To the west of the forebulge was a slope, and the deep Woodruff Basin in which debris-flow carbonates, turbidites, siltstones, mudstones, and cherts accumulated (Sandberg et al., 2003). During the Famennian and Early Carboniferous, the Antler Orogeny caused basinal sediments to be thrust eastwards onto the slope and outer shelf, resulting in tectonic interfingering of shallow and deep-water deposits along the Roberts

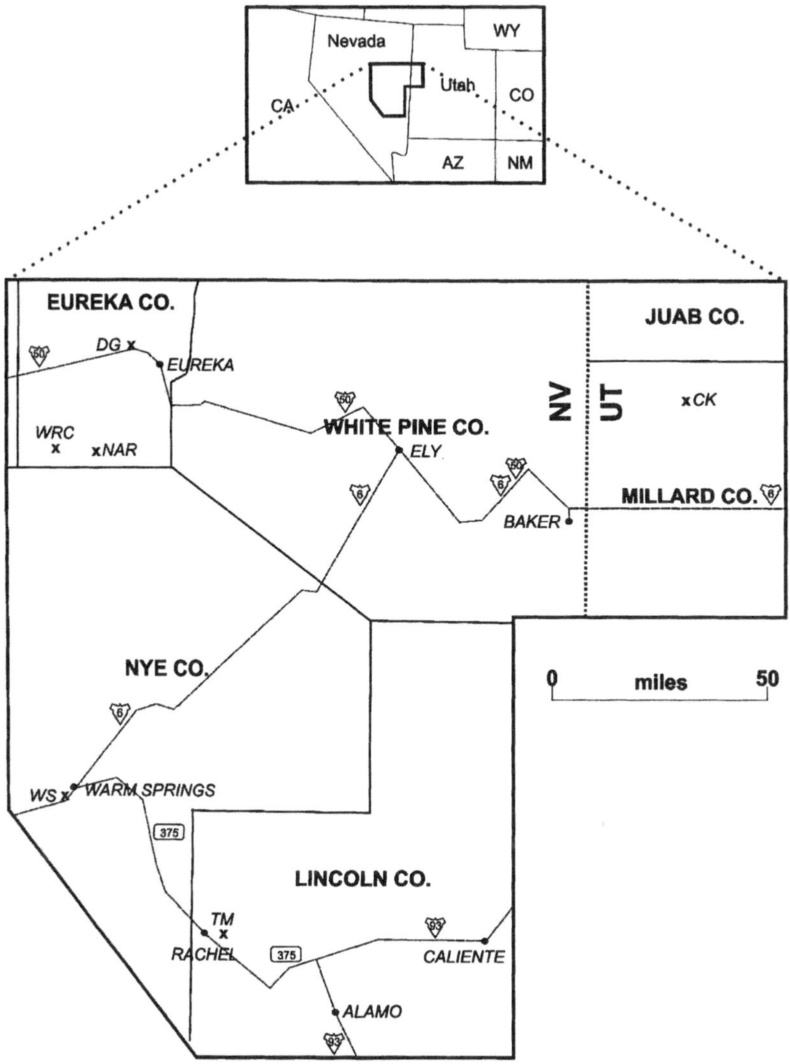


*Figure 1.* Locality map of study sections in the western United States, showing principal towns and highways. CK = Coyote Knolls; DG = Devils Gate; NAR = Northern Antelope Range; TM = Tempiute Mountain; WRC = Whiterock Canyon; WS = Warm Springs.

Mountains Thrust, which runs along the western edge of the Pilot Basin (Sandberg and Poole, 1977; Sandberg et al., 1989). The studied sections are representative of a variety of depositional settings, and are presented from east to west, from the intra-shelf Pilot Basin, across the proto-Antler forebulge, to the slope and basinal sites of the Woodruff Basin (Fig. 2; Appendix A). Slope deposits are well exposed in the well-known Devils Gate section while further west the Northern Antelope Range and Tempiute Mountain sections expose middle to basal slope/Basin floor deposits. Warm Springs and Whiterock Canyon expose the deepest water sections of this study, from the Basin floor of the Woodruff Basin

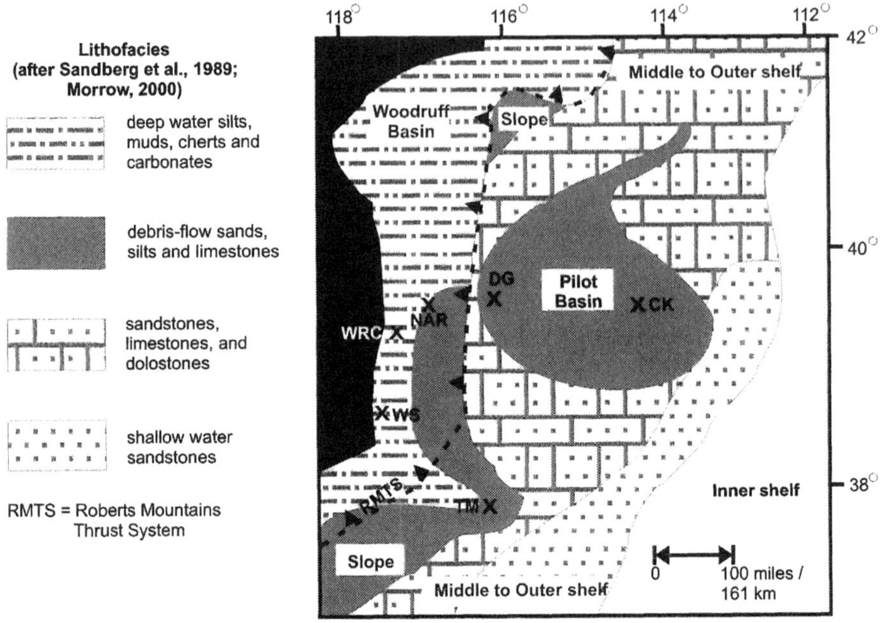


*Figure 2.* Lithofacies and palaeogeography of the Great Basin during the Frasnian-Famennian interval showing location of sections studied in this chapter. Modified from Sandberg et al. (1989) and Morrow (2000). Note movement on the Roberts Mountains thrust (hachured) has not been fully restored, thus foreshortening the shelf–basin transition in some areas (section abbreviations as in Fig. 1).

(Sandberg et al., 1997, 2002, 2003; Morrow, 2000; see Fig. 2). The sections vary considerably in thickness, depending on their depositional environment: basinal sections of the Woodruff Basin being thinner than both the slope sections and the expanded Pilot Basin section of Coyote Knolls (Fig. 3).

## 3. Analytical procedure and rationale

Oxygenation levels recorded in the F-F boundary sections were assessed using two independent criteria: sediment fabric (i.e. presence of lamination and/or degree of bioturbation) and pyrite framboid size populations. Boundary sections were logged in detail with care taken to identify the presence (or absence) of fine lamination/bioturbation features. Field determinations were supplemented by petrographic analysis of thin sections and examination of polished blocks. Burrow size is known to decrease along a transition from oxic to dysoxic conditions, and burrows disappear completely in anoxic conditions (Savrda and Bottjer, 1986). Euxinic conditions (defined by the presence of a sulphidic lower water column) are also associated with laminated sediment, but it is not possible to distinguish anoxic conditions, where anoxia is only developed at the seabed, from euxinic conditions using sediment fabric alone. This is better achieved from analysis of pyrite framboid populations.

Polished blocks were examined under backscatter SEM at a magnification of × 2670, to determine the pyrite content and, where present, the size distribution of pyrite framboid

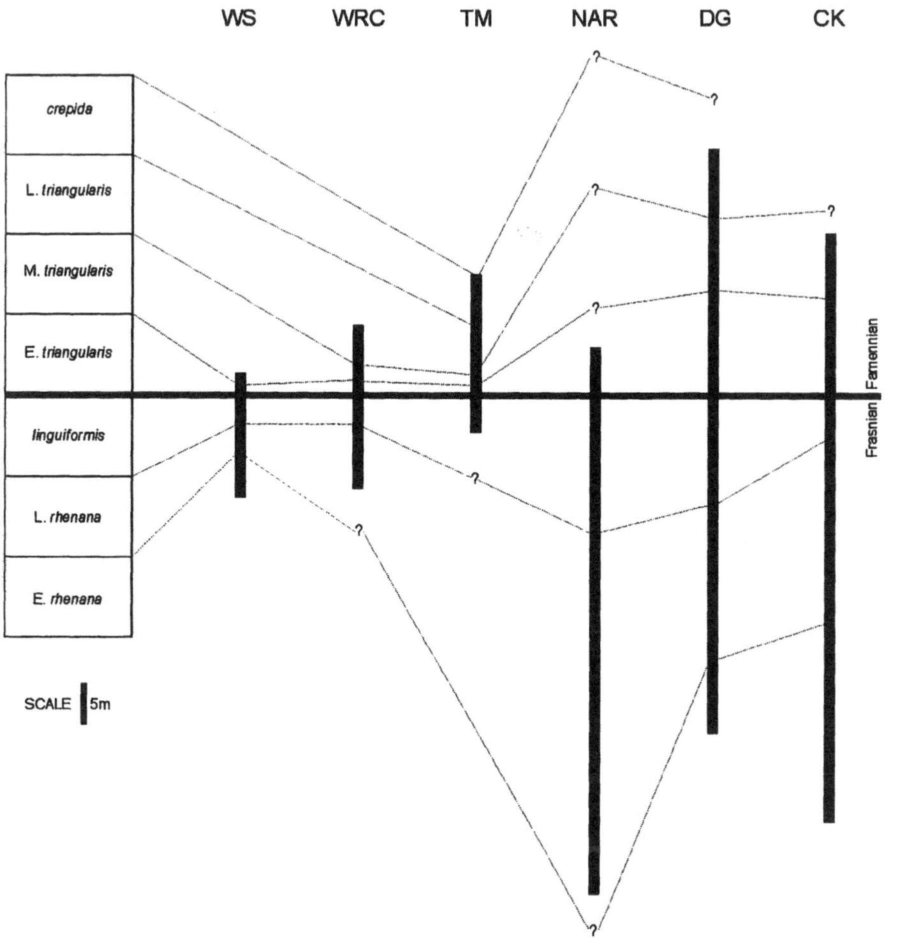


*Figure 3.* Correlation of study sections (abbreviations as in Fig. 1), showing the development of thinner, basinal sections in the western outcrops, and thicker, slope sections to the east.

populations (Fig. 4). Where possible, at least 100 framboids were measured per sample although in some of the more carbonate-rich lithologies the pyrite content is low and the density of framboids consequently too low to achieve the desired 100 framboid sample size (see Appendix B).

Framboids form as an aggregate of iron monosulphide microcrysts, and Wilkin and Barnes (1996, 1997) proposed a four-stage model for their formation. Firstly, iron monosulphide microcrysts nucleate. These then react to form greigite ($Fe_3S_4$), followed by aggregation of greigite microcrysts (due to magnetic attraction) to form spherical framboids. Finally, greigite is replaced by pyrite to form pyrite framboids. The reaction of iron monosulphide to form greigite requires weakly oxidising conditions, in order to supply elemental sulphur, whereas the first and fourth stages both require reducing conditions (Raiswell, 1982; Canfield and Thamdrup, 1994). Thus, the environment where pyrite framboids can form is tightly constrained to the close proximity of the redox boundary.

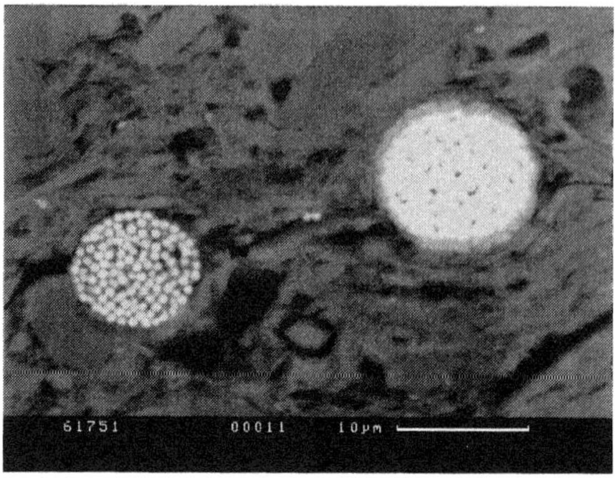


*Figure 4.* SEM photograph of pyrite framboids from the Devils Gate section (bed 4). Note that the framboid on the right of the picture exhibits an oxidised rim but still retains the framboid morphology.

Framboids cease to grow in the more intensely anoxic conditions of the underlying sulphate-reduction zone, where more slowly growing crystalline pyrite forms (Wilkin and Barnes, 1997). If bottom waters become anoxic (i.e. euxinic conditions develop), then framboids develop in the water column but are unable to achieve diameters much larger than 5 µm before they sink below the Fe-reduction zone and cease to grow (Wilkin et al., 1996). For example, in modern Black Sea sediments, deposited in the type example of a euxinic basin, framboids rarely exceed 6 µm in diameter (Muramoto et al., 1991). Thus, euxinicity produces populations of tiny framboids with a narrow size range. In contrast, in dysoxic environments, where framboids form in surficial sediments, size is largely governed by the local availability of reactants and the framboids are larger and more variable in size (Wilkin et al., 1996).

In a detailed study of a range of modern depositional environments, Wilkin et al. (1996) found that framboids formed in a euxinic water column (such as the Black Sea) have a mean diameter, standard deviation, and maximum framboid diameter that are smaller than those framboids formed under oxygenated or dysoxic water columns. Wilkin et al. (1996) suggested that, because framboid size is preserved throughout diagenesis and lithification, the technique could also be applied to ancient sediments. Subsequently, Wignall and Newton (1998) found a close correspondence between the framboid size parameters noted above and water column oxygenation levels determined using palaeoecological criteria. They confirmed that pyrite framboid analyses can additionally provide a subtle indicator of environmental stability (Wignall and Newton, 1998).

Many lower dysaerobic biofacies record variable oxygenation regimes in which long-term euxinicity is punctuated by brief seafloor oxygenation events that allowed colonisation by a benthic pioneer fauna (Hallam, 1980; Wignall and Myers, 1988). Thus, there is often a dichotomy between geochemical evidence from lower dysaerobic facies which records long-term (average) depositional conditions and palaeoecological evidence which can record shorter-term (atypical) events (Raiswell et al., 2001). Such facies have a distinctive pyrite framboid size signature dominated by small, syngenetic populations formed in euxinic conditions but with the addition of rare, larger (>5 μm diameter) framboids formed within the sediment during brief intervals of seafloor oxygenation (Wignall and Newton, 1998; Table 1). It is therefore useful to measure maximum framboid diameter (MFD), provided a sufficiently large sample size is obtained, as this can provide a further subtle indicator of transient oxygenation events.

A great advantage of pyrite framboid analysis over other geochemical proxies for ancient redox levels is the ability of framboids to withstand both diagenetic alteration and weathering. Once formed, framboids are not observed to grow or accrete in modern environments, and anoxic conditions are generally characterised by the growth of crystalline or amorphous forms of pyrite. On exposure, framboids and other sedimentary pyrite are weathered to iron oxy-hydroxides, but this only involves the pseudomorphing of the framboids; it is still possible to measure framboid size even in intensely weathered samples (Fig. 4).

## 4. Great Basin sections

### *4.1. Coyote Knolls*

#### *4.1.1. Lithological and faunal variations*

Over 70 m of F-F sediments from the deep-water, intra-shelf Pilot Basin were studied (Fig. 5; see Appendix 1 for locality details). Sandberg et al. (1988, 1997) provide the most detailed account of the section, and have placed the F-F boundary precisely using conodont biostratigraphy. The base of the section is in the Early *rhenana* Zone, which contains the topmost part of the Guilmette Formation (beds 1–6). Above this, the remainder of the sequence comprises part of the lower member of the Pilot Shale, and continues up into the Middle *triangularis* Zone (beds 7–43).

The Guilmette Formation (beds 1–6) comprises well-bedded, thoroughly bioturbated, pale-grey micrites, with a rich fauna including corals, brachiopods, homoctenids, bivalves, goniatites, "spiny" ostracods, and trilobites. Micrite nodules and pyrite lumps are also present. The micrites are intercalated with a few brown calcareous siltstone and silty limestone horizons, which are also shown to be intensely bioturbated (Fig. 5).

The succeeding Pilot Shale (beds 7–43) is dominated by fine-grained, laminated shales, siltstones, and calcisiltites, with common pyrite blebs. These hemipelagic strata are interbedded with thin, siltstone turbidites, which contain shelf benthos and occasionally have bioturbated tops (e.g. bed 19, Fig. 5). Some beds (e.g. bed 15) have undergone slumping. Bed 18 records a change from quartz-rich calcisilts to a black, crinoidal wackestone, overlain again by quartz-rich siltstones. The F-F boundary lies at the top of bed 33, a finely laminated, brown, coarse calcisiltite. The basal Famennian, bed 34, is a graded conglomerate with many ooids and peloids and rare grapestones, floating in a calcisilt matrix. The early Famennian is characterised by an increased proportion of clastics, and a number of coarse siltstone and sandstone turbidites, which contain calcareous peloids and shelf benthos. There are also a number of additional conglomeratic debris-flow beds (Fig. 5).

Autochthonous benthic fauna in the Pilot Shale is scarce, but allochthonous fossils occur in the turbidites, and are dominated by a shelf fauna of crinoids and brachiopods. Rare, fragmented gastropods, and uniserial foraminifera are also observed in the Late

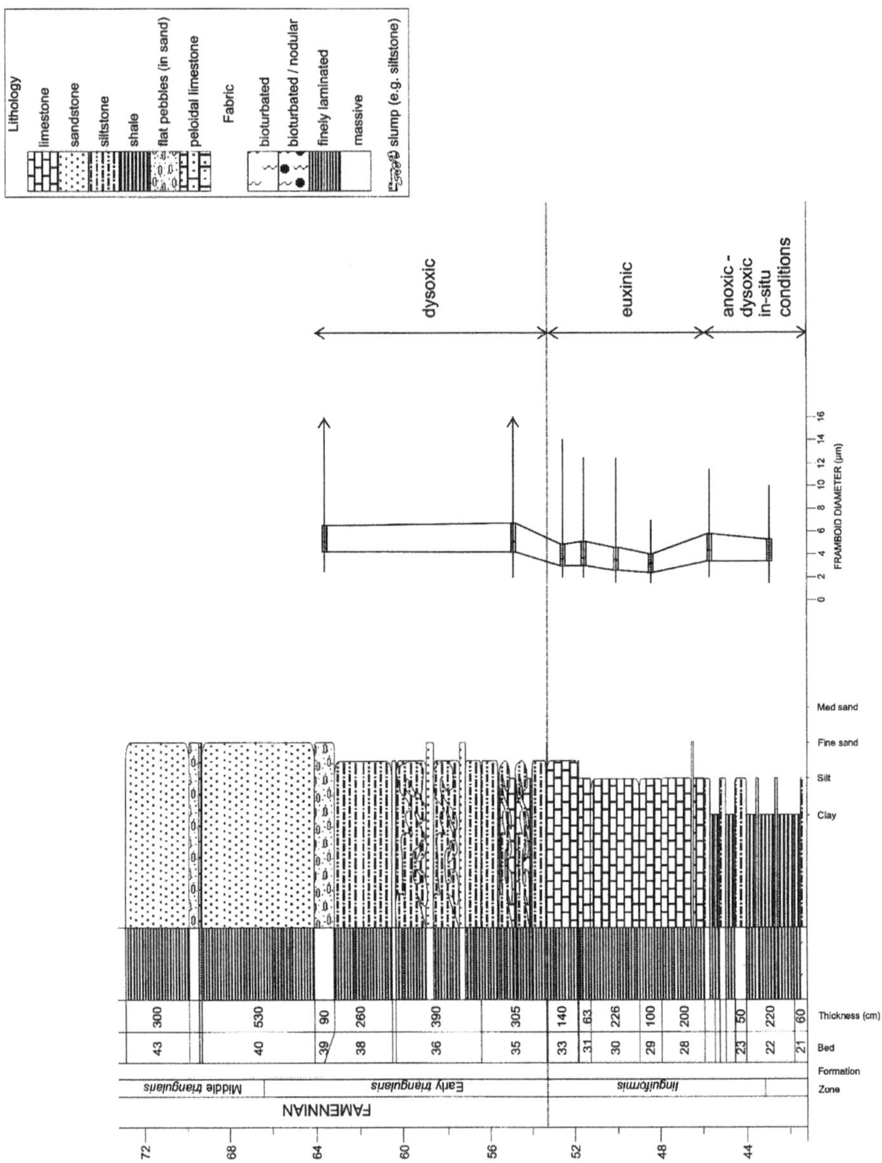

Lithology
limestone
sandstone
siltstone
shale
flat pebbles (in sand)
peloidal limestone
Fabric
bioturbated
bioturbated / nodular
finely laminated
massive
slump (e.g. siltstone)
dysoxic
euxinic
anoxic - dysoxic in-situ conditions
FRAMBOID DIAMETER (µm)
Med sand
Fine sand
Silt
Clay
Thickness (cm)
Bed
Formation
Zone
FAMENNIAN
Middle triangularis
Early triangularis
linguiformis

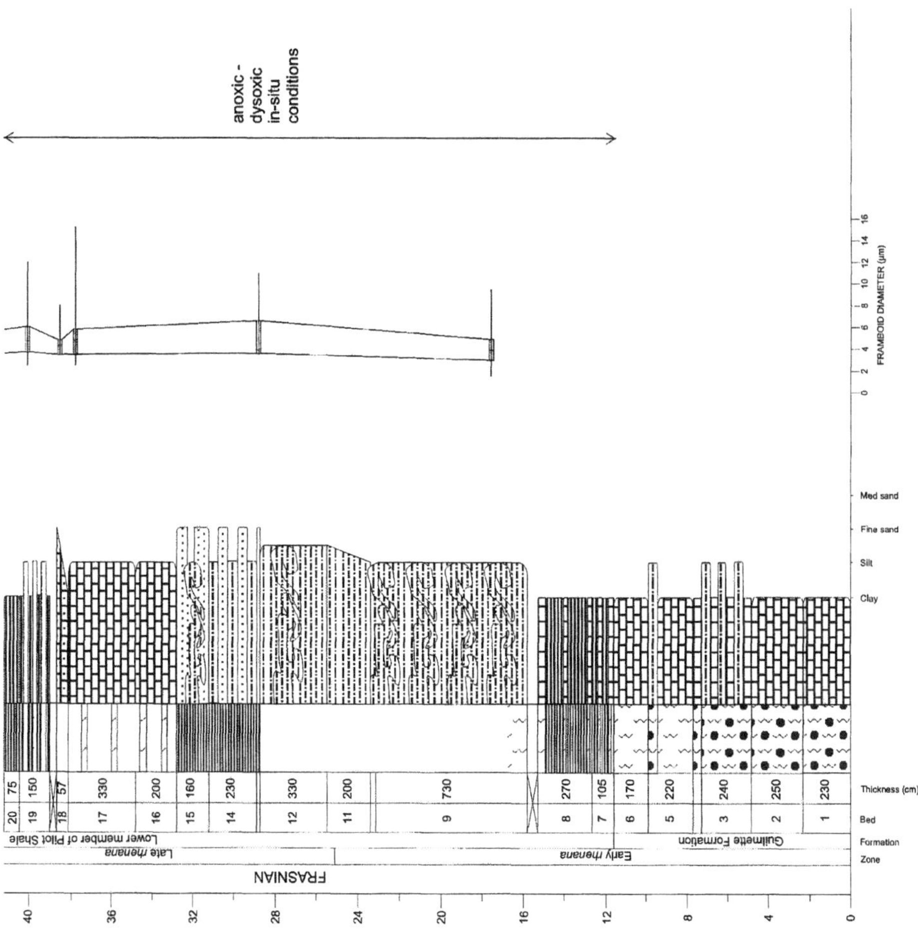


*Figure 5.* Log of Coyote Knolls section, including pyrite framboid "box and whisker" data (box shows 25th, 50th, and 75th percentile, with horizontal line representing minimum and maximum framboid diameter). Note that grade (e.g. clay) refers only to grain size, and not lithology, since this is a mixed clastic/and carbonate sequence. Conodont zonation based on Sandberg et al. (1997).

*rhenana* Zone. Homoctenids are occasionally observed within the Frasnian, but not above bed 28 in the *linguiformis* Zone. Within the topmost Frasnian, thin turbidites in beds 31 and 32 contain ostracods, crinoids, and abundant bioclasts. The early Famennian conglomerates contain crinoids, brachiopods, bryozoans, echinoderms, foraminifera, and other shell debris. Near the top of the section, in the Middle *triangularis* Zone, bed 41 contains abundant, small, prismatic-walled, circular fossils that are probably calcispheres. These are overgrown with a calcite cement.

### 4.1.2. *Pyrite framboid analysis*

Framboids are absent within the pale limestones of the Guilmette Formation, but in the overlying Pilot Shale, framboids become common (Fig. 6). Within the Early *rhenana* Zone, there are abundant, small framboids with a mean diameter of 4.3 μm (Fig. 6; Appendix B). In the Late *rhenana* Zone framboid mean diameter increases (ranging from 4.8 μm to 5.6 μm) and they are common in the finer-grained sediments. Bed 18, a black crinoidal wackestone, yielded only rare framboids in comparison to other beds. The overlying *linguiformis* Zone sediments also contain abundant framboids, although of smaller mean diameter (Fig. 6; Appendix B). Bed 29 is particularly rich in tiny framboids with a very small mean diameter of only 3.7 μm. In the Early *triangularis* Zone the framboid populations become less abundant and distinctly larger, with diameters up to 23 μm.

### 4.1.3. *Depositional style and oxygenation history*

Pyrite framboid, faunal, and lithological data all suggest that the Guilmette Formation was deposited in a well-oxygenated, shelf environment. The overlying Pilot Shale records a major deepening in the Early *rhenana* Zone (termed the *semichatovae* transgression, Sandberg et al., 2002), and a distinct change in environmental conditions. The finely laminated sediments, common turbidites, and slumped horizons indicate a basinal or distal base-of-slope environment. The lamination and common framboids within the calcisiltites suggest that oxygen-poor conditions prevailed with an anoxic seafloor developed throughout the Early to Late *rhenana* zones. The post-event bioturbation following turbidity flows suggests the turbidites are likely to have been sourced from the better-oxygenated shelf, from where they transported a living infauna that was able to briefly colonise and bioturbate the seabed, as is observed in some modern dysoxic basins (Follmi and Grimm, 1990). Not all of the turbidity flows caused oxygenation events, however, as bioturbation is rare in the upper part of the Late *rhenana* to *linguiformis* zones. This may reflect deteriorating oxygen levels on the shelf, although allochthonous shelf benthos remains present during the Late *rhenana* Zone. Alternatively, an intensely anoxic lower water column at this time may have been more immune to transient reoxygenation events.

The upper part of the Late *rhenana* Zone records a lithological change to more persistent and darker shales, with a corresponding decrease in framboid diameter, indicative of a decrease in benthic oxygen levels within the basin (Fig. 6). Pyrite framboid and lithological evidence suggest further deterioration in oxygen levels by the middle part of the *linguiformis* Zone, as fine lamination becomes persistent and the mean framboid diameter drops to 3.7 μm. Truly euxinic conditions probably developed at this time, and continued up to the F-F boundary. This is the level of the Upper Kellwasser Horizon in Europe. A

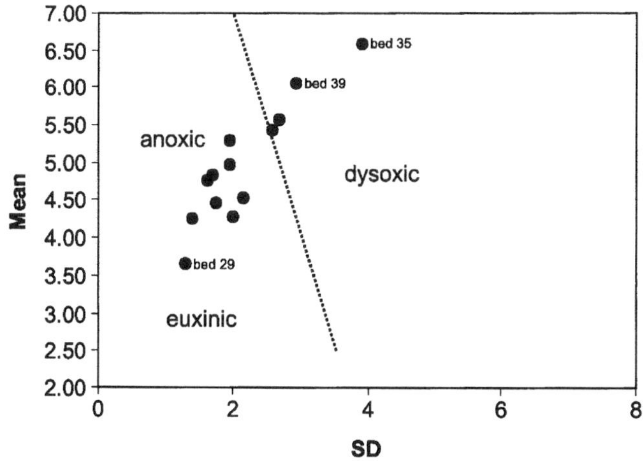


*Figure 6.* Mean framboid diameter (microns) vs. standard deviation plot for pyrite framboids at Coyote Knolls. Note that bed 29 (linguiformis Zone) plots as the most intensely anoxic bed, and that by beds 35 and 39 (Famennian), the framboid distribution is characteristic of dysoxic conditions.

small amount of allochthonous benthic shelf fauna is observed, even within this level, indicating that anoxia probably did not develop in the sediment source area.

The basal Famennian records an improvement in oxygenation levels, with the framboid distribution of bed 35 suggestive of a dysoxic setting. There is also the appearance of high-energy event beds, probably debris flows, that brought coarser material to the base-of-slope. Despite this improvement, the majority of the beds in the Early *triangularis* Zone remain finely laminated. Interestingly, allochthonous fauna becomes more abundant in the basal Famennian, with crinoids and brachiopods dominating, suggesting normal oxygen levels on the shelf.

### 4.2. Devils Gate

#### 4.2.1. Lithological and faunal variations

This highly accessible section has been intensively studied (Sandberg and Poole, 1977; Sandberg et al., 1988, 1989, 1997, 2003; Casier et al., 1996; Morrow, 2000; Joachimski et al., 2002). It is regarded as one of the most important F–F reference sections (Sandberg et al., 1988, 2002), and the F–F boundary is well constrained by conodont dating. The sequence is a composite of three closely spaced sections (Fig. 7). In total a sequence of over 70 m in thickness, comprising 42 beds, has been investigated here.

This expanded sequence belongs to the upper member of the Devils Gate Limestone Formation (Fig. 7) and it ranges in age from the lower part of the Early *rhenana* Zone up to the Late *triangularis* Zone (Fig. 7). The base of the logged sequence corresponds to the "*semichatovae* transgression" (Sandberg et al., 1997, 2002, 2003) and the base of transgressive-regressive cycle IId of Johnson et al. (1985). This is manifest as a transition from

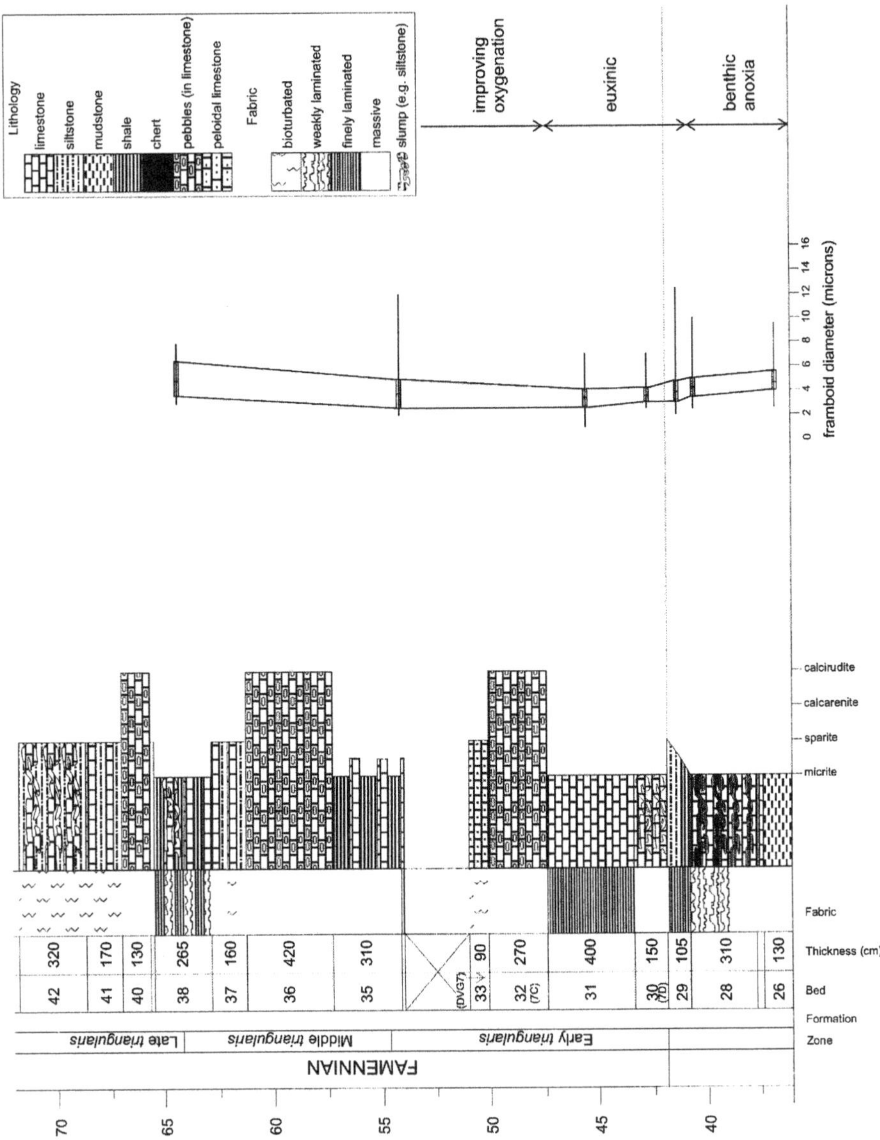
Lithology
limestone
siltstone
mudstone
shale
chert
pebbles (in limestone)
peloidal limestone
Fabric
bioturbated
weakly laminated
finely laminated
massive
slump (e.g. siltstone)
improving oxygenation
euxinic
benthic anoxia
framboid diameter (microns)
calcirudite
calcarenite
sparite
micrite
Fabric
Thickness (cm)
Bed
Formation
Zone
Late triangularis
Middle triangularis
Early triangularis
FAMENNIAN
(DVG7)
(7C)
(7D)

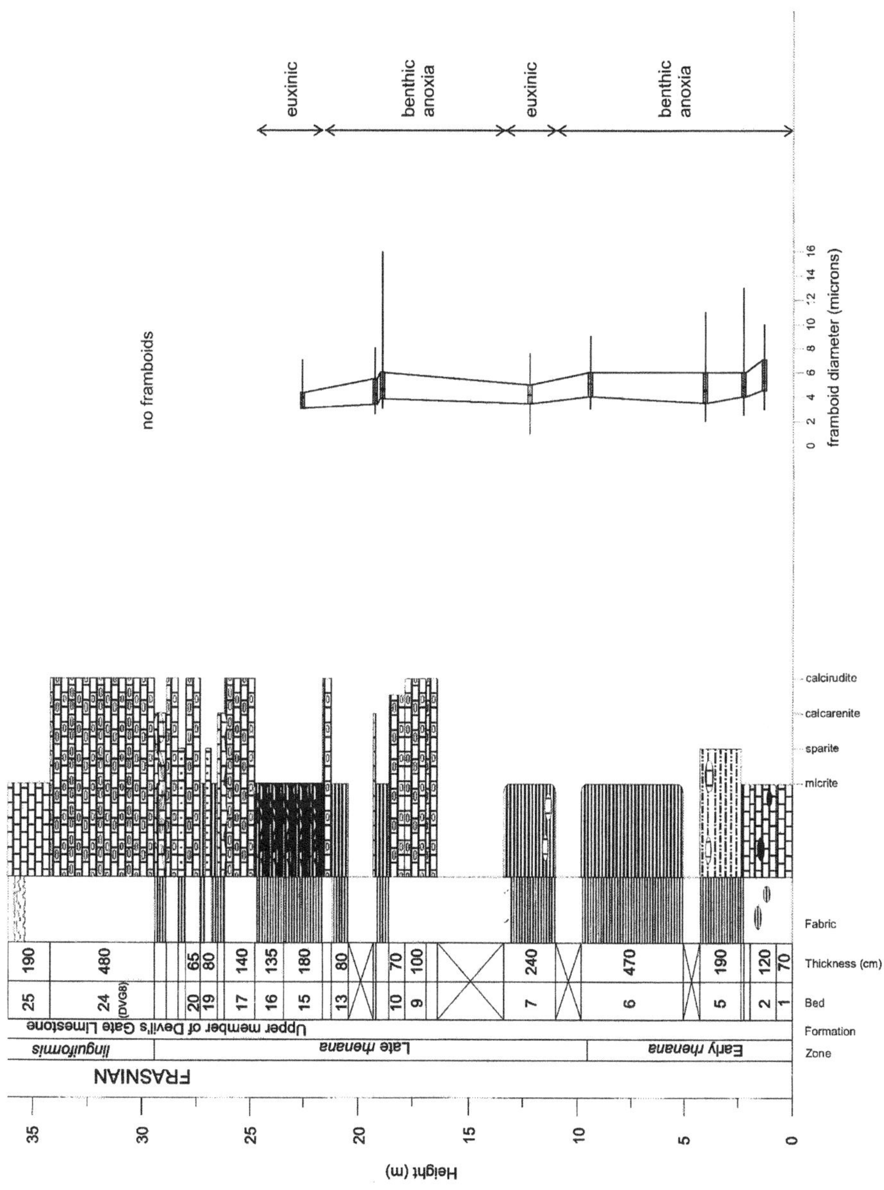


*Figure 7.* Log of Devils Gate section including pyrite framboid "box and whisker" data. Sample numbers from Sandberg et al. (1988) are given in brackets in bed column, where appropriate. Conodont zonation based on Sandberg et al. (1988).

medium-grey, massive micrites (beds 1–3) to organic-rich, dark, finely laminated radiolarian chert (bed 4), finely laminated siltstone (bed 5), and finely laminated, calcareous shale (bed 6).

Hemipelagic sedimentation during the Late *rhenana* to *triangularis* zones (beds 7–42) comprises cherts, silty shales, micrites, and calcareous siltstones, many of which are finely laminated. These often display slumping and soft sediment deformation. The fine-grained sediments are frequently interbedded with thick beds of limestone breccia and conglomerate, which are usually clast supported and lack clast orientation. The clasts include peloidal limestones, large flat pebbles, and chert pebbles (e.g. beds 8–9, 14, 17, 20, 22, 24, 32, 36, and 40). A number of calcarenite turbidites are also present. During the upper part of the *linguiformis* Zone, there was a brief pause of this allodapic deposition (bed 24 is the highest Frasnian sediment-gravity flow), allowing persistent hemipelagic sedimentation across the F-F boundary. The rocks at this level comprise massive to weakly laminated mudstones and shales. The F-F boundary, defined by conodonts, is placed at the top of bed 29, a badly weathered, finely laminated, silty shale, which coarsens upwards into siltstone (Sandberg et al., 1988; Fig. 7). The basal Famennian is characterised by turbidites, interbedded with siltstones and shales. There are numerous thick conglomerates in the Early and Middle *triangularis* zones, which again are interbedded with siltstones. The topmost part of the sequence (Late *triangularis* Zone) becomes dominated by a "ribbon facies" of interbedded micritic limestones and siltstones on a scale of a few centimetres.

The non-conodont fauna in the Devils Gate section is varied, but not abundant, and is mostly restricted to allochthonous shelf fauna within the turbidites and coarser beds (Fig. 7). Other than abundant conodonts, the Early *rhenana* Zone hemipelagic sediments contain only occasional calcispheres, and radiolarians within some chert beds. Bioturbation is absent except for that seen within silty, sparry concretions in bed 5, where the fine lamination has been partly disrupted by burrowing. In the Late *rhenana* Zone the thick calcirudites contain a rich shelf fauna of crinoids, silicified rugose corals, *Syringopora*, stromatoporoids, echinoderm fragments, gastropods, brachiopods, ostracods, several species of foraminifera, calcareous algae (possibly gymnocodiaceans), and calcispheres. In contrast, background sediments generally remain finely laminated, and without fossils. Bed 24 marks the base of the *linguiformis* Zone, and contains a rich shelf fauna similar to that of the beds below. In addition, Sandberg et al. (1988) recorded common *Iowatrypa* and *Tabulophyllum* within this bed. Indeed bed 24 marks the last occurrence of both these genera (Fig. 7). The *linguiformis* Zone hemipelagic-style sediments (beds 25–29) contain small, unidentified bioclasts, and calcispheres. At this level the beds are massive or weakly bioturbated although the topmost Frasnian, bed 29, records a distinct change to unfossiliferous, finely laminated silty shales, from the more massive or weakly bioturbated beds beneath (Fig. 7).

The basal Famennian turbidites contain a fauna including brachiopods, (probably *Leiorhynchus*) and abundant calcispheres. These are interbedded with unfossiliferous hemipelagic sediments. The conglomerates in the Early and Middle *triangularis* zones contain a diverse, allochthonous fauna of solitary rugose corals, foraminifera (belonging to different taxa to those seen in the Frasnian strata), crinoids, brachiopods, ostracods, calcified cyanobacteria, large orthocone nautiloids, and calcispheres. The intervening sediments are mostly massive or bioturbated. The Late *triangularis* Zone is generally bioturbated and contains thin-shelled brachiopods, ostracods, calcispheres, and fusilinid foraminifera (Fig. 7).

In a study of the ostracods from F-F boundary strata at the Devils Gate section, Casier and Lethiers (1998) noted the presence of an abundant, but allochthonous, Frasnian and Famennian fauna. Seventy ostracod species were identified within the Frasnian, belonging to an association typical of well-oxygenated, marine conditions (Casier and Lethiers, 1998). This undoubtedly records conditions in the source area of the allodapic limestones and not at the depositional site. Of these species, 16 pass through the F-F boundary, indicating a severe extinction rate in excess of 75%.

### *4.2.2. Pyrite framboid analysis*

SEM analysis of the fine-grained, hemipelagic strata of the Early *rhenana* to Late *triangularis* zones reveals that framboidal pyrite is common throughout much of the Devils Gate sequence. In the Early *rhenana* Zone, bed 2 contains uncommon framboids with a relatively large mean diameter (Fig. 8; Appendix B). Above this bed, the silty shales of beds 4–7 (the record of the *semichatovae* transgression), contain abundant, small framboids. The framboids become even smaller at the base of the Late *rhenana* Zone, in a finely laminated, black, silty shale. Above this (bed 11) there is an increase in mean diameter and during the middle part of the Late *rhenana* Zone (beds 12 and 15) framboids are uncommon and somewhat smaller. Framboids were not found in samples from the upper part of the Late *rhenana* Zone.

Framboids become common once again in the upper part of the *linguiformis* Zone. In the topmost Frasnian, the middle part of bed 29 contains abundant, small framboids, which record a decrease in mean diameter to only 4.4 μm. The Early *triangularis* Zone hemipelagic sediments also contain common, small framboids with very similar mean diameters, but by the Late *triangularis* Zone, only rare, large framboids are present (Fig. 8; Appendix B).

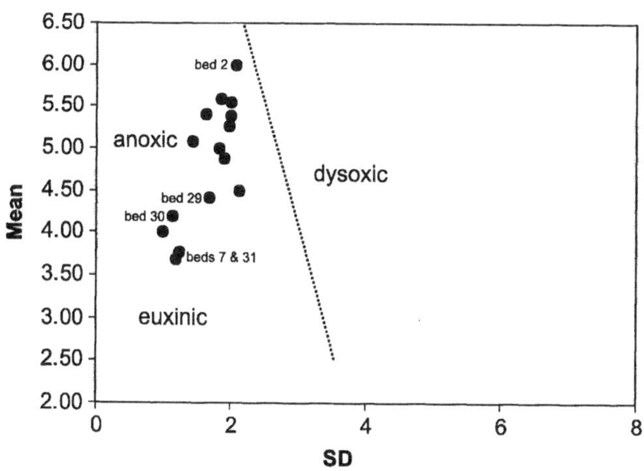


*Figure 8.* Mean framboid diameter (microns) vs. standard deviation plot for pyrite framboids at Devils Gate. Note that the sequence as a whole plots within the anoxic/euxinic fields, with bed 7 (basal Late rhenana Zone) and beds 29–31 (F-F boundary beds) plotting as the most intensely anoxic beds.

#### *4.2.3. Depositional style and oxygenation history*

The interbedding of fine-grained strata and thick, limestone breccias, conglomerates, and turbidites together with the common slumped horizons and soft sediment deformation is indicative of slope deposition at Devils Gate. As at Coyote Knolls, the material moving down-slope was sourced from an oxygenated area where a diverse shelly benthos thrived. However, the various gravity flows clearly came to rest in an anoxic setting as testified by the abundant framboids and fine lamination of the interbedded fine-grained strata. Peak intensity of anoxia (i.e. euxinic conditions), characterised by the smallest framboid populations, was encountered in the early to middle part of the Late *rhenana* Zone and the late part of the *linguiformis* Zone. The intervening interval is recorded by strata that are massive or bioturbated and lack pyrite. This suggests that oxygen levels at the depositional site increased slightly at this time, although they probably remained dysoxic, as some beds are laminated. The basal Famennian hemipelagic sediments (beds 30 and 31) also contain common framboids, with size distributions indicating that the latest Frasnian euxinic pulse continued into the Early *triangularis* Zone. However, the presence of rare, dysaerobic brachiopods in bed 30 (*Leiorhynchus*) suggests dysoxic interludes. Later in the *triangularis* Zone an increase of the size and a decline in abundance of framboids indicates an improvement in benthic oxygen levels. During the Late *triangularis* Zone, beds 41 and 42 record the return of weak bioturbation.

The fluctuating intensity of bottom water oxygen-restriction may reflect relative sea-level changes in the Devils Gate section. Thus, the onset of anoxia in the Early *rhenana* Zone may correlate with the *semichatovae* transgression (Event 8 of Sandberg et al. (1997) and Event 4 of Sandberg et al. (2002)). Improvement of benthic oxygenation in the Late *rhenana* to early *linguiformis* zones may reflect subsequent shallowing, although this was rapidly superceded by euxinic-style deposition and probably rapid deepening across the F-F boundary. After a relatively quiet period of deposition during the late *linguiformis* Zone, the early part of the Famennian is characterised by high-energy events. Bed 32 is a large debris flow, interpreted to be a tsunamite by Sandberg et al. (1988) because, unlike other debris flows at Devils Gate, it contains abundant clasts of nearshore carbonates and reworked Frasnian conodonts at its top. It is also markedly erosive at its base and locally erodes down to a level in the uppermost Frasnian strata (Sandberg et al., 2003; Morrow and Sandberg, 2003).

### ***4.3. Northern Antelope Range***

#### *4.3.1. Lithological and faunal variations*

This highly expanded, composite section (Fig. 3; Appendix A) has previously been studied by Johnson et al. (1980, 1996), Hose et al. (1982), Morrow (2000), Morrow and Sandberg (2003), and Sandberg et al. (2003). At the base of the studied section several metres of finely laminated cherts and shales occur, interbedded with thin beds of turbiditic calcarenites. This is the upper part of the lower tongue of the Woodruff Formation (beds 1–6, Late *rhenana* Zone) and the calcarenites are composed of well-sorted, well-rounded grains of peloids (Fig. 9). This material dominates the overlying Fenstermaker Wash Formation (beds 7–26, Late *rhenana* to Early *triangularis* zones), which consists of thick beds of massive, peloidal,

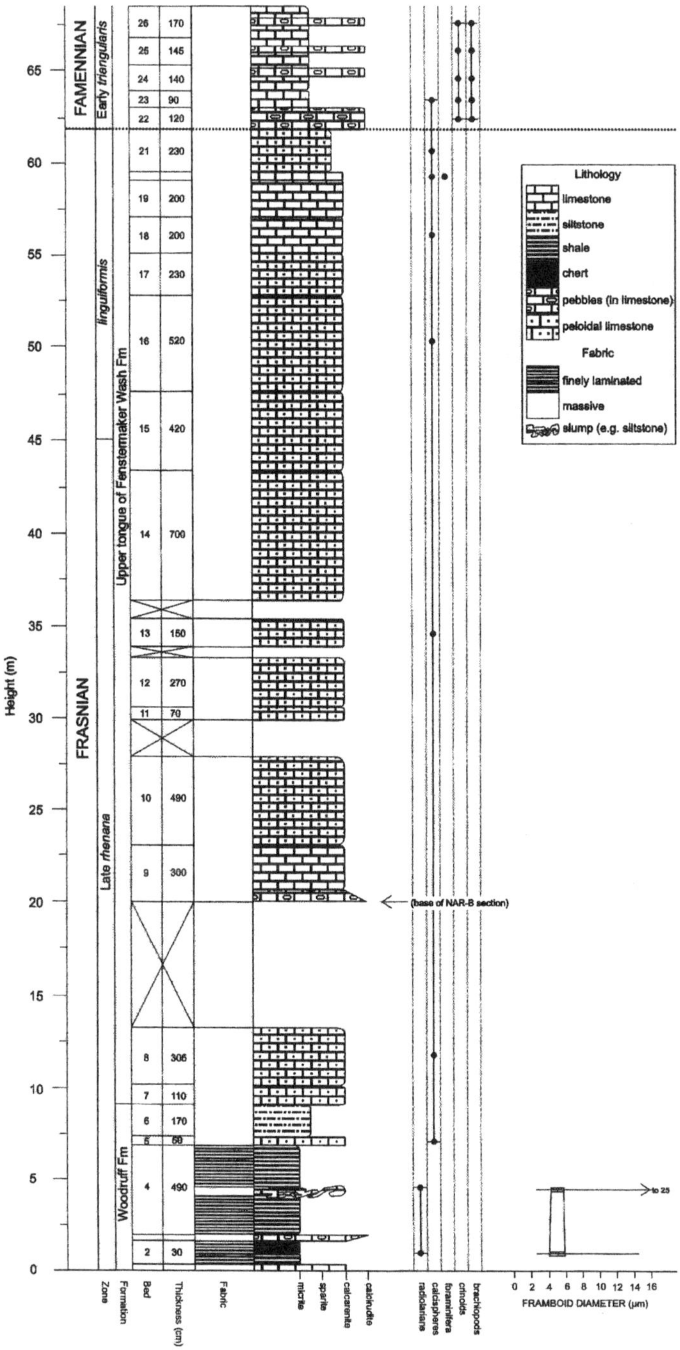


*Figure 9.* Log of Northern Antelope Range section including pyrite framboid "box and whisker" data. Conodont zonation based on Morrow (2000).

sandy calcarenites with significant quartz grain content. The peloids are strikingly well-sorted and rounded (Fig. 10). There is little lithological variation within the Fenstermaker Wash Formation until the *linguiformis* Zone, when the calcarenites become finer grained and have a markedly reduced quartz content (Morrow, 2000; Morrow and Sandberg, 2003). The lower Famennian (bed 22) is a distinctive calcirudite containing rounded, 5-cm-diameter pebbles of micrite, and silicified bioclasts. Above this, microsparites and peloidal calcarenites extend to the top of the section (Fig. 9).

Fauna in the Woodruff Formation is extremely sparse and limited to radiolarians, conodonts, and sponge spicules. The Fenstermaker Wash Formation is equally, poorly fossiliferous. Calcispheres are scattered throughout, although they are rare in the topmost Frasnian. Other fossils include silicified brachiopods and crinoids in beds 22–26, but generally the high degree of rounding and sorting of peloids suggests poor preservational conditions for fossils.

### *4.3.2. Pyrite framboid analysis*

The cherty beds of the Woodruff Formation contain common framboids, with small mean diameters between 5–6 μm, but relatively large MFD values up to 25 μm (Fig. 9). No framboids are observed in the Fenstermaker Wash Formation although bed 20 of the *linguiformis* Zone contains common pyrite blebs, in contrast to the surrounding beds that lack pyrite.

### *4.3.3. Depositional style and oxygenation history*

The finely laminated cherts and shales of the Woodruff Formation indicate a deep-water environment and the presence of calcareous turbidites suggests a base-of-slope depositional environment. The considerable bed thickness and lack of grading displayed by the calcarenites of the succeeding Fenstermaker Wash suggests that they are not typical turbidites. Instead, they may record high-concentration turbidity flows from close to source or even grain flows. Morrow (2000) and Sandberg et al. (2003) record deep-water conodonts within the Fenstermaker Wash Formation, in keeping with a slope interpretation. However, they interpreted the sediments to have formed on a shallow-water carbonate bank setting in which case the sediment-gravity flows must have entrained conodonts while they flowed down-slope, or alternatively the conodonts accumulated in the slope setting between gravity-flow events.

The absence of hemipelagic sediments in the Fenstermaker Wash Formation renders the reconstruction of oxygen levels at the depositional site extremely difficult. However, the cherts and shales of the underlying Woodruff Formation, within the Late *rhenana* Zone, are finely laminated and exhibit pyrite framboids with a size distribution characteristic of an anoxic environment. The high MFD in these beds suggests that bottom waters were occasionally dysoxic thereby allowing large framboids to grow within the surface layers of the sediment. The appearance of pyrite blebs in bed 20, in the later part of the *linguiformis* Zone, probably reflects a subtle decrease in oxygen levels although assessing any redox variations in such rapidly deposited, allochthonous carbonates is problematic. The scarcity of fossils in the Fenstermaker Wash Formation reflects a taphonomic control (i.e. a high energy source area inimical to the preservation of fossils) and does not allow inferences about oxygen levels at the depositional site. Allochthonous benthos is not recorded during the Frasnian but becomes common in the early Famennian. This may again be preservation controlled, but also might reflect an improvement in oxygen levels on the shelf source area at this time.

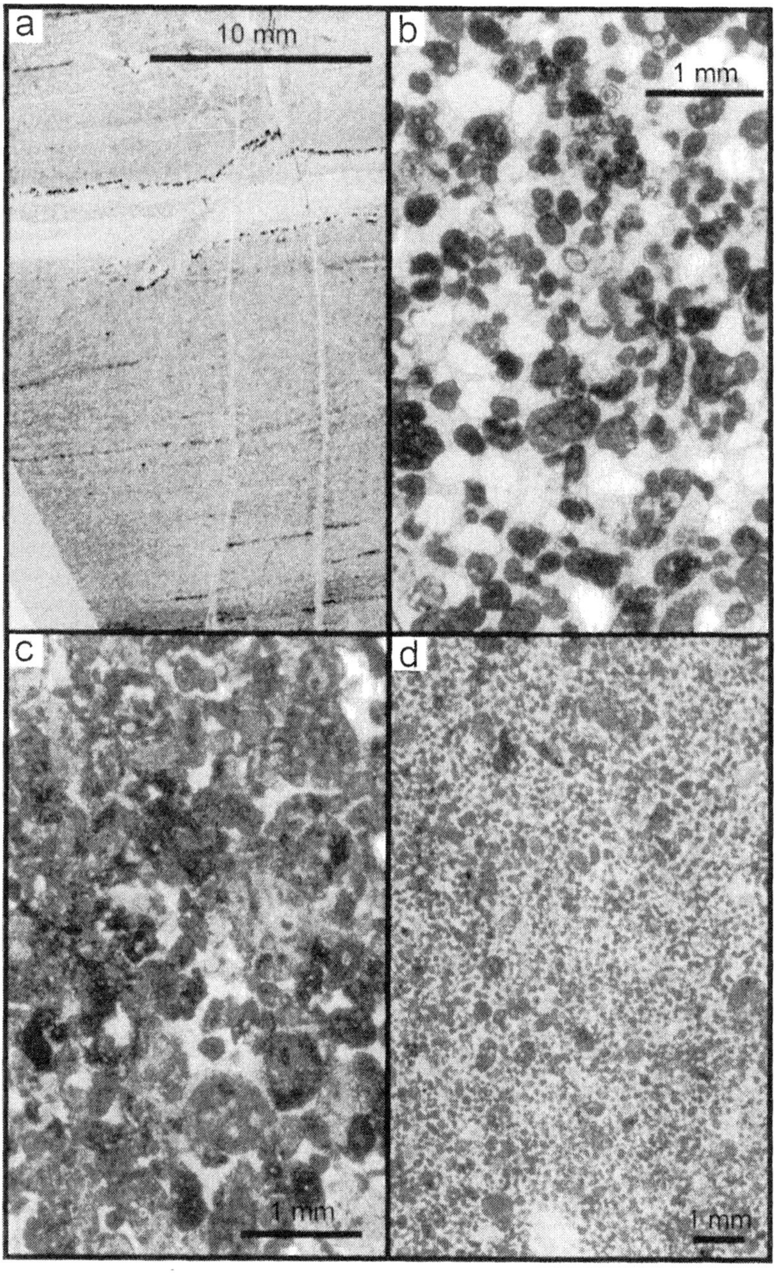


*Figure 10.* Photomicrograph of Northern Antelope Range rocks: (a) Lower part of bed 4 (Late *rhenana* Zone), finely laminated, pyritic calcareous shale of the Woodruff Formation; (b) Bed 8 (Late *rhenana* Zone), well-rounded, peloidal calcarenite, consisting of approximately 60% peloids and 40% quartz grains; (c) Bed 20 (*linguiformis* Zone), fine-grained calcarenite with notably reduced clastic content; and (d) Lower Famennian bed 23, peloidal microsparite.

The dramatic facies change between the two formations is interpreted to be the result of proto-Antler forebulge migration, which caused large amounts of sediment to be washed down-slope (Sandberg et al., 2003). The unusual facies of peloidal calcarenites (see Fig. 13), supplemented in the boundary beds by common, abraded bioclasts are similar to those described from the Famennian-age Palliser Formation of western Canada (Peterhaensel and Pratt, 2001). Although the source is unclear, these authors argued that excess nutrients in the ocean led to unusually high levels of microendolithic bioerosion of the skeletal grains deposited on the carbonate platform. Consequently, bioclasts (predominantly crinoids in the Canadian example) were intensely bored to form micritic peloids, and cortoids. The destruction of bioclasts clearly would mask the diversity and abundance of shelf benthos.

### 4.4. Tempiute Mountain

#### 4.4.1. Lithological and faunal variations

No specimens of *Palmatolepis linguiformis* have been found in this 20 m-thick mountainside section (Fig. 11); however, late evolutionary variants of *rhenana* Zone conodonts are present, suggesting that the *linguiformis* Zone is present (Morrow, 2000; Morrow, pers. comm. 2002). The studied section begins in this probable *linguiformis* Zone strata and continues up into the Early *crepida* Zone, and belongs entirely to the Devils Gate Limestone. Here, the Devils Gate consists of thin-bedded, laminated to bioturbated micrites, microsparites, and calcisiltites (Fig. 11). Interbedded with these sediments are common, massive calcisiltite, calcarenite, and sandstone turbidite beds of varying composition, which show common synsedimentary deformation (Morrow and Sandberg, 2003). The upper Frasnian turbidites are carbonate dominated, whereas those in the lower Famennian are mostly sandstones. The F-F boundary has not been precisely constrained, but is thought to lie within a short, unexposed interval between beds 8 and 9. The lower Famennian strata consist of sandstone turbidites that contain flat pebbles of limestone. At the top of the section, beds 14 and 15 are laminated, pyritic, medium-to-dark-grey micrites (Fig. 11).

Fauna in the main Tempiute Mountain boundary section is sparse. Samples recovered from the *punctata* to *jamieae* Zone strata below the study section shown in Fig. 11 contain poorly preserved tentaculitoids and ostracods. Much of the fauna is contained within the turbidite beds, and includes crinoids, tentaculitoids (in bed 4), ostracods, bivalves, conodonts, and calcispheres. No fossils, other than conodonts, were observed in the *triangularis* Zone, but in the *crepida* Zone, bed 15 contains abundant siliceous sponge spicules, calcispheres, and rare brachiopod fragments.

#### 4.4.2. Pyrite framboid analysis

Most samples from Tempiute Mountain do not yield framboids. However, the fine-grained sediments occasionally reveal very small numbers of framboids, often with small diameters (Fig. 11; Appendix B), in particular in the upper part of the *linguiformis* Zone. Much of the sequence around the boundary contains pyrite blebs, but only bed 14, in the Late *triangularis* Zone, contains common framboids, with a mean diameter of 6.8 μm (see Fig. 15).

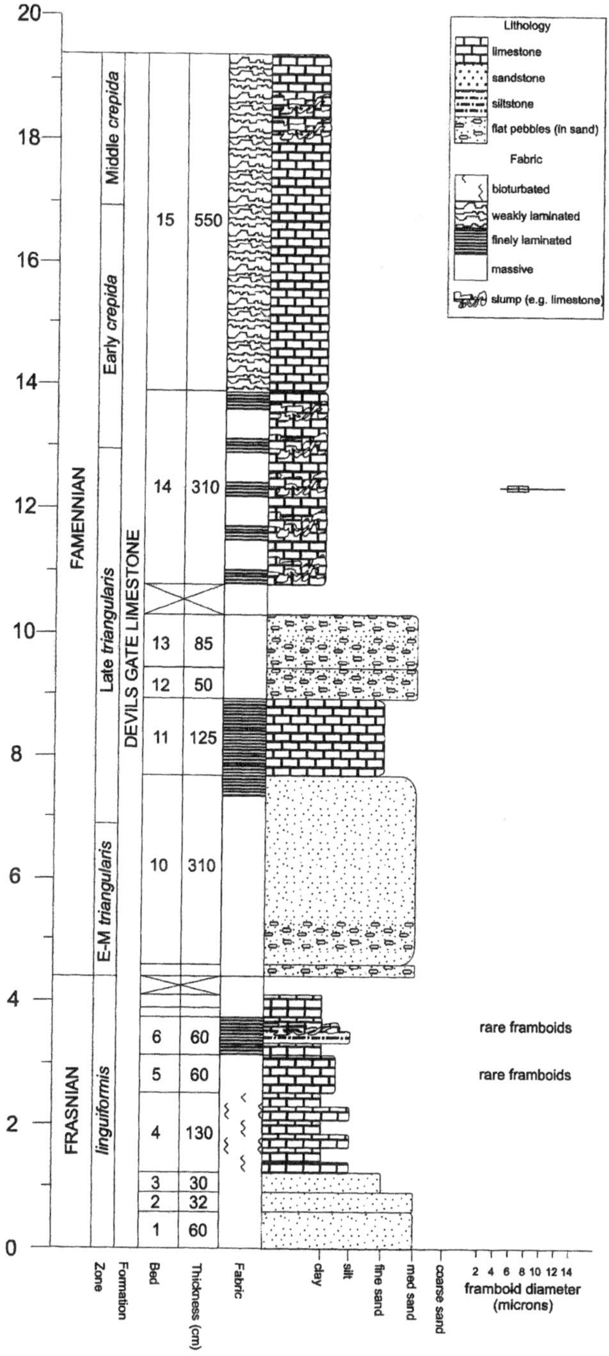


*Figure 11.* Log of Tempiute Mountain section including pyrite framboid "box and whisker" datum. Note that grade (e.g. clay) refers only to grain size, and not lithology, since this is a mixed clastic/and carbonate sequence. Conodont zonation based on Morrow (2000).

#### *4.4.3. Depositional style and oxygenation history*

Turbidites and background hemipelagic strata typical of a base-of-slope or basin floor depositional setting dominate this section. However, elsewhere in the F-F stratigraphy (*linguiformis* to Early *crepida* zones) slump folding is common indicating a slope setting (Morrow, 2000; Morrow and Sandberg, 2003). There is a noticeable change in sediment source across the F-F boundary from carbonate-sourced to clastic-sourced turbidity flows.

Pyrite framboids provide tentative evidence for anoxia at Tempiute Mountain but their rarity is rather enigmatic. Most of the turbidite beds are not framboidal, and contain a benthic fauna sourced from a shelf. The lack of an in-situ benthic fauna and presence of lamination suggests oxygen-depleted benthic conditions. However, conditions were not persistently anoxic, resulting in occasional bioturbation. Bed 6 exhibits persistent, fine laminations, and those framboids that have been found are very small, suggesting that euxinic conditions may have developed during the late part of the *linguiformis* Zone.

As in other sections, the early Famennian witnessed the appearance of abundant intraclasts, most notably of the "flat pebble" variety, at Tempiute Mountain. A similar appearance of flat-pebble conglomerates is also seen in equatorial sections in the immediate aftermath of the end-Permian mass extinction (Wignall and Twitchett, 1999). They record the early lithification of thin beds of carbonate that have not been bioturbated. The lack of such burrowing may be a signal of oxygen-restriction in the shallower-water source area of such intraclasts.

Finely laminated sediments and relatively small framboids in the Late *triangularis* to Early *crepida* zones provide clear evidence for anoxia, although the presence of sponge spicules in the *crepida* Zone suggests that benthic conditions improved to at least dysoxic levels allowing this benthic fauna to develop. The F-F boundary was not seen at Tempiute Mountain and it is not possible to conclusively state whether an anoxic event occurred at the boundary. However, lithological, faunal, and pyrite analyses suggest that conditions were probably oxygen-poor throughout the F-F interval.

### *4.5. Whiterock Canyon*

#### *4.5.1. Lithological and faunal variations*

This canyon section from near Horse Heaven Mountain, Nevada, comprises 20 m of section divided into 13 beds. Deposited in the Woodruff Basin, the section spans the Early *rhenana* to Late *triangularis* zones, and is entirely developed within the Woodruff Formation (Bratton et al., 1999; Morrow, 2000; Morrow and Sandberg, 2003). This comprises dark, finely laminated cherts and shales, interbedded with thin siltstones, some of which are calcareous (Fig. 12). Several of the *linguiformis* Zone beds are notably organic-rich. There is evidence for minor slumping in bed 3, but the majority of beds are undisturbed. The topmost Frasnian level (bed 6) is a finely laminated dark-grey to black, silty, organic-rich chert with common pyrite, that becomes siltier in its upper part. The directly overlying Famennian beds consist of siltstone rather than chert. The silt occurs as thin, graded laminae.

Bioturbation is absent in the Whiterock Canyon section and fossils are sparse (Fig. 12), other than abundant radiolarians in the Frasnian, especially in the topmost *linguiformis* Zone (bed 6). No fossils other than conodonts are found in the Famennian.

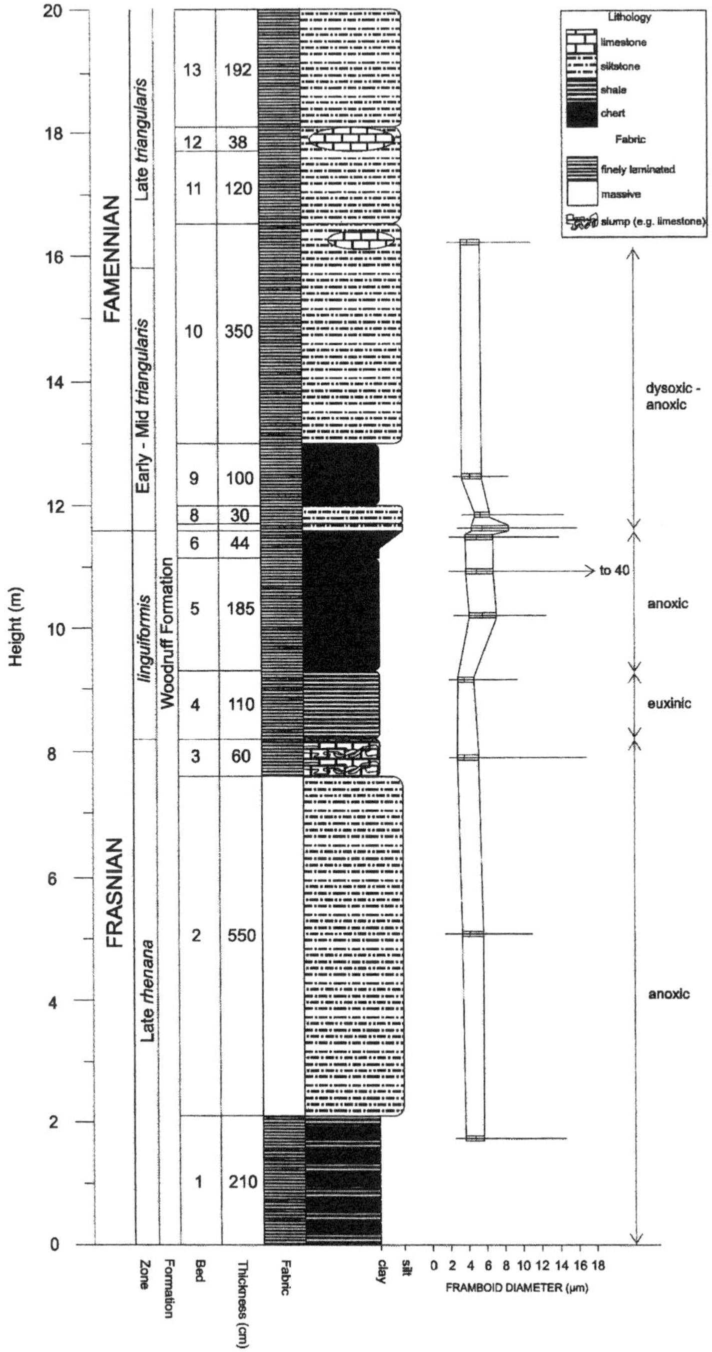


*Figure 12.* Log of Whiterock Canyon section, including pyrite framboid "box and whisker" data. Conodont zonation from Morrow (2000).

### 4.5.2. *Pyrite framboid analysis*

Polished blocks examined under the SEM reveal that almost the entire Whiterock Canyon sequence contains framboidal pyrite (Fig. 13; Appendix B). Within the Late *rhenana* Zone, beds 1 and 2 contain common framboids with consistent, small mean diameters. Bed 3 records a decrease in mean framboid diameter, and there is a further decrease in the *linguiformis* Zone (bed 4). In the middle and upper part of the *linguiformis* Zone, beds 5 and 6 contain abundant framboids, which record a small rise in mean diameter. The top of bed 5 (upper *linguiformis* Zone) has a much larger maximum framboid diameter of 40 μm (Fig. 13; Appendix B). By the topmost Frasnian, bed 6, framboids are again abundant and record a decrease in mean diameter, although they are slightly larger than those in the lower part of the sequence. There is an increase in mean framboid diameter in the basal Famennian, coincidental with the increased input of detrital silts within turbidites. This may in part be due to winnowing and reworking of framboids in the silt turbidites, resulting in the smallest framboid fraction being lost. Beds 7 and 8 contain less-abundant framboids with an increased mean diameter (Fig. 13; Appendix B). During the Middle *triangularis* Zone, framboids again become abundant, and record another drop in mean diameter to small sizes.

### 4.5.3. *Depositional style and oxygenation history*

The fine-grained nature of the sediment suggests deposition within a low energy, deep-water setting, although the thin, graded siltstones of the Famennian may record the distal-most development of a prograding turbidite succession. Sedimentation rates were clearly substantially lower than those seen in the more proximal slope sections to the east (Fig. 3).

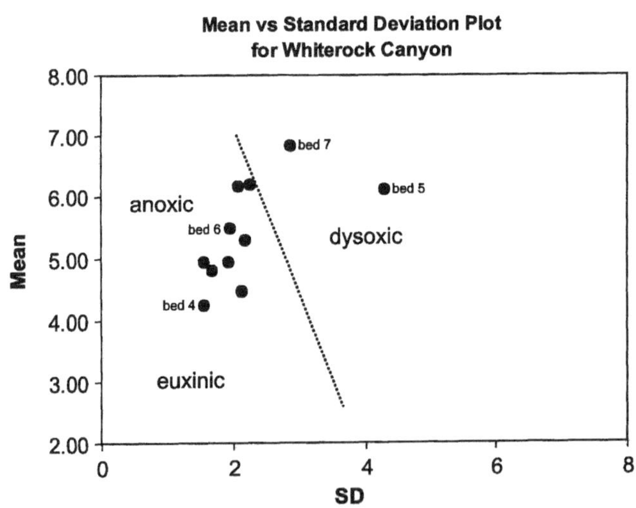


*Figure 13.* Mean framboid diameter (microns) vs. standard deviation plot for pyrite framboids at Whiterock Canyon. Note that most beds plot within the anoxic field, with beds 5 (linguiformis Zone) and 7 (Early triangularis Zone) plotting in the dysoxic field.

Pyrite framboid data and the fine laminae suggest that depositional conditions were persistently anoxic in this deep-water setting within the Woodruff Basin. However, there was subtle variation on this theme with small sizes and low size variability of early *linguiformis* Zone framboids indicating euxinic conditions. In the higher parts of the *linguiformis* Zone, the larger mean framboid diameter suggests conditions improved slightly to anoxic-dysoxic levels.

In the Early *triangularis* Zone, beds 7 and 8 both record a higher mean framboid diameter, characteristic of dysoxic conditions. These beds contain a succession of thin, distal turbidites, which might have led to a brief improvement in benthic oxygen levels, or alternatively modification (winnowing) of the framboid populations during transport. However, by the Middle *triangularis* Zone, the framboid distribution indicates that conditions again deteriorated to anoxic levels that prevailed to the top of the section.

### *4.6. Warm Springs*

#### *4.6.1. Lithological and faunal variations*

The 15-m-thick Warm Springs section, which is developed within the Woodruff Formation (Sandberg et al., 1997), has been divided into eight beds and spans the Early *rhenana* Zone to earliest Famennian. The base of the section comprises black, finely laminated cherts, and light-grey, finely laminated, pyritic siltstones, interbedded on a 5 mm–5 cm scale (Fig. 14). There is also a thin chert breccia consisting of chert clasts within a chert matrix that may be the product of tectonic cataclasis. The Late *rhenana* to *linguiformis* zones expose weakly bioturbated, pyritic, red, calcareous siltstone, with lenticular lenses of black, laminated chert. Part of bed 6 has undergone considerable deformation, possibly during slumping. Bed 7 comprises laminated, pyritic, calcareous siltstones, which weather red but are dark-grey when fresh. The F-F boundary has not been precisely placed using conodonts, but is constrained to within bed 7 (Sandberg et al., 1997; Morrow, pers. comm. 2002). Thus, the topmost Frasnian records a distinct lithological change from bioturbated to laminated silts. The early Famennian section shows intense tectonic deformation and consists of black cherts (Fig. 14).

As at Whiterock Canyon, fossils are extremely rare in the Warm Springs section apart from the abundant radiolarians in the chert beds, and microfaunas recovered from rare carbonate concretions. Also similar is the disappearance of radiolaria towards the F-F boundary, although in this section they return in the Famennian.

#### *4.6.2. Pyrite framboid analysis*

Warm Springs samples contain very rare pyrite framboids except around the F-F boundary where they become more common (Fig. 14). Very rare, but quite small framboids occur in the sediments of the Early *rhenana* and Late *rhenana* zones (Appendix B).

#### *4.6.3. Depositional style and oxygenation history*

This is a condensed section of cherts and fine siltstones in which pelagic and hemipelagic deposition dominated, although there is possible evidence for minor slumping during the

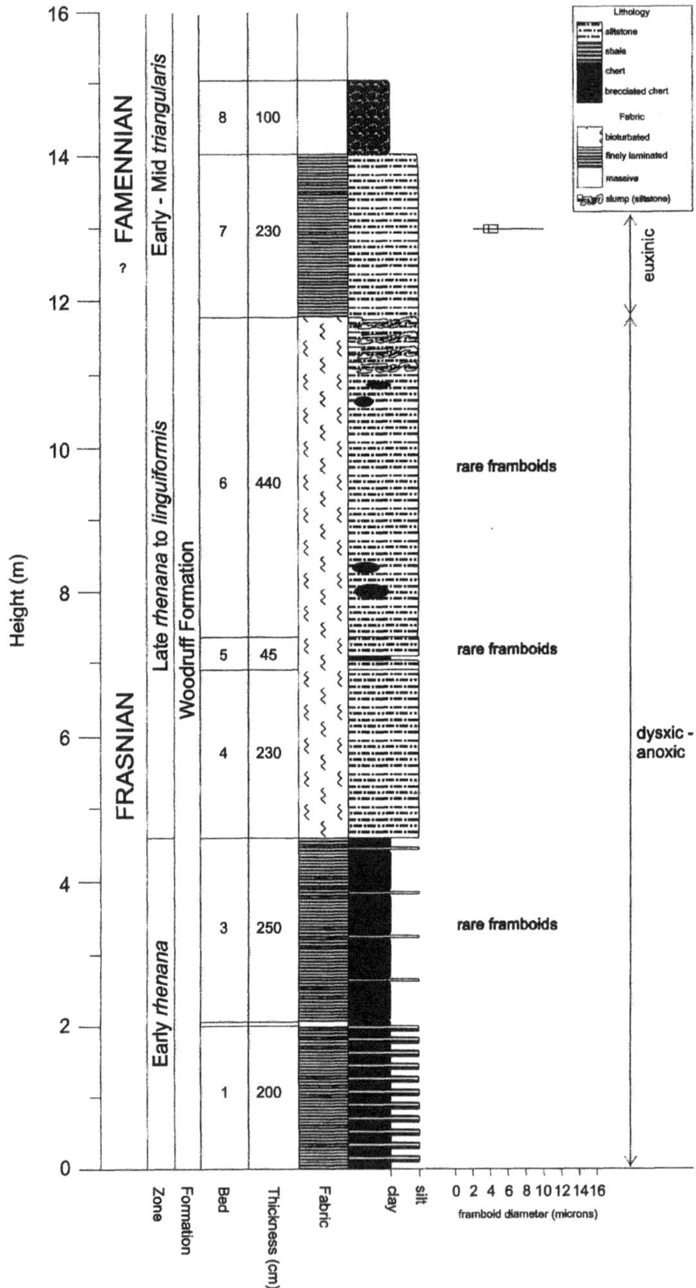


*Figure 14.* Log of Warm Springs section including pyrite framboid "box and whisker" data. Conodont zonation from Sandberg et al. (1997).

*linguiformis* Zone. A basin floor setting appears appropriate, although Sandberg et al. (1997) suggested a "transitional" location between lower slope and basin based on the occurrence of olistoliths (not seen in this section) derived from the carbonate platform to the east. It is of course possible for olistoliths to travel considerable distances along a basin floor. Whatever the precise location within the basin, Warm Springs is clearly a deep-water distal section.

Oxygen-restricted deposition clearly dominated deposition at Warm Springs section although the higher parts of the Late *rhenana* Zone display small burrows (Ichnofabric Index 2–3, Droser and Bottjer, 1986) and a decline in framboidal pyrite abundance indicating improvement of benthic oxygen levels. The most intensely anoxic conditions are recorded in the F-F boundary and lowest Famennian strata where small-diameter framboids are abundant.

## 5. Discussion

### *5.1. Summary of oxygenation history*

To the east of the proto-Antler forebulge, the slope and basin floor sections of the relatively shallow intra-shelf, Pilot basin (e.g. Devils Gate and Coyote Knolls) records persistent anoxic conditions during most of the studied F-F interval, probably due to the semi-restricted nature of deposition in this intracontinental basin (Figs. 15 and 16). Compared with the Devils Gate section, the intensity of anoxia is generally greater and more persistent in the Coyote Knolls section. This probably reflects its deeper-water location. Thus, persistent euxinic conditions are developed throughout the late *rhenana* to end-*linguiformis* interval at Coyote Knolls while euxinicity at Devils Gate began later (late *linguiformis* Zone) and persisted a little longer (Early *triangularis* Zone).

Anoxic sedimentation on the slope of the Pilot basin was frequently interrupted by turbidites and debris flows that contain a diverse shelf fauna, indicating oxygenated conditions in the source area which was probably the crest of the proto-Antler forebulge. Casier

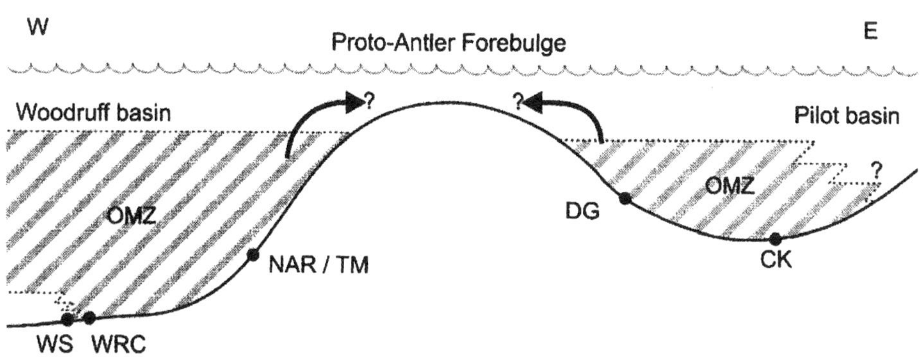


*Figure 15.* Schematic cross-section showing estimated palaeogeographic positions of Great Basin Frasnian-Famennian sections, and inferred oxygen minimum zone (OMZ) during the latest Frasnian (WRC = Whiterock Canyon; WS = Warm Springs; NAR = Northern Antelope Range; DG = Devils Gate; TM = Tempiute Mountain; CK = Coyote Knolls). Arrows indicate potential spilling of anoxic waters onto shallow shelf areas during latest Frasnian.

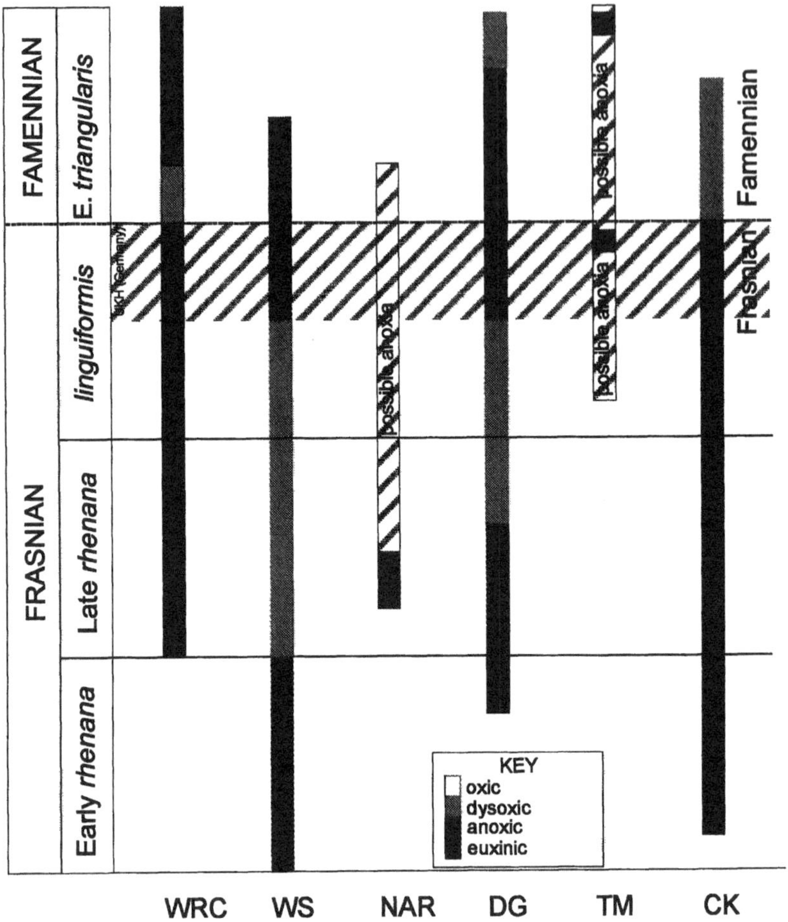


*Figure 16.* Timing of development of oxygen-poor conditions in the Great Basin (section abbreviations as in Fig. 1). Note that true palaeo-oxygenation conditions across the F-F boundary in NAR and TM sections are probably masked by turbidite events. The hachured shaded band relates to the development of anoxia in the Devils Gate and Coyote Knolls sections and is contemporaneous with the Upper Kellwasser Horizon in Germany.

and Lethiers (1998) identified a diverse ostracod assemblage within the Frasnian at Devils Gate indicative of a well-oxygenated environment but no assemblage characteristic of oxygen-depleted conditions has been identified close to the boundary. However, the ostracods are allochthonous, because they are found within turbidites and debris flow beds derived from the shelf, and these data lend further support to the suggestion that the shelf remained oxygenated. The ostracod evidence does not, however, preclude anoxic conditions developing at the depositional site.

To the west of the Pilot Basin and the adjacent proto-Antler forebulge, the eastern slopes of the Woodruff Basin were dominated by rapid deposition of allochthonous strata (e.g. Northern Antelope Range and Tempiute Mountain). These preserve a poor record of palaeo-oxygenation levels. However, interbedded hemipelagic sediments of the Late *rhenana* and *linguiformis* zones are often laminated and framboidal, indicating that anoxic conditions were developed. Allochthonous fauna is contained within the common, thick sediment-gravity flows sourced from the nearby, better-oxygenated shelf.

Interestingly, the deep-water sections of the Woodruff Basin record marginally better oxygenation conditions than the deepest section of the Pilot Basin (Coyote Knolls). Thus dysoxic conditions, with weakly bioturbated strata, are recorded in the Late *rhenana* and *linguiformis* zones at the deep-water Warm Springs section (Figs. 15 and 16). The most intensely anoxic conditions in the basinal sections are recorded from the Early *rhenana* and latest Frasnian strata. If the Warm Springs section is the deepest record, it is possible that the floor of the Woodruff Basin was, for much of the F-F interval, around the lower limit of a mid-water oxygen minimum zone (Fig. 15). Intensification and expansion of this zone around the F-F boundary may be responsible for the basin-wide spread of anoxia at this time. Support for this oxygenation history comes from the even deeper water record of the Shoshone Mountains of northern Nevada. Here, the Slaven Chert records the quiet accumulation of biogenic silica in a "relatively distal or isolated location" (Boundy-Sanders et al., 1999). A Late *rhenana* Zone section of the Chert consists of abundant siliceous sponge spicules that indicate ample seafloor oxygen levels at this time (Boundy-Saunders et al., 1999). This is in marked contrast with the intensely oxygen-restricted deposition of contemporaneous slope sections studied here.

### *5.2. Comparison with inorganic geochemical data*

The oxygenation history for the Pilot and Woodruff basins contrasts with that suggested by Bratton et al. (1999). They investigated the Whiterock Canyon and Coyote Knolls sections and estimated palaeo-oxygen levels on the basis of trace metal concentrations, principally vanadium. Based on a decrease in V concentration from 2663 ppm to just 284 ppm at Whiterock Canyon, Bratton et al. (1999, p. 282) suggested that within the *linguiformis* Zone "there is a dramatic shift to fully oxic Zone 1 conditions" around 40 cm below the F-F boundary. However, the rocks at Whiterock Canyon are intensely and deeply weathered, and it is likely that this has affected the trace metal content. Such weathering has no effect neither on the pyrite framboid distribution, nor the sediment fabric, which remains finely laminated throughout. The inferred palaeo-oxygenation history of Bratton et al. (1999) is not supported by the lithological and pyrite framboid data in this study, which indicates that anoxic conditions prevailed up to the F-F boundary in the Woodruff Basin.

Similarly, the oxygenation history inferred by Bratton et al. (1999) for the Pilot Basin (Coyote Knolls) contradicts that suggested by data in this study. On the basis of V concentrations, Bratton et al. (1999) suggested that the main anoxic pulse in the Pilot Basin ended 6 m below the F-F boundary, during the *linguiformis* Zone, and that conditions were fully oxygenated at the boundary. The presence of tiny framboids and fine lamination, and the absence of fossils in the boundary beds are contrary to this interpretation, and indicate that anoxia persisted up to the F-F boundary. Bratton et al. (1999) have a low sampling density in the *linguiformis* Zone, and it may be possible that their "oxic" samples were taken from the turbidite beds, and do not reflect conditions at the depositional site itself.

The clearest record of depositional conditions comes from those sections where high sedimentation rates and turbidite deposition have not hindered the assessment of background oxygen levels. In summary, conditions were oxygen restricted during much of the

Early *rhenana* to *triangularis* zones in both the Woodruff and Pilot basins (Fig. 16). Anoxia appears to have been most intense around the Early–Late *rhenana* Zone boundary, and again in the latest *linguiformis* zone, when euxinic conditions developed. These euxinic pulses are age-equivalent with the Kellwasser anoxic horizons in Europe and may be considered the Great Basin manifestation of these phenomena.

### *5.3. Oxygenation history on the shelf*

No sections recording deposition in a shelf environment have been studied here and inferences can only be made on the basis of the allochthonous clasts sourced from this setting. The presence of a varied benthic shelf fauna entrained within sediment-gravity flows in the Devils Gate and Coyote Knolls sections suggests that, in general, the source area supplying carbonate clasts to the Pilot Basin was well-oxygenated. However, turbidites are not present in the critical, topmost *linguiformis* Zone interval at Devils Gate, and those from this interval at Coyote Knolls are not fossiliferous. This tentative evidence for shelfal anoxia provides a kill mechanism for the 75% extinction recorded amongst ostracod species from the Devils Gate sediment-gravity flows (Casier and Lethiers, 1998). Further indirect evidence for shallow-water anoxia near the F-F boundary is provided by the basal Famennian flat-pebble conglomerates in the Tempiute Mountain section, which are thought to have formed in response to the suppression of bioturbation by anoxia. The resultant thin-bedded carbonates have been eroded by storms to source the distinctive and unusual “flat pebbles” seen in the allodapic carbonates (see Wignall and Twitchett, 1999).

### *5.4. Event stratigraphy and sea-level change*

Sandberg et al. (1997) identified a 20-event stratigraphy in the Middle and Late Devonian of Nevada and Utah. Of these, events 8 (major eustatic rise during the Early *rhenana* Zone), 9 (rapid eustatic rise and fall during the *linguiformis* Zone), and 10 (mass extinction at the F-F boundary) fall within the time frame of this study. Within the Early *rhenana* Zone in the Pilot Basin (Coyote Knolls), the carbonate platform is drowned and the Guilmette Formation is replaced by the Pilot Shale, evidence of a major deepening. This corresponds to a major eustatic rise, Event 8 of Sandberg et al. (1997), but could equally be the result of regional tectonic subsidence at this locality. The Devils Gate, Northern Antelope Range (Sandberg et al., 2003), and Tempiute Mountain (Morrow and Sandberg, 2003) sections also record evidence of deepening during the Early *rhenana* Zone, which gives evidence of the *semichatovae* transgression throughout the region.

Based on this study, there is little lithological evidence for Event 9 of Sandberg et al. (1997) because the lithology and facies around this level change little, or this crucial interval is not preserved. The sea-level curve of Sandberg et al. (2002) shows a sea-level fall throughout the late part of the *linguiformis* Zone. This is primarily based on the increased abundance of the normally shallow-water conodont genus *Icriodus*, from a minor component to as high as 30% of the total conodont fauna within previous deeper water settings (Sandberg et al., 1988).

Several sections record an increase in clastic content at or near to the F-F boundary (Fig. 17). At Coyote Knolls this is manifest as a change from calcisiltite to coarse siltstone. At Devils Gate, micrites and silty shales grade into siltstones in the boundary bed, but the lithology abruptly returns to carbonates for the remainder of the Early *triangularis* Zone. The Whiterock Canyon section records an influx of silt in the uppermost Frasnian and basal Famennian (Fig. 17). At Tempiute Mountain there is a marked change in lithology as micritic background sedimentation is replaced by thick, sandstone turbidites at the F-F boundary. These lithological changes are not seen everywhere, for example the clastic content in the Northern Antelope Range declines in the same interval (Fig. 17), but they have been used as evidence for a major sea-level fall beginning in the later part of the *linguiformis* Zone (Sandberg et al., 2002; Morrow and Sandberg, 2003). In this case, the intensification and peak of anoxia at the F-F boundary in the Great Basin would occur during a major regressive interval; this is an unusual occurrence because other major anoxic events of the geological record occur during transgression or highstand (e.g. Hallam and Wignall, 1997, 1999). However, the increased clastic content might not be related to sea-level changes at all, but could arise from a reduction in carbonate productivity on the shelf during the mass extinction or a climate change to increased humidity and the development of a mixed clastic-carbonate depositional system. We would argue that the evidence for a major regression at the end of the *linguiformis* Zone is not compelling.

Event 10 of Sandberg et al. (1997), the F-F mass extinction, is difficult to recognise in many of the studied sections (although the allochthonous fossils, such as corals (e.g. Sandberg et al., 1988), ostracods (e.g. Casier and Lethiers, 1998), and conodont faunas (e.g. Sandberg et al., 1988) clearly record major extinction steps at Devils Gate). The most notable crisis in the basinal sections is the temporary loss of radiolaria in the F-F boundary interval.

## 6. Anoxia as an extinction mechanism

Anoxia has long been regarded as one of the most likely kill mechanisms of the F-F mass extinction (e.g. House, 1985; Casier, 1987; Sandberg et al., 1988; Walliser et al., 1989; Goodfellow et al., 1989; Buggisch, 1991; Becker, 1993; Joachimski and Buggisch, 1993; Joachimski et al., 2001; Bond et al., 2004). Implicit in the anoxia kill hypothesis is the need for globally synchronous anoxia. The lack of a demonstrable synchronicity has been one of the key arguments against the mechanism (Becker et al., 1991; McGhee, 1996; Copper, 1998; Bratton et al., 1999; House, 2002). The principal support for the detractors has come from the Nevada and Utah sections where, as noted above, improved oxygenation has been reported from the F-F boundary (Bratton et al., 1999). However, our evidence suggests that anoxia prevailed and often intensified leading up to the F-F boundary in the Great Basin. The Upper Kellwasser anoxic event is thus seen to be present in the western United States, considerably strengthening the likelihood that it was the cause of the F-F mass extinction. However, the same cannot be said of the Lower Kellwasser event. Based on lithology and framboid analyses, the Lower Kellwasser Horizon in Germany is within the early part of the Late *rhenana* Zone, whereas at Kowala Quarry in Poland, this anoxic event occurs much later, near the top of the Late *rhenana* Zone (Bond et al., 2004). Feist and Schindler

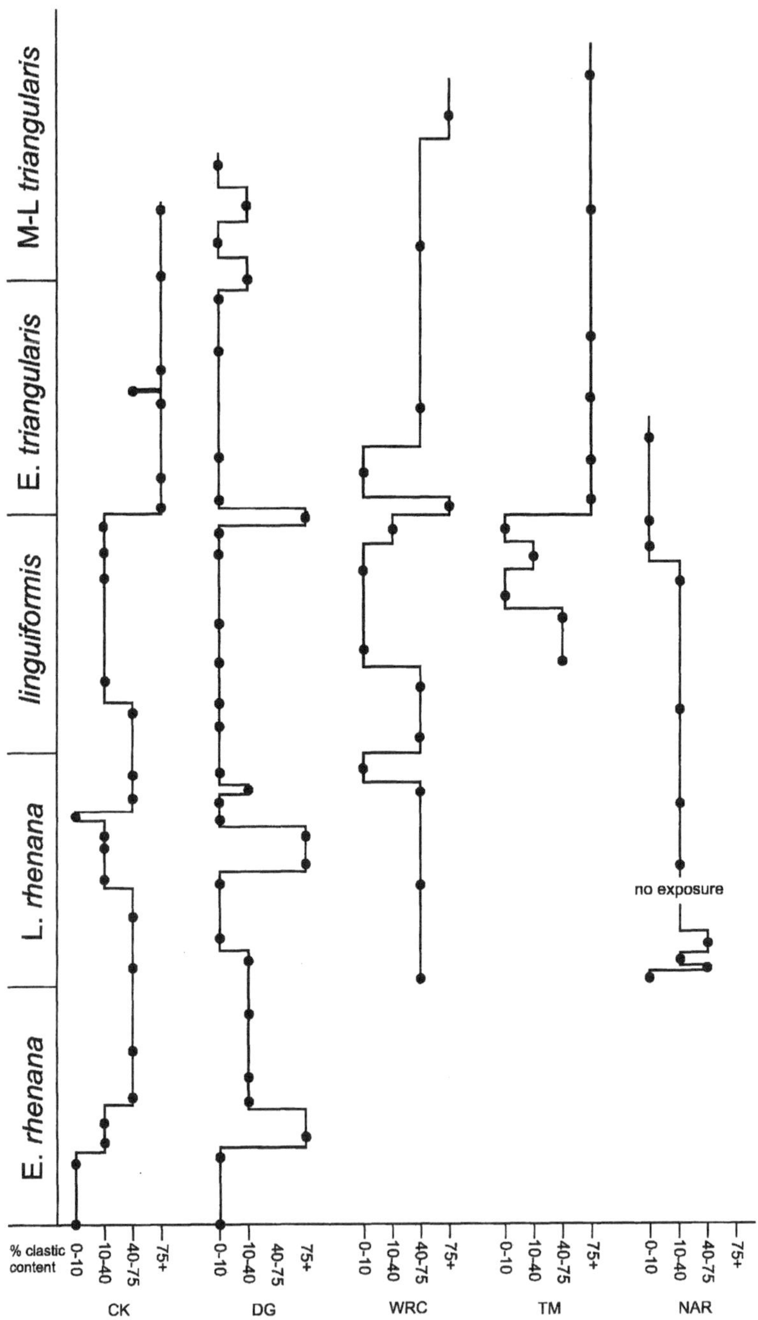


*Figure 17.* Summary of changes in clastic content in the F-F boundary sections of this study. These values are based on petrographic analyses. Data points are represented by nodes.

(1994) have previously noted a non-synchroneity between French and German developments of Lower Kellwasser "dark facies" based on lithology, as have Crick et al. (2002), based on magnetic susceptibility correlation data. Similarly, in the Great Basin, there is no clear, basin-wide increase in anoxic intensity at any level in the Late *rhenana* Zone and several deep-water sections seem rather well oxygenated at this time. Synchroneity is therefore the key difference between these two anoxic events – and an explanation as to why only the anoxic event during the latest *linguiformis* Zone saw catastrophic faunal losses, because at this time the extent of anoxia was at its peak extent both vertically and laterally within the water column.

## Acknowledgements

We thank Jared Morrow for his invaluable field guidance and discussions during this project. He is also thanked, together with Michael Joachimski for their comments on an earlier version of the manuscript. Bond's research was funded by a PhD studentship from the Natural Environment Research Council.

## References

Becker, R.T., 1993. Stratigraphische Gliederung und Ammonoideen-Faunen im Nehdenium (Oberdevon II) von Europa und Nord-Afrika. Cour. Forsch.-Inst. Sencken. 155, 1–405.

Becker, R.T., House, M.R., Kirchgasser, W.T., Playford, P.E., 1991. Sedimentary and faunal changes across the Frasnian/Famennian boundary in the Canning Basin of Western Australia. Hist. Biol. 5, 183–196.

Bond, D., Wignall, P.B., Racki, G., 2004. Extent and duration of marine anoxia during the Frasnian-Famennian (Late Devonian) mass extinction in Poland, Germany, Austria and France. Geol. Mag. 141, 1–29.

Boundy-Sanders, S.Q., Sandberg, C.A., Murchey, B.L., Harris, A.G., 1999. A late Frasnian (Late Devonian) radiolarian, sponge spicule, and conodont fauna from the Slaven Chert, northern Shoshone Range, Roberts Mountains Allochthon, Nevada. Micropaleontology 45, 62–68.

Bratton, J.F., Berry, W.B.N., Morrow, J.R., 1999. Anoxia pre-dates Frasnian-Famennian boundary mass extinction horizon in the Great Basin, USA. Palaeogeogr. Palaeoclimatol. Palaeoecol. 154, 275–292.

Buggisch, W., 1972. Zur Geologie und Geochemie der Kellwasserkalke und ihrer begleitenden Sedimente (Unteres Oberdevon). Geology and geochemistry of the lowermost Devonian Kellwasser Limestone and its associated sediments. Abhand. Hess. Landesamtes Bodenforsch. 62, 1–67.

Buggisch, W., 1991. The global Frasnian-Famennian 'Kellwasser Event'. Geol. Rund. 80, 49–72.

Canfield, D.E., Thamdrup, B., 1994. The production of $^{34}S$-depleted sulphide during bacterial disproportionation of elemental sulfur. Science 266, 1973–1975.

Casier, J.-G., 1987. Etude biostratigraphique et paléoécologique des ostracodes du récif de marbre rouge du Hautmont à Vodelée (partie supérieure du Frasnien, Bassin de Dinant, Belgique). Rev. Paléobiol. 6, 193–204.

Casier, J.-G., Lethiers, F., 1998. Les Ostracodes du Frasnien terminal (zone a linguiformis des Conodontes) de la coupe du col de Devils Gate (Nevada, USA). Latest Frasnian (linguiform conodont zone) ostracods from the Devils Gate Pass section, Nevada, USA. Bull. L'Inst. Royal Sci. Nat. Belg. Sci. Terre 68, 77–95.

Casier, J.-G., Lethiers, F., Claeys, P., 1996. Ostracod evidence for an abrupt mass extinction at the Frasnian /Famennian boundary (Devils Gate, Nevada, USA). Compt. Rend. l'Acad. Sci., Ser. II. Sci. Terre Planet. 322(5), 415–422.

Copper, P., 1998. Evaluating the Frasnian-Famennian mass extinction: comparing brachiopod faunas. Acta Palaeont. Polon. 43, 137–154.

Crick, R.E., Ellwood, B.B., Feist, R., El Hassani, A., Schindler, E., Dreesen, R., Over, D.J., Girard, C., 2002. Magnetostratigraphy susceptibility of the Frasnian/Famennian boundary. Palaeogeogr. Palaeoclimatol. Palaeoecol. 181(1–3), 67–90.
Droser, M.L., Bottjer, D.J., 1986. A semiquantitative field classification of ichnofabric. Journ. Sed. Pet. 56, 558–559.
Feist, R., Schindler, E., 1994. Trilobites during the Frasnian Kellwasser Crisis in European Late Devonian cephalopod limestones. Cour. Forsch.-Inst.. Sencken. 169, 195–223.
Follmi, K.B., Grimm, K.A., 1990. Doomed pioneers: Gravity flow deposition and bioturbation in marine oxygen-deficient environments. Geology, 18, 1969–1072.
Goodfellow, W.D., Geldsetzer, H.H.J., McLaren, D.J., Orchard, M.J., Klapper, G., 1989. Geochemical and isotopic anomalies associated with the Frasnian-Famennian extinction. Hist. Biol. 2, 51–72.
Hallam, A., 1980. Black shales. J.. Geol. Soc. London 137, 123–124.
Hallam, A., Wignall, P.B., 1997. Mass extinctions and their aftermath. Oxford University Press, Oxford.
Hallam, A., Wignall, P.B., 1999. Mass extinctions and sea-level changes. Earth-Sci. Rev. 48, 217–250.
Hose, R.K., Armstrong, A.K., Harris, A.G., Mamet, B.L., 1982. Devonian and Mississippian rocks of the northern Antelope Range, Eureka County, Nevada. US Geol. Surv. Prof. Paper 1182, 19p.
House, M.R., 1985. Correlation of mid-Palaeozoic ammonoid evolutionary events with global sedimentary perturbations. Nature 313(5997), 17–22.
House, M.R., 2002. Strength, timing, setting and cause of mid-Palaeozoic extinctions. Palaeogeogr. Palaeoclimatol. Palaeoecol. 181(1–3), 5–26.
Joachimski, M.M., Buggisch, W., 1993. Anoxic events in the late Frasnian – causes of the Frasnian-Famennian faunal crisis? Geology 21, 675–678.
Joachimski, M.M., Ostertag-Henning, C., Pancost, R.D., Strauss, H., Freeman, K.H., Littke, R., Damste, J.S.S., Racki, G., 2001. Water column anoxia, enhanced productivity and concomitant changes in $\delta^{13}C$ and $\delta^{34}S$ across the Frasnian-Famennian boundary (Kowala – Holy Cross Mountains/Poland). Chem. Geol. 175(1–2), 109–131.
Joachimski, M.M., Pancost, R.D., Freeman, K.H., Ostertag-Henning, C., Buggisch, W., 2002. Carbon isotope geochemistry of the Frasnian-Famennian transition. Palaeogeogr. Palaeoclimatol. Palaeoecol. 181, 91–109.
Johnson, J.G., Klapper, G., Elrick, M., 1996. Devonian transgressive-regressive cycles and biostratigraphy, Northern Antelope Range, Nevada: establishment of reference horizons for global cycles. Palaios 11, 3–14.
Johnson, J.G., Klapper, G., Sandberg, C.A., 1985. Devonian eustatic fluctuations in Euramerica. Geol. Soc. Am. Bull. 96, 567–587.
Johnson, J.G., Klapper, G., Trojan, W.R., 1980. Brachiopod and conodont successions in the Devonian of the northern Antelope Range, central Nevada. Geol. Palaeont. 14, 77–116.
Lüning, S., Kolonic, S., Loydell, D., Craig, J., 2003. Reconstruction of the original organic richness in weathered Silurian shale outcrops (Murzuq and Kufra basins, southern Libya). Geoarabia 8, 299–308.
McGhee, G.R., 1996. The Late Devonian Mass Extinction. Columbia University Press, New York.
Morrow, J.R., 1997. Shelf-to-basin event stratigraphy, conodont paleoecology, and geologic history across the Frasnian-Famennian (F-F mid-Late Devonian) boundary mass extinction, central Great Basin, western U.S. Boulder, Colorado. University of Colorado, Ph.D. dissertation, 355pp.
Morrow, J.R., 2000. Shelf-to-basin lithofacies and conodont paleoecology across Frasnian-Famennian (F-F, mid–Late Devonian) boundary, central Great Basin (Western U.S.A.). Cour. Forsch.-Inst. Sencken. 219, 1–57.
Morrow, J.R., Sandberg, C.A., 2003. Late Devonian sequence and event stratigraphy across the Frasnian-Famennian (F-F) boundary, Utah and Nevada. In: Harries, P.J. (Ed.), High-Resolution Approaches in Stratigraphic Paleontology. Kluwer Dordrecht, pp. 351–419.
Muramoto, J.A., Honjo, S., Fry, B., Hay, B.J., Howarth, R.W., Cisne, J.L., 1991. Sulfur, iron and organic carbon fluxes in the Black Sea: sulfur isotopic evidence for origin of sulfur fluxes. Deep-Sea Res. 38, S1151–S1187.
Peterhaensel, A., Pratt, B.R., 2001. Nutrient-triggered bioerosion on a giant carbonate platform masking the postextinction Famennian benthic community. Geology 29(12), 1079–1082.
Racki, G., Racka, M., Matyja, H., Devleeschouwer, X., 2002. The Frasnian/Famennian boundary interval in the South Polish-Moravian shelf basins: integrated event-stratigraphical approach. Palaeogeogr. Palaeoclimatol. Palaeoecol. 181, 251–297.
Raiswell, R., 1982. Pyrite texture, isotopic composition and the availability of iron. Am. J. Sci. 282, 1244–1263.
Raiswell, R., Newton, R., Wignall, P.B., 2001. An indicator of water-column anoxia: resolution of biofacies variations in the Kimmeridge Clay (Upper Jurassic, UK). J. Sed. Res. 71, 286–294.

Raup, D.M., Sepkoski, J.J., 1982. Mass extinction in the marine fossil record. Science 215, 1501–1503.
Sandberg, C.A., Morrow, J.R., Poole, F.G., Ziegler, W., 2003. Middle Devonian to Early Carboniferous event stratigraphy of Devils Gate and Northern Antelope Range sections, Nevada, U.S.A. Cour. Forsch.-Inst. Sencken. 242, 187–207.
Sandberg, C.A., Morrow, J.R., Warme, J.E., 1997. Late Devonian Alamo impact event, global Kellwasser events, and major eustatic events, eastern Great Basin, Nevada and Utah. Brigham Young Univ. Geol. Stud. 42, 129–160.
Sandberg, C.A., Morrow, J.R., Ziegler, W., 2002. Late Devonian sea-level changes, catastrophic events, and mass extinctions. In: Koeberl, C., MacLeod, K.G., (Eds), Catastrophic Events and Mass Extinctions: impacts and Beyond. Geol. Soc. Am. Spec. Pap. 356, 473–487.
Sandberg, C.A., Poole, F.G., 1977. Conodont biostratigraphy and depositional complexes of Upper Devonian cratonic-platform and continental-shelf rocks in the wester United States. In: Murphy, M.A., Berry, W.B.N., Sandberg, C.A. (Eds.), Devonian of North America. Camp. Mus. Contrib. 4, 144–182.
Sandberg, C.A., Poole, F.G., Johnson, J.G., 1989. Upper Devonian of Western United States. In: McMillan, N.J., Embry, A.F., Glass, D.J. (Eds), Devonian of the World. Can. Soc. Petrol. Geol. Memoir 14(III), 183–220.
Sandberg, C.A., Ziegler, W., Dreesen, R., Butler, J.L., 1988. Part 3: Late Frasnian mass extinction: conodont event stratigraphy, global changes, and possible causes. Cour. Forsch.-Inst. Sencken. 102, 263–307.
Savrda, C.E., Bottjer, D.J., 1986. Trace-fossil model for reconstruction of paleo-oxygenation in bottom waters. Geology 14(1), 3–6.
Schindler, E., 1990. Die Kellwasser-Krise (hohe Frasne-Stufe, Ober Devon). Gottinger Arbeiten zur Geol. Paläont. 46, 1–115.
Scotese, C.R., McKerrow, W.S., 1990. Revised world maps and introduction. Mem. Geol. Soc. Lond. 12, 1–21.
Sepkoski, J.J., 1996. Patterns of Phanerozoic extinction: a perspective from global data bases. In: Walliser, O.H. (Ed.), Global Events and Event Stratigraphy, Springer-Verlag, Berlin, PP. 35–52.
Walliser, O.H., 1996. Global events in the Devonian and Carboniferous. In: Walliser, O.H. (Ed.), Global Events and Event Stratigraphy. Springer-Verlag, Berlin, pp. 225–250.
Walliser, O.H., Groos-Uffenorde, H., Schindler, E., Ziegler, W., 1989. On the Upper Kellwasser Horizon (boundary Frasnian/Famennian). Cour. Forsch.-Inst. Sencken. 110, 247–255.
Wendt, J., Belka, Z., 1991. Age and depositional environment of Upper Devonian (early Frasnian to early Famennian) black shales and limestones (Kellwasser Facies) in the eastern Anti-Atlas, Morocco. Facies 25, 51–90.
Wignall, P.B., Myers, K.J., 1988. Interpreting benthic oxygen levels in mudrocks – a new approach. Geology 16, 452–455.
Wignall, P.B., Newton, R.J., 1998. Pyrite framboid diameter as a measure of oxygen deficiency in ancient mudrocks. Am. J. Sci. 298, 537–552.
Wignall, P.B., Twitchett, R.J., 1999. Unusual intraclastic limestones in Lower Triassic carbonates and their bearing on the aftermath of the end-Permian mass extinction. Sedimentology 46(2), 303–316.
Wilkin, R.T., Barnes, H.L., 1996. Pyrite formation by reactions of iron monosulfides with dissolved inorganic and organic sulfur species. Geochim. Cosmo. Acta 60, 4167–4179.
Wilkin, R.T., Barnes, H.L., 1997. Formation processes of framboidal pyrite. Geochim. Cosmo. Acta 61, 323–339.
Wilkin, R.T., Barnes, H.L., Brantley, S.L., 1996. The size distribution of framboidal pyrite in modern sediments: an indicator of redox conditions. Geochim. Cosmo. Acta 60, 3897–3912.
Yudina, A.B., Racki, G., Savage, N.S., Racka, M., Malkowski, K., 2002. The Frasnian-Famennian events in a deep-shelf succession, Subpolar Urals: biotic, depositional and geochemical records. Acta Palaeont. Polon. 47, 355–372.

## Appendix A Locality details

Coyote Knolls (39.47° N, 113.62° W)

This section is exposed in and above a T-shaped gully in Tule Valley, Millard County, western Utah (Fig. 5). This remote section can be accessed via a dirt track, 14.3 km north of the Cowboy Pass, which is a gravel road 50 km to the north of US Highway 50. The gully section lies at the bottom of a well-bleached scree slope, 5.5 km due west from Coyote Knolls (for more details see Sandberg et al., 1997). The gully exposes an almost unbroken F-F sequence of over 70 m thickness, dipping westwards, with the F-F boundary in the western wall of the gully. In total, our study records 43 beds.

Devils Gate (39.57° N, 116.07° W)

The base of the section is reached following Highway 50 for 13 km west from the town of Eureka. Shortly before the road passes through "Devils Gate", a track (the old highway) branches off on the north side. After following this track for 300 m, the section begins in the hillside immediately to the north (for more details see Sandberg et al., 1997). The bed numbers of Sandberg et al. (1988) are painted on several of the beds, but are not followed here. In total, our study comprises 42 beds.

Northern Antelope Range (NAR-A = 39.24° N, 116.24° W)

This location is a composite of two separate sections, NAR-A and -B. NAR-B, a steep slope section, is stratigraphically lower and is the easiest section to access. It can be reached following the main dirt road on the eastern side of the Antelope Valley, which runs several miles south of US Highway 50. Shortly before a cattle grid and fence meet the road, a rough track leads off to the southeast, into the range. After approximately 2 km this track ends, and the NAR-B section is exposed in a cliff and hillside. NAR-A is exposed in a cliff section some 1000 m to the north. In total the two sections expose a F-F boundary sequence almost 70 m thick, comprising 26 beds. A map of the locality is also given in Morrow (1997).

Tempiute Mountain (37.62° N, 115.64° W)

The section can be accessed by exiting Highway 375 on a dirt road 4.8 km northwest of Coyote Summit, to the east of the town of Rachel. Driving north, after 2.2 km, take the left fork, and continue northwest for 2.2 km to a gate in the canyon between Tempiute Mountain and Chocolate Drop. After parking by this gate, the section can be reached by walking up the canyon and joining an old mine footpath which runs to the southwest. Skirting round the summit, the section lies on the western side of the mountain at an elevation of approximately 7800 ft. In total, our study includes 15 beds. A map of the locality is also given in Morrow (1997).

Whiterock Canyon (37.62° N, 115.64° W)

This section is accessed by exiting Highway 50 southwards on the Antelope Valley road, 32 km west of Eureka. This road runs along the valley floor towards Segura Ranch. After 36 km, shortly before the ranch, take Forest Road 025 westwards into the Monitor Range. After a further 7 km, a track leads off the Forest Road southwards into Copenhagen Canyon. After 5 km take a fork westwards, past Rabbit Hill to the north, and into Whiterock Canyon, towards Horse Heaven Mountain. Following this track to the end, the section can be reached by walking a further kilometre west. The section, which comprises 13 beds, lies in trenches dug into a scree slope on the north side of the canyon. A map of the locality is also given in Morrow (1997).

Warm Springs (38.18° N, 116.39° W)

The section is reached by following Highway 6 for 4.1 km south from the T-junction by the abandoned Warm Springs station. Parking by the highway, behind a knob of volcanic rocks, walk 500 m to the north along a track into a small, narrow canyon. The studied section, which comprises eight beds, is exposed in the western wall of the canyon. The bed numbers of Sandberg et al. (1997) are painted on several of the beds, but are not followed here.

**Appendix B**

Pyrite framboid data ($N$ = number in sample, SD = standard deviation, FD = framboid diameter (μm)).

| Zone | Bed | $N$ | Mean | SD | Max. FD | Min. FD |
|---|---|---|---|---|---|---|
| Coyote Knolls | | | | | | |
| Early *rhenana* | 9 | 110 | 4.3 | 1.4 | 9.5 | 1.5 |
| Late *rhenana* | 13 | 7 | 5.6 | 2.7 | 11.0 | 3.5 |
| | 17 | 104 | 5.4 | 2.6 | 15.5 | 2.5 |
| | 18 | 6 | 4.8 | 1.7 | 8.0 | 3.5 |
| | 19 | 95 | 5.3 | 2.0 | 12.0 | 2.5 |
| | 22 | 103 | 4.8 | 1.6 | 10.0 | 1.5 |
| *linguiformis* | 27 | 88 | 5.0 | 2.0 | 11.5 | 2.0 |
| | 29 | 106 | 3.7 | 1.3 | 7.0 | 1.5 |
| | 30 | 99 | 4.3 | 2.0 | 12.5 | 1.5 |
| | 31 | 67 | 4.5 | 1.8 | 12.5 | 2.0 |
| | 33 | 85 | 4.5 | 2.2 | 14.0 | 2.0 |
| Early *triangularis* | 35 | 64 | 6.6 | 3.9 | 22.0 | 2.5 |
| | 39 | 65 | 6.1 | 2.9 | 23.0 | 2.5 |
| Devils Gate | | | | | | |
| Early *rhenana* | 2 | 17 | 6.0 | 2.1 | 10.0 | 3.0 |
| | 4 | 90 | 5.6 | 2.0 | 13.0 | 2.5 |
| | 5 (top) | 77 | 5.3 | 2.0 | 11.0 | 2.0 |
| | 6 (top) | 24 | 5.6 | 1.9 | 9.0 | 3.0 |
| Late *rhenana* | 7 (mid) | 103 | 3.8 | 1.2 | 7.5 | 1.0 |
| | 11 | 69 | 5.4 | 2.0 | 16.0 | 3.0 |
| | 12 | 12 | 5.0 | 1.8 | 8.0 | 2.5 |
| | 15 | 13 | 4.2 | 1.1 | 7.0 | 3.0 |
| *linguiformis* | 26 | 95 | 5.1 | 1.4 | 9.5 | 2.5 |
| | 28 (top) | 28 | 4.9 | 1.9 | 10.0 | 2.5 |
| | 29 (mid) | 114 | 4.4 | 1.7 | 12.5 | 2.0 |
| Early *triangularis* | 30 (top) | 47 | 4.0 | 1.0 | 7.0 | 2.5 |
| | 31 | 47 | 3.7 | 1.2 | 7.0 | 1.0 |
| | 34 | 40 | 4.5 | 2.1 | 12.0 | 2.0 |
| Late *triangularis* | 38 | 17 | 5.4 | 1.6 | 8.0 | 3.0 |
| Tempiute Mountain | | | | | | |
| *linguiformis* | 5 | 4 | 3.5 | 0.4 | 4.0 | 3.0 |
| | 6 | 3 | 4.0 | 1.0 | 5.0 | 3.0 |
| | 7 | 8 | 5.1 | 1.7 | 8.5 | 3.5 |
| Early triangularis | 9 | 2 | 10.3 | 3.9 | 13.0 | 7.5 |
| Late *triangularis – crepida* | 14 | 57 | 6.8 | 1.9 | 12.5 | 4.0 |

Appendix B (*Continued*)

| Zone | Bed | *N* | Mean | SD | Max. FD | Min. FD |
|---|---|---|---|---|---|---|
| Whiterock Canyon | | | | | | |
| Late rhenana | 1 (top) | 81 | 5.3 | 2.2 | 14 | 2.5 |
| | 2 (mid) | 80 | 4.9 | 1.9 | 11 | 1.5 |
| | 3 (mid) | 95 | 4.5 | 2.1 | 17 | 2 |
| linguiformis | 4 (top) | 98 | 4.2 | 1.6 | 9.5 | 2 |
| | 5 (mid) | 74 | 6.2 | 2.3 | 12.5 | 2.5 |
| | 5 (top) | 87 | 6.1 | 4.3 | 40 | 2 |
| | 6 (top) | 100 | 5.5 | 2.0 | 14 | 2 |
| Early–Mid triangularis | 7 | 85 | 6.8 | 2.9 | 16 | 3 |
| | 8 (mid) | 40 | 6.2 | 2.1 | 14.5 | 3.5 |
| | 9 | 52 | 4.9 | 1.6 | 8.5 | 2.5 |
| Mid–Late triangularis | 10 (top) | 106 | 4.8 | 1.7 | 11 | 2 |
| Warm Springs | | | | | | |
| Early rhenana | 3 | 4 | 6.4 | 1.4 | 7.5 | 4.5 |
| Late rhenana–linguiformis | 5 | 17 | 5.9 | 1.3 | 8.5 | 3.5 |
| | 6 | 10 | 5.7 | 1.2 | 7.5 | 4.5 |
| F-F boundary | 7 | 118 | 4.6 | 1.6 | 10.0 | 2.0 |

Understanding Late Devonian and Permian-Triassic Biotic and Climatic Events:
Towards an Integrated Approach
edited by D.J. Over, J.R. Morrow and P.B. Wignall



# *Chapter 10*

# *Late Permian double-phased mass extinction and volcanism: an oceanographic perspective*

*Grzegorz Racki[a,]* and Paul B. Wignall[b]*

[a]*Department of Earth Sciences, Silesian University, 41-200 Sosnowiec, Poland*
[b]*School of Earth Sciences, University of Leeds, Leeds LS2 9JT, Great Britain*

**Abstract**

The two mass extinctions of the Late Permian, end-Guadalupian and end-Changhsingian events, coincide with intense volcanic paroxysms located both in subpolar and tropical realms. For the Siberian Traps eruptions in particular, thermal destabilization of methane hydrates in extensive bipolar reservoirs initiated catastrophic release of $CH_4$ that may have triggered a subsequent runaway greenhouse culminated through the Permian–Triassic (P–Tr) boundary interval. In addition, increased geothermal heat and volcano–ice interactions since the early Late Permian resulted finally in unusual high-latitude warmth via catastrophic methane release. This combination of factors suggests excessive permafrost melting with a consequent inflow of heated and low-salinity waters. This is postulated to have provoked shutdown of the ocean thermohaline system, and thus to have caused the superanoxic event of the P–Tr boundary timespan, even if many global feedbacks and interactions were acting earlier in concert towards progressive warming and destabilized oceanic circulation and biogeochemical cycling. Both larger than hitherto assumed, Chinese and Siberian volcanic cataclysms are considered as the main driving force of profound P–Tr oceanic stagnation and superanoxia, nutrient stress and bioproductivity crash, in action in the geodynamically activated activity (due to plate boundary reorganization and accelerated rifting processes), fragile ecosystem. This end-Permian crisis acme was recorded in the Late Changhsingian-Griesbachian Chert Gap and overall 'fossilization low' through ~10 Ma, and the consequent biogeochemical and ecological peculiarities of the 'dead'/survival intervals. In contrast, a specific volcanogenic/tectonoeustatic scenario for the end-Guadalupian crisis may be focused on the Phanerozoic sea-level minimum and/or an extraordinary concomitant uplift of the Gondwana supercontinent.

*Keywords:* Permian–Triassic boundary; mass extinction; volcanism; thermohaline circulation; anoxia; palaeoclimate; mantle plumes

## 1. Introduction

The cause of the largest known mass extinction event, at the end of the Permian period, is still a matter of intense debate, with a variety of catastrophic, rapid and gradual scenarios all in contention (Hallam and Wignall, 1997; Erwin et al., 2002; White, 2002; Benton and Twitchett, 2003). Recent studies have highlighted the more complex and gradual (stepwise), primarily Earth-bound aspects of the major crisis (e.g. Berner, 2002; De Wit et al., 2002; Rees, 2002; Shen and Shi, 2002; Gruszczyński et al., 2003; Weidlich et al., 2003; Wignall and Newton, 2003; Dorritie, 2004; Kidder and Worsley, 2004b; Korte et al., 2004; Krull et al.,

*Corresponding author. Fax: +48-32-291-58-65. *E-mail address:* racki@us.edu.pl (G. Racki).

2004; Newton et al., 2004; Payne et al., 2004; Twitchett et al., 2004), and its exceptional evolutionary–ecological significance (Bambach et al., 2002; Retallack et al., 2003; Stanley, 2003; Benton et al., 2004; Falkowski et al., 2004; McGhee et al., 2004). Matters of its magnitude and duration, and a purported poor quality fossil record, are also important (Metcalfe et al., 2001; Twitchett, 2001; Erwin et al., 2002; Foote, 2003; Mundil et al., 2004).

Many workers have implicated Siberian Traps volcanism as a key trigger in a cascade of environmental deterioration (see summary in Wignall, 2001 and Dorritie, 2004). Recent works of Reichow et al. (2002) and Saunders et al. (2005) indicate that this igneous province continues beneath the West Siberian basin located nearly 1000 km to the west, as previously postulated by many workers (e.g. Campbell et al., 1992; Dobretsov, 1997; Kozur, 1998, p. 259). Therefore, the volume of volcanics of the Earth's largest continental flood basalt province was probably *at least* twice as extensive as previously assumed ($>3 \times 10^6$ km$^3$). Dobretsov (1997), taking into account the probable northern (Taymyr) and northwestern (Kara and Barents Seas) continuation of the province, estimated the volume of the traps as $16 \times 10^6$ km$^3$. Many eruptive ages require accurate confirmation (Saunders et al., 2005), but the province may even extend further southward into the Transbaikal Zone of central Kazakhistan and Uzbekistan (see Fig. 3 in Nikishin et al., 2002; also Kruk et al., 1999, Lyons et al., 2002; Yakubchuk and Nikishin, 2004). Thus, Dobretsov and Vernikovsky's (2001) suggestion of a total volume of igneous rocks (intrusives and extrusives) of 30–40 $\times 10^6$ km$^3$, with the lava volume comprising 20–30% of the total volume, may not be far from the truth at least for a broad Late Permian to Early Triassic timespan.

The temporal link of the largest igneous province with the biggest extinction event is now very strong, but the feedback-forcing mechanisms remain conjectural, especially in the context of extraterrestrial impact versus volcanism debate (e.g. Alvarez, 2003; Keller, 2003; Vermeij, 2004; White and Saunders, 2005). However, an abrupt mass extinction, such as that which terminated the end of the Cretaceous, has yet to be demonstrated for the end of the Permian. Wignall (2001) stressed that paroxysmal volcanism of immense scale may trigger drastic global environmental extremes, mostly by a positive feedback loop involving global warming (i.e. runaway greenhouse; Kerr, 1998, Hallam, 1999, White, 2002). In particular, the buildup of volcanically derived $CO_2$ is envisaged to initially trigger a destabilization of highly volatile seafloor methane hydrates in marine sediments with the dramatic release of $CH_4$ (as interpreted from rapid large-scale negative shifts of $\delta^{13}C$; Krull et al., 2000, 2004; Dorritie, 2002, 2004; Gruszczyński et al., 2003; Ryskin, 2003; Kidder and Worsley, 2004b; but see Grard et al., 2005), originally suggested for the Permian–Triassic (P–Tr) biospheric turnover by Erwin (1993).

There were two discrete pulses of extinctions in the later Permian, and both coincide with major igneous provinces (Wignall, 2001; see Fig. 1 and Tables 1–2). As noted above, one occurred exactly at the P–Tr boundary (~251 Ma; see Section 1.2), and was followed by a long-delayed Early Triassic ecosystem recovery, and a second, somewhat smaller biotic crisis that took place ~9 Ma earlier at the end of the Guadalupian age, and was succeeded by a moderate Lopongian radiation (Stanley and Yang, 1994; Hallam and Wignall, 1997; Kozur, 1998; Erwin et al., 2002; Shen and Shi, 2002). Thus, two time-scales of the large-scale biospheric perturbation are distinguished: (1) a prolonged 15 Ma Late Permian to Early Triassic *biocrisis* as a whole, and (2) the brief, less than 0.5 Ma P–Tr boundary *mass extinction*.

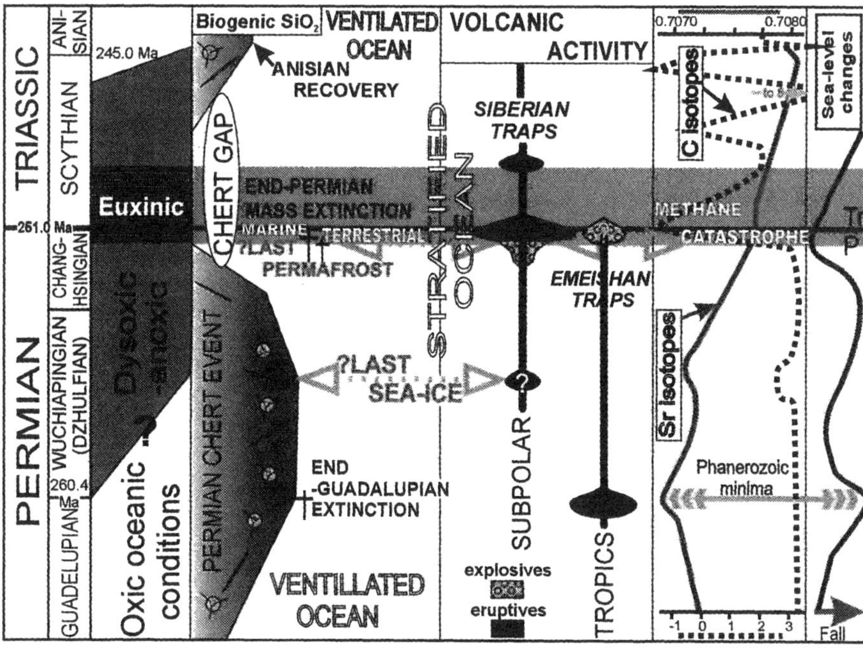


*Figure 1.* A scheme of the Permian–Triassic biotic crisis in terms of major oceanographic, magmatic (flood basalt), geochemical and eustatic changes, based on Isozaki (1997, Fig. 3), Wignall (2001, Fig. 5) and Dorittie (2004), with supplementary data from Holser and Magiritz (1987, Figs. 5, 7 and 8), Kakuwa (1996, Fig. 11), Hallam and Wignall (1997, Fig. 5.13), Kozur (1998, Fig. 1), Hallam and Wignall (1999, Fig. 11), Beauchamp and Baud (2002), Erwin et al. (2002), Reichow et al. (2002, Fig. 3), Wignall and Twitchett (2002a), Courtillot and Renne (2003), Kamo et al. (2003), Korte et al. (2003, 2004; Fig. 6), Payne et al. (2004; Fig. 3) and Kaiho et al. (2005); stage duration after Gradstein et al. (2004). Only a tentative correlation is presented because of the uncertain date of the P–Tr boundary (250–251 or 253 Ma; Metcalfe et al., 2001; Kamo et al., 2003; Mundil et al., 2004). Note the supposed two-step key deglaciation–volcanism and methane release links (arrows ending dotted line; cf. Dorittie, 2004), and also that the major input of young non-radiogenic low-ratio strontium occurred well before the climax of Siberian Traps. This perhaps reflects the weathering of Chinese volcanic sequences (Huang and Zhou, 1997) and/or oceanic massive seafloor eruptions (Veevers et al., 1994; Hallam, 1999).

### *1.1. Volcanism and catastrophic kill mechanisms*

The *causal* link between volcanism and extinction magnitudes seems to be complex and variable (cf. Erwin, 1993; Condie, 2001; Wignall, 2001; Courtillot and Renne, 2003; Pálfý, 2003; White and Saunders, 2005), especially in light of the new somewhat contradictory compilations of superplume events by Abbott and Isley (2002) and Vaughan and Scarrow (2003), and recent summaries of possible dynamic interactions in volcanism–climate system (Stothers, 1999; Robock, 2000; Zielinski, 2002; Adams et al., 2003; Blake, 2003; Grattan, 2005). In addition to the often proposed effects of volcanic eruptions, such as massive methane release, $CO_2$-induced warmth, oxygen-depleted oceans and atmosphere, $SO_2$-derived cooling and acid rain, the oceanographic consequences of the high-latitude lava outpouring needs to be considered, as well as a potentially crucial impact of tropospheric sulphate aerosols localized over one hemisphere (White and Saunders, 2005). Thawing of polar ice could potentially have a major impact on the ocean thermohaline

*Table 1.* Chronostratigraphic subdivision of the later Permian to earliest Triassic, showing correspondence between the international (proposed by Subcommission on Permian Stratigraphy of IUGS during the Beijing International Meeting; Jin et al., 1997), Western Tethys and Russian scales (modified after Fig. 1 in Crasquin-Soleau, 2003, and International Stratigraphic Chart, International Commission of Stratigraphy; Gradstein et al., 2004, see Mundil et al., 2004 for different dates).

| Proposed world scale | | | | Weatern Tethys | Urals and Russian Platform | |
|---|---|---|---|---|---|---|
| Triassic | | | Induan | Induan | Induan | 251.0 Ma |
| PERMIAN | LATE | Lopingian | Changsingian | Dorashamian | Tatarian | 253.8 Ma |
| | | | Wuchiapingian | Dzhulfian | | 260.4 Ma |
| | MIDDLE | Guadalupian | Capitanian | Midian | | 265.8 Ma |
| | | | Wordian | Murgabian | Kazanian | |
| | | | Roadian | Kubergandian | Ufimian | 268.0 Ma |
| | | | Kungurian/ Leonardian | Bolorian | Kungurian | |

system and climate (e.g. Conaghan et al., 1994; Veevers et al., 1994; Krhovský and Cejchan, 1995; Isozaki, 1997; Dorritie, 2002, 2004; Courtillot and Renne, 2003). In this context, it is important to note that the northern Boreal Ocean has been seen as a key area in understanding the catastrophic stagnation that affected the oceans at this time (e.g. Wignall and Twitchett 1996, 2002a; Kozur, 1998; Dorritie, 2002, 2004).

Two key questions are addressed here regarding changes in Late Permian oceanic circulation and its possible volcanic trigger:

(1) Are there any depositional and biotic signatures of a circulation reordering?
(2) Was the possible igneous trigger limited only to the P–Tr boundary interval and northern circumpolar zone?

### *1.2. Principal timing concerns*

Although there is clearly a close temporal correlation of the P–Tr mass extinction and Siberian Traps eruptions, precise synchroneity has not yet been demonstrated (see discussion in Metcalfe et al., 2001; Peng et al., 2001; Erwin et al., 2002; Reichow et al., 2002; Courtillot and Renne, 2003). In fact, some radiometric dates suggest a mostly Triassic age of the volcanism (Wignall, 2001; Erwin, 2003). However, the most recent, refined geochronological analyses show, despite different P–Tr boundary dates, a good coincidence of the major eruptions with the mass extinction event (251.4±0.3 Ma, Kamo et al., 2003; 252.6±0.2

*Table 2.* Four major areas of Late Permian to Early Triassic magmatic activity (see Fig. 4), with emphasis on its magnitude, and palaeogeographic and geotectonic location.

| Name of igneous province (geographic location) | Main varieties of magmatic products | Palaeogeo-graphic/ geodynamic contexts | Age in Ma | Original size: surface area ($10^6$ km$^2$)/ volume ($10^6$ km$^3$) | Basic references [additional references] |
|---|---|---|---|---|---|
| Siberian Traps (Russia): Tung uska (main centre), Taymyr, Petchora-Barents Sea, Kuznetsk, Verkhoyansk | Basalt, tuffs, (alkaline granitoids, bimodal magmatic rocks) | N subpolar zone/super-plume province in N Pangea, including several plume sets | Since ?258 to 247 (to ?229), peak 251.7 to 251.1 | >4/>3 (with intrusives 16) | Kamo et al. (2003), White and Saunders (2005) [Dobretsov, 1997; Kozur, 1998; Courtillot and Renne, 2003] |
| Emeishan Traps (China) | Basalt, tuffs (trachyte, rhyolite, gabbro) | Subtropical zone/super-plume province in S China | 259±3 (to ?236) | 2 to 5/0.3 to >1 (with intrusives 8.9) | Courtillot and Renne (2003), Ali et al. (2005), Izokh et al. (2005) [Zhu et al., 2003] |
| Oman–India (Panjal Traps)-Western Australia | Basalt-andesite-rhyolite-granite | Temperate zone/rift zone to the divergent Neo-Tetyhan S margin | >268-245 | ?0,05 (?4000 km long zone)*/ unknown | Veevers and Tewari (1995) [Wignall, 2001] |
| New England-Brook Street (Australia-New Zealand to ?West Antarctica) | Granite-rhyolite-basalt-andesite-tuffs | S subpolar to polar zone/ magmatic arc at the Panthalassan–Pangean margin | Effusives (268) 258-250 (major intrusion: 255-245) | unknown (>3000 km long Pacific-type margin*) | Veevers et al. (1994) [Michaelsen and Henderson, 2000; Campbell and Conaghan, 2001; Stewart, 2001] |

*Approximated mostly from Figures 1b and 2b in Veevers and Tewari (1995); size of the Indian Panjal volcanics is only estimated (12,000 km$^2$, maximum thickness of 2.5 km; see Wignall, 2001).

Ma, Mundil et al., 2004). This confirms previous palynological data from crater–lacustrine deposits in the Siberian Traps Province (Naumov and Luzina, 1997). Dircct biostratigraphical dating of the volcanism can often provide the most decisive evidence for comparing eruption timings and comparing with bioevents. Thus, conchostracan and sporomorph evidence indicates that precursory magmatism commenced in the Wuchiapingian (= Dzhulfian) over the Siberian Platform and Timan–Petchora Basin (Kozur, 1998, Table 2 therein). Therefore, the Siberian superplume activity as a whole was a *more prolonged phenomenon* (Fig. 1 and Table 2) than is usually presumed (2 Ma in Saunders et al., 2005; see also Vernikovsky et al., 2003; Gurevitch et al., 2004), in accordance with overall duration of flood basalt volcanism lasting at least 5–10 Ma (see Table 2). Unfortunately, precise biostratigraphical data on the position of P–Tr boundary within the volcanic succession remains somewhat controversial (see Kozur, 1998; Lozovsky, 1998; Mogutcheva and Betekhtina, 1998; Yaroshenko and Lozovsky, 2004). If abnormal gymnosperm pollens provide an evidence of global atmospheric perturbations due to volcanic cataclysm, this environmental stress culminated indeed in the latest Permian (Vischer et al., 2004).

## 2. Circulation in the Panthalassa superocean

Palaeoceanographic circulation styles are broadly classified into two basic types: thermohaline circulation, characterized by cold, deep water, which originates at high latitudes and ventilates the deep oceans and halothermal (also termed haline or reversed thermohaline) circulation, involving the formation of warm, saline, deep water in the tropics (Horne, 1999). The latter mode is often implicated to produce oxygen-poor deep waters although it should be noted that the biggest halothermal circulation system seen today, in the Mediterranean, produces extremely well-oxygenated deep waters (see also Bice and Marotzke, 2001).

Computer-simulated circulation in the Panthalassa Ocean indicates relatively vigorous meridional and longitudinal thermohaline overturning with deep water formed by cooling in polar oceans, and subordinately by evaporation in the subtropics (Kutzbach et al., 1990; see also Hotinski et al., 2001, 2002; Winguth et al., 2002; Zhang et al., 2002; Kidder and Worsley, 2004b).

> Early scenarios of P–Tr oceanography invoked major changes around the P–Tr boundary, including a sudden mixing event (e.g. Malkowski et al., 1989). However, the role of salinity changes in these events has yet to be fully explored. Holser and Magaritz (1987) supposed that instabilities and rapid swings between the two major circulation states would be expected when continental glaciers and any sea ice melted during glacial waning and retreat, with decreasing surface salinity resulting in density stratification and euxinic oceanic conditions (see the modern concept of self-sustained salt oscillator summarized in Krhovský and Cejchan, 1995 and Seidov and Haupt, 2003). The 'injection events' from polar areas during climatic oscillations should produce a fertility crisis in the freshened and/or warmed surface waters, and sporadic bloom conditions (Thierstein and Berger, 1978). Potentially, extreme low-salinity evidence in surface waters comes from the so-called 'fungal spike' of terminal Permian age now reinterpreted by some as a bloom of freshwater algae (Afonin et al., 2001). However, some recent views on the P–Tr climates (see Section 4) rather preclude the above scenarios. Kidder and Worsley (2004b) claimed that euxinia in the deep ocean would be established by the chain of global events, which led to global warming, low atmospheric oxygen, evaporite deposition, and increased injection of warm, saline bottom water. The sustainable mid-latitude sinking of the brines is inferred by these authors from the necessity of continued poleward heat transport to maintain the warm,

moist high-latitude vegetation (cf. Gibbs et al., 2002). In their scenario, the cessation of polar sinking as polar surface waters, which became progressively warmer would cause the deep oceans to become rapidly filled with elevated salinity waters that would become increasingly anoxic and finally euxinic by the P–Tr boundary interval.

## 3. End-Permian plankton collapse and setting of the Chert Gap

A landward and southward expansion of silica "factories" occurred during the Middle Permian, and was especially extensive during the early Late Permian. This phenomenon has been termed the Permian Chert Event (PCE) by Murchey and Jones (1992) and Beauchamp and Baud (2002). In the Wuchiapingian, a succession from warm-water to impoverished cold-water carbonate producers to siliceous communities is seen, manifest in spiculitic and/or radiolarian cherty facies (Murchey and Jones, 1992).

A dramatic crash of biosiliceous productivity occurred across the P–Tr boundary (Chert Gap; Fig. 1), a unique feature of this mass extinction (Racki, 1999). As first shown by Isozaki (1997), in the deep-water pelagic sections of Japan, a remarkable symmetrical lithologic change, from bedded red, hematitic chert to siliceous mudstones to carbonaceous shale (with a total organic carbon content of 4–10%), and back again, occurs during the P–Tr transition. This symmetric succession is observed in pelagic sequences of accreted oceanic terranes of Japan and North America (British Columbia, Nevada), as well as in deep-water facies in Russia (Sikhote-Alin), Oman and South China (Noble and Renne, 1990; Blome and Reed, 1995; Isozaki, 1997; Kozur, 1998; Cordey et al., 1999; Bragin, 2000; Bragin and Krylov, 2002; Kato et al., 2002; Filippov and Kemkin, 2003; Sperling, 2004). Radiolarian-free carbonaceous black shale, and red to grey to variegated grey shale occur exactly in the P–Tr position in the eastern Palaeo-Tethys as well (Thailand; Sashida et al., 2000, Sashida and Salyapongse, 2002). In the Boreal shelf seas, there is a similar loss of chert deposition as spiculites are abruptly replaced by shales in the latest Permian (Wignall et al., 1998).

The sedimentary break, most prominent in low latitudes (Kozur, 1998), is arranged in two steps: (1) a likely diachronous decline of siliceous productivity initiated in the Wujiapingian, followed by (2) gradual end-Permian (Changhsingian) collapse. The mid-latitude deep-water continental margin sections of British Columbia indicate a more abrupt collapse of radiolarian productivity in the Late Changhsingian, somewhat before the loss of sponge spicules (Wignall and Newton, 2003). Chert deposition resumed in the distal oceanic areas during the early Middle Triassic (Anisian) 10 Ma later (Kakuwa, 1996; Isozaki, 1997; Kato et al., 2002). According to the diversity compilations of Vishnevskaya and Kostyuchenko (2000), 85% of radiolarian species became extinct in the Late Permian. Many radiolarian morphologies disappeared, except for primitive spherical spumellarians and some simple spicular or even skeleton-free taxa (cf. Kozur, 1998; Bragin, 2000; Yao and Kuwahara, 2000; De Wever et al., 2003).

Beauchamp and Baud (2002) emphasized that the abrupt end-Permian collapse of the PCE was associated with a major transgression and a return to much warmer oceanic and continental climatic conditions. In their scenario, closely comparable to that proposed earlier by Casey (1993), the conditions necessary for the onset, expansion and climax of the PCE were provided by the vigorous thermohaline circulation of nutrient-rich cold waters.

By implication, these conditions must have ceased rapidly in the Late Changhsingian, with the demise of effective silica production caused by the breakdown of northerly derived water influx, perhaps due to rapid and complete melting of sea ice. In this scenario, this termination of thermohaline conveyor eventually led to warm and sluggish oceanic settings unfavourable to the production of siliceous oozes (Figs. 1, 2 and 3). Kidder and Worsley (2004b) implied alternatively only nutrient limitation and possible surface water freshening as the cause of the P–Tr chert and phosphorite gap (cf. Dorritie, 2004). Wignall and Twitchett (2002a) showed that the onset of oceanic anoxia–dysoxia was near synchronous in latest Permian followed by a mid-Griesbachian peak of superanoxia, when dysoxic conditions developed in all but shallowest of marine settings. Much more variable recovery of bottom-water oxygenation continued from the latest Griesbachian until the Middle Triassic.

## 4. Late Permian climate

The Late Permian is usually thought of as an undoubted post-glacial time (see Hallam and Wignall, 1997), with the last extensive continental ice sheets of the Permo–Carboniferous

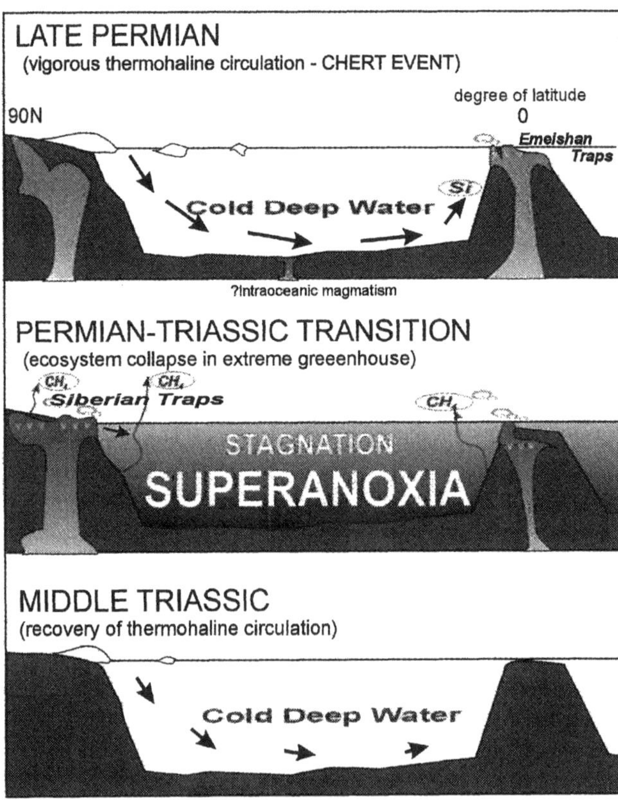


*Figure 2.* Idealized cross-section presentation of Late Permian to Triassic oceanic ecosystems to show contrasting forms of changing deep ocean circulation (cf. Horne, 1999), and twofold impact of volcanism in polar and tropical regions.

disappearing in the Middle Permian (González-Bonorino and Eyles, 1995; Crowell, 1999). However, the evidence varies from region to region. For example, sediments from the margins of the newly opening Neotethyan Ocean reveal a prominent shift towards warmer humid conditions through the Early to Middle Permian (Angiolini et al., 2003). In the southern high latitudes (Australia), cryogenic structures, stone rolls in palaeosols and dropstones were recorded by Veevers et al. (1994) and Conaghan et al. (1994) in the terminal Permian. However, the main uncertainty in climatic evaluation is the interpretation placed upon the extensive Late Permian Boreal chert record considered by Beauchamp and Baud (2002) to have formed beneath polar sea ice, although contemporaneous dropstone evidence, presented by these authors, is essentially conjectural. Nevertheless, Late Permian (Tatarian) siliceous–diamictite deposits are known in northeastern Asia (Omolon massif; Fig. 2 in Byakov, 2000).

Climate proxies (compiled in Table 3) also indicate some climatic trends in the Late Permian. The early Wujiapingian warming was followed by a cooling trend that began towards the end of this age (Kozur, 1998), confirmed by the oxygen isotopic trend (Gruszczyński et al., 2003, p. 173) and palaeontological data (Chumakov and Zharkov, 2003; Chumakov, 2004). The latest Permian cooling is also implied from carbon isotopes in pedogenic carbonates that indicate a reduction in atmospheric $CO_2$ (cf. Kidder and Worsley, 2004b, p. 216). This was reversed in the latest Changhsingian by a rapid warming of maximally ~6°C (Kidder and Worsley, 2004b), for which the first evidence comes from the latest Permian of Greenland (Wignall and Twitchett, 2002b), and finally culminated in the earliest Triassic 'postapocalyptic' greenhouse (e.g. Sheldon and Retallack, 2002; Chumakov and Zharkov, 2003; Chumakov, 2004; but see Korte et al., 2004). On the other hand, Kozur (1998, p. 264) and Dorritie (2004) hypothesized strong cooling pulse(s) in low latitudes promoted by volcanism (see Table 3). These recurring volcanogenic winters were probably indeed brief-term phenomena (Robock, 2000; Thordarson et al., 2001; Adams et al., 2003; see also Blake, 2003) and as a result have left little geological evidence. Significantly, no historical eruptions have caused any long-term cooling. Even ultra-violent, volcanic mega-eruptions, like the Toba event 75 K years ago, that produced

*Table 3.* Recent interpretations of the Late Permian to Early Triassic climates, with emphasis on a cooling phase in the terminal Permian (cf. Chumakov and Zharkov, 2003). The data are derived from diversity of biotic and lithologic proxies (Kozur, 1998; Beauchamp and Baud, 2002), conodonts (Mei et al., 2002) and carbonate platform evolution on the Gondwana shelf (Pakistan; Mertmann, 2003); carbonate oxygen isotope signature based on Hallam and Wignall (1997, Fig. 5.14, p. 129; see also Gruszczyński et al., 2003, Fig. 2 therein) and White (2002, Fig. 3 therein versus Korte et al., 2004). Correlation with major volcanic events and the zenith of Permian Chert Event (see Fig. 1) is marked.

| Ages | | Kozur (1998) | Beauchamp and Baud (2002) | Mei et al. (2002) | Mertmann (2003) | $\delta^{18}O_{carb.}$ |
|---|---|---|---|---|---|---|
| Scythian | Olenekian | no data | (supergreenhouse) very warm dry | no data | no data | |
| | Induan | very warm / (?volcanic winter) | | | Tr | Siberian |
| Changhsingian | | (volcanic winter) continuous cooling | very cool | cool | cool P | PCE acme |
| Wuchiapingian | | warm | | very warm | warm | Emeishan |
| Capitanian | | no data | cool dry | warm | | -6 -4 -2 |

up to 3000 $km^3$ of magma, had no perceptible change on climate (see summary in Wignall, 2001), as well as the largest Phanerozoic explosive eruption in Middle Ordovician (Bergström et al., 2004).

Proposed episodic ice sheet growth and ice rafting in both the Permian circumpolar humid belts may be the result of northward drift of the supercontinent (e.g. Crowell, 1999; Zharkov and Chumakov, 2001). Chumakov (1994) attributed glaciomarine diamictites in the Verhhoyansk–Kolyma block to the early Tatarian, but new biostratigraphical dating by atomodesmid bivalves (Byakov, 2000) has revealed also a *Late* Tatarian age of the partly glaciomarine Khivach Horizon. According to the modern correlation scheme, however, these strata are unlikely to be of terminal Permian age (Table 1), but could be certainly a glacial record of the late Wuchiapingian cooling pulse (Table 3). Consequently, Chumakov and Zharkov (2003) reasonably concluded that the Late Permian (Tatarian) cooling "was probably of global scale and relatively strong" (p. 369), and responsible for the glacial and glaciomarine facies in the circumpolar belts "with small-scale permanent or intermittent glaciation centers" (p. 362). The bipolar climatic pattern is presented in Figure 4.

Regardless of the detailed climatic changes, the concept of Beauchamp and Baud (2002) of pervasive sea ice lacks evidence, but a seasonal sea ice growth is still possible due to a set of plausible palaeoceanographic conditions in the northern hemisphere. The permafrost features are known from south polar humid areas only, but equivalent palaeosol studies are required in the northern domain (see summary in Naugolnykh, 2004). However, it is noteworthy that the Korvunchaya Svita, associated with the Siberian Traps, is marked by floras predominated by ferns and quillworts, and so certainly may represent a tundra- or taiga-type vegetation (G. Retallack, e-mail comm., 2004).

The majority of Northern Hemisphere permafrost present today in latitudes of 60 and 68°N reaches thicknesses approaching 1500 m. Further north, a layer of permafrost can even be traced under the Arctic Ocean on the shallow continental shelf exemplified by the Beaufort Sea, an area of widespread offshore permafrost (see Dorritie, 2004). Therefore, if similar thicknesses of permafrost were present in the Changhsingian northern polar regions, a large-scale hydrological perturbation may have been caused by rapid total ice melting in the volcanism-affected northern regions (~60°N; Gurevitch et al., 2004), especially if the pole was located very close to the Siberian continent (Fig. 16 in Golonka et al., 2003), i.e. occupied by thick permafrost and for the most part frozen Boreal Ocean (see Fig. 4). Such a complete and rapid deglaciation event is reliably identified in the Gondwanan subpolar belt (65–70°S), where terminal Permian permafrost and broadleaf deciduous swamp woodlands were replaced by earliest Triassic frost-free domains with conifer forests in Australia and Antarctica (Conaghan et al., 1994; Retallack and Krull, 1999; Krull et al., 2000).

Coeval lava flows, intercalated with shallow-marine, predominantly volcaniclastic, sediments emplaced under periglacial conditions, are also reported from the Gondwana shelf (e.g. Campbell and Conaghan, 2001; Carr and Jones, 2001). Ice–lava contacts would have occurred mostly during the initiation phase of Siberian flood magmatism (sensu Jerram and Widdowson, 2005), if the above quoted re-dating of Verhhoyansk–Kolyma dropstones and other worldwide cooling evidence are accepted (Table 3). This initiation took place through the Wuchiapingian age (Kozur, 1998), although earlier (Kazanian) volcanotectonic activation was substantiated in the more western domain by basaltic clasts in the Kama region (Chaikin and Meskhi, 2003).

In summary, the most recent dates for the age of (1) the latest ice rafting and permafrost in both the circumpolar humid domains, combined with (2) magmatic activation stages, provide a starting point for considering Late Permian volcanic–oceanographic causal links (Fig. 1).

### *4.1. Continental signature of the P–Tr climat es*

A synthesis of terrestrial climate change during the P–Tr boundary has yet to be undertaken, however, there were clearly major ecological and sedimentological changes. The loss of peat swamps and consequent cessation of coal formation everywhere in the world, until the Middle Triassic, is one of the more remarkable features of the extinction (Retallack and Krull, 1999; Krull et al., 2000; Michaelsen, 2002; Sheldon and Retallack, 2002; Kidder and Worsley, 2004b). This floral extinction has been attributed to rapid and intense warming mostly by water stress and frequent wildfires (see discussion in Wang and Chen, 2001), and the final deforestation, recorded in the P–Tr 'fungal' spike, was seen by Visscher et al. (1996, 2004) as caused by widespread volcanic paroxysms (via atmospheric chemistry changes, ozone depletion and acidification). Soils in wet, cold environments were particularly vulnerable to the volcanically forced acid contamination and extreme weather changes (Grattan, 2005).

> Ward et al. (2000) noted, in sections from the Karoo region of South Africa, a rapid change in fluvial style from meandering to braided rivers. Neatly combining these two observations of vegetation loss and fluvial change, Ward and coworkers suggested that the former caused the latter. Thus, the loss of vegetation would have caused a decline in riverbank stability and thus a switch from meandering to braided-style channels. However, these two observations do not come from the same regions. The coal-gap evidence comes mostly from continents situated at high palaeolatitudes (Australia, Antarctica; Retallack and Krull, 1999; Sheldon and Retallack, 2002), whereas the fluvial change is seen in South Africa and numerous other lower latitude Pangea locations (e.g. India, southern Urals; Newell et al., 1999; Sarkar et al., 2003). However, it is not a ubiquitous facies change and it is significant that in the Madagascan part of the Karoo depositional system, meandering river-style deposition persists across the P–Tr boundary (Wescott and Diggens, 1998).

Braided river development is favoured in a semi-arid climatic regime, where rainfall is erratic and often concentrated during brief intervals that are interspersed with long dry phases. The P–Tr boundary-braided rivers could therefore be a climatic signal indicating a brief increase in humidity in central Pangean locations otherwise characterized by more truly arid conditions (Korte et al., 2003, 2004; see also Kump et al., 2000).

## 5. Volcanogenic breakdown of the thermohaline system

This review of Late Permian palaeoclimates from the interval preceding the end-Permian crisis reveals clear evidence for a marked latitudinal temperature gradient with cold/freezing polar areas. The resultant volcanic influence in these regions is postulated to have been a key factor in the crisis. As reliably reviewed and/or modelled by Weaver et al. (1999), Ganopolski and Rahmstorf (2001) and Bigg et al. (2003), a change in the modern Arctic freshwater budget, driven by anthropogenic factors, should easily cause a shutdown of the North Atlantic Deep Water production, despite a number of competing effects (Fig. 3). Seidov et al. (2001) simulated a multifaceted pattern of the circulation overturns, because, among other factors, meltwater events in one hemisphere may lead to intensification of the thermohaline conveyor in the opposite hemisphere that in turn may cause a warming of abyssal water masses. Similar effects are revealed for the two most recent deglaciations when Southern Ocean warming promoted an abrupt re-commencement of more vigorous Atlantic overturning circulation (Knorr and Lohmann, 2003). We propose that Late Permian oceans could have been similarly susceptible to major changes of circulation caused by modest changes in high-latitude freshwater run-off.

The Siberian Traps are the result of enormous fissure eruptions with the major feeder zone located in the rifted West Siberian Basin, previously elevated by a hot mantle plume, along the northwestern border of the rapidly subsiding Siberian Platform (Callot et al., 2004; Saunders et al., 2005; see also Czamanske et al., 1998). The volcanic succession, up to 6.5-km thick, is composed of lava flows of 30–40 m thick with flow lengths of up to 70–80 km (Sharapov et al., 2003). Most of the volume of overheated basaltic magma (Zolotukhin and Podgornykh, 1998), of a few million cubic kilometres in volume, was probably extruded over ~0.6 Ma in the critical P–Tr mass extinction interval (Kamo et al., 2003; see also Gurevitch et al., 2004; Saunders et al., 2005). Common values for emplacement rates in large igneous provinces are between 2 and 10 $km^3$ per year, and individual flows were probably of the order of 100–1000 $km^3$ (up to 10,000 $km^3$; Jerram and Widdowson, 2005) with eruptions lasting for minimum periods of between 5 and 15 years (Coffin and Eldholm, 1994, Fig. 6 therein; Self et al., 2000; Condie, 2001, p. 59). However, note that the size of individual flows has rarely been measured directly assessed in any province; for instance, the gigantic Roza lava flow field of the Columbia River flood

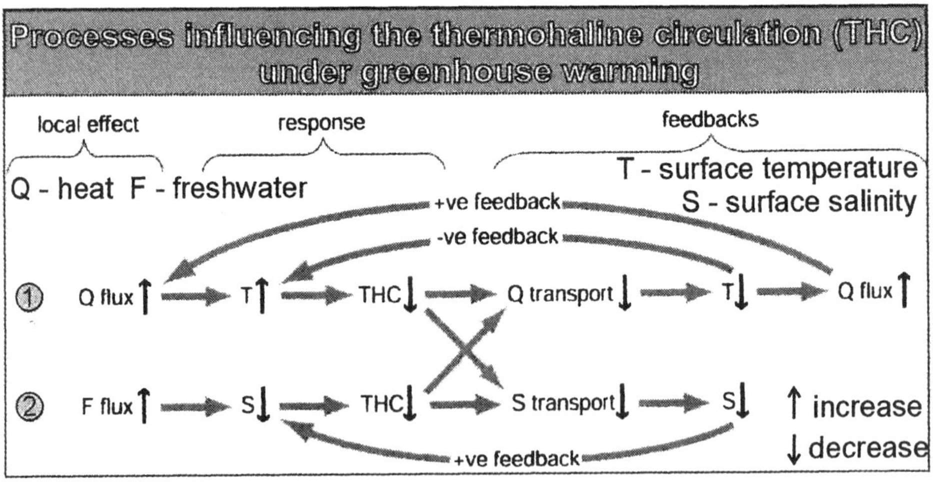


*Figure 3.* Scheme of key response to local effects and associated feedback influences on the global thermohaline system under greenhouse forcing (modified Fig. 14 from Bigg et al., 2003). Warming and freshwater fluxes refer to atmosphere–ocean interactions in high latitudes; Q – heat, F – freshwater, T – surface temperature, S – surface salinity. Copyright ©2003 by The Royal Meteorological Society.

basalts, with an area of ~40,300 $km^2$ and a volume of ~1300 $km^3$, is estimated to have been constructed over a 14-years period on ~150-km long vent system (Thordarson and Self, 1998). To put the estimate into perspective, the recent *global* eruption rate in continental volcanic systems (magma and lava) is 1.6 $km^3$ per year (Fig. 2.2 in Schmincke, 2004), and the largest historic eruption, Eldgjá in southern Iceland, 934 A.D., produced 19.6 $km^3$ of basaltic lavas during 3–8 years (Thordarson et al., 2001).

The basaltic flows may have been dominated by extremely thermally efficient, inflated and compound (Hawaiian pahoehoe) thick sheets that could have had an especially protracted environmental impact (Self et al., 1998, 2000). Heat fluxes of the order $10^4$ $W/m^2$ are likely (Emanuel et al., 1995, p. 13,765). Alternatively, faster rates of emplacement for the pahoehoe flow fields were advocated by Anderson et al. (1999), and predicted by Kilburn (2004) after quantitative analysis of fracturing styles. In fact, study of lava flow dynamics, that probably vary significantly, requires knowledge of lava viscosity, eruption rate and duration (Blake and Bruno, 2000). For example, the emplacment style of the Deccan Trap extrusives is thought to have varied considerably (Bondre et al., 2004).

In addition, Svensen et al. (2004) have noted the presence of a vast hydrothermal vent system associated with the eruption of the North Atlantic Igneous Province around the Palaeocene–Eocene boundary. A similar phenomenon may certainly have occurred in the P–Tr northern Eurasian igneous domain, marked by extensive rift systems (Nikishin et al., 2002; Saunders et al., 2005), where elevated geothermal gradients are set up in the crust above the active superplume, promote a large-scale circulation of hydrothermal fluids, which generated a diversity of ore deposits (see Gatinskii and Rundkvist, 2003; Yakubchuk and Nikishin, 2004; also in the S-Chinese domain; Izokh et al., 2005). Excessive emission of hydrothermal organohalogens and a resultant severe disruption of the stratospheric ozone balance have been mooted at this time. This is advanced as a cause of the chronic environmental mutagenesis recorded by anomalous ('mutant') pollen grains and unseparated tetrads of lycopsid microspores that become common in the extinction interval (Visscher et al., 2004; Yaroshenko and Lozovsky, 2004).

### *5.1. Immediate lava–ice interactions*

This is a field requiring numerical simulation, but the vast flood basalt flows attributable to the Siberian superplume are likely to have caused direct thawing of permafrost in this polar region (cf. Dorritie, 2002, 2004; Fig. 2). The thermodynamics of recent subglacial eruptions are very poorly understood (Höskuldsson and Sparks, 1997; Smellie, 2000). However, the best analogue, although of negligible scale, is contemporary activity in the plume-related Icelandic rift zone system (Gudmundsson, 2000), where eruptions in contact with ice can rapidly generate large quantities of meltwater (known as megaflood or jökulhlaup events), and even large subglacial lakes (Björnsson, 2002; Gíslason et al., 2002).

### *5.2. Longer-term geothermal effects*

In addition to subglacial and ice-margin eruptions, igneous activity no doubt enhanced geothermal gradients to drive hydrothermal circulation. Thus, the prolonged increased magmatic heat inflow and persistently elevated geothermal gradients and geothermal systems, are also suspects in the shutdown of the P–Tr thermohaline system (Dorritie, 2004). As to subsurface meltdown conditions, model calculations by Kukkonen and Safanda (2001) indicated that doubling of the temperature gradient thins the permafrost by about 40–50% with a consequent major hydrologic change. Björnsson (2002) described vigorous hydrothermal subglacial systems at volcanic sites in Iceland that continuously affect drainage areas of up to 200 $km^2$, with a total geothermal output of the order of 10,000 MW and a production rate of meltwater between 0.06 and 0.2 $km^3$ per year. Perturbance effects for key conveyor areas, which resulted from such large-scale activity over vast regions, require numerical simulation.

### 5.3. *Volcanogenic greenhouse*

The main source of global stress, caused by the huge expelling of volcanic $CO_2$, has long been identified as a prime reason of polar warming (Conaghan et al., 1994; Veevers et al., 1994). In addition, intrusion of voluminous mantle-derived melts in carbon-rich sedimentary strata may have caused an explosive release of methane through thousands of hydrothermal vent complexes, as implied for the Palaeocene–Eocene boundary (Svensen et al., 2004). Such a scenario is postulated here for the northern peripolar domain (*cf.* Dorritie, 2002, 2004), where mantle plume(s) affected coal-bearing deposits in Permian basin (cf. Czamanske et al., 1998; Kamo et al., 2004; Visscher et al., 2004).

Methane emission rates are affected by several factors, particularly the extent of ice sheets and permafrost and eustatic sea level, but ocean bottom-water temperatures are the most important single constraint (*cf.* the Clathrate Gun Hypothesis of Kennett et al., 2003). For example, as calculated by Buffett and Archer (2004), the predicted steady-state methane inventory decreases by 85% in response to 3°C of warming, caused for example by a doubling of present day concentration of $CO_2$ (cf. Stouffer and Manabe, 2003); thus, the global hydrate reservoirs were perhaps declined in progressively warmer Permian world. Methane derived directly from permafrost is however unlikely to have contributed a significant component of total methane release in the P–Tr boundary interval, which must instead have come from reservoirs such as the Arctic shelf. The modern Arctic store of methane hydrate is in the order of ~400 Gt of C, associated with permafrost regions but not necessarily buried in permafrost as hydrates can exist below the permafrost (Kvenvolden, e-mail comm., 2004; see also discussion in Kvenvolden, 1998); this may represent only a few percent of the hydrate-bound C inventory in shelf margins (cf. also Dorritie, 2004). Nevertheless, all the estimates of present day methane oceanic inventories are *highly speculative* and hold between 500 and 21,000 Gt of C, with the most widely cited amount of ~10,000 Gt (Kvenvolden, 1998; Judd et al., 2002; Buffett and Archer, 2004; Dorritie, 2004; Milkov, 2004). Evidence from the magnitude of the P–Tr boundary negative C isotope excursion suggests relatively high values between 4200 and 4800 Gt of $^{13}C$-depleted carbon (Berner, 2002; Gruszczyński et al., 2003).

Additionally, in the case of peatlands underlain by permafrost, as proved in the Permian Gondwanan domain but not yet studied in the Boreal domain, degradation of the dynamic system alters peat hydrology and ecological characters, all of which affect gaseous C emissions (see review in Kennett et al., 2003). Consequently, permafrost melt is associated regionally today with a 1.6- and 30-fold increase in $CO_2$ and $CH_4$ emissions, respectively, as a response to progressive surface global warming (Turestsky et al., 2002; Christensen et al., 2004). Of the recent natural fluxes of methane, 35% (= 0.03 Gt C/year; see Table 11.2 in Schlesinger, 1997) are from wetlands in northern latitudes, but this contribution could be far more significant in the terminal Permian from high-latitude productive peat swamps.

Poleward heat transport and involved thermohaline instabilities are highly sensitive to forcing by switches in atmospheric $pCO_2$ (Winguth et al., 2002; Stouffer and Manabe, 2003; Archer et al., 2004). However, in modelling studies, the (volcanically forced) temperature rise alone only causes weakening of the circulation, and other factors are required to complete the shutdown of the ocean conveyor belt (Ganopolski and Rahmstorf, 2001; Piotrowski et al., 2004).

## 6. Stepwise volcanic modulation in different climatic belts

Great attention has been focused on the major end-Permian biotic turnover and its temporal coincidence with the high-latitude Siberian Traps. However, the earlier end-Guadalupian extinction has also been linked with the tropical flood basalt eruptions of the Emeishan province in southwestern China (Fig. 4; Chung and Jahn, 1995; Kozur, 1998; Hallam and Wignall, 1997; Wignall, 2001; Lo et al., 2002; Ali et al., 2005).

### *6.1. Low-latitude igneous province(s)*

Initially considered a relatively small flood basalt event, recent work indicates the Emeishan igneous province is more extensive (Xiao et al., 2003; Ali et al., 2005), and also includes lavas in northwestern Vietnam (Izokh et al., 2005). This plume activity seems to have been associated with continental break-up and rifting of the western margin of the South China Craton to create the Songpan–Ganzi ocean basin (Song et al., 2004). Preserved fragments of the lava sequences are more than 4-km thick. Their entire volume, including intrusives, is recently calculated to be almost $9 \times 10^6$ km$^3$ (Zhu et al., 2003), but the volcanic and upper-intrusive portion seems to be relatively small (~$1 \times 10^6$ km$^3$, Courtillot and Renne, 2003; ~$0.3 \times 10^6$ km$^3$, Ali et al., 2005).

Radiometric dating of the Emeishan Traps indicates two phases of volcanism: (1) an initial outpouring of lava formation, coincident with a proposed end-Guadalupian extinction event at 256–260 Ma (Thompson et al., 2001; Ali et al., 2002, 2005; Zhou et al., 2002a), and (2) later magmatism that may have produced much of the widespread deposition of acidic tuffs in the Upper Permian–Early Triassic sections of South China (Kozur, 1998; Lo et al., 2002; Nikishin et al., 2002; Guo et al., 2004; see also Izokh et al., 2005). This history poses something of a dilemma, because magnetostratigraphic data indicate the rapid emplacement of the province (1–2 Ma; see summary in Ali et al., 2005). Even if Lo et al. (2002) indeed dated two discrete episodes of this extrusion separated by a 4–7 Ma timespan, these ages remain questionable due to Ar–Ar monitor standard calibration (Courtillot and Renne, 2003; Ali et al., 2005). Whilst He et al. (2003, Fig. 5a therein) indicated supplementary Middle and Late Permian basaltic sheets in addition to the main end-Guadalupian volcanic rock body, Xiao et al. (2003) described an exclusively Late Permian, thick (~5 km) low- to high-Ti basalt succession (Binchuan section) underlying the Lower Triassic

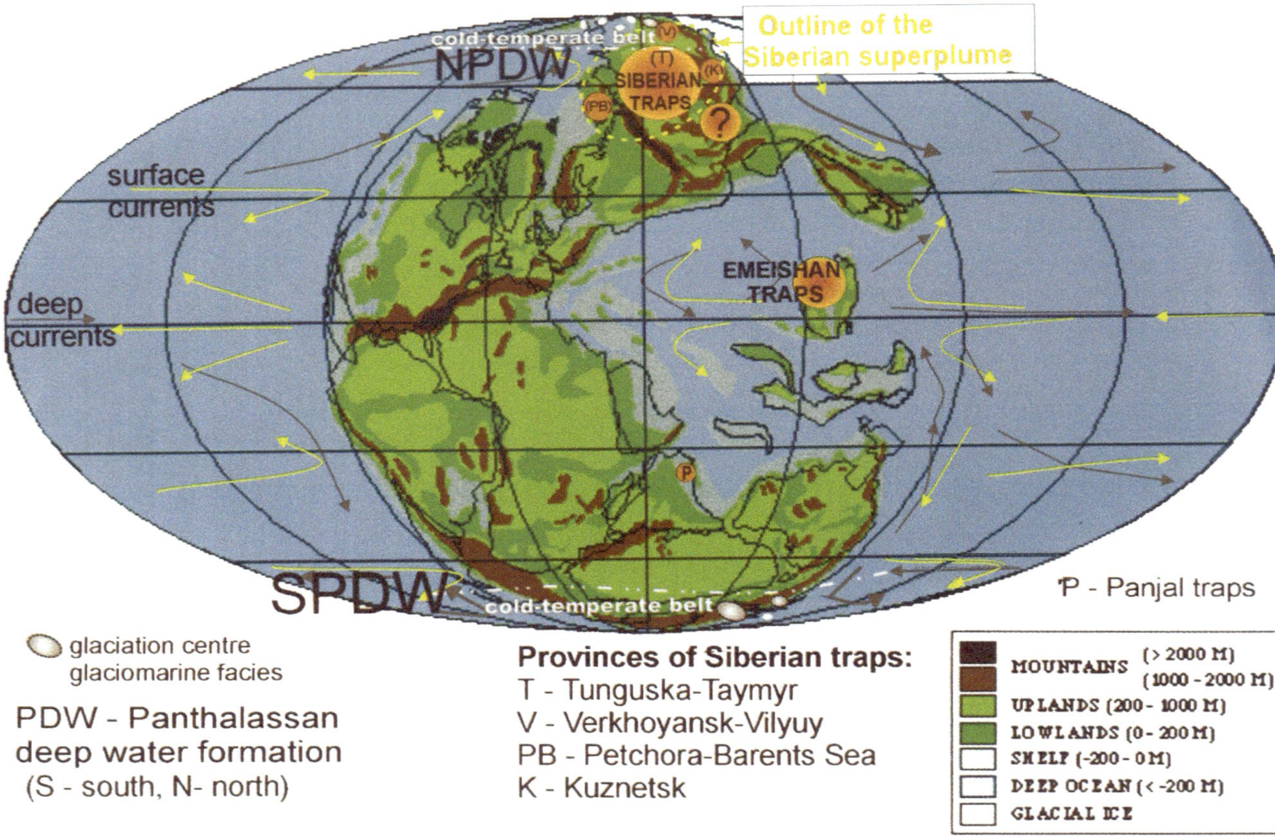


*Figure 4.* Main Late Permian to Early Triassic igneous provinces, shown as centres of mantle superplume activity (based on Fig. 2 in Yakubchuk and Nikishin, 2004, Fig. 3 in Nikishin et al., 2002 and Fig. 68 in Zharkov, 2004, with supplementary data from Fig. 1 in Dobretsov, 1997), against oceanic circulation pattern (based on Winguth et al., 2002, Fig. 5 therein; see also Fig. 7 in Kiessling et al., 1999), and glaciation record (based on Fig. 1 in Chumakov and Zharkov, 2003 and Fig. 93 in Chumakov, 2004). The base world map taken from "The Paleogeographic Atlas Project. University of Chicago" (http://pgap.uchicago.edu/PGAPhome.html).

siliciclastics. Chung and Jahn (1995) stressed earlier a notable complexity of the plume–lithosphere interactions within the Emeishan flood magmatism. This unusual temporal pattern of volcanicity needs further study.

A range of long-duration explosive volcanic activity is undoubtedly recorded in the sections of South China (Thompson et al., 2001). Massive silicic extrusions, similar or larger in magnitude to the Toba mega-eruption (see Zielinski, 2002), characterized the critical Changhsingian timespan (Isozaki et al., 2004). Altered ash layers (tonsteins), dispersed in Late Permian coal-bearing formation, corroborate a transition from mostly calc-alkalic, silica-poor magma, towards more siliceous and K-rich, rhyolitic source at the P–Tr boundary (Zhou et al., 2000). In addition, geochemical (e.g. platinum group elements) and mineralogical anomalies of the coals indicate hydrothermal activity and volcanic ash input into mire during peat deposition (Dai et al., 2003). Thick piles of acid (trachyte and rhyolite) rocks also form an important member of the upper Emeishan (high-Ti) basaltic succession (Lo et al., 2002; Xiao et al., 2003). However, Kozur (1998) supposed the continuous distribution of the P–Tr 'boundary tuff' layers from southeastern Siberia, up to 25-m thick, to South China over $\sim 2 \times 10^6$ $km^2$ that correspond to a final peak of the initial highly explosive phase of Siberian volcanism (cf. also Dobretsov and Vernikovsky, 2001; Courtillot and Renne, 2003). Alternatively, large pyroclastic eruptions from a southerly volcanic arc that collided with the South China continent are postulated by Newkirk et al. (2002). In fact, the highly explosive source remains unknown (Isozaki et al., 2004). Remarkably, the P–Tr air falls are absent in nearby North China (Enkin et al., 1992), even though this block was affected by coeval collisional event (Jia et al., 2004), and mafic intrusive activity is recorded in the palaeo-Asian oceanic domain (Zhou et al., 2002b).

A tholeiitic basalt province in India (Panjal Traps) also erupted near the Middle–Late Permian boundary (see Wignall, 2001), and jointly with the Late Permian to earliest Triassic volcanic series in Oman and western Australia, which were associated with the incipient opening of the Neo-Tethys ocean (Veevers and Tewari, 1995). In addition, voluminous subduction-related volcanism is known along Proto-Cordilleran-Gondwana and Proto-Altaid orogens (Nikishin et al., 2002). When combined with the Siberian volcanism, the data provide some evidence of a temporal link between violent volcanism and both Late Permian mass extinctions (Courtillot, 1999; Hallam and Wignall, 1999; Isozaki, 2002), even if the site of the explosive and extrusive processes remains somewhat unclear (Wignall, 2001).

### *6.2. Magmatic events in the subpolar Gondwanan domain*

The proposed oceanographic scenario triggered by the Siberian Traps in north circumpolar regions (Fig. 2) may be applied to the high southern regions as well. Veevers et al. (1994, p. 189) stressed: 'Antipodal to Siberian Traps and no less voluminous are the tuffs, ignimbrites, and parent granites erupted in the magmatic arc along the Panthalassan margin of the Gondwananland province'. This seems to be an overstated claim (cf. Nikishin et al., 2002, p. 24), but the distinctly *accelerated* volcanism and orogenic activity along extensive cratonward-subduction zone (Pangean Rim of Fire, Golonka, 2002; Jenkins et al., 2002; Nikishin et al., 2002) cannot be omitted in the P–Tr global change scenario.

Michaelsen and Henderson (2000) reported uniform Late Permian eruptive activity over a period of some 9 Ma in eastern Australia, exemplified by the extensive Platypus

Tuff Bed (258.9 ± 2.7 Ma), a tephra event unit with a pyroclastic volume above 10 $km^3$ (Michaelsen et al., 2001). The 2.5-km-thick sequence in northeastern New South Wales (Emmaville Volcanics) consists of diverse ignimbrites, volcaniclastics and minor lavas (Stewart, 2001) and moderate-sized compound basaltic to andesitic flows in lava sequences up to 48 $km^3$ in volume (Campbell and Conaghan, 2001). Some volcanic units in the southern Sydney Basin flowed into cold water, shallow shelf environments with water depths of ~10–25 m (Carr and Jones, 2001).

## 7. Consequences of subaerial versus seafloor volcanism

As recognized by Coffin and Eldholm (1994) and Courtillot (1999, p. 99), environmental consequences of subaqueous and subaerial volcanic activity are different. Sea-floor eruptions may affect the oceanic biota due to fertilization (Vermeij, 1995; Racki and Cordey, 2000), but Vermeij (1995, p. 144) suspected that warming induced by undersea volcanism is a key evolutionary stimulus that is more important than increased nutrient release.

The Late Permian magmatism is usually considered as limited to continental subaerial settings (e.g. Nikishin et al., 2002), but this could be biased due to a non-preservation of subducted oceanic plateaus (see review in Saunders et al., 1996). However, pillow lavas are also found in older parts of the Emeishan Traps within a carbonate shelf setting (Lo et al., 2002; see also Xiao et al., 2003; Song et al., 2004), whilst later flows are developed in marine clastic domains (He et al., 2003). As proved by fossils found in tuffs and pillow structures (Czamanske et al., 1998, pp. 108–109; Fedorenko in Tanton and Hager, 2000), depositional setting of the lower Tunguska explosive-to-extrusive volcanic sequence also appears to be subaqueous probably in lagoons and/or lakes. Non-marine bivalve coquinas characterize the oldest tuffaceous series, for example, the Tutonchan Horizon (Mogutcheva and Betekhtina, 1998). New data from the northern Pangea shelf also point to partial connection of the Siberian superplume activity with the Boreal marine basins (cf. Dorritie, 2002, 2004). The basic lava surges certainly extended into the deltaic and marine clastic sequences of the Petchora–Barents Sea basins, affected by intense rifting activity (Dobretsov, 1997; Nikishin et al., 2002; Golonka et al., 2003). In particular, Shipilov (2003) showed the development of Barents–Kara continental margin characterized by the Late Permian–Early Triassic basaltic magmatism in various tectonic settings. The northwestern geographical extent of the subaqueous extrusion remains poorly constrained and more reliable radiometric datings are needed (Saunders et al., 2005).

Even though the P–Tr first-order sea-level lowstand timespan was distinguished by an overall low spreading/subduction and mantle outgassing rates due to supercontinental stasis (Holser and Magiritz, 1987; Gibbs et al., 1999), some workers have nonetheless suggested massive intra-oceanic magmatism and increased hydrothermal fluxes resembling the mid-Cretaceous superplume event recorded in vast basaltic plateaus (Veevers et al., 1994; Carpenter and Lohmann, 1999). In particular, the rapid sea-level rise beginning in the latest Permian (Hallam and Wignall, 1999) could be a consequence of oceanic plateau emplacement (Hallam, 1999). This flood volcanism should be preserved as extensive accreted oceanic plateaus, such as those found in the obducted terranes of western North America. Early to Mid-Triassic accreted oceanic plateau and/or seamount associations dominated by basic pyroclastic sediments (Pickett and Robertson, 2004) are known in northern Turkey (Nilüfer unit), with a volume of mafic pillow lava of $2 \times 10^5$ km$^3$ (Genç, 2004); biostratigraphic datings are scarce (Okay and Göncüoğlu, 2004), and "it is quite likely that the age of the Nilüfer formation spans the Permian-Triassic boundary" (A.I. Okay, e-mail comm., 2004). In this regard, it is noteworthy that Nikishin et al. (2002, p. 20–21) described considerable magmatic activity in the Late Permian and Triassic Tethyan rifts, but again more reliable dating is needed. The hotspot activity in the eastern Panthalassa, described from accreted terranes of the Canadian Cordillera, is also speculated to be an echo of the end-Permian mantle plume in Siberia (Tardy et al., 2003).

### *7.1. Geothermally modified circulation?*

Can the pulsed massive volcanogenic heating have contributed, at least episodically, to the unusual vigorous circulation advocated for the Wuchiapingian zenith of the Chert Event by Beauchamp and Baud (2002)? The widespread development of biogenic cherts around the margins of Panthalassa coincides with the emplacement of the Emeishan plume in an area that was probably characterized by strong poleward and eastward flowing currents along the Tethys coast (Kutzbach et al., 1990; Winguth et al., 2002; Fig. 4). On the other hand, the *long-lasting* (even ~8 Ma; Fig. 1) Siberian magmatic–hydrothermal system as a whole probably existed since the Wuchiapingian in a direct proximity of the Panthalassan deep water formation, occurring seasonally in the north-eastern, Boreal part (see Winguth et al., 2002, p. X-7; Fig. 4); possible interactions during the two-step collapse of Chert Event needs further consideration.

The question of possible influence of both effective 'heat factories' remains unanswered (see Section 5.2), but the importance of geothermal forcing (sensu lato) in driving deep zonal circulation is poorly known even in modern oceans (Kadko et al., 1995). This energy source is currently neglected in numerical modelling. A typical value of the geothermal fluxes on abyssal plains is 0.05 W/m$^2$, whilst air–sea heat fluxes are of the order of 100 W/m$^2$. Adcroft et al. (2001, p. 1738) claimed that this naive comparison is inappropriate, because the heating of a fluid from below appears to be essential in creating deep circulation in their numerical simulations. This volcanogenic/hydrothermal factor could be significant during emplacement of large igneous fields, especially in (semi) enclosed marine basins (Kadko et al., 1995, p. 456; Coffin and Eldholm, 1994, p. 26), as seen in the modelling of the dynamics of effluent plumes and sea surface temperature anomalies by Vogt (1989), Emanuel et al. (1995) and Speer (1997).

Hydrothermal fluids remove a significant portion of heat from the oceanic lithosphere for plates typically up to about 65-Ma old (Kadko et al., 1995; Villinger et al., 2002; Alt, 2003; German and Von Damm, 2003; Fisher et al., 2003; Johnson and Pruis, 2003). Advective heat output between 0.1 and 0.3 W/m$^2$ commonly occurs on mid-ocean ridge flanks (Table 1 in Kadko et al., 1995), but anomalies up to 1 W/m$^2$, i.e. 200 times above the geothermal abyssal background, are reported from some basaltic fields (Fisher et al., 2003). This energy transfer is exemplified by the warm (25°C) ridge-flank hydrothermal springs on 3.5 Ma crust of the Juan de Fuca Ridge, northeast Pacific Ocean (Mottl et al., 1998).

The heating is parameterized in the numerical model of recent oceans as a spatially uniform energy flux of 0.05 W/m$^2$ (Adcroft et al., 2001), and through an idealized flat ocean floor (Scott et al., 2001). A magnitude of the advective response is inversely proportional to the deep stratification, and therefore the effect of additional geothermal heating is strongest in the relatively sluggish Pacific Ocean. The overturn changes might influence biogeochemical cycling and nutrient transport although they have little impact on total oceanic heat transfer. Adcroft et al. (2001, p. 1738) concluded that 'geothermal heating induces a substantial change in the deep circulation which is larger than previously assumed and subsequently, the warming of the deep ocean is only a quarter of that suggested by the diffusive limit'.

### 7.2. *Volcanogenic fertilization?*

Growing numbers of observations, and some numerical simulations, underline significance of volcano- and tectono-biogeochemical feedbacks (Carpenter and Lohmann, 1999; Stanley and Hardie, 1999; Condie, 2001; Jones and Jenkyns, 2001; Cather et al., 2003; German and Von Damm, 2003; Handoh and Lenton, 2003; Siemann, 2003; Weissert and Erba, 2004). As discussed by Racki and Cordey (2000), tectono-volcanic processes might contribute in several ways to the PCEvent acme.

Very high fertilization potential of newly erupted volcanic ash is remarkable (Vermeij, 1995; Frogner et al., 2001; Felitsyn and Kirianov, 2002; Cather et al., 2003), and a trap-derived nutrification is predicted for neighbouring seas (e.g. Sansone et al., 2002; Pálfy, 2003; Dorritie, 2004). The results of Gíslason et al. (2002) show that modern subglacial eruptions affect near-surface oceanic layers, where either Mn, Fe, Si or Cu are rate-determining for the biomass growth. The biologically mediated degradation via microbes of basaltic glass (i.e. 10–20% of the crust) may also add significantly to the nutrient cycling (Staudigel et al., 1995; cf. Stefánsson and Gíslason, 2001). The Late Permian to Early Triassic magmatic activity, in particular Emeishan Traps, may have achieved at least some of these effects, because they were erupted near sea level onto a previously submerged marine platform.

The flux and bioavailability of many elements can be strongly increased by fluid discharge from hydrothermal vent fields (Resing and Sansone, 2002), as calculated for the global Fe input (Mottl et al., 1998); the fluxes could be significant in presumably Fe-limited P–Tr oceans (Kidder and Worsley, 2004a), but only prior to the superanoxic event. Sinton and Duncan (1997) suggested that hydrothermalism, associated with extensive mid-Cretaceous submarine volcanism, was responsible for the dissolved oxygen depletion (cf. also Cathles and Hallam, 1991, p. 669; Sinninghe Damsté and Köster, 1998; Racki and Cordey, 2000; Leckie et al., 2002). The mantle–hydrosphere chemical budget is as yet far from understood (see summary in Alt, 2003; Johnson and Pruis, 2003), especially its control by geological processes and variability during the course of Earth's history (German and Von Damm, 2003, p. 216).

## 8. End-Guadalupian crisis and a potential volcanic cause

A major acidic (rhyo-dacitic) tuff, ~2-m thick, has been recognized at the Middle–Upper Permian boundary horizon in China (Isozaki et al., 2004) and in a paleoatoll limestone within an accretionary complex in Japan, which at that time was probably more than 3000 km away from the eastern Pangea (Isozaki and Ota, 2001); this suggests a key role of the explosive acidic volcanism in the end-Guadalupian biosphere crisis (Isozaki et al., 2004; Ali et al., 2005).

The Chinese basalt eruptions may have triggered a similar series of environmental changes to the Siberian Traps, namely oceanic warming and destabilization of clathrate reservoirs (Lo et al., 2002) confirmed by negative C-isotopic shift (Kaiho et al., 2005). This may have been facilitated by sea-level fall, because eustatic sea level reached perhaps one of the lowest points of the Phanerozoic at this time (Hallam and Wignall, 1999). This coincides with a Phanerozoic minimum of the seawater $^{87}Sr/^{86}Sr$ curve (Korte et al., 2003; Fig. 1).

Several causal links can be postulated between volcanism and devastation of warm-water faunas during major regression. The profound sea-level fall was almost certainly not of the glacioeustatic origin supposed, among others, by Courtillot and Renne (2003, p. 128), because polar ice caps were in an overall retreat since the Early Permian (Hallam and Wignall, 1997; see Section 4). Tectonoeustasy could be evoked, such as the rapid rifting phenomena modelled by Cathles and Hallam (1991), especially if the seawater $^{87}Sr/^{86}Sr$ negative anomaly records worldwide tectonism, active volcanism and amplified hydrothermal fluxes (cf. Nikishin et al., 2002). As an alternative, one might surmise a major uplift of the Pangea supercontinent (Erwin, 1993, p. 164; Zharkov and Chumakov, 2001), connected with exhumation of newly erupted mantle-derived basaltic series (Huang and Zhou, 1997; Crowell, 1999, p. 39; Lo et al., 2002; see also Dessert et al., 2001 for the Deccan Trap modelling).

## 9. Discussion

Both the end-Guadalupian and end-Permian extinctions coincide broadly with the zenith of accretion and onset of fragmentation of Pangea and so are likely to coincide with a new mode of mantle convention, which could include a plate boundary reorganization, accelerated rifting activity and the Siberian superplume event (e.g. Holser and Magiritz, 1987; Courtillot et al., 1999; Abbot and Isley, 2002; Golonka, 2002; Nikishin et al., 2002; Gatinskii and Rundkvist, 2003; Shipilov, 2003; Bottjer, 2004; Yakubchuk and Nikishin, 2004). The nature of the configuration of Pangea and Panthalassia may have made a global ecosystem that was particularly unable to cope with these unusual volcanic triggers (Bottjer, 2004). In particular, the extensive zone of aridity within the supercontinent interior would have made any silicate-weathering feedback mechanisms that can normally counter-balance any increased atmospheric $CO_2$ levels, unusually inefficient (see discussion in Kidder and Worsley, 2004b). The generally low values of seawater $^{87}Sr/^{86}Sr$ are probably significant in this regard.

The climatic/oceanographic causal links during the Late Permian two-step biosphere reorganization partly fit scenarios proposed for the mid-Cretaceous superplume-related oceanic anoxic events (Vogt, 1989; Kerr, 1998; Leckie et al., 2002; Weissert and Erba, 2004), albeit with no corresponding major extinction during the later interval despite recurrent biocalcification crises. The Cretaceous interval is a prime example of the geological period of unusually active tectonism and pulses of ophiolite obduction (Vaughan and Scarrow, 2003) that drove enhanced mantle outgassing and liberation of $CH_4$ by superplume-driven submarine volcanism (Jahren, 2002; Wissler et al., 2003), as well as phosphorus supply due to weathering rate and/or *benthic regeneration under anoxic conditions* (Handoh and Lenton, 2003; Nederbragt et al., 2004). A cascade of causal linkages, triggered by excess hydrothermal activity associated with sea-level rise, is presented in comprehensive model for oceanic anoxia by Jones and Jenkyns (2001). Why these events were not as lethal as the P–Tr superanoxia is not clear, but we would advance the observations that no Cretaceous events were as prolonged or as severe as the P–Tr superanoxic event. Thus, Cretaceous anoxic facies are rarely seen in waters shallower than deep shelf locations, whereas the P–Tr anoxia extended to inner shelf water depths (Wignall and Twitchett, 1996, 2002a). In fact, some of the Cretaceous oceanic anoxic events could be surprisingly dynamic phenomena promoted by variations in the sea surface temperature and intensified surface productivity, i.e. not simply by ocean stagnation (Wilson and Norris, 2001). We also reiterate the sea-floor location of the mid-Cretaceous volcanism and the arid nature of the Pangea supercontinent, and its inefficiency of $CO_2$ drawdown due to lowered silicate weathering rate (Kidder and Worsley, 2004b).

We propose that the severity of the anoxia, highlighted by biomarker evidence for widespread photic-zone euxinic conditions (Grice et al., 2005), may at least in part relate to the site of Siberian volcanism at high latitudes, where permafrost melt water would have been able to critically retard the sinking of deep waters and thus slow down global oceanic overturn from the latest Wuchiapingian onwards (Fig. 1). This immense volcanism probably caused additional methane outbursts (Dorritie, 2002, 2004), particularly from a large bipolar permafrost and continental margin reservoirs (Krull et al., 2000, 2004; Retallack et al., 2003), resulting finally in atmospheric oxygen drop (Weidlich et al., 2003), although it is a moot point whether this was contributory to the terrestrial extinctions.

Furthermore, instead of the lack of sufficient ocean heat transport into polar latitudes, an increased geothermal heat inflow to the volcanically active regions offers an alternative for the temperate high-latitude climatic paradox during the Late Permian ('a cold bias of the climate model'), as discussed by Gibbs et al. (2002). According to Kidder and Worsley (2004b), increased upwelling by cyclonic storm activity and heat trapping polar cloudiness supplemented poleward heat transport by warm deep-ocean brines. The simulation of Ufnar et al. (2004) indicates that the intensification of the hydrologic cycle during greenhouse periods probably plays an essential role in the warming of high latitudes due to augmented poleward heat transfer by $H_2O$ vapour. However, this effect could be weakened if volcanogenic sulphate aerosols accumulate in the troposphere (White and Saunders, 2005). Future quantitative modelling has to address the enigma of sustainable and efficient halothermal dynamics (cf. Zhang et al., 2002) in the unusual superstagnant/euxinic P–Tr oceans, advocated by these authors. However, the first fully coupled comprehensive climate model of Kiehl and Shields (2005) simulates high-latitude warming and a oceanic stagnation as a result of elevated carbon dioxide levels.

## 10. Conclusions and implications

The key to the P–Tr mass extinction lay in the development of a series of successive intrusive and extrusive/explosive eruptions within a ~10 Ma time interval, both in subpolar and tropical realms (Figs. 2 and 4, Table 2) of both subaerial and subaqueous nature. This was an exceptional feature of the P–Tr mass extinction (see also Dorritie, 2002, 2004; Isozaki,

2002; Racki, 2003), and is considered particularly important due to the effect on high-latitude surface water and ocean deep-water generation. The polar location and volume of Siberian Traps volcanism are also held responsible for perhaps one of the greatest methane release events of the Phanerozoic (Dorritie, 2004; but see contradicting modeling results of Grard et al., 2005).

Three-step temporal scenario of the end-Permian ecosystem collapse is proposed (Fig. 5):

1. Long-term evolution of global ecosystem after Permian vanishing of the low-latitude Alleghenian/Variscan orogenesis in the Pangean supercontinent, marked by many feedbacks and interactions acting in concert towards progressive warming and destabilized oceanic circulation and biogeochemical cycling (Chumakov and Zharkov, 2003; Chumakov, 2004; Kidder and Worsley, 2004a, b; Zharkov, 2004).

2. Late Permian to Early Triassic pulsed magmatic activity worldwide in the geodynamically reactivated plate-tectonic setting, which resulted in amplified climatic and

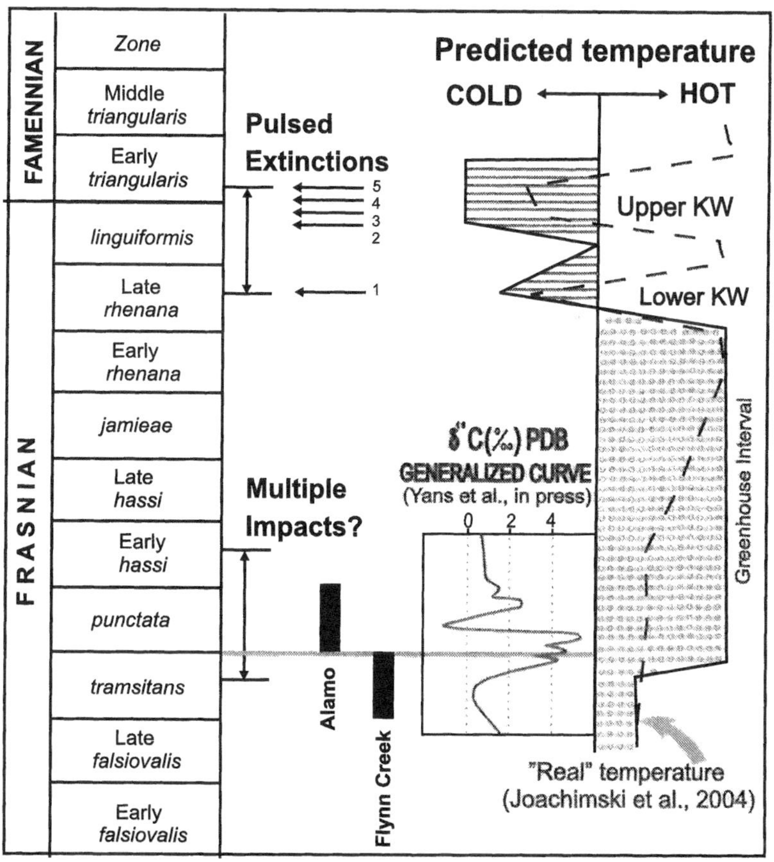


*Figure 5.* A scheme of the three-order temporal–causal fabric for the Permian–Triassic mass extinction: for details of overall palaeogeographic and climatic–oceanographic setting, see Chumakov and Zharkov (2003), Kidder and Worsley (2004b) and Zharkov (2004), and for scenario of methane catastrophe, see Dorittie (2004).

ecological instability (= P–Tr biotic crisis), helping also *finally* to slow thermohaline circulation.

3. End-Permian methane catastrophe triggered mainly by the major cataclysmic pulse of the Siberian superplume in the northern polar domain. This provoked a closing cascade of devastating effects in the overheated globe, which resulted in the P–Tr ecosystem collapse (Dorritie, 2004).

During the Late Olenekian and Middle Triassic, i.e. several million years after volcanicity drop, a gradual recovery of plankton productivity took place (Figs. 1 and 2). This was due to initiation of poleward transport of heat and moisture, and progressive Panthalassan ventilation thanks to revived thermohaline activity. The recovery processes are poorly known, but could possibly include the inflow of warm salty waters from the Tethys (cf. Bigg et al., 2003). The equator-to-pole temperature gradient increased chiefly due to restored cooling in high latitudes, although Haupt and Seidov (2001) suggest this was in one hemisphere only, but its sedimentary signature is conjectural (Price, 1999; Zharkov and Chumakov, 2001). The climatic change was partly due to an effective $CO_2$ sink resulting from the weathering of giant basaltic series in gradually more humid settings (e.g. Dessert et al., 2001, 2003; Stefánsson and Gíslason, 2001; Felitsyn, 2002; Chadwick, et al., 2003).

This critical review of Late Permian records identifies several area requiring further study and rigorous testing beginning within more refined temporal pattern, and probabilistic but environmentally justified event correlation (Fig. 1). As concluded by Wignall (2001), Courtillot and Renne (2003) and Palfý (2003), the overall synchrony of cataclysmic flood

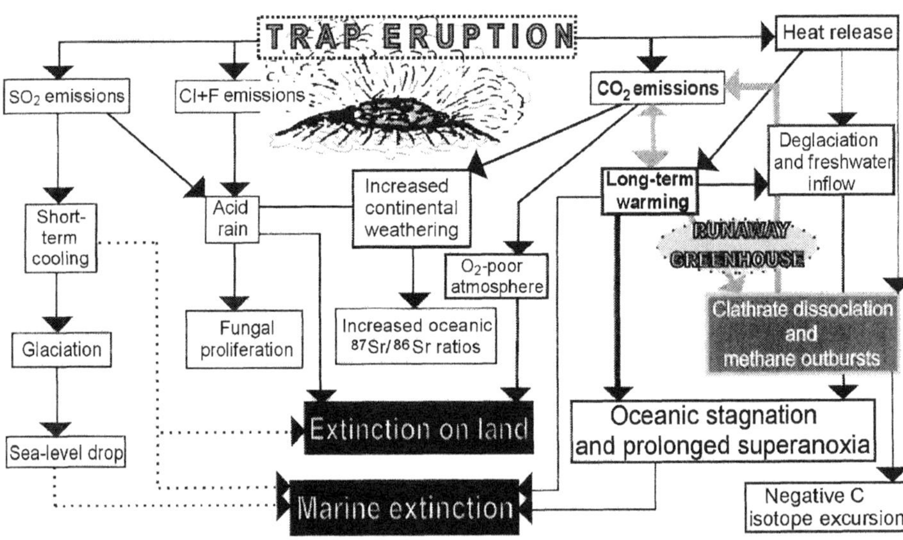


*Figure 6.* Postulated chain of events caused by the eruption of the Siberian Traps (based on Fig. 6 in Wignall, 2001), with marked different causal role of cooling processes (not supported by geological record; dotted line) and greenhouse warming (a key feedback during end-Permian and probably several other mass extinctions; thick continuous line). The flow chart is limited to continental igneous provinces, and omits a possible influence of intra-oceanic volcanism discussed in the text (especially geothermally promoted changes in circulation, and a volcanogenic fertilization; see Fig. 3 in Pálfy, 2003).

basalt eruptions and major mass extinctions permits causal links both brief- and long-term, although these links are to some extent still unknown (cf. Jerram and Widdowson, 2005; White and Saunders, 2005), which in each case may be mediated by catastrophic climatic perturbations (mostly excessive warming) and induced stagnation of ocean water (Fig. 6).

## Acknowledgements

We thank Jason Ali and Tom Worsley for their comments on an early version of this article. We appreciate thorough reviews by Andy Saunders and Sara Pruss, which have led to final improvements in the manuscript. This work benefitted from discussions and information exchange with Jan Golonka, Jeffery R. Scott, Greg Retallack, Keith A. Kvenvolden and Aral Okay.

## References

Abbott, D.H., Isley, A.E., 2002. The intensity, occurrence, and duration of superplume events and eras over geological time. J. Geodyn. 34, 265–307.

Adams, J.B., Mann, M.E., Ammann, C.M., 2003. Proxy evidence for an El Niño-like response to volcanic forcing. Nature 426, 274–278.

Adcroft, A., Scott, J.R., Marotzke J., 2001. Impact of geothermal heating on the global ocean circulation. Geophys. Res. Lett. 28, 1735–1738.

Afonin, S.A., Barinova, S.S., Krassilov, V., 2001. A bloom of *Tympanicysta* Balme (green algae of zygnematalean affinities) at the Permian–Triassic boundary. Geodiversitas 23, 481–487.

Ali, J.R., Thompson, G.M., Song, X.Y., Wang, Y.L., 2002. Emeishan Basalts (SW China) and the 'end-Guadalupian' crisis: magnetobiostratigraphic constraints. J. Geol. Soc. Lond. 159, 21–29.

Ali, J.R., Thompson, G.M., Zhou, M.F., Song, X.Y., 2005. Emeishan large igneous province, SW China. Lithos 79, 475–489.

Alt, J.C., 2003. Hydrothermal fluxes at mid-ocean ridges and on ridge flanks. C. R. Geosci. 335, 853–864.

Alvarez, W.D.H., 2003. Comparing the evidence relevant to impact and flood basalt at times of major mass extinctions. Astrobiology 3, 153–161.

Anderson, S.W., Stofan, E.R., Smrekar, S.E., Guest, J.E., Wood, B., 1999. Pulsed inflation of pahoehoe lava flows: implications for flood basalt emplacement. Earth Planet. Sci. Lett. 168, 7–18.

Angiolini, L., Balini, M., Garzanti, E., Nicora, A., Tintori, S., Crasquin, S., Muttoni, G., 2003. Permian climatic and paleogeographic changes in Northern Gondwana: the Khuff Formation of Interior Oman. Palaeogeogr., Palaeoclimatol., Palaeoecol. 191, 269–300.

Archer, D., Martin, P., Buffett, B., Brovkin, V., Rahmstorf, S., Ganopolski, A., 2004. The importance of ocean temperature to global biogeochemistry. Earth Planet. Sci. Lett. 222, 333–348.

Bambach, R.K., Knoll, A.H., Sepkoski, J.J., 2002. Anatomical and ecological constraints on Phanerozoic animal diversity in the marine realm. Proc. Natl. Acad. Sci. USA. 99, 6854–6859.

Beauchamp, B., Baud, A., 2002. Growth and demise of Permian biogenic chert along northwest Pangea: evidence for end-Permian collapse of thermohaline circulation. Palaeogeogr., Palaeoclimatol., Palaeoecol. 184, 37–63.

Benton, M.J., Tverdokhlebov, V.P., Surkov, M.V., 2004. Ecosystem remodelling among vertebrates at the Permian–Triassic boundary in Russia. Nature 432, 97–100.

Benton, M.J., Twitchett, R.J., 2003. How to kill (almost) all life: the end-Permian extinction event. Trends Ecol. Evol. 18, 358–365.

Bergström, S.M., Huff, W.D., Saltzman, M.R., Kolata, D.R., Leslie, S.A., 2004. The greatest volcanic ash falls in the Phanerozoic: Trans-Atlantic relations of the Odroviciam Millbrig and Kinnekulle K-bentonites. Sedim. Record 4, 4–7.

Berner, R.A., 2002. Examination of hypotheses for the Permo–Triassic boundary extinction by carbon cycle modeling. Proc. Natl. Acad. Sci. USA. 99, 4172–4177.

Bice, K.L., Marotzke, J., 2001. Numerical evidence against reversed thermohaline circulation in the warm Paleocene/Eocene ocean. J. Geophys. Res. 106(C6), 1529–11542.
Bigg, R., Jickells, T.D., Liss, P.S., Osborn, T.J., 2003. The role of the oceans in climate. Int. J. Climatol. 23, 1127–1159.
Björnsson, H., 2002. Subglacial lakes and jökulhlaups in Iceland. Global Planet. Change 35, 255–271.
Blake, S., 2003. Correlations between eruption magnitude, $SO_2$ yield, and surface cooling. In: Oppenheimer, C., Pyle, D.M., Barclay, J. (Eds), Volcanic Degassing. Geol. Soc. Lond. Spec. Publ., Vol. 213, pp. 371–380.
Blake, S., Bruno, B.C., 2000. Modelling the emplacement of compound lava flows. Earth Planet. Sci. Lett. 184, 181–197.
Blome, C.D., Reed, K.M., 1995. Radiolarian biostratigraphy of the Quinn River Formation, Black Rock Terrane, north-central Nevada: correlations with eastern Klamath Terrane geology. Micropaleontology 41, 49–68.
Bondre, N.R., Duraiswami, R.A., Dole, G., 2004. Morphology and emplacement of flows from the Deccan Volcanic Province, India. Bull. Volcanol. 66, 29–45.
Bottjer, D.J., 2004. Early Triassic and the supercontinent cycle. Geol. Soc. Am., Abstr. Pr. 36(5), 334.
Bragin, N.Y., 2000. The Permian–Triassic crisis in the biosphere as manifested in the Paleo-Pacific deep-water sequences. Stratigr. Geol. Correlation 8, 232–242.
Bragin, N.Y., Krylov, K.A., 2002. Horizons of Carbonaceous rocks in Triassic and Lower Jurassic cherty deposits of the Dal'negorsk reference section (Primor'e). Stratigr. Geol. Correlation 10, 503–509.
Buffett, B., Archer D., 2004. Global inventory of methane clathrate: sensitivity to changes in the deep ocean. Earth Planet. Sci. Lett. 227, 185–199.
Byakov, A.S., 2000. Bivalve zonation in Permian deposits of northeastern Asia. Stratigr. Geol. Correlation 8, 30–48.
Callot, J.P., Gurevitch, E., Westphal, M., Pozzi, J.P., 2004. Flow patterns in the Siberian traps deduced from magnetic fabric studies. Geophys. J. Int. 156, 426–430.
Campbell, L.M., Conaghan, P.J., 2001. Flow-field and palaeogeographic reconstruction of volcanic activity in the Permian Gerringong Volcanic Complex, southern Sydney Basin, Australia. Austr. J. Earth Sci. 48, 357–375.
Campbell, I.H., Czamanske, G.K., Fedorenko, V.A., Hill, R.I., Stepanov, V., 1992. Synchronism of the Siberian Traps and the Permian–Triassic boundary. Science 258, 1760–1763.
Carpenter, S.J., Lohmann, K.C., 1999. Reply to the comment by S.T. Petsch on carbon isotope ratios of Phanerozoic marine cements: re-evaluating global carbon and sulfur systems. Geochim. Cosmochim. Acta 63, 761–766.
Carr, P.F., Jones, B.G., 2001. The influence of palaeoenvironment and lava flux on the emplacement of submarine, near-shore Late Permian basalt lavas, Sydney Basin (Australia). J. Volcanol. Geotherm. Res. 112, 247–266.
Casey, R.E., 1993. Radiolaria. In: J.H. Lipps (Ed.), Fossil Prokaryotes and Protists. Blackwell, Boston, pp. 249–284.
Cather, S.M., Dunbar, N.W., McDowell, F.W., McIntosh, W.C., Scholle P.A., 2003. Early Oligocene global cooling, volcanic iron fertilization, and the ignimbrite flare-up of southwestern North America. Geol. Soc. Am., Abstr. Pr. 35(6), 255.
Cathles, L.M., Hallam, A., 1991. Stress-induced changes in plate density, Vail sequences, epeirogeny, and short-lived global sea level fluctuations. Tectonics 10, 659–671.
Chadwick, O.A., Gavenda, R.T., Kelly, E.F., Ziegler, K., Olson, C.G., Elliott, W.C., Hendricks, D.M., 2003. The impact of climate on the biogeochemical functioning of volcanic soils. Chem. Geol. 202, 195–223.
Chaikin, V.G., Meskhi, A.M., 2003. Late Permian volcanism in the Kama region. Dokl. Earth Sci. 391, 634–635.
Christensen, T.R., Johansson, T., Åkerman, H.J., Mastepanov, M., Malmer, N., Friborg, T., Crill, P., Svensson, B.H., 2004. Thawing sub-arctic permafrost: effects on vegetation and methane emissions. Geophys. Res. Lett. 31, L04501, doi 10.1029/2003GL018680.
Chumakov, N.M., 1994. Evidence of Late Permian glaciation in the Kolyma River basin: a repercussion of the Gondwana glaciation in Northeast Asia. Stratigr. Geol. Correlation 2, 130–150.
Chumakov, N.M., 2004. Climate and climate zones of the Permian and Early Triassic (in Russian). In: Semikhatov, M.A., Chumakov, P.M. (Eds), Klimat v Epokhi Krupnykh Biosfernykh Peresroyek. Geol. Inst. RAS Trans. No. 550, pp. 230–256.
Chumakov, N.M., Zharkov, M.A., 2003. Climate during the Permian–Triassic biosphere reorganizations. Article 2. Climate of the Late Permian and Early Triassic: general inferences. Stratigr. Geol. Correlation 11, 361–375.
Chung, S.L., Jahn, B.M., 1995. Plume–lithosphere interaction in generation of the Emeishan flood basalts at the Permian–Triassic boundary. Geology 23, 889–892.
Coffin, M.F., Eldholm, O., 1994. Large igneous provinces: crustal structure, dimensions, and external consequences. Rev. Geophys. 32, 1–36.

Conaghan, P.J., Shaw, S.E., Veevers, J.J., 1994. Sedimentary evidence of the Permian/Triassic global crisis induced by the Siberian hotspot. In: Embry, A.F., Beauchamp, B., Glass, D.J. (Eds), Pangea: Global Environments and Resources. Can. Soc. Petr. Geol. Memoir, Vol.17, pp. 785–795.

Condie, K.C., 2001. Mantle Plumes and their Record in Earth History. Cambridge University Press, Cambridge, 306pp.

Cordey, F., Marcoux, J., Bucher, H., Girard, C., Lécuyer, C., Crasquin-Soleau, S., Baud, A., Orchard, M., 1999. Record of the Permo–Triassic Crisis in oceanic environments: new targets in Oman, Nevada, and British Columbia. In: EUG99, Terra Abstr., Strasbourg, p. 275.

Courtillot, V., Jaupart, C., Manighetti, I., Tapponnier, P., Besse, J., 1999. On causal links between flood basalts and continental breakup. Earth Planet. Sci. Lett. 166, 177–195.

Courtillot, V.E., 1999. Evolutionary Catastrophes. The Science of Mass Extinction. Cambridge University Press, Cambridge, 173pp.

Courtillot, V.E., Renne, P.R., 2003. On the ages of flood basalt events. C.R. Geosci. 335, 113–140.

Crasquin-Soleau, S., 2003. Middle and Late Permian stratigraphic correlations on the Russian Platform: significance of ostracodes. Rev. Micropaléontol. 46, 23–33.

Crowell, J.C., 1999. Pre-Mesozoic Ice Ages: their bearing on understanding the climate system. Geol. Soc. Am. Mem. 192, 1–106.

Czamanske, G.K., Gurevitch, A.B., Fedorenko, V., Simonov, O., 1998. Demise of the Siberian Plume: paleogeographic and paleotectonic reconstruction from the prevolcanic and volcanic record, North-Central Siberia. Int. Geol. Rev. 40, 95–115 (http://www.mantleplumes.org/).

Dai, S.F., Ren, D.Y., Zhang, J.Y., Hou, X.Q., 2003. Concentrations and origins of platinum group elements in Late Paleozoic coals of China. Int. J. Coal Geol. 45, 59–70.

Dessert, C., Dupré, B., François, L.M., Schott, J., Gaillardet, J., Chakrapani, G., Bajpai, S., 2001. Erosion of Deccan Traps determined by river geochemistry: impact on the global climate and the $^{87}Sr/^{86}Sr$ ratio of seawater. Earth Planet. Sci. Lett. 188, 459–474.

Dessert, C., Dupré, B., Gaillardet, J., François, L.M., Allègre, C.J., 2003. Basalt weathering laws and the impact of basalt weathering on the global carbon cycle. Chem. Geol. 202, 257–273.

De Wever, P., O'Dogherty, L., Caridroit, M., Dumitrica, P., Guex, J., Nigrini, C., Caulet, J.P., 2003. Diversity of radiolarian families through time. Bull. Soc. Géol. Fr. 174, 453–469.

De Wit, M.J., Ghosh, J.G., de Villiers, S., Rakotosolofo, N., Alexander, J., Tripathi, A., Looy, C., 2002. Multiple organic carbon isotope reversals across the Permo–Triassic boundary of terrestrial Gondwana sequences: clues to extinction patterns and delayed ecosystem recovery. J. Geol. 110, 227–246.

Dobretsov, N.L., 1997. Permian–Triassic magmatism and sedimentation in Eurasia as a result of a superplume. Dokl. Earth Sci. 354, 497–500.

Dobretsov, N.L., Vernikovsky V.A., 2001. Mantle plumes and their geologic manifestations. Int. Geol. Rev. 43, 771–787.

Dorritie, D., 2002. Consequences of Siberian Traps volcanism. Science 297, 1808–1809.

Dorritie, D., 2004. Killer in Our Midst. Methane Catastrophes in Earth's Past and Near Future. http://www.killerinourmidst.com/.

Emanuel, K.A., Speer, K., Rotunno, R., Srivastava, R., Molina, M., 1995. Hypercanes: a possible link in global extinction scenarios. J. Geophys. Res. 100(D7), 13,755–13,765.

Enkin, R.J., Yang, Z.Y., Chen, Y., Courtillot, V., 1992. Paleomagnetic constraints on the geodynamic history of the major blocks of China from the Permian to the present. J. Geophys. Res. 97(B10), 13,953–13,989.

Erwin, D.H., 1993. The Great Paleozoic Crisis. Life and Death in the Permian. Columbia University Press, New York, 327pp.

Erwin, D.H., Bowring, S.A., Jin, Y.G., 2002. End-Permian mass extinctions: a review. In: Koeberl, C., MacLeod, K.G. (Eds), Catastrophic Events and Mass Extinction: Impacts and Beyond. Geol. Soc. Am. Spec. Pap., Vol. 356, pp. 363–383.

Falkowski, P.G., Schofield, O., Katz, M.E., van de Schootbrugge, B., Knoll, A.H., 2004. Why is the land green and the ocean red? In: Thierstein, H., Young, J. (Eds), Coccolithophorids. Springer, Berlin, pp. 429–453.

Felitsyn, S.B, 2002. The redistribution of phosphorus in basic volcanic rocks. Lithol. Miner. Resour. 37, 107–109.

Felitsyn, S.B., Kirianov, V.Y., 2002. Mobility of phosphorus during the weathering of volcanic ashes. Lithol. Miner. Resour. 37, 275–278.

Filippov, A.N., Kemkin, I.V., 2003. Clastic rocks from Permian and Triassic cherty sequences in Sikhote Alin and Japan. Lithol. Miner. Resour. 38, 36–47.

Fisher, A.T., Davis, E.E., Hutnak, M., Spiess, V., Zuhlsdorff, L., Cherkaoui, A., Christiansen, L., Edwards, K., Macdonald, R., Villinger, H., Mottl, M.J., Wheat, C.G., Becker, K., 2003. Hydrothermal recharge and discharge across 50 km guided by seamounts on a young ridge flank. Nature 421, 618–621.

Foote, M., 2003. Origination and extinction through the Phanerozoic: a new approach. J. Geol. 111, 125–148.

Frogner, P., Gíslason, S.R., Óskarsson, N., 2001. Fertilizing potential of volcanic ash in ocean surface water. Geology 29, 487–490.

Ganopolski, A., Rahmstorf, S., 2001. Stability and variability of the thermohaline circulation in the past and future: a study with a coupled model of intermediate complexity. In: Seidov, D., Bernd J., Haupt, B.J., Maslin, M., (Eds), The Ocean and Rapid Climate Change: Past, Present and Future. Am. Geophys. Union, Geophys. Monogr., Vol. 126, pp. 261–275.

Gatinskii, Y.G., Rundkvist, D. V., 2003. Mesozoic–Cenozoic metallogenic provinces, Wilson cycle, and plume tectonics. Russ. J. Earth Sci. 5, 31–44.

Genç, Ş.C., 2004. Triassic large igneous province in the Pontides, northern Turkey: geochemical data for its tectonic setting. J. Asian Earth Sci. 22, 503–516.

German, C.R., Von Damm, K.L., 2003. Hydrothermal processes. In: H. Elderfield (Ed.), Vol. 6: The Oceans and Marine Geochemistry; Holland, H.D., Turekian, K.K. (Eds), Treatise on Geochemistry. Elsevier–Pergamon, Amsterdam, pp. 181–222.

Gibbs, M.T., Bluth, G.J.S., Fawcett, P.J., Kump, L.R., 1999. Global chemical erosion over the last 250 my: variations due to changes in paleogeography, paleoclimate, and paleogeology. Am. J. Sci. 299, 611–651.

Gibbs, M.T., Rees, P.M., Kutzbach, J.E., Ziegler, A.M., Behling, P.J., Rowley, D.B., 2002. Simulations of Permian climate and comparisons with climate-sensitive sediments. J. Geol. 110, 33–55.

Gíslason, S.R., Snorrason, Á., Kristmannsdóttir, H.K., Sveinbjörnsdóttir, Á.E., Torsander, P., Ólafsson, J., Castet, S., Dupré, B., 2002. Effects of volcanic eruptions on the $CO_2$ content of the atmosphere and the oceans: the 1996 eruption and flood within the Vatnajökull Glacier, Iceland. Chem. Geol. 190, 181–205.

Golonka, J., 2002. Plate-tectonic maps of the Phanerozoic. In: Kiessling, W., Flügel, E., Golonka, J. (Eds), Phanerozoic Reef Patterns. Soc. Econ. Paleontol. Mineral. Spec. Publ., Vol. 72, pp. 21–75.

Golonka, J., Bocharova, N.Y., Ford, D., Edrich, M.E., Bednarczyk, J., Wildharber, J., 2003. Paleogeographic reconstructions and basins development of the Arctic. Mar. Petr. Geol. 20, 211–248.

González-Bonorino, G., Eyles, N., 1995. Inverse relation between ice extent and the late Paleozoic glacial record of Gondwana. Geology 23, 1015–1018.

Gradstein, F., Ogg, J., Smith, A., 2004. A Geological Time Scale 2004. Cambridge University Press, Cambridge.

Grard, L.M.F., Dessert, C., Dupré, B., Goddéris, Y., 2005. Basaltic volcanism and mass extinction at the Permo-Triassic boundary: environmental impact and modeling of the global carbon cycle. Earth Planet. Sci. Lett. 234, 207–221.

Grattan, J., 2005. Pollution and paradigms: lessons from Icelandic volcanism for continental flood basalt studies. Lithos 79, 343–353.

Grice, K., Cao, C., Love, G.D., Böttcher, M.E., Twitchett, R.J., Grosjean, E., Summons, R.E., Turgeon, S.C., Dunning, W., Jin, Y., 2005. Photic zone euxinia during the Permian-Triassic superanoxic event. Science 307, 706–709.

Gruszczyński, M., Małkowski, K., Szaniawski, H., Wang C.Y., 2003. The carbon biogeochemical cycle across the Permian–Triassic boundary strata and its implications: isotope record from the Changhsingian Stage at Meishan, south China. Acta Geol. Polon. 53, 167–179.

Gudmundsson, A., 2000. Dynamics of volcanic systems in Iceland: example of tectonism and volcanism at juxtaposed hot spot and mid-ocean ridge systems. Annu. Rev. Earth Planet. Sci. 28, 107–140.

Guo, F., Fan, W.M., Wang, Y.J., Li, C.W., 2004. When did the Emeishan mantle plume activity start? Geochronological and geochemical evidence from ultramafic–mafic dikes in southwestern China. Int. Geol. Rev. 46, 226–234.

Gurevitch, E.L., Heunemann, C., Rad'ko, V., Westphal, M., Bachtadse, V., Pozzi, J.P., Feinberg, H., 2004. Palaeomagnetism and magnetostratigraphy of the Permian–Triassic northwest central Siberian Trap Basalts. Tectonophysics 379, 211–226.

Hallam, A., 1999. Discussion on oceanic plateau formation – a cause of mass extinction and black shale deposition around the Cenomanian–Turonian boundary. J. Geol. Soc. Lond. 156, 208.

Hallam, A., Wignall, P.B., 1997. Mass Extinctions and Their Aftermath. Oxford University Press, Oxford, 320 pp.

Hallam, A., Wignall, P.B., 1999. Mass extinctions and sea-level changes. Earth-Sci. Rev. 48, 217–250.

Handoh, I.C., Lenton, T.M., 2003. Periodic mid-Cretaceous oceanic anoxic events linked by oscillations of the phosphorus and oxygen biogeochemical cycles. Global Biogeochem. Cycles 17, 1092, doi 10.1029/2003GB002039.

Haupt, B.J., Seidov, D., 2001. Warm deep-water ocean conveyor during Cretaceous time. Geology 29, 295-298

He, B., Xu, Y.G., Chung, S.L., Xiao, L., Wang, Y., 2003. Sedimentary evidence for a rapid, kilometer-scale crustal doming prior to the eruption of the Emeishan flood basalts. Earth Planet. Sci. Lett. 213, 391–405.

Herrmann, A.D., Patzkowsky, M.E., Pollard, D., 2003. Obliquity forcing with 8–12 times preindustrial levels of atmospheric $pCO_2$ during the Late Ordovician glaciation. Geology 31, 485–488.

Holser, W.T., Magiritz, M., 1987. Events near the Permian–Triassic boundary. Mod. Geol. 11, 155–180.

Horne, D.J., 1999. Ocean circulation modes of the Phanerozoic: implications for the antiquity of deep-sea benthonic invertebrates. Crustaceana 72, 999–1018.

Höskuldsson, A., Sparks, R.S.J., 1997. Thermodynamics and fluid dynamics of effusive subglacial eruptions. Bull. Volcanol. 59, 219–230.

Hotinski, R.M., Bice, K.L., Kump L.R., Bice, K.L., Kump, L.R., Najjar, R.G., Arthur, M.A., 2001. Ocean stagnation and end-Permian anoxia. Geology 29, 7–10.

Hotinski, R.M., Kump. L.R., Bice, K.L. 2002. Comment on Could the Late Permian deep ocean have been anoxic? Paleoceanography 17, 1052, doi 10.1029/2001PA000680.

Huang, S.J., Zhou, S., 1997. Carbon and strontium isotopes of late Palaeozoic marine carbonates in the upper Yangtze Platform, Southwest China. Acta Geol. Sin. (English Edition) 71, 282–292.

Isozaki Y., 1997. Permo–Triassic superanoxia and stratified superocean: records from lost deep sea. Science 276, 235–238.

Isozaki, Y., 2002. End-Permian double-phased mass extinction and killer volcanism. Geol. Soc. Am., Abstr. Pr. 34(6), 360.

Isozaki, Y., Ota, A., 2001. Middle–Upper Permian (Maokouan–Wuchiapingian) boundary in mid-oceanic paleo-atoll limestone of Kamura and Akasaka, Japan. Proc. Japan Acad. Sci. B77, 104–109.

Isozaki, Y., Ota, A., Matsuda T., Sakai H., Ji, Z.S., Shimizu, N., Kobayashi, N., Kawahata, H., Nishi, H., Takano, M., Kubo, T., 2004. Stratigraphy of the Middle–Upper Permian and Lowermost Triassic at Chaotian, Sichuan, China. Record of Late Permian double mass extinction event. Proc. Japan Acad. B80, 10–16.

Izokh, A.E., Polyakov, G.V., Hoa, T.T., Balykin, P.A., Phuong, N.T., 2005. Permian-Triassic ultramafic-mafic magmatism of Northern Vietnam and Southern China as expression of plume magmatism. Russ. Geol. Geophys. 46, 942–951.

Jahren, A.H., 2002. The biogeochemical consequences of the mid-Cretaceous superplume. J. Geodyn. 34, 177–191.

Jenkins, R.B., Landenberger, B., Collins, W.J., 2002. Late Palaeozoic retreating and advancing subduction boundary in the New England Fold Belt, New South Wales. Austr. J. Earth Sci. 49, 467–489.

Jerram, D.A., Widdowson, M., 2005. The anatomy of continental flood basalt provinces: geological constraints on the processes and products of flood volcanism. Lithos 79, 385–405.

Jia, D.C., Hu, R.Z., Lu, Y., Qiu, X.L., 2004. Collision belt between the Khanka block and the North China block in the Yanbian region, northeast China. J. Asian Earth Sci. 23, 211–219.

Jin, Y.G., Wardlaw, B.R., Glenister, B.F., Kotlyar, G.V., 1997. Permian chronostratigraphic subdivisions. Episodes 20, 10–15.

Johnson, H.P., Pruis, M.J., 2003. Fluxes of fluid and heat from the oceanic crustal reservoir. Earth Planet. Sci. Lett. 216, 565–574.

Jones, C.E., Jenkyns, H.C., 2001. Seawater strontium isotopes, oceanic anoxic events, and seafloor hydrothermal activity in the Jurassic and Cretaceous. Am. J. Sci. 301, 112–149.

Kadko, D., Baross, J., Alt, J., 1995. The magnitude and global implications of hydrothermal flux. In: Humphris, S.E., Zierenberg, R.A., Mullineaux, L.S., Thomson, R.E. (Eds), Seafloor Hydrothermal Systems: Physical, Chemical, Biological, and Geological Interactions. Am. Geophys. Union, Geophys. Monogr., Vol. 91, pp. 446–466.

Kaiho, K., Chen, Z.Q., Ohashi, T., Arinobu, T., Sawada, K., Cramer, B.S., 2005. A negative carbon isotope anomaly associated with the earliest Lopingian (Late Permian) mass extinction. Palaeogeogr. Palaeoclimatol. Palaeoecol. 225, 172–180.

Kakuwa, Y., 1996. Permian–Triassic mass extinction event recorded in bedded chert sequence in southwest Japan. Palaeogeogr. Palaeoclimatol. Palaeoecol. 121, 35–51.

Kamo, S.L., Czamanske, G.K., Amelin, Y., Fedorenko, V.A., Davis, D.W., Trofimov, V.R., 2003. Rapid eruption of Siberian flood-volcanic rocks and evidence for coincidence with the Permian–Triassic boundary and mass extinction at 251 Ma. Earth Planet. Sci. Lett. 214, 75–91.

Kato, Y., Nakao, K., Isozaki, Y., 2002. Geochemistry of Late Permian to Early Triassic pelagic cherts from southwest Japan: implications for an oceanic redox change. Chem. Geol. 182, 15–34.

Keller, G., 2003. Biotic effects of impacts and volcanism. Earth Planet. Sci. Lett. 215, 249–264.

Kennett J.P., Cannariato, K.G., Hendy, I.L., Behl, R.J., 2003. Methane Hydrates in Quaternary Climate Change: The Clathrate Gun Hypothesis. Am. Geophys. Union, Spec. Publ. Vol. 54, pp. 1–216.

Kerr, A.C., 1998. Ocean plateau formation: a cause of mass extinction and black shale deposition around the Cenomanian–Turonian boundary? J. Geol. Soc. Lond. 155, 619–626.

Kidder, D.L., Worsley, T.R., 2004a. Biologic iron limitation in the Late Permian and Early Triassic ocean. Geol. Soc. Am., Abstr. Pr. 36(5), 335.

Kidder, D.L., Worsley, T.R., 2004b. Causes and consequences of extreme Permo–Triassic warming to globally equable climate and relation to the Permo–Triassic extinction and recovery. Palaeogeogr. Palaeoclimatol. Palaeoecol. 203, 207–237.

Kiehl, J.T., Shields, C.A., 2005: Climate simulation of the latest Permian: Implications for mass extinction. Geology 33, 757–760.

Kiessling, W., Flügel, E., Golonka, J., 1999. Paleoreef maps: evaluation of a comprehensive database on Phanerozoic reefs. Bull. Am. Assoc. Petrol. Geol. 83, 1552–1587.

Kilburn, C.R.J., 2004. Fracturing as a quantitative indicator of lava flow dynamics. J. Volcanol. Geotherm. Res. 132, 209–224.

Knorr, G., Lohmann, G., 2003. Southern Ocean origin for the resumption of Atlantic thermohaline circulation during deglaciation. Nature 424, 532–536.

Korte, C., Kozur, H.W., Bruckschen, P., Veizer, J., 2003. Strontium isotope evolution of Late Permian and Triassic seawater. Geochim. Cosmochim. Acta 67, 47–62.

Korte, C., Kozur, H.W., Joachimski, M.M., Strauss, H., Veizer, J., Schwark, L., 2004. Carbon, sulfur, oxygen and strontium isotope records, organic geochemistry and biostratigraphy across the Permian/Triassic boundary in Abadeh, Iran. Int. J. Earth Sci. 93, 565–581.

Kozur, H.W., 1998. Some aspects of the Permian–Triassic boundary (PTB) and of the possible causes for the biotic crisis around this boundary. Palaeogeogr. Palaeoclimatol. Palaeoecol. 143, 227–272.

Krhovský, J., Cejchan, P., 1995. Cascade of causally linked effects of rapid glaciation–deglaciation events: a possible cause of non-selectivity of mass extinctions. Geolines (Praha) 3, 27–37.

Kruk, N.N., Plotnikov, A.V., Vladimirov, A.G., Kutolin, V.A., 1999. Geochemistry and geodynamic conditions of the trap rock formation in the Kuznetsk Basin. Dokl. Earth Sci. 369A, 1387–1390.

Krull, E.S., Lehrmann, D.J., Druke, D., Kessel, B., Yu, Y.Y., Li, R.X., 2004. Stable carbon isotope stratigraphy across the Permian–Triassic boundary in shallow marine carbonate platforms, Nanpanjiang Basin, south China. Palaeogeogr. Palaeoclimatol. Palaeoecol. 204, 297–315.

Krull, E.S., Retallack, G.J., Campbell, H.J., Lyon, G.L., 2000. $\delta 13C_{org}$ chemostratigraphy of the Permian–Triassic boundary in the Maitai Group, New Zealand: evidence for high-latitudinal methane release. New Zealand J. Geol. Geophys. 43, 21–32.

Kukkonen, I.T., Safanda, J., 2001. Numerical modelling of permafrost in bedrock in northern Fennoscandia during the Holocene. Global Planet. Change 29, 259–273.

Kump, L.R, Brantley, S.L., Arthur, M.A., 2000. Chemical, weathering, atmospheric $CO_2$, and climate. Annu. Rev. Earth Planet. Sci. 28, 611–667.

Kutzbach, J.E., Guetter, P.J., Washington, W.M., 1990. Simulated circulation of an idealized ocean for Pangean time. Paleoceanography 5, 299–317.

Kvenvolden, K.A., 1998. A primer on the geological occurrence of gas hydrate. In: Henriet, J.P., Mienert, J. (Eds), Gas Hydrates: Relevance to World Margin Stability and Climate Change. J. Geol. Soc. Lond., Spec. Publ. 137, p. 9–30.

Leckie, R.M., Bralower, T.J., Cashman, R., 2002. Oceanic anoxic events and plankton evolution: Biotic response to tectonic forcing during the mid-Cretaceous. Paleoceanography 17, 1041, doi 10.1029/2001PA000623.

Lo, C.H., Chung, S.L., Lee, T.Y., Wu, G., 2002. Age of the Emeishan flood magmatism and relations to Permian–Triassic boundary events. Earth Planet. Sci. Lett. 198, 449–458.

Lozovsky, V.R., 1998. The Permian–Triassic boundary in the continental series of Eurasia. Palaeogeogr. Palaeoclimatol. Palaeoecol. 143, 273–283.

Lyons, J.J., Coe, R.S., Zhao, X., Renne, P.R., Kazansky, A.Y., Izokh, A.E., Kungurtsev, L.V., Mitrokhin, D.V., 2002. Paleomagnetism of the early Triassic Semeitau igneous series, eastern Kazakstan. J. Geophys. Res. 107 (B7), doi 10.1029/2001JB000521.

Malkowski, K., Gruszczynski, M., Hoffman, A., Hałas, S., 1989. Oceanic stable isotope composition and a scenario for the Permo–Triassic crisis. Histol. Biol. 2, 289–309.

McGhee, G.R., Sheehan, P.H., Bottjer, D.J., Droser, M.L., 2004. Ecological ranking of Phanerozoic biodiversity crises: ecological and taxonomic severities are decoupled. Palaeogeogr. Palaeoclimatol. Palaeoecol. 211, 289–297.

Mei, S.H., Henderson, C.M., Wardlaw, B.R., 2002. Evolution and distribution of the conodonts Sweetognathus and Iranognathus and related genera during the Permian, and their implications for climate change. Palaeogeogr. Palaeoclimatol. Palaeoecol. 180, 57–91.

Mertmann, D., 2003. Evolution of the marine Permian carbonate platform in the Salt Range (Pakistan), Palaeogeogr. Palaeoclimatol. Palaeoecol. 191, 373–384.

Metcalfe, I., Nicoll, R.S., Mundil, R., Foster, C., Glen, J., Lyons, J., Wang, X.F., Wang, C.Y., Renne, P.R., Black, L., Xun, Q., Mao, X.D., 2001. The Permian–Triassic boundary mass extinction in China. Episodes 24, 239–244.

Michaelsen, P., 2002. Mass extinction of peat-forming plants and the effect on fluvial styles across the Permian–Triassic boundary, northern Bowen Basin, Australia. Palaeogeogr. Palaeoclimatol. Palaeoecol. 179, 173–188.

Michaelsen, P., Henderson, R.A., 2000. Sandstone petrofacies expressions of multiphase basinal tectonics and arc magmatism: Permian–Triassic north Bowen Basin, Australia. Sediment. Geol. 136, 113–136.

Michaelsen, P., Henderson, R.A., Crosdale, P.J., Fanning, C.M., 2001. Age and significance of the Platypus Tuff Bed, a regional reference horizon in the Upper Permian Moranbah Coal Measures, north Bowen Basin. Austr. J. Earth Sci. 48, 183–192.

Milkov, A.V., 2004. Global estimates of hydrate-bound gas in marine sediments: how much is really out there? Earth-Sci. Rev. 66, 183–197.

Mogutcheva, N.K., Betekhtina, O.A., 1998. Urgent problems of the stratigraphy of the continental Siberian Triassic. Geol. Geofiz. 39, 287–296.

Mottl, M.J., Wheat, G., Baker, E., Becker, N., Davis, E., Feely, R., Grehan, A., Kadko, D., Lilley, M., Massoth, G., Moyer, C., Sansone F., 1998. Warm springs discovered on 3.5 Ma oceanic crust, eastern flank of the Juan de Fuca Ridge. Geology 26, 51–54.

Mundil, R., Ludwig, K.R., Metcalfe, I., Renne, P.R., 2004. Age and timing of the Permian mass extinctions: U/Pb dating of closed-system zircons. Science 305, 1760–1763.

Murchey, B.L., Jones, D.L., 1992. A Mid-Permian Chert Event: widespread deposition of biogenic siliceous sediments in coastal, island arc and oceanic basins. Palaeogeogr. Palaeoclimatol. Palaeoecol. 96, 161–174.

Naugolnykh, S.V., 2004. Permian and Triassic paleosols (in Russian). In: Semikhatov, M.A., Chumakov, P.M. (Eds), Klimat v Epokhi Krupnykh Biosfernykh Peresroyek. Geol. Inst. RAS Trans. No. 550, pp. 221–229.

Naumov, V.A., Luzina, I.V., 1997. Palynoassemblages from deposits of the crater lakes of the Tunguska paleovolcanic area. Stratigr. Geol. Correlation 5, 319–333.

Nederbragt, A.J., Thurow, J., Vonhof, H., Brumsack, H.J., 2004. Modelling oceanic carbon and phosphorus fluxes: implications for the cause of the late Cenomanian Oceanic Anoxic Event (OAE 2). J. Geol. Soc. Lond. 161, 721–728.

Newell, A.J., Tverdokhlebov, V.P., Benton, M.J., 1999. Interplay of tectonics and climate on a transverse fluvial system, Upper Permian, Southern Uralian Foreland Basin, Russia. Sediment. Geol. 127, 11–29.

Newkirk, T.T., Lehrmann, D., Hudak, G., 2002. Tephrastratigraphy and analysis of tectonic setting of Triassic intermediate volcanic strata: Nanpanjiang basin, South China. Geol. Soc. Am., Abstr. Pr. 34(6), 512.

Newton, R.J., Pevitt, E.L., Wignall, P.B., Bottrell, S.H., 2004. Large shifts in the isotopic composition of seawater sulphate across the Permo–Triassic boundary in northern Italy. Earth Planet. Sci. Lett. 218, 331–345.

Nikishin, A.M., Ziegler, P.A., Abbott, D., Brunet, M.F., Cloetingh, S., 2002. Permo–Triassic intraplate magmatism and rifting in Eurasia: implications for mantle plumes and mantle dynamics. Tectonophysics 351, 3–39.

Noble, P., Renne, P., 1990. Paleoenvironmental and biostratigraphic sgnificance of siliceous microfossils of the Permo–Triassic Redding Section, Eastern Klamath Mountains, California. Marine Micropaleontol. 15, 379–391.

Okay, A.I., Göncüoğlu, M.C., 2004. The Karakaya Complex: a review of data and concepts. Turkish J. Earth Sci. 13, 75–95.

Palfý, J., 2003. Volcanism of the Central Atlantic Magmatic Province as a potential driving force in the end-Triassic mass extinction. In: Hames, W., Mchone, J.G., Renne, P., Ruppel, C. (Eds), The Central Atlantic Magmatic Province: Insights from Fragments of Pangea, Am. Geophys. Union, Geophys. Monogr., Vol. 136, pp. 255–267.

Payne, J.L., Lehrmann, Wei, J.Y., Orchard, M.J., Schrag, D.P., Knoll, A.H., 2004. Large perturbations of the carbon cycle during recovery from the end-Permian extinction. Science 305, 506–509.

Peng, Y., Tong, J., Shi, G. R., Hansen, H. J., 2001. The Permian–Triassic boundary stratigraphic set: characteristics and correlation. Newslett. Strat. 39, 55–71.

Pickett, E.A., Robertson, A.H.F., 2004. Significance of the volcanogenic Nilüfer Unit and related components of the Triassic Karakaya Complex for Tethyan subduction/accretion processes in NW Turkey. Turkish J. Earth Sci. 13, 97–143.

Piotrowski, A.M., Goldstein, S.L., Hemming, S.R., Fairbanks, R.G., 2004. Intensification and variability of ocean thermohaline circulation through the last deglaciation. Earth Planet. Sci. Lett. 225, 205–220.

Price, G.D., 1999. The evidence and implications of polar ice during the Mesozoic. Earth-Sci. Rev. 48, 183–210.

Racki, G., 1999. Silica-secreting biota and mass extinctions: survival patterns and processes. Palaeogeogr. Palaeoclimatol. Palaeoecol.154, 107–132.

Racki, G., 2003. End-Permian mass extinction: oceanographic consequences of double catastrophic volcanism. Lethaia 36, 171–173.

Racki, G., Cordey, F., 2000. Radiolarian palaeoecology and radiolarites: is the present the key to the past? Earth-Sci. Rev. 52, 83–120.

Rees, P.M., 2002. Land-plant diversity and the end-Permian mass extinction. Geology 30, 827–830.

Reichow, M.K., Saunders, A.D., White, R.V., Pringle, M.S., Al'Mukhamedov, A.I., Medvedev, A.I., Kirda, N.P., 2002. $^{40}Ar/^{39}Ar$ dates from the West Siberian Basin: Siberian flood basalt province doubled. Science 296, 1846–1849.

Resing, J.A., Sansone, F.J., 2002. The chemistry of lava–seawater interactions II: the elemental signature. Geochim. Cosmochim. Acta 66, 1925–1941.

Retallack, G.J., Krull, E.S., 1999. Landscape ecological shift at the Permian–Triassic boundary in Antarctica. Austr. J. Earth Sci. 46, 785–812.

Retallack, G.J., Smith, R.M.H., Ward, P.D., 2003. Vertebrate extinction across Permian–Triassic boundary in Karoo Basin, South Africa. Geol. Soc. Am. Bull. 115, 1133–1152.

Robock, A., 2000. Volcanic eruptions and climate. Rev. Geophys. 38, 191–219.

Ryskin, G., 2003. Methane-driven oceanic eruptions and mass extinctions. Geology 31, 741–744.

Sansone, F.J., Benitez-Nelson, C.R., Resing, J.A., DeCarlo, E.H., Vink, S.M., Heath, J.A., Huebert B.J., 2002. Geochemistry of atmospheric aerosols generated from lava–seawater interactions. Geophys. Res. Lett. 29, 1335, doi 10.1029/2001GL013882.

Sarkar, A., Yoshioka, H., Ebihara, M., Naraoka, H., 2003. Geochemical and organic carbon isotope studies across the continental Permo–Triassic boundary of Raniganj Basin, eastern India. Palaeogeogr. Palaeoclimatol. Palaeoecol. 191, 1–14.

Sashida, K., Igo, H., Adachi, S., Ueno, K., Kajiwara, Y., Nakornsri, N., Sardsud, A., 2000. Late Permian to Middle Triassic radiolarian faunas from northern Thailand. J. Paleontol. 74, 789–811.

Sashida, K., Salyapongse, S., 2002. Permian radiolarian faunas from Thailand and their paleogeographic significance. J. Asian Earth Sci. 20, 691–701.

Saunders, A.D., England, R.W., Reichow, M.K., White, R,V., 2005. A mantle plume origin for the Siberian traps: uplift and extension in the West Siberian Basin, Russia. Lithos 79, 407–424.

Saunders, A.D., Tarney, J., Kerr, A.C., Kent, R.W., 1996. The formation and fate of large oceanic igneous provinces. Lithos 37, 81–95.

Schlesinger, W.H., 1997. Biogeochemistry: An Analysis of Global Change. Academic Press, San Diego, 588pp.

Schmincke, H.U., 2004. Volcanism. Springer-Verlag, Berlin. 324 pp.

Scott, J.R., Marotzke, J., Adcroft, A., 2001. Geothermal heating and its influence on the meridional overturning circulation. J. Geophys. Res. 106(C12), 31,141–31,154.

Seidov, D., Barron, E., Haupt B.J., 2001. Meltwater and the global ocean conveyor: northern versus southern connections. Global Planet. Change 30, 257–270.

Seidov, D., Haupt B.J., 2003. On sensitivity of ocean circulation to sea surface salinity. Global Planet. Change 36, 99–116.

Self, S., Keszthelyi, L., Thordarson T., 1998. The importance of pahoehoe. Annu. Rev. Earth Planet. Sci. 26, 81–110.

Self, S., Thordarson T., Keszthelyi, L., 2000. Eruption dynamics in flood basalt volcanism. Geol. Soc. Am., Abstr. Pr. 32(7), 396.

Sharapov, V.N., Cherepanov, A.N., Popov, V.N., 2003. Hydrodynamic parameters of volcanic trap formation at the Siberian Platform. Dokl. Earth Sci. 388, 41–44.

Sheldon, N.D., Retallack, G.J., 2002. Low oxygen levels in earliest Triassic soils. Geology 30, 919–922.

Shen, S.H., Shi, G.R., 2002. Paleobiogeographical extinction patterns of Permian brachiopods in the Asian—western Pacific region. Paleobiology 28, 449–463.

Shipilov, E.V., 2003. Interference of tectonic–geodynamic regimes in Permian–Triassic evolution of the Arctic periphery of North Eurasia. Dokl. Earth Sci. 393, 1215–1220.

Siemann, M.G., 2003. Extensive and rapid changes in seawater chemistry during the Phanerozoic: evidence from Br contents in basal halite. Terra Nova 15, 243–248.

Sinninghe Damsté, J.S., Köster, J., 1998. A euxinic southern North Atlantic Ocean during the Cenomanian/Turonian oceanic anoxic event. Earth Planet. Sci. Lett. 158, 165–173.

Sinton, C.W., Duncan, R.A., 1997. Potential links between ocean plateau volcanism and global ocean anoxia at the Cenomanian–Turonian boundary. Econ. Geol. 92, 836–842.

Smellie, J., 2000. Terrestrial subglacial eruptions – a personal perspective. Comm. Volcanogenic Sediments Newslett. 19, 2–3.

Song, X.Y., Zhou, M.F., Cao, Z.M., Robinson, P.T., 2004. Late Permian rifting of the South China Craton caused by the Emeishan mantle plume? J. Geol. Soc. Lond. 161, 773–781.

Speer, K.G., 1997. Could buoyant plumes penetrate across the ocean's thermocline? Philos. Trans. Roy. Soc. Lond. A355, 443–457.

Sperling, E.A., 2004. A possible Permian–Triassic boundary section at Quinn River Crossing, NW Nevada. Geol. Soc. Am., Abstr. Pr. 36(4), 19.

Stanley, G.R., 2003. The evolution of modern corals and their early history. Earth Sci. Rev. 60, 195–225.

Stanley, S.M., Hardie, L.A., 1999. Hypercalcification: paleontology links plate tectonics and geochemistry to sedimentology. GSA Today 9, 1–6.

Stanley, S.M., Yang, X., 1994. A double mass extinction at the end of the Paleozoic era. Science 266, 1340–1344.

Staudigel, H., Chastain, R.A., Yayanos, A., Bourcier, W., 1995. Biologically mediated dissolution of glass. Chem. Geol. 126, 147–154.

Stefánsson, A., Gíslason, S.R., 2001. Chemical weathering of basalts, southwest Iceland: effect of rock crystallinity and secondary minerals on chemical fluxes to the ocean. Am. J. Sci. 301, 513–556.

Stewart, A.L., 2001. Facies in an Upper Permian volcanic succession, Emmaville Volcanics, Deepwater, north-eastern New South Wales. Austr. J. Earth Sci. 48, 929–942.

Stothers, R.B., 1999. Volcanic dry fogs, climate cooling, and plague pandemics in Europe and the Middle East. Clim. Change 42, 713–723.

Stouffer, R.J., Manabe, S., 2003. Equilibrium response of thermohaline circulation to large changes in atmospheric $CO_2$ concentration. Clim. Dynam. 20, 759–773.

Svensen, H., Planke, S., Malthe-Sorenssen, A., Jamtveit, B., Myklebust, R., Eidem, T.R., Rey, S.S., 2004. Release of methane from a volcanic basin as a mechanism for initial Eocene global warming. Nature 429, 542–545.

Tanton, L.T.E., Hager, B.H., 2000. Melt intrusion as a trigger for lithospheric foundering and the eruption of the Siberian flood basalts. Geophys. Res. Lett. 27, 3937–3940.

Tardy, M., Lapierre, H., Bosch, D., Cadoux, A., Narros, A., Struik, L.C., Brunet, P., 2003. Le terrane de Slide Mountain (Cordillères canadiennes): une lithosphère océanique marquée par des points chauds. Can. J. Earth Sci. 40, 833–852.

Thierstein, H.R., Berger, W.H., 1978. Injection events in ocean history. Nature 276, 461–466.

Thompson, G.M., Ali, J.R., Song, X., Jolley, D.W., 2001. Emeishan basalts, SW China: reappraisal of the formation's type area stratigraphy and a discussion of its significance as a large igneous province. J. Geol. Soc. Lond. 158, 593–599.

Thordarson, T., Miller, D.J., Larsen, G., Self, S., Sigurdsson, H., 2001. New estimates of sulfur degassing and atmospheric mass-loading by the 934 A.D. Eldgjá eruption, Iceland. J. Volcanol. Geotherm. Res. 108, 33–54.

Thordarson, T., Self, S., 1998. The Roza member, Columbia river basalt group: a gigantic pahoehoe lava flow field formed by endogenous processes? J. Geophys. Res. 103(B11), 27,411–27,445.

Turetsky, M.R., Wieder, R.K., Vitt, D.H., 2002. Boreal peatland C fluxes under varying permafrost regimes. Soil Biol. Biochem. 34, 907–912.

Twitchett, R.J., 2001. Incompleteness of the Permian–Triassic fossil record: a consequence of productivity decline? Geol. J. 36, 341–353.

Twitchett, R.J., Krystyn, L., Baud, A., Wheeley, J.R., Richoz, S., 2004. Rapid marine recovery after the end-Permian mass-extinction event in the absence of marine anoxia. Geology 32, 805–808.
Ufnar, D.F., González, L.A., Ludvigson, G.A., Brenner, R.L., Witzke, B.J., 2004. Evidence for increased latent heat transport during the Cretaceous (Albian) greenhouse warming. Geology 32, 1049–1052.
Vaughan, A.P.M., Scarrow, J.H., 2003. Ophiolite obduction pulses as a proxy indicator of superplume events? Earth Planet. Sci. Lett. 213, 407–416.
Veevers, J.J., Conaghan, P.J., Shaw, S.E., 1994. Turning point in Pangean environmental history at the Permian/Triassic (P/Tr) boundary. In: Klein, G.D. (Ed.), Pangea: Paleoclimate, Tectonics and Sedimentation during Accretion, Zenith and Breakup of a Supercontinent. Geol. Soc. Am. Spec. Pap., Vol. 288, pp. 187–196.
Veevers, J.J., Tewari, R.C., 1995. Permian–Carboniferous and Permian–Triassic magmatism in the rift zone bordering the Tethyan margin of southern Pangea. Geology 23, 467–470.
Vermeij, G.J., 1995. Economics, volcanoes, and Phanerozoic revolutions. Paleobiology 21, 125–152.
Vermeij, G.J., 2004. Ecological avalanches and the two kinds of extinction. Evol. Ecol. Res. 6, 315–337.
Vernikovsky, V.A., Pease, V.L., Vernikovskaya, A.E., Romanov, A.P., Gee, D.G., Travin, A.V., 2003. First report of early Triassic A-type granite and syenite intrusions from Taimyr: product of the northern Eurasian superplume? Lithos 66, 23–36.
Villinger, H., Grevemeyer, I., Kaul, N., Hauschild, J., Pfender, M., 2002. Hydrothermal heat flux through aged oceanic crust: where does the heat escape? Earth Planet. Sci. Lett. 202, 159–170.
Vishnevskaya, V.S., Kostyuchenko, A.S., 2000. The evolution of radiolarian biodiversity. Paleontol. J. 34, 124–130.
Visscher, H., Brinkhuis, H., Dilcher, D.L., Elsik, W.C., Eshet, Y., Looy, C.V., Rampino, M.R., Traverse, A., 1996. The terminal Paleozoic fungal event: evidence of terrestrial ecosystem destabilization and collapse. Proc. Natl. Acad. Sci. USA. 93, 2155–2158.
Visscher, H., Looy, C.V., Collinson, M.E., Brinkhuis, H., van Konijnenburg-Van Cittert, J.H., Kurschner, W.M., Sephton, M.A., 2004. Environmental mutagenesis during the end-Permian ecological crisis. Proc. Natl. Acad. Sci. USA 101, 12952–12956.
Vogt, P.R., 1989. Volcanogenic upwelling of anoxic, nutrient-rich water, a possible factor in carbonate-bank/reef demise and benthic faunal extinctions? Geol. Soc. Am. Bull. 101, 1225–1245.
Wang, Z.Q., Chen, A.S., 2001. Traces of arborescent lycopsids and dieback of the forest vegetation in relation to the terminal Permian mass extinction in North China. Rev. Palaeobot. Palynol. 117, 217–243.
Ward, P.D., Montgomery, D.R., Smith, R., 2000. Altered river morphology in South Africa related to the Permian–Triassic extinction. Science 289, 1740–1743.
Weaver, A.J., Bitz, C.M., Fanning, A.F., Holland, M.M., 1999. Thermohaline circulation: high-latitude phenomena and the difference between the Pacific and Atlantic. Annu. Rev. Earth Planet. Sci. 27, 231–285.
Weidlich, O., Kiessling, W., Flügel, E., 2003. Permian–Triassic boundary interval as a model for forcing marine ecosystem collapse by long-term atmospheric oxygen drop. Geology 31, 961–964.
Weissert, H., Erba, E. 2004. Volcanism, $CO_2$ and palaeoclimate: a Late Jurassic–Early Cretaceous carbon and oxygen isotope record. J. Geol. Soc. Lond. 161, 695–702.
Wescott, W.A. Diggens, J.N., 1998. Depositional history and stratigraphical evolution of the Sakamena Group (Middle Karoo Supergroup) in the southern Morondava Basin, Madagascar. J. African Earth Sci. 27, 461–479.
White, R.V., 2002. Earth's biggest 'whodunnit': unravelling the clues in the case of the end-Permian mass extinction. Philos. Trans. Roy. Soc. Lond. A360, 2963–2985.
White, R.V., Saunders, A.D., 2005. Volcanism, impact and mass extinctions: incredible or credible coincidences? Lithos 79, 299–316.
Wignall, P.B., 2001 Large igneous provinces and mass extinctions. Earth Sci. Rev. 53, 1–33.
Wignall, P.B., Hallam, T., 1992. Anoxia as a cause of the Permian–Triassic mass extinction: facies evidence from northern Italy and the western United States. Palaeogeogr., Palaeoclimatol., Palaeoecol. 93, 21–46.
Wignall, P.B., Morante, R., Newton, R., 1998. The Permo–Triassic transition in Spitsbergen: $^{13}C_{org}$ chemostratigraphy, Fe and S geochemistry, facies, fauna and trace fossils. Geol. Mag. 135, 47–62.
Wignall, P.B., Newton, R., 2003. Contrasting deep-water records from the Upper Permian and Lower Triassic of south Tibet and British Columbia: evidence for a diachronous mass extinction. Palaios 18, 153–167.
Wignall, P.B., Twitchett, R.J., 1996. Oceanic anoxia and the end-Permian mass extinction. Science 272, 1155–1158.
Wignall, P.B., Twitchett, R.J., 2002a. Extent, duration, and nature of the Permian–Triassic superanoxic event. In: Koeberl, C., MacLeod, K.G. (Eds), Catastrophic Events and Mass Extinction: Impacts and Beyond. Geol. Soc. Am. Spec. Pap., Vol. 356, pp. 395–413.

Wignall, P.B., Twitchett, R.J., 2002b. Permian–Triassic sedimentology of Jameson Land, East Greenland: incised submarine channels in an anoxic basin. J. Geol. Soc. Lond. 159, 691–703.
Wilson, P.A., Norris, R.D., 2001. Warm tropical ocean surface and global anoxia during the mid-Cretaceous period. Nature 412, 425–429.
Winguth, A. M. E., Heinze, C., Kutzbach, J. E., Maier-Reimer, E., Mikolajewicz, U., Rowley, D., Rees, A., Ziegler, A.M., 2002. Simulated warm polar currents during the middle Permian. Paleoceanography 17, 1057, doi 10.1029/2001PA000646.
Wissler, L., Funk, H.P., Weissert, H., 2003. Response of Early Cretaceous carbonate platforms to changes in atmospheric carbon dioxide levels. Palaeogeogr., Palaeoclimatol., Palaeoecol. 200, 187–205.
Xiao, L., Xu, Y.G., Chung, S.L., He, B., Mei, H.J., 2003. Chemostratigraphic correlation of Upper Permian lavas from Yunnan Province, China: extent of the Emeishan large igneous province. Int. Geol. Rev. 45, 753–766.
Yakubchuk, A., Nikishin, A., 2004. Noril'sk–Talnakh Cu–Ni–PGE deposits: a revised tectonic model. Miner. Deposita 39, 125–142.
Yao, A., Kuwahara, K., 2000. Permian and Triassic radiolarians from the southern Guizhou Province, China. J. Geosci., Osaka City University 43, 1–19.
Yaroshenko, O.P., Lozovsky, V.R., 2004. Palynological assemblages of continental Lower Triassic in Eastern Europe and their interregional correlation. Paper 1: Palynological assemblages of the Induan Stage. Stratigr. Geol. Correlation 12, 275–285.
Zhang, R., Follows, M., Marshall J., 2002. Mechanisms of thermohaline mode switching with application to warm equable climates. J. Clim. 15, 2056–2072.
Zharkov, M.A., 2004. Paleogeographic rearrangements and depositional environments in the Permian and Early Triassic (in Russian). In: Semikhatov, M.A., Chumakov, P.M. (Eds), Klimat v Epokhi Krupnykh Biosfernykh Peresroyek. Geol. Inst. RAS Trans. No. 550, pp. 158–180.
Zharkov, M.A., Chumakov, N.M., 2001. Paleogeography and sedimentation settings during Permian-Triassic reorganizations in biosphere. Strat. Geol. Correlation 9, 340–363.
Zhou, M.F., Malpas, J., Song, X.Y., Robinson, P.T., Sun, M., Kennedy, A.K., Lesher, C.M., Keays, R.R., 2002a. A temporal link between the Emeishan large igneous province (SW China) and the end-Guadalupian mass extinction. Earth Planet. Sci. Lett. 196, 113–122.
Zhou, M.F., Yang, Z.X., Song, X.Y., Lesher, C.M., Keays, R.R., 2002b. Magmatic Ni–Cu–(PGE) sulfide deposits in China. In: Cabri, L.J. (Ed.), The Geology, Geochemistry, Mineralogy, Mineral Beneficiation of the Platinum-Group Elements. Canadian Institute of Mining, Metallurgy and Petroleum, Special Volume 54, pp. 619–636.
Zhou, Y.P., Bohor, B.F., Ren, Y.L., 2000. Trace element geochemistry of altered volcanic ash layers (tonsteins) in Late Permian coal-bearing formations of eastern Yunnan and western Guizhou Provinces, China. Int. J. Coal Geol. 44, 305–324.
Zhu, D., Luo, T.Y., Gao, Z.M., Zhu, C.M., 2003. Differentiation of the Emeishan flood basalts at the base and throughout the crust of southwest China. Int. Geol. Rev. 45, 471–477.
Zielinski, G.A., 2002. Climatic impact of volcanic eruptions. The Sci. World 2, 869–884.
Zolotukhin, V.V., Podgornykh, N.M., 1998. On possible overheating of the Permo–Triassic trap melts of the Siberian platform. Dokl. Earth Sci. 363, 226–229.

Understanding Late Devonian and Permian-Triassic Biotic and Climatic Events:
Towards an Integrated Approach
edited by D.J. Over, J.R. Morrow and P.B. Wignall



# Chapter 11

# Fossil preservation during the aftermath of the end-Permian mass extinction: taphonomic processes and palaeoecological signals

*Margaret L. Fraiser* *, *David J. Bottjer*

*Department of Earth Sciences, University of Southern California, Los Angeles, CA 90089-0740, USA*

**Abstract**

A high number of Lazarus taxa and an alleged rarity of silicified faunas during the Early Triassic have been interpreted previously as indications that the fossil record following the end-Permian mass extinction is poor and unreliable for palaeontologic studies. This preliminary examination of Early Triassic fossil preservation reveals that silicified faunas are actually moderately common in Lower Triassic strata. Furthermore, an evaluation of environmental and ecological characteristics of the aftermath of the end-Permian mass extinction indicates that the numerous Early Triassic molds and recrystallized calcareous fossils are at least suitable for palaeoecologic studies. Within this context, future tests with the aim of continuing to examine the potential Early Triassic preservation bias are proposed.

*Keywords*: Early Triassic; end-Permian mass extinction; silicification; taphonomy; Lazarus taxa

## 1. Introduction: the potential Early Triassic preservation bias

The end-Permian mass extinction played a pivotal role in shaping the broadscale pattern of taxonomic diversity through the later Phanerozoic. Two mass extinctions near and at the end of the Permian eliminated 71% and 80% of marine invertebrate species, comprising the largest decreases in Phanerozoic marine biodiversity (Stanley and Yang, 1994). The end-Permian mass extinction triggered the abrupt phyletic switch from the rhynchonelliform brachiopod-dominated Palaeozoic Fauna to the bivalve- and gastropod-dominated Modern Fauna (Gould and Calloway, 1980; Sepkoski, 1981).

Mounting research indicates that the aftermath of the end-Permian mass extinction was as crucial as the mass extinction itself in shaping the evolutionary history of life on Earth. Sedimentologic (Woods et al., 1999; Pruss et al., 2004) and isotopic (e.g. Marenco et al., 2003; Payne et al., 2004) evidence indicates that the physiologically stressful environmental conditions likely related to the cause of the end-Permian mass extinction persisted through the Early Triassic, for possibly 5–6 million years (Martin et al., 2001; Mundil et al., 2004). Because of these global environmental perturbations, biotic recovery was suppressed in Early Triassic oceans: global taxonomic diversity remained very low (Ciriacks, 1963; Schubert and Bottjer, 1995; Erwin and Pan, 1996) and opportunistic

*Corresponding author. Fax: +1-414-229-5452. *E-mail address:* mfraiser@uwm.edu (M.L. Fraiser).

organisms proliferated (Rodland and Bottjer, 2001; Fraiser and Bottjer, 2004). In addition to the phyletic switch from the Palaeozoic Fauna to the Modern Fauna, the prolonged deleterious environmental conditions during the Early Triassic facilitated a host of temporary and permanent structural changes in benthic level-bottom marine palaeocommunities (Fraiser and Bottjer, 2005).

Thus, the vast majority of studies focusing on the end-Permian mass extinction and its Early Triassic aftermath indicate that, during this geologically brief interval in Earth's history, broadscale taxonomic and palaeoecologic patterns were drastically and permanently altered. However, a variety of literature on the mode of preservation of Early Triassic skeletonized invertebrate marine fossils and on Early Triassic taxonomic patterns questions the quality of the Lower Triassic fossil record and, therefore, the notion that the Early Triassic was as significant to the course of evolution as previously determined.

The Lower Triassic fossil record has been described as a "fossilization low" (Twitchett, 2001) because a large portion of Early Triassic taxa are Lazarus taxa (Batten, 1973; Erwin and Pan, 1996; Twitchett, 2001), taxa that temporarily disappear from the fossil record but reappear later unchanged (Flessa and Jablonski, 1983). Aftermaths of mass extinctions are typified by high numbers of Lazarus taxa from several phyla; these taxon outages may be the result of migration of taxa to refugia, population sizes below the level of detection, lack of sampling, or taphonomic bias (Jablonski, 1986a, b; Erwin and Pan, 1996; Kauffman and Harries, 1996; Wignall and Benton, 1999; Twitchett, 2001). Nevertheless, the more Lazarus taxa present in a particular interval, the less complete the fossil record is considered to be (e.g. Twitchett, 2001).

The high number of Early Triassic Lazarus taxa recently has been attributed to a taphonomic bias rather than any ecological cause because of an alleged dearth of silicified Early Triassic invertebrate marine faunas. The Early Triassic is also considered to represent a long "chert gap" when very little biogenic chert was produced and/or preserved (Racki, 1999; Beauchamp and Baud, 2002) because of latest Permian global warming and subsequent shutdown of thermohaline circulation that eliminated conditions favorable to the production, accumulation, and preservation of biogenic silica (Beauchamp and Baud, 2002). [However, a Lower Triassic chert has been discovered recently (Hori et al., 2003; Takemura et al., 2003), and the well-entrenched Early Triassic "chert gap" is more likely an Early Triassic "chert eclipse". Furthermore, global Early Triassic patterns of biogenic silica and silica-secreting organisms deserve reexamination (cf. Hori et al., 2003; Takemura et al., 2003).] Erwin (1996) and Erwin and Pan (1996) propose that poor preservation makes many gastropod taxa only appear to be absent during the Early Triassic because several genera are known almost exclusively from silicified material, others are difficult to identify without silicification, and many had robust, easily preserved shells. It was thus postulated that many Early Triassic Lazarus taxa may have actually been present in large numbers but were not preserved because of a lack of extensive early silicification of Early Triassic fossils; the Lazarus taxa represent organisms whose calcareous body fossils were simply dissolved away, leaving no record of their existence (Erwin, 1996; Erwin and Pan, 1996). More recent studies revealing that silicified faunas have a higher fidelity of fossil preservation than non-silicified faunas (Cherns and Wright, 2000; Wright et al., 2003) lend credence to these interpretations about the Early Triassic fossil record. It is contended that a lack of fossil silicification may bias our understanding of recovery events (Schubert et al., 1997) and skew palaeoecological interpretations (Cherns and Wright, 2000; Erwin and Kidder, 2000).

Certainly, a lack of silicified fossils may indicate that a preservation bias threatens previous perceptions of the taxonomic patterns during the aftermath of the end-Permian mass extinction. Taxonomic studies require excellent fossil preservation, such as that provided by early silicification. Increasingly, however, taxonomic approaches, like counting the number of Lazarus taxa, are considered to be one-dimensional and insufficient for elucidating a complete picture of evolution if used alone. Palaeoecological approaches, in which organisms and their interactions with each other and the environment are evaluated, are required to fully understand evolutionary patterns and processes, including aftermaths and biotic recoveries from mass extinctions (e.g. Bottjer, 2001). Excellent fossil preservation such as silicification is not a requirement in palaeoecological studies because interpretations can be made using a wide range of data available from the rock record (e.g. Bottjer, 2001) [although taphonomic analyses are vital to palaeoecological studies to ensure that palaeoecological patterns are not artifacts of preservation processes (e.g. Parsons-Hubbard et al., 1999; Márquez-Aliaga and Ros, 2002)]. Furthermore, a recent study on partially silicified Early Triassic gastropods revealed only one Lazarus genus despite the fauna's presumed increased preservation potential, suggesting other controls on the abundance of Lazarus taxa (Wheeley and Twitchett, 2005). Doubts about the quality of the Lower Triassic fossil record must be addressed with regard to palaeoecologic as well as taxonomic data not only because the end-Permian mass extinction and its aftermath are considered to be a defining interval in evolutionary history, but because taxonomic and ecologic patterns during the aftermaths of mass extinctions can be decoupled and differ in severity, and the most significant result of a mass extinction can be the new ecological patterns that arise during its aftermath (Droser et al., 1997, 2000).

Very few studies have addressed the Early Triassic Lazarus phenomenon (q.v. Wignall and Benton, 1999) or have examined early diagenetic processes that potentially affected Early Triassic faunas. The purpose of this pilot study was to test the hypothesis that the Early Triassic fossil record is not taphonomically biased and that it actually records primary palaeoecologic signals during the aftermath of the end-Permian mass extinction. Until the usefulness of the Lower Triassic fossil record is determined, all taxonomic and palaeoecologic studies of the aftermath of the end-Permian mass extinction will be precluded by the possibility of preservation bias (*sensu* Erwin and Pan, 1996). Presented here is: an assessment of Lower Triassic fossil preservation; a discussion of environmental and ecological effects upon Early Triassic early diagenetic processes; and proposed future tests of the usefulness of the Lower Triassic fossil record for palaeoecologic and taxonomic studies.

## 2. Methods

Several studies have focused on preservation of select Lower Triassic invertebrate marine fossils of particular regions (e.g. Assereto and Rizzini, 1975; Boyd and Newell, 1976; Nice, 1985; Boyd et al., 1999; Moffat and Bottjer, 1999), but the general nature of global Early Triassic fossil preservation has not been synthesized or evaluated. To gain a basic understanding of Early Triassic fossil preservation around the world for the purposes of this study, fossiliferous Lower Triassic strata deposited in level-bottom marine environments in western Palaeotethys, eastern Panthalassa, and open-ocean Panthalassa were examined in the field where they are exposed today in the western United States, northern Italy, and Japan (Fig. 1). The members and formations examined range from approximately 10 m (e.g. Kamura Formation, Kyushu, Japan) to nearly 700 m (e.g. Werfen Formation) in thickness, and numerous localities tens to hundreds of kilometers apart were selected for each member and formation. The examined strata span the entire Early Triassic and represent shallow marine subtidal to slope deposition and continuous accumulation on open-ocean seamounts (Kamura Formation). The field observations were supplemented with an evaluation of literature-based observations and data.

## 3. Preservation of Early Triassic skeletonized invertebrate benthic marine fossils

Silicified and partially silicified faunas were found in several beds deposited in present-day western U.S.A. (Figs 1 and 2): Smithian lower member, Union Wash Formation; Spathian upper member, Thaynes Formation; and the Spathian Virgin Limestone Member, Moenkopi Formation. Partially silicified faunas also were discovered in Lower Triassic sections of northern Italy; no silicified faunas were found in Japan. A review of the literature supports the field-based observations and reveals that despite being patchy, silicified faunas are moderately common in Lower Triassic strata (Fig. 1) and are not non-existent as previously maintained (Erwin, 1996). Silicified and partially silicified faunas have been discovered and documented in other sections from Panthalassan, Palaeotehthyan, and Neotethyan sections (Boyd and Newell, 1997, 2002; Newell and Boyd, 1995, 1999; Krystyn et al., 2003; Wheeley and Twitchett, 2005) (Fig. 2).

Nevertheless, Early Triassic bivalves and microgastropods around the world were most commonly preserved as internal and external molds (Fig. 3) or were recrystallized. In the Spathian Virgin Limestone Member of Nevada, bivalve shell voids were commonly filled by sparry calcite or were left open. Less commonly, in the Spathian upper member of the Thaynes Formation in Utah, voids created by dissolution of bivalve shells were filled by unlithified clastic material piped in from conduits (Boyd et al., 1999) and, in the lower Dinwoody Formation in northeastern Nevada, fossils of the bivalve *Claraia* cf. *stachei* were thickened diagenetically by uniform addition of calcite over the shell exterior before dissolution of aragonitic inner layers (Boyd and Newell, 1976). Microgastropods from the Werfen Formation are commonly coated with iron-oxide, giving them a reddish-pink color in outcrop (Assereto and Rizzini, 1975). The linguliform brachiopod *Lingula* is most commonly represented by original shell material in Griesbachian strata around the world (Xu and Grant, 1992; Rodland and Bottjer, 2001). Rhynchonelliform brachiopods, echinoid spines, and crinoids in Lower Triassic strata around the world were most commonly recrystallized. Preservation often precludes identification of fossils to the species level (e.g. Fraiser and Bottjer, 2004), but the modes of life of most Early Triassic benthic marine invertebrate organisms have been identified easily as epifaunal and shallow infaunal (Schubert and Bottjer, 1995). Though molds and recrystallized shells are not the ideal modes of preservation for taxonomic studies, they can be used to determine relative abundances of higher groups of taxa, number of guilds, and tiering, and are therefore sufficient for palaeoecological studies.

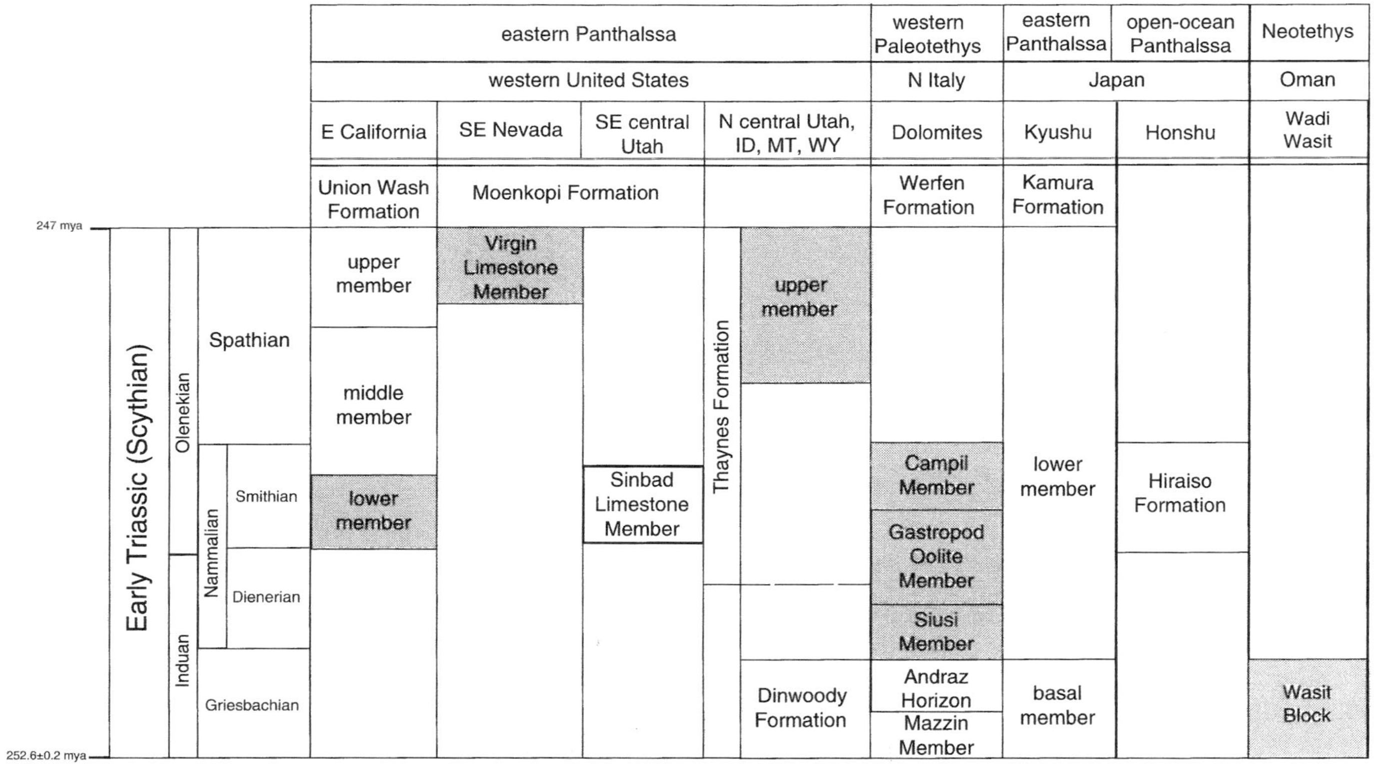


*Figure 1.* Generalized stratigraphy of fossiliferous Lower Triassic sections containing silicified and partially silicified faunas (indicated by shading). The Union Wash, Moenkopi, Thaynes, Dinwoody, Werfen, Kamura, and Hiraiso Formations were examined in the field (modified from Kimura et al., 1991; Stone et al., 1991; Schubert and Bottjer, 1995; Twitchett and Wignall, 1996; Sano and Nakashima, 1997; Martin et al., 2001; F. Corsetti, pers. comm., 2003; Mundil et al., 2004); Oman data from Krystyn et al. (2003).

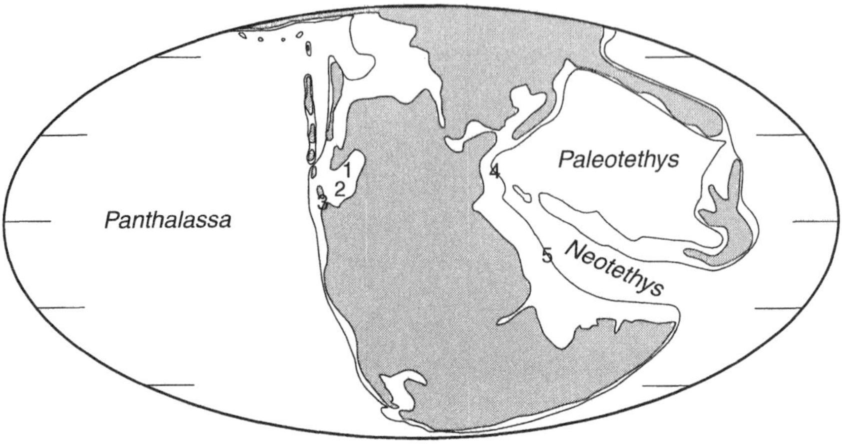


*Figure 2.* Early Triassic palaeogeographic map modified from Scotese (1994) indicating where silicified and partially silicified faunas occur: (1) upper member, Thaynes Formation, western USA; (2) Virgin Limestone Member, Moenkopi Formation, western USA; (3) lower member, Union Wash Formation, western USA; (4) Siusi, Gastropod Oolite, and Campil Members, Werfen Formation, northern Italy; (5) Wasit Block, Wadi Wasit, Oman.

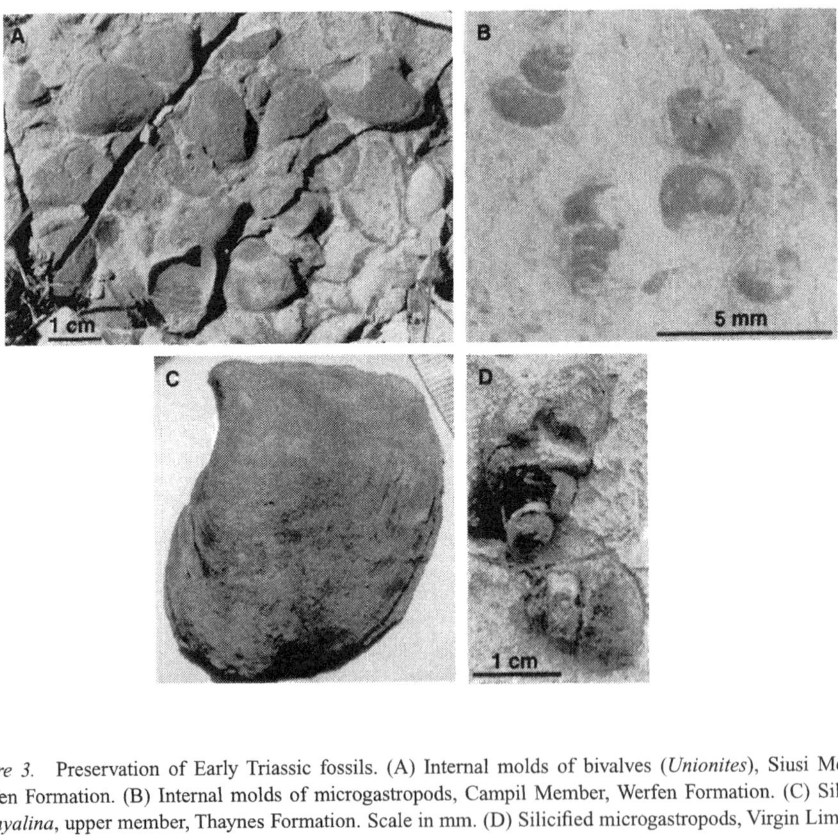


*Figure 3.* Preservation of Early Triassic fossils. (A) Internal molds of bivalves (*Unionites*), Siusi Member, Werfen Formation. (B) Internal molds of microgastropods, Campil Member, Werfen Formation. (C) Silicified *Promyalina*, upper member, Thaynes Formation. Scale in mm. (D) Silicified microgastropods, Virgin Limestone Member, Moenkopi Formation.

## 4. Environmental and ecological effects upon Early Triassic early diagenetic processes

Silicified faunas have been considered to be vital, perhaps even essential, for palaeontologic studies (e.g. Schubert et al., 1997), and numerous well-known taxonomic studies have been based on silicified faunas (e.g. Cooper and Grant, 1972). Silicification preserves intricate details of calcareous invertebrate shells and makes them relatively casy to extract from rock matrix in the laboratory. Major differences can exist between coeval silicified and non-silicified faunas when aragonitic shells are not preserved by early silicification, resulting in non-silicified faunas that originally comprised only calcitic shells (Cherns and Wright, 2000; Wright et al., 2003). According to Schubert et al. (1997), good preservation of calcareous fossils does not compensate for the lack of silicified faunas.

While it has been shown here that silicified fossils are actually moderately common in Lower Triassic strata, many Early Triassic faunas are preserved as molds and recrystallized calcareous shells and a taphonomic bias still may be credible for describing the condition of the Lower Triassic fossil record in some regions. Under actualistic conditions, originally aragonitic shells typically dissolve and are replaced as a result of meteoric or burial diagenetic processes during early diagenesis because aragonite is less stable than calcite at surface temperatures and pressures, even during "aragonite seas" (e.g. Tucker and Wright, 1990). Microbially mediated reactions associated with the decay of organic matter (e.g. Walter and Burton, 1990) and high levels of atmospheric and hence ocean concentrations of $CO_2$ (Berner and Kothavala, 2001; Sabine et al., 2004) are processes that increase the acidity of seawater and can contribute to dissolution of calcareous shells (Hautmann, 2004). However, an evaluation of the non-actualistic environmental and ecological characteristics of Early Triassic oceans indicates that the Lower Triassic fossil record does not suffer from a taphonomic "megabias" (*sensu* Schubert et al., 1997) and that useful, unbiased palaeontologic data were in fact preserved.

Rapid dissolution of biogenic carbonate is pronounced in the upper part of the sediment column under actualistic conditions. Biogenic reworking of sediments increases oxygen levels in the mixed layer and exasperates oxidative decay of organic matter, thereby increasing acidity near the sediment–water interface and causing pore waters to become undersaturated with respect to both aragonite and calcite and enhancing the dissolution of calcareous shells (Aller, 1981; Walter and Burton, 1990). Shell dissolution is therefore greatest in areas with well-developed infaunal benthic communities, and maximum shell preservation occurs in regions with low bioturbation (Aller, 1981). Under actualistic conditions, epifaunal, shallow infaunal, and small organisms are expected to be preferentially dissolved while calcareous shells of deeper-burrowing and larger organisms should be less prone to dissolution (e.g. Wright et al., 2003). Sediment reworking by infaunal organisms also disrupts mold space left after shells dissolve (Cherns and Wright, 2000). The majority of Early Triassic fossils were originally aragonite and are from organisms that had epifaunal or shallow infaunal modes of life (Schubert and Bottjer, 1995); they are therefore predicted to have been preferentially dissolved. However, bioturbation during the Early Triassic resembled the Late Cambrian/Early Ordovician (Bottjer et al., 1996): bioturbation was primarily horizontal (Twitchett, 1999; Pruss and Bottjer, 2004a), depth of infaunal trace fossils never exceeded 10 cm (Schubert and Bottjer, 1995; Twitchett, 1999; Pruss and Bottjer, 2004b), and bioturbation intensity was low (Schubert and Bottjer, 1995; Twitchett, 1999). The anachronistic characteristics of bioturbation during the Early Triassic prevented destruction and dissolution of calcareous shells and may have actually aided calcareous fossil preservation. Even if after death benthic marine invertebrate calcareous shells were dissolved on the seafloor in some regions due to decreased saturation state of seawater (Berner and Kothavala, 2001; Sabine et al., 2004), the low levels of bioturbation would have ensured that molds were not disturbed and may have even aided preservation because of high alkalinity at the sediment–water interface. The preservation of Early Triassic fossil molds also may be an indication that skeletal aragonite survived shallow burial and that dissolution took place during deeper burial (*sensu* Wright et al., 2003). Furthermore, the decrease in ichnogeneric diversity and in depth and extent of bioturbation during the Early Triassic are excellent proxies for ecological degradation during the aftermath of the end-Permian mass extinction independent of fossilization processes and falsify the hypothesis that taxonomic and ecologic patterns during the Early Triassic are more apparent than real (*sensu* Erwin, 1996; Erwin and Pan, 1996; Wheeley and Twitchett, 2005).

In addition to the lack of extensive bioturbation, evidence in the form of large $\delta^{34}S$ excursions (Marenco et al., 2003), submarine carbonate fans (Woods et al., 1999; Baud and Richoz, 2004), and subtidal stromatolites (Pruss and Bottjer, 2004b) indicate that alkalinity periodically built up in Early Triassic oceans due to bacterial sulfate reduction. Though this phenomenon varied regionally and temporally throughout the Early Triassic, increased oceanic alkalinity would have aided carbonate fossil preservation in some regions during the Early Triassic.

The prevalence of molds and recrystallized fossils of originally aragonitic organisms only millimeters to 2–3 cm in size (Fraiser and Bottjer, 2004) is another indication of the fidelity of the Lower Triassic fossil record and supports the hypothesis that the Lower Triassic fossil record reflects a primary ecological signal. Small aragonitic shells would be expected to dissolve before larger ones (e.g. Wright et al., 2003); the hypothesis that incompletely silicified larger fossils may result in micromorphic faunas (Erwin and Kidder, 2000) is not supported. The abundance of small originally phosphatic shells (e.g. *Lingula*) (Rodland and Bottjer, 2001; Price-Lloyd and Twitchett, 2002) lends further support to the hypothesis proposed herein.

Even though various aspects of the Early Triassic biotic crisis were manifested at different times at different places around the world, palaeocommunities in the Panthalassa, Palaeotethys, Neotethys, and Boreal Oceans are remarkably similar in terms of biodiversity and palaeoecology (c.f. Fraiser and Bottjer, 2005). That silicified and nonsilicified benthic skeletonized invertebrate marine faunas in various oceans and latitudes around the world share taxonomic and palaeoecologic characteristics (i.e. low biodiversity, low levels of tiering, numerical dominance by the same genera) is another line of evidence that refutes the hypothesis of an extensive Early Triassic taphonomic bias.

Thus, based on several lines of evidence, the Early Triassic "fossilization low" appears actually to be a palaeoecological indicator of the severity of the physio-chemically harsh environmental conditions that persisted long after the end-Permian mass extinction (*sensu* Wignall and Benton, 1999) rather than merely poor preservation. Furthermore, the fossil record of Early Triassic ammonoids is considered suitable for determining their taxonomic and morphologic recovery patterns after the end-Permian mass extinction (McGowan, 2004a, b), even though their originally aragonitic shells, once settled on the seafloor after death of the animal, would have been subjected to the same early diagenetic processes as originally aragonitic skeletonized benthic organisms.

## 5. Testing the quality and utility of the Lower Triassic fossil record of skeletonized invertebrate benthic marine fossils

This preliminary examination of the preservation of Early Triassic skeletonized benthic invertebrate marine fossils has revealed not only that silicified faunas are moderately common in Lower Triassic strata, but also that molds and recrystallized body fossils of Early Triassic skeletonized benthic invertebrate marine organisms are adequate at least for palaeoecologic studies. This is likely because environmental characteristics unique to the Early Triassic prevented originally calcareous shells of benthic marine invertebrate organisms and their molds from being preferentially dissolved or destroyed. Although these preliminary observations indicate that reservations about the quality of the Lower Triassic fossil record are not completely valid, future testing is required to more accurately determine the quality and utility of the nonsilicified Lower Triassic fossil record for taxonomic and palaeoecologic purposes, because the majority of Early Triassic fossils are nonsilicified (Fig. 3) and palaeoecologic analyses of the aftermath of the end-Permian mass extinction have focused on nonsilicified faunas (e.g. Rodland and Bottjer, 2001; Fraiser and Bottjer, 2003, 2004). Three tests of the quality and utility of the Lower Triassic fossil record, one literature-based and two field-based, are proposed here.

Determining how the number and extent of Early Triassic silicified faunas compares with the number and extent of silicified faunas from the Permian, Middle, and Late Triassic will further test the assumption that silicified faunas are rare in Lower Triassic strata deposited during the aftermath of the end-Permian mass extinction. Preliminary evidence presented here indicates that Early Triassic silicified faunas are actually moderately common in Lower Triassic strata, and the number of late Early Triassic silicified faunas may even be larger than that of the Middle Triassic (C. McRoberts, pers. comm., 2004).

Secondly, the biodiversity, relative abundances, sizes, and extent of secondary tiering (epibiont cover) of Early Triassic silicified faunas should be compared to the biodiversity, relative abundances, sizes, and extent of secondary tiering of coeval nonsilicified faunas; similar methods have been used previously (Cherns and Wright, 2000). Comparisons of the biodiversities and relative abundances of the two groups will determine if nonsilicified Early Triassic faunas are significantly biased because originally aragonitic shells were preferentially dissolved (*sensu* Cherns and Wright, 2000; Wright et al., 2003). Comparisons of ranges in fossil size and in extent of epibiont cover will assess the extent of any shell dissolution. These comparisons will test the model proposed here that the Lower Triassic fossil record is adequate for discerning palaeoecology.

Taxonomic and palaeoecologic characteristics of different Early Triassic marine environments must also be compared and contrasted. Early dissolution of aragonite can be a major process in offshore settings regardless of the depths of the aragonite or calcite compensation, where a source of organic matter is available upon which microbial processes can act and increase acidity (Wright et al., 2003). Shell dissolution and preservation therefore can be dependent on facies-controlled factors.

## 6. Conclusions

The high number of Lazarus taxa during the aftermath of the end-Permian mass extinction previously has been attributed to an alleged dearth of silicified faunas that also has been used previously as an indication that the entire Early Triassic fossil record is afflicted by a taphonomic bias (Erwin, 1996; Erwin and Pan, 1996; Twitchett, 2001). However, silicified faunas are moderately common in Lower Triassic strata and preservation of calcareous benthic skeletonized invertebrate marine shells during the Early Triassic was possibly aided by environmental, ecological, and geochemical characteristics in the early diagenetic environment unique to the aftermath of the end-Permian mass extinction. Early Triassic fossil molds and recrystallized shells reflect primary ecologic signals, if not taxonomic ones. Early silicification of calcareous organisms may be important for determining taxonomic patterns, but it is not required for determining palaeoecological patterns.

This preliminary examination indicates that the Early Triassic "fossilization low" is largely due to ecological and environmental factors that persisted during the aftermath of the end-Permian mass extinction (*sensu* Wignall and Benton, 1999), not because of a preservation bias as previously posited (Erwin, 1996; Erwin and Pan, 1996). Consideration of all aspects of the fossil record for critical intervals in the history of life is thus a crucial task necessary for evaluating the preservation and hence utility of particular fossil faunas to provide answers to large-scale evolutionary and palaeoecological questions.

## Acknowledgments

MLF thanks T. Oji, T. Kudo, H. Yamagishi, and A. Yamamoto, University of Tokyo, and H. Sano and T. Igawa, Kyushu University, for field assistance in Japan; D. Boyd, University of Wyoming, and B. Hussaini and N. Newell, the AMNH, for discussions and assistance in identifying Early Triassic bivalves; C. McRoberts, State University of New York at Cortland, for an unpublished list of Early Triassic bivalves; D. Erwin, NMNH, for discussions about the potential Early Triassic preservation bias; and C. Jamet and P. Marenco, University of Southern California, for field assistance in the western U.S. This research was supported by grants to MLF from the American Museum of Natural History Lerner-Gray Fund for Marine Research, the Geological Society of America, the Paleontological Society, the USC Graduate and Professional Student Senate, the USC Department of Earth Sciences, and the Wrigley Institute for Environmental Studies, and by a grant to DJB from the USC Women in Science and Engineering Program (WISE). Reviews by C. McRoberts and P. Wignall improved an earlier version of this manuscript.

## References

Aller, R.C., 1981. Carbonate dissolution in nearshore terrigenous muds: the role of physical and biological reworking. J. Geol. 90, 79–95.

Assereto, R.L., Rizzini, A., 1975. Reworked Ferroan Dolomite grains in the Triassic "oolite a gasteropodi" of Camoniche Alps (Italy) as indicators of early diagenesis. Neues Jahrbuch Geol. Palaontol Abhandlungen, 148(2), 215–232.

Batten, R.L., 1973. The vicissitudes of the gastropods during the interval of Guadalupian-Ladinian time. In: Logan, A., Hills, L. V. (Eds), The Permian and Triassic Systems and their Mutual Boundary. Canadian Society of Petroleum Geologists Memoir, Vol. 2, pp. 596–607.

Baud, A., Richoz, S., 2004. Anachronistic facies after mass extinctions: tethyan basal Triassic calcimicrobial cap rocks. GSA Abstr. Progr. 35, 336.

Beauchamp, B., Baud, A., 2002. Growth and demise of Permian biogenic chert along northwest Pangea: evidence for end-Permian collapse of thermohaline circulation. Palaeogeogr. Palaeoclimatol. Palaeoecol. 184, 37–63.

Berner, R.A., Kothavala, Z., 2001. GEOCARB III: a revised model of atmospheric $CO_2$ over Phanerozoic time. Am. J. Sci. 301, 182–204.
Bottjer, D.J., 2001. Biotic recovery from mass extinctions. In: Briggs, D.E.G., Crowther, P.R. (Eds), Palaeobiology II. Blackwell Science, Oxford, pp. 202–206.
Bottjer, D.J., Schubert, J.K., Droser, M.L., 1996. Comparative evolutionary palaeoecology: assessing the changing ecology of the past. In: Hart, M. B. (Ed.), Biotic Recovery from Mass Extinction Events. Geological Society Special Publication No. 102, pp. 1–13.
Boyd, D.W., Newell, N.D., 1976. Diagenetic image reversal in a Triassic pelecypod. Univ. Wyoming Contrib. Geol. 14, 65–68.
Boyd, D.W., Newell, N.D., 1997. A reappraisal of Trigoniacean families (Bivalvia) and a description of two new Early Triassic species. American Museum Novitates, No. 3216, p. 14.
Boyd, D.W., Newell, N.D., 2002. A unique pterioid bivalve from the Early Triassic of Utah. American Museum Novitates, No. 3375, p. 9.
Boyd, D.W., Nice, D.E., Newell, N.D., 1999. Silt injection as a mode of fossilization: a Triassic example. Palaios 14, 545–554.
Cherns, L., Wright, V.P., 2000. Missing molluscs as evidence of large-scale, early skeletal aragonite dissolution in a Silurian sea. Geology 28(9), 791–794.
Ciriacks, K.W., 1963. Permian and Eotriassic bivalves of the Middle Rockies. Bull. Am. Mus. Nat. Hist. 125, 1–100.
Cooper, G.A., Grant, R.E., 1972. Permian brachiopods of west Texas, I. Smithsonian Contributions to Paleobiology, p. 231.
Droser, M.L., Bottjer, D.J., Sheehan, P.M., 1997. Evaluating the ecological architecture of major events in the Phanerozoic history of marine invertebrate life. Geology 25(2), 167–170.
Droser, M.L., Bottjer, D.J., Sheehan, P.M., McGhee, G.R., Jr., 2000. Decoupling of taxonomic and ecologic severity of Phanerozoic marine mass extinctions. Geology 28, 675–678.
Erwin, D.H., 1996. Understanding biotic recoveries: extinction, survival, and preservation during the end-Permian mass extinction. In: Jablonski, D., Erwin, D.H., Lipps, J. (Eds), Evolutionary Paleobiology. The University of Chicago Press, Chicago, pp. 398–418.
Erwin, D.H., Kidder, D.L., 2000. Depositional controls on selective silicification of Permian fossils, southwestern United States. In: Wardlaw, B.R., Grant, R.E., Rohr, D.M. (Eds), The Guadalupian Symposium. Smithsonian Contributions to the Earth Sciences, No. 32, pp. 407–415.
Erwin, D.H., Pan, H., 1996. Recoveries and radiations: gastropods after the Permo-Triassic mass extinction. In: Hart, M.B. (Ed.), Biotic Recovery from Mass Extinction Events. Geological Society Special Publication No. 102, pp. 223–229.
Flessa, K.W., Jablonski, D., 1983. Extinction is here to stay. Paleobiology, 9, 315–321.
Fraiser, M.L., Bottjer, D.J., 2003. Paleoecology of bivalves during their initial rise to dominance. Geol. Soc. Am. Ann. Meet. Abstr. Progr. 35(6), 417.
Fraiser, M.L., Bottjer, D.J., 2004. The non-actualistic Early Triassic gastropod fauna: a case study from the Lower Triassic Sinbad Limestone Member. Palaios 19, 259–275.
Fraiser, M.L., Bottjer, D.J., 2005. Restructuring in benthic level-bottom shallow marine communities due to prolonged environmental stress following the end-Permian mass extinction. Comptes Rendus Palevol 4, in press.
Gould, S.J., Calloway, C.B., 1980. Clams and brachiopods-ships that pass in the night. Paleobiology 6(4), 383–396.
Hautmann, M., 2004. Effect of end-Triassic $CO_2$ maximum on carbonate sedimentation and marine mass extinction. Facies 50, 257–261.
Hori, R., Sakakibara, M., Takemura, A., Takemura, S., Sporli, K.B., Sakamoto, S., Ogane, K., Nakamura, Y., Campbell, H., 2003. Early Triassic radiolarians from Arrow Rocks in the Waipapa Terrane, North Island, New Zealand. Tenth Meeting of the International Association of Radiolarian Palaeontologists, Abstracts and Programs, University of Lausanne.
Jablonski, D., 1986a. Background and mass extinctions: the alternation of macroevolutionary regimes. Science 231, 129–133.
Jablonski, D., 1986b. Causes and consequences of mass extinctions: a comparative approach. In: Elliot, D.K. (Ed.), Dynamics of Extinction. Wiley, New York, pp. 183–230.
Kauffman, E.G., Harries, P.J., 1996. The importance of crisis progenitors in recovery from mass extinction. In: Hart, M.B. (Ed.), Biotic Recovery from Mass Extinction Events. Geological Society Special Publication No. 102, pp. 15–39.

Kimura, T., Hayami, I., Yoshida, S., 1991. Geology of Japan. University of Tokyo Press, Tokyo.

Krystyn, L., Richoz, S., Baud, A., Twitchett, R., 2003. A unique Permian-Triassic boundary section from the Neotethyan Hawasina Basin, central Oman Mountains. Palaeogeogr. Palaeoclimatol. Palaeoecol. 191(3–4), 329–344.

Marenco, P.J., Corsetti, F.A., Bottjer, D.J., Kaufman, A.J., 2003. Killer oceans of the Early Triassic. Geol. Soc. Am. Annu. Meet. Abstr. Progr. 35(6), 386.

Márquez-Aliaga, A., Ros, S., 2002. Taphonomy richness of the scarce Spanish bivalves Triassic record: some examples to be discussed. In: De Renzi, M., Alonso, M.V.P., Belinchón, M., Peñalver, E., Montoya, P., Márquez-Aliaga, A. (Eds), Current Topics on Taphonomy and Fossilization, International Conference Taphos 2002. 3rd Meeting on Taphonomy and Fossilization, pp. 199–206.

Martin, M.W., Lehrmann, D.J., Bowring, S.A., Enos, P., Ramezani, J., Wei, J., Zhang, J., 2001. Timing of the Lower Triassic carbonate bank buildup and biotic recovery following the end-Permian mass extinction across the Nanpanjiang Basin, south China. Geol. Soc. Am. Abstr. Prog. 33(6), 201.

McGowan, A.J., 2004a. Ammonoid taxonomic and morphologic recovery patterns after the Permian-Triassic. Geology 32, 665–668.

McGowan, A.J., 2004b. The effect of the Permo-Triassic bottleneck on Triassic ammonoid morphological evolution. Paleobiology 30, 369–395.

Moffat, H.A., Bottjer, D.J., 1999. Echinoid concentration beds: two examples from the stratigraphic spectrum. Palaeogeogr. Palaeoclimatol. Palaeoecol. 149, 329–348.

Mundil, R., Ludwig, K. R., Metcalfe, I., Renne, P. R., 2004. Age and timing of the Permian mass extinctions: U/Pb dating of closed-system zircons. Science 305, 1760–1763.

Newell, N.D., Boyd, D.W., 1995. Pectinoid bivalves of the Permian-Triassic crisis. Bull. Am. Mus. Nat. Hist. 227, 1–95.

Newell, N.D., Boyd, D.W., 1999. A new Lower Triassic *Permophorus* from the Central Rocky Mountains. American Museum Novitates, 3263, 1–5.

Nice, D.E., 1985. The Role of Skeletal Diagenesis in Carbonate Fabric Development, Thaynes Formation (Lower Triassic), Northeast Utah and Southeast Idaho. M.S. Thesis, University of Wyoming, Laramie.

Parsons-Hubbard, K.M., Callender, W.R., Powell, E.N., Brett, C.E., Walker, S.E., Raymond, A.L., Staff, G.M., 1999. Rates of burial and disturbance of experimentally-deployed molluscs: implications for preservation potential. Palaios 14(4), 337–351.

Payne, J.L., Lehrmann, D.J., Wei, J., Orchard, M.J., Schrag, D.P., Knoll, A.H., 2004. Large perturbations of the carbon cycle during recovery from the end-Permian extinction. Science 305, 506–509.

Price-Lloyd, N., Twitchett, R.J., 2002. The Lilliput Effect in the aftermath of the end-Permian mass extinction event. Geol. Soc. Am. Annu. Meet. Abstr. Programs 34(6), 355.

Pruss, S.B., Bottjer, D.J., 2004a. Early Triassic trace fossils of the western United States and their implications for prolonged environmental stress from the end-Permian mass extinction, Palaios 19, 551–564.

Pruss, S.B., Bottjer, D.J., 2004b. Late Early Triassic microbial reefs of the western United States: a description and model for their deposition in the aftermath of the end-Permian mass extinction. Palaeogeogr. Palaeoclimatol. Palaeoecol. 211, 127–137.

Pruss, S.B., Fraiser, M.L., Bottjer, D.J., 2004. The proliferation of Early Triassic wrinkle structures: implications for environmental stress following the end-Permian mass extinction. Geology, 32, 461–464.

Racki, G., 1999. Silica-secreting biota and mass extinctions: survival patterns and processes. Palaeogeogr. Palaeoclimatol. Palaeoecol. 154, 107–132.

Rodland, D.L., Bottjer, D.J., 2001. Biotic recovery from the end-Permian mass extinction: behavior of the inarticulate brachiopod *Lingula* as a disaster taxon. Palaios 16(1), 95–101.

Sabine, C.L. et al., 2004. The ocean sink for anthropogenic $CO_2$. Science 305, 367–371.

Sano, H., Nakashima, K., 1997. Lowermost Triassic (Griesbachian) microbial bindstone-cementstone facies, southwest Japan. Facies 36, 1–24.

Schubert, J.K., Bottjer, D.J., 1995. Aftermath of the Permian-Triassic mass extinction event: paleoecology of Lower Triassic carbonates in the western USA. Palaeogeogr. Palaeoclimatol. Palaeoecol. 116, 1–39.

Schubert, J.K., Kidder, D.L., Erwin, D.H., 1997. Silica-replaced fossils through the Phanerozoic. Geology 25(11), 1031–1034.

Scotese, C.R., 1994. Early Triassic paleogeographic map. In: Klein, G.D. (Ed.), Pangea: Paleoclimate, Tectonics, and Sedimentation During Accretion, Zenith, and Breakup of a Supercontinent. Geological Society of America Special Paper 288, Boulder, CO.

Sepkoski, J.J., Jr., 1981. A factor analytic description of the Phanerozoic marine fossil record. Paleobiology 7, 36–53.

Stanley, S.M, Yang, X., 1994. A double mass extinction at the end of the Paleozoic era. Science 266, 1340–1344.

Stone, P., Stevens, C.H., Orchard, M.J., 1991. Stratigraphy of the Lower and Middle (?) Triassic Union Wash Formation, East-Central California. U.S.G.S. Bull. 1928, 1–26.

Takemura, A., Aita, Y., Sakai, T., Kamata, V., Suzuki, N., Iiori, R.S., Famakita, S., Sakakibara, M., Campbell, H., Fujiki, T., Ogane, K., Takemura, S., Sakamoto, S., Kodama, A., Nakamura, Y., 2003. Conodont-based age determination for a radiolarian-bearing Lower Triassic chert sequence in Arrow Rocks, New Zealand. Tenth Meeting of the International Association of Radiolarian Palaeontologists, Abstracts and Programs, University of Lausanne.

Tucker, M.E., Wright, V.P., 1990. Diagenetic Processes, Products and Environments. Carbonate Sedimentology. Blackwell Science, Oxford, pp. 314–364.

Twitchett, R.J., 1999. Palaeoenvironments and faunal recovery after the end-Permian mass extinction. Palaeogeogr. Palaeoclimatol. Palaeoecol. 154, 27–37.

Twitchett, R.J., 2001. Incompleteness of the Permian-Triassic fossil record: a consequence of productivity decline? Geol. J. 36, 341–353.

Twitchett, R.J., Wignall, P.B., 1996. Trace fossils and the aftermath of the Permo-Triassic mass extinction: evidence from northern Italy. Palaeogeogr. Palaeoclimatol. Palaeoecol. 124, 137–151.

Walter, L.M., Burton, E.A., 1990. Dissolution on platform carbonate sediments in marine pore fluids. Am. J. Sci. 290, 601–643.

Wheeley, J.R., Twitchett, R.J., 2005. Palaeoecological significance of a new Griesbachian (Early Triassic) gastropod assemblage from Oman. Lethaia 38, 1–9.

Wignall, P.B., Benton, M.J., 1999. Lazarus taxa and fossil abundance at times of biotic crisis. J. Geol. Soc. London, 156, 453–456.

Woods, A.D., Bottjer, D.J., 2000. Distribution of ammonoids in the Lower Triassic Union Wash Formation (eastern California): evidence for paleoceanographic conditions during recovery from the end-Permian mass extinction. Palaios 15, 535–545.

Woods, A.D., Bottjer, D.J., Mutti, M., Morrison, J., 1999. Lower Triassic large seafloor carbonate cements: their origin and a mechanism for the prolonged biotic recovery from the end-Permian mass extinction. Geology 27, 645–648.

Wright, P., Cherns, L., Hodges, P., 2003. Missing molluscs: field testing taphonomic loss in the Mesozoic through early large-scale aragonite dissolution. Geology 31(3), 211–214.

Xu, G., Grant, R.E., 1992. Permo-Triassic brachiopod successions and events in South China. In: Sweet, W.C., Zunyi, Y., Dickens, J.M., Hongtu, Y. (Eds), Permo-Triassic Events in the Eastern Tethys. Cambridge University Press, Cambridge, pp. 98–108.

Understanding Late Devonian and Permian-Triassic Biotic and Climatic Events:
Towards an Integrated Approach
edited by D.J. Over, J.R. Morrow and P.B. Wignall



# *Chapter 12*

# *Environmental trends of Early Triassic biofabrics: implications for understanding the aftermath of the end-Permian mass extinction*

*Sara B. Pruss*, Frank A. Corsetti and David J. Bottjer*

*Department of Organismic and Evolutionary Biology, Harvard University, 26 Oxford Street, Cambridge, MA 02138*

**Abstract**

The Lower Triassic Virgin Limestone Member of the Moenkopi Formation provides a holistic picture of the lag phase of the biotic recovery in the aftermath of the end-Permian mass extinction. This unit records deposition in siliciclastic and carbonate nearshore to mid-shelf palaeoenvironments in the western United States during latest Early Triassic (Spathian) time. Nearshore siliciclastic-dominated units are characterized by a relatively diverse trace fossil assemblage, including vertical traces such as *Rhizocorallium* and *Arenicolites*, but record low ichnofabric indices. Middle shelf carbonate-dominated units, however, contain a low diversity trace fossil assemblage dominated by simple horizontal traces and tend to record high ichnofabric indices. Subtidal microbial structures occur as wrinkle structures in siliciclastic strata and stromatolitic and thrombolitic buildups in carbonate strata. The strong variations in trace fossil assemblages, ichnofabric indices, and microbial development across the onshore–offshore transect demonstrate that an examination of a variety of facies in different palaeoenvironments is required to reveal the true picture of the biotic recovery. In addition, the prolonged absence of deep bioturbation during the Early Triassic and the distribution of microbial fabrics in subtidal carbonates and siliciclastics indicates a return to Early Phanerozoic-style substrates and suggests that latest Early Triassic shelf environments continued to experience environmental stress, perhaps episodically, for 4–7 million years after the end-Permian mass extinction.

*Keywords*: Early Triassic, Virgin Limestone, biofabric, microbialite, biotic recovery

## 1. Introduction

Following the end-Permian mass extinction, the diversity of marine organisms suffered the most severe reduction in the history of metazoan life. Approximately 80% of marine species became extinct during this event (Stanley and Yang, 1994) and the recovery of organisms did not occur for 4–7 million years after the extinction (Hallam, 1991; Martin et al., 2001). The Early Triassic encompasses an unusually long lag phase from the post-extinction biotic recovery (Hallam, 1991; Erwin, 1993), and marine communities are low in diversity (Hallam and Wignall, 1997) and commonly characterized by biotic recovery opportunists (Rodland and Bottjer, 2001; Fraiser and Bottjer, 2004). Changes in biofabric, fabrics imparted to the rocks though biologic activity (e.g. ichnofabric and microbialites), have also been observed. For example, trace fossils undergo a reduction in diversity, size, and depth of penetration into the sediment after the end-Permian mass extinction

*Corresponding author. Fax: +1-213-740-8801. *E-mail address:* spruss@fas.harvard.edu (S.B. Pruss).

(Twitchett and Wignall, 1996; Twitchett, 1999; Ausich and Bottjer, 2002; Pruss and Bottjer, 2004a) and normal marine microbialites and flat-pebble conglomerates show an anachronistic resurgence (e.g. Wignall and Twitchett, 1999; Pruss et al., 2004).

The Virgin Limestone Member of the Moenkopi Formation is one of the few Lower Triassic examples of a mixed siliciclastic-carbonate succession deposited in intertidal to mid-shelf palaeoenvironments that allows for trends in biofabric to be observed in coeval siliciclastic and carbonate environments. Analysis of biofabric trends across an onshore–offshore transect has revealed that the distribution of microbial structures and trace fossils varies when comparing nearshore siliciclastic deposits to more offshore carbonates. The siliciclastic biofabrics are characterized predominantly by vertical trace fossils, low ichnofabric indices, and wrinkle structures while the carbonate biofabrics are dominated by horizontal trace fossils, high ichnofabric indices and stromatolites/thrombolites. Analysis of biofabric variations from nearshore to offshore settings and between siliciclastic and carbonate palaeoenvironments suggests that a true picture of the biotic recovery can only be gleaned from examination of all facies. Here, we report a detailed description and interpretation of biofabrics from the Virgin Limestone Member of the Moenkopi Formation in order to better understand how lateral heterogeneities in facies associations can have an effect on palaeobiological interpretations during the biotic rebound from the end-Permian mass extinction.

## 2. Geologic setting

The Moenkopi Formation records the waxing and waning of an epicontinental sea during Smithian–Spathian time (late Early Triassic) and consists of non-marine redbeds and nearshore to middle-shelf siliciclastics and carbonates (e.g. Larson, 1966; Shorb, 1983). The sea flooded southwestern North America and extended eastward to Utah. The Virgin Limestone Member, the subject of this report, is a mixed siliciclastic-carbonate unit in the Moenkopi Formation that was deposited during the last stage of the Early Triassic and is exposed from eastern California to southwestern Utah; palaeoenvironments become increasingly shallower to the east (e.g. Shorb, 1983). The Virgin Limestone consists of limestones, dolomitic limestones, siltstones, and sandstones (Belnap, 1971; Shorb, 1983). The Virgin Limestone crops out as limestone ledges with intervening beds of poorly exposed siliciclastic strata. Siliciclastic strata dominate the nearshore sections of the Virgin Limestone whereas the shelfal sections are composed mostly of carbonate with intervening siliciclastic interbeds. Fossils found in the Virgin Limestone include bivalves, gastropods, echinoderms, and rare ammonoids (e.g. Shorb, 1983; Schubert and Bottjer, 1995). Microbial reefs and other anachronistic facies have also been documented from the Virgin Limestone (e.g. Pruss and Bottjer, 2004b; Pruss et al., 2004).

## 3. Methods

Biofabric trends were examined across a nearshore–middle shelf transect to investigate the lateral distribution of trace fossils, ichnofabric indices, and microbial fabrics between these palaeoenvironments and rock types. Investigating both siliciclastic and carbonate facies can illustrate more about potential palaeoenvironmental biases of uppermost Lower Triassic strata than studying one or the other in isolation.

The Virgin Limestone Member was measured at 4 localities in eastern California and southern Nevada (Fig. 1). These include two localities in the Muddy Mountains (Muddy Mountains Overton and Muddy Mountains Ute), one locality at Mountain Pass, and one locality in the Spring Mountains (Lost Cabin Springs) (for detailed locality information, see Pruss, 2004). At each locality, detailed stratigraphic sections were measured; fossil content, sedimentary structures, and grain size were recorded to ascertain palaeoenvironments. All beds containing trace fossils were noted, and ichnofabric indices were assigned (Droser and Bottjer, 1986). Because exposure of beds containing trace fossils varied greatly, ichnofabric indices were only assigned to beds where the amount of bioturbation could accurately be scored. Microbial structures and buildups were noted and sampled. In addition to detailed field analyses, thin-sections were made from collected samples of trace fossils and microbial structures. Microbial fabrics were also examined in thin-section to determine if any microbial communities were preserved.

## 4. Palaeoenvironmental analysis

The four localities investigated here were deposited on a mixed siliciclastic-carbonate ramp in latest Early Triassic time (e.g. Shorb, 1983) and can be generally placed in two categories: siliciclastic-dominated nearshore and carbonate-dominated inner to middle shelf. In general, the two Muddy Mountains sections are examples of siliciclastic-dominated nearshore sections and the Mountain Pass and Lost Cabin Springs sections represent carbonate-dominated shelf palaeoenvironments.

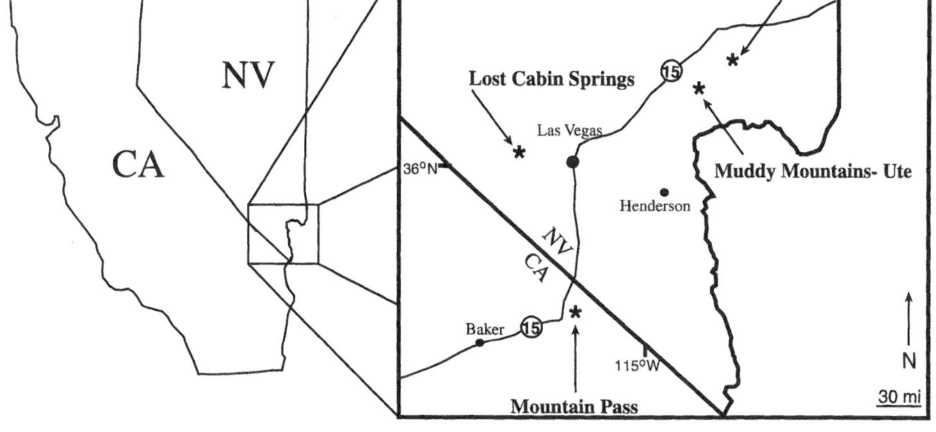


*Figure 1.* Locality map showing sites where Lower Triassic Virgin Limestone was studied. Modified from Pruss and Bottjer (2004a).

### 4.1. Muddy Mountains Overton

The Muddy Mountains Overton locality is dominated by siltstone and calcareous siltstone and contains some thin limestone beds (Fig. 2). The thin limestone beds consist primarily of bivalve-gastropod packstones, intraclastic grainstones and oolitic grainstones

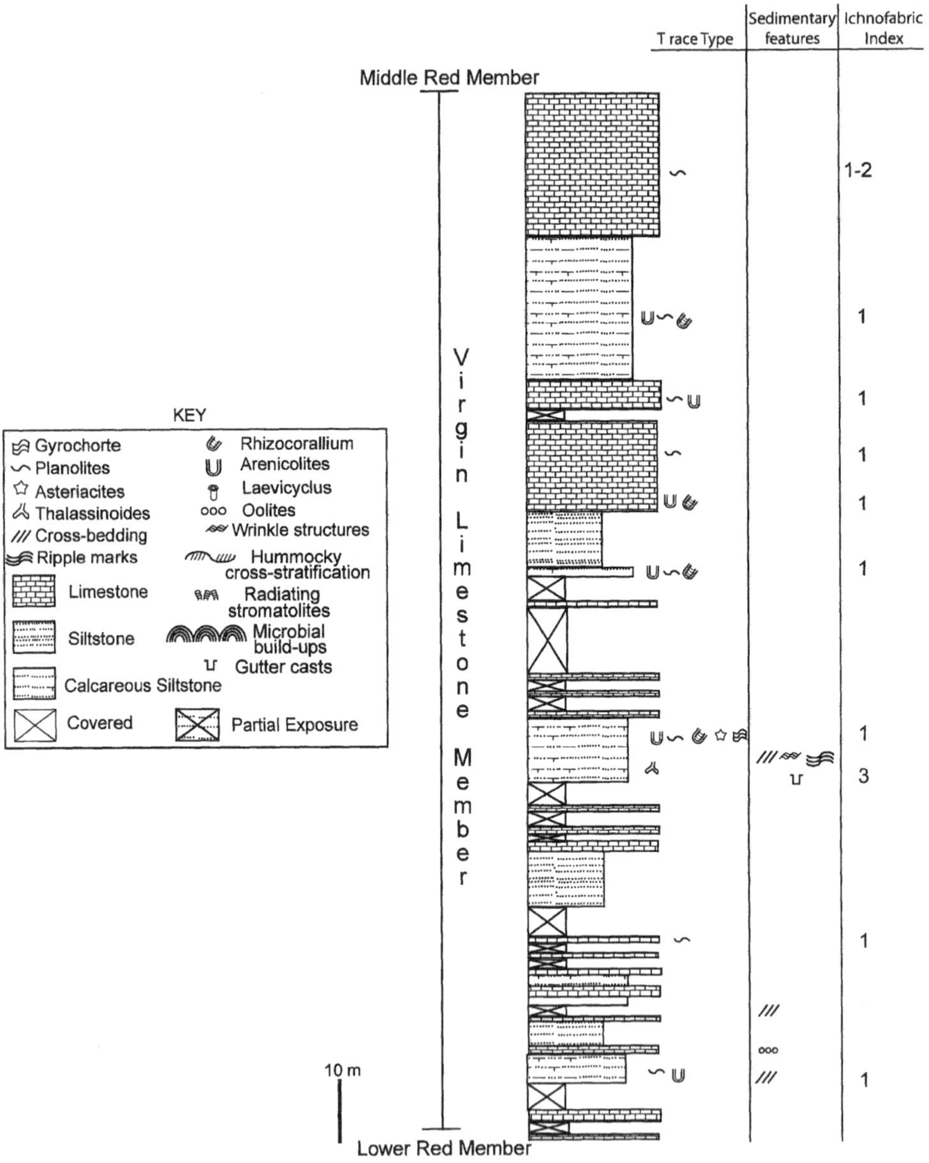


*Figure 2.* Measured stratigraphic section from the Muddy Mountains Overton locality. Trace fossils, sedimentary features including sedimentary and microbial structures, and ichnofabric indices are shown. Modified from Pruss and Bottjer (2004a).

and are particularly common in the basal portion of the section. The limestone beds thicken up-section. Common sedimentary structures include interference ripple marks, trough and tabular cross-stratification, and gutter casts. The intraclastic grainstones and gutter casts suggest that this unit was deposited in a subtidal storm-dominated palaeoenvironment.

### *4.2. Muddy Mountains Ute*

Although this locality is within a few kilometers of the Muddy Mountain Overton locality, there are some clear lithologic differences between the sections (Fig. 3). Oolite dominates the basal portion of the Virgin Limestone Member at this locality. The remaining section consists of limestone, siltstone and calcareous siltstone, with limestone beds becoming increasingly abundant up-section. Many of the limestones in the basal portion of this section are dominated by ooid grainstones and bioclastic packstones; the carbonate mudstones become more prominent up-section. Carbonate and siliciclastic intraclasts are found throughout the section in both the limestones and siliciclastics. Like the carbonates, a change in siliciclastic grain size also occurs. This change is from fine-grained sandstones and siltstones in the basal part of the section to fine-grained siltstones and mudstones in the middle and upper portion. Sedimentary structures include gutter casts and cross-stratification (including hummocky cross-stratification in the upper part of the section). The Muddy Mountains Ute section was likely deposited in a subtidal storm-dominated mixed siliciclastic-carbonate ramp near and basinward of an oolite shoal.

### *4.3. Mountain Pass*

The palaeoenvironments at the Mountain Pass locality are the most difficult to ascertain because much of this section is covered (Fig. 4). Systematic changes in the exposed limestone ledges do, however, provide information. The basal bed of the Virgin Limestone Member at the Mountain Pass locality is oolitic. The limestone ledges become increasingly micritic up-section. Because these carbonate ledges are thick, and show a decrease in grain size up-section, the Virgin Limestone at this locality is interpreted to represent deposition in a carbonate shelf palaeoenvironment; however, no information can be gleaned from the abundant covered intervals. It is interesting to note that the Mountain Pass locality is much thinner than the other localities. Here the Virgin Limestone rests unconformably on the conglomeratic Timpoweap Member, which is composed of clasts of the Lower Permian Kaibab Formation (e.g. Shorb, 1983), rather than the Lower Red Member. It is possible that the Lower Red Member is missing because of faulting (common in the area). Alternatively, it is possible that the Mountain Pass section was deposited on a basinal high and the Lower Red Member was never present.

### *4.4. Lost Cabin Springs*

The Lower Red Member is not exposed at the section described here from Lost Cabin Springs, which was chosen for its superior exposure of the remaining section, but it was

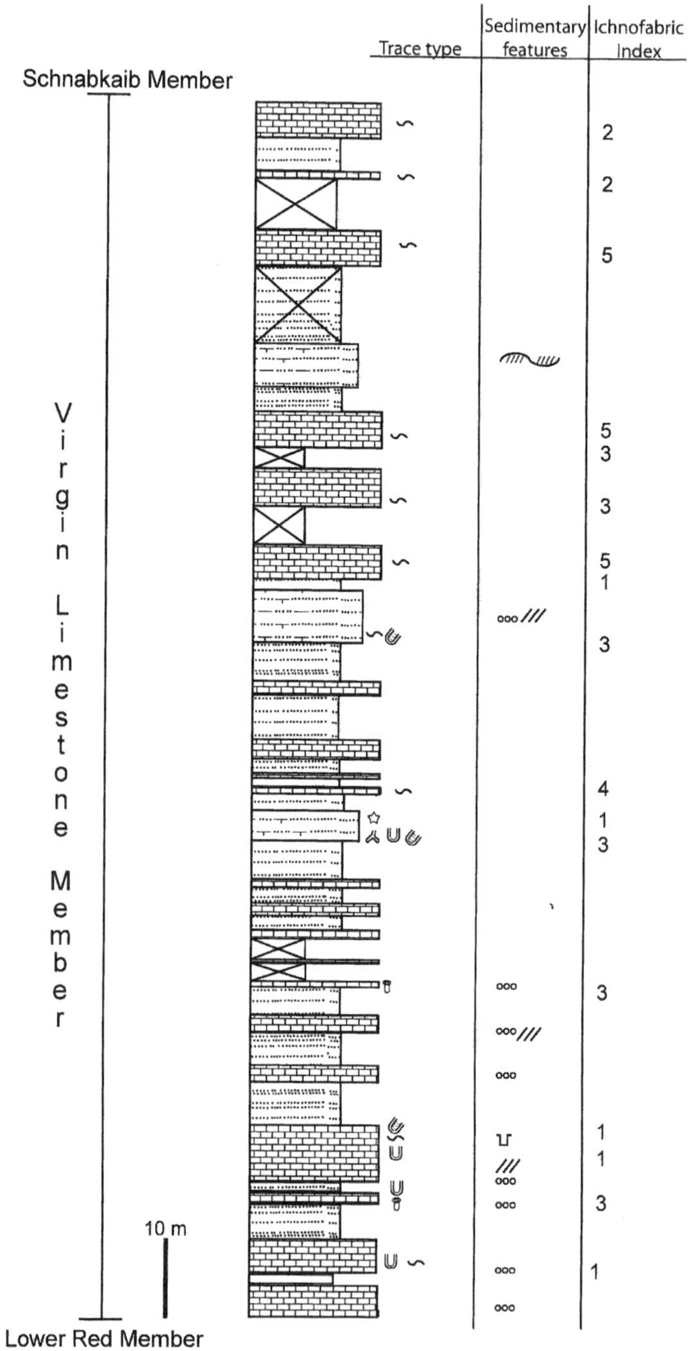


*Figure 3.* Measured stratigraphic section from the Muddy Mountains Ute locality. Trace fossils, sedimentary features including sedimentary and microbial structures, and ichnofabric indices are shown. Key for symbols is in Figure 2. Modified from Pruss and Bottjer (2004a).

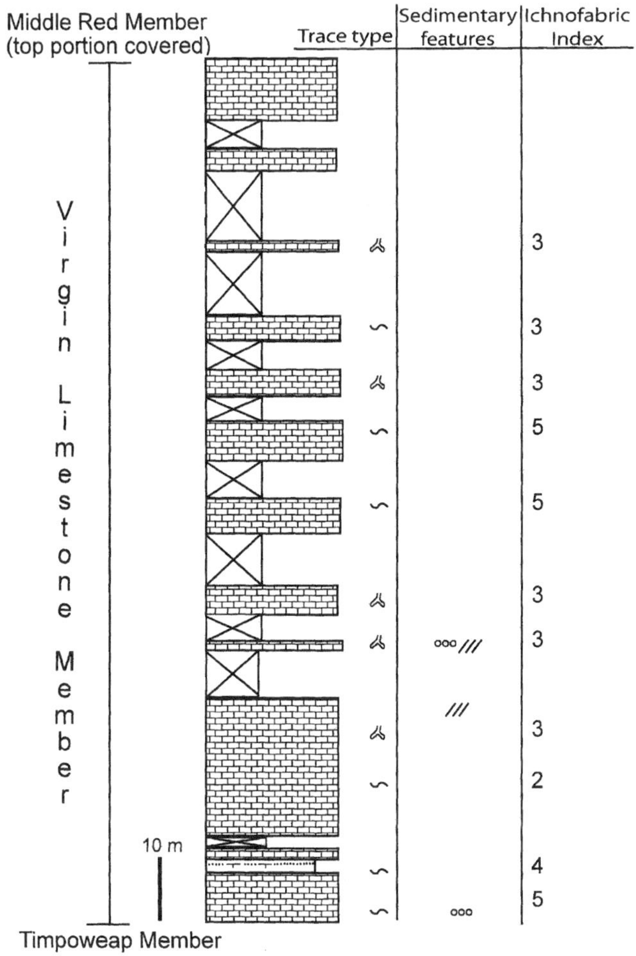


*Figure 4.* Measured stratigraphic section from Mountain Pass locality. Trace fossils, sedimentary features including sedimentary and microbial structures, and ichnofabric indices are shown. Key for symbols is in Figure 2. Modified from Pruss and Bottjer (2004a).

present laterally, and we estimate that as much as 50 m of siliciclastics and thin limestone beds are missing from the basal part of our measured section. The lowermost beds measured at the Lost Cabin Springs section are mostly calcareous siltstones and partially covered siliciclastic-rich intervals (Fig. 5). Limestone units become more prominent upsection, and these become increasingly fine-grained through the middle of the section. The presence of hummocky cross-stratification in the upper part of the section would indicate deposition on a storm-dominated carbonate ramp. Because the Lost Cabin Springs section contains the largest proportion of fine-grained carbonate units, it is interpreted to represent the deepest palaeoenvironments along the onshore–offshore transect.

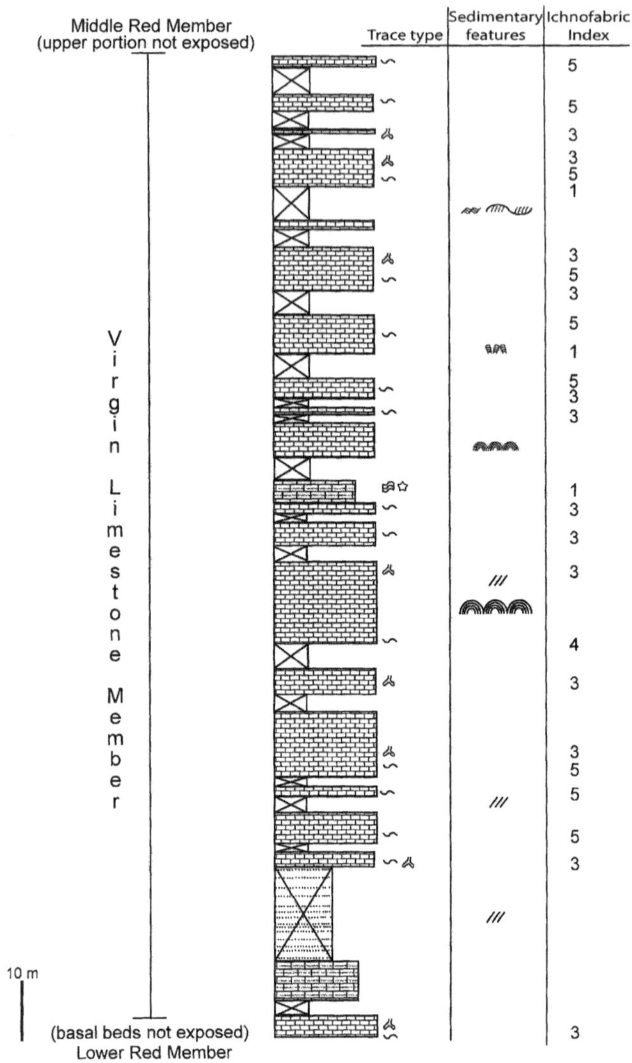


*Figure 5.* Measured stratigraphic section from the Lost Cabin Springs locality. Trace fossils, sedimentary features including sedimentary and microbial structures, and ichnofabric indices are shown. Key for symbols is in Figure 2. Modified from Pruss and Bottjer (2004a).

## 5. Trace fossils and ichnofabric indices

Trace fossils and ichnofabric were examined across an onshore–offshore transect of the Virgin Limestone Member. In this study, results from Pruss and Bottjer (2004) were used to assess the changes in ichnofabric index from nearshore to middle shelf settings. The highest ichnofabric indices (ii=4–6) were reported from thinly bedded limestones dominated by the horizontal trace *Planolites* (Pruss and Bottjer, 2004a) (Fig. 6, brace A). Thinly

*Figure 6.* Photograph of limestone unit within the Virgin Limestone Member showing variations in ichnofabric index from 1 (brace A) to 5 (brace B) from Muddy Mountains Ute locality.

bedded limestones also occasionally showed no bioturbation (ii=1) (Fig. 6, brace B). The siliciclastic beds, containing vertical trace fossils such as *Arenicolites* and *Rhizocorallium*, limestones containing *Thalassinoides* (ii=3), and oolites with *Laevicyclus* (ii=3) had low ichnofabric indices (ii=1–3) (Pruss and Bottjer, 2004a).

### *5.1. Muddy Mountains Overton and Ute*

The Muddy Mountains Overton and Ute localities contain vertical trace fossils such as *Arenicolites* and *Rhizocorallium* that are preserved primarily in calcareous siltstones and limestones (Pruss and Bottjer, 2004a). *Planolites*, a horizontal ichnofossil, occurs in limestones and calcareous siltstones, but *Thalassinoides* traces are rare at both localities (preserved in only two beds). *Gyrochorte* and *Asteriacites* are also found in only two beds, but these are uncommon at all localities (Pruss and Bottjer, 2004a). The vertical trace fossil *Laevicyclus* is only found in two oolitic beds at the Muddy Mountains Ute locality. At the Muddy Mountains Overton locality the range in ichnofabric indices is 1–3 ($n$=10 beds, 8 had ii of 1, 1 had ii 1–2, 1 had ii 3). The range in ichnofabric indices at the Muddy Mountains Ute locality is 1–5 ($n$=17 beds, 5 had ii of 1, 2 beds had ii of 2, 6 beds had ii of 3, 1 bed had ii of 4, 3 beds had ii of 5). It is important to reiterate that the ichnofabric indices presented here only represent the indices of scored beds, and the number of scored beds was determined by exposure. At both localities, beds scored with lower ichnofabric indices were more common than those at Mountain Pass and Lost Cabin Springs.

### *5.2. Mountain Pass and Lost Cabin Springs*

*Arenicolites* and *Rhizocorallium*, vertical trace fossils, are rarely preserved at the Lost Cabin Springs and Mountain Pass localities unlike the Muddy Mountains Overton and Ute localities. The vertical trace fossil *Thalassinoides* is, however, commonly preserved in carbonate units at Mountain Pass and Lost Cabin Springs. *Planolites* is the most common trace fossil at these localities. *Gyrochorte* and *Asteriacites* are found in only one bed at Mountain Pass and Lost Cabin Springs. The range in ichnofabric indices at Mountain Pass is 2–5 ($n$=11 beds, 1 bed had ii of 2, 6 beds had ii of 3, 1 bed had ii of 4, 3 beds had ii of 5). The range of ichnofabric indices from the Lost Cabin Springs locality is 1–5 ($n$=25, 3 beds had ii of 1, 12 beds had ii of 3, 1 bed had ii of 4, 9 beds had ii of 5). These results indicate that the nearshore environments contain fewer scored beds with high levels of bioturbation than the more offshore sections.

## 6. Microbial structures

The distribution of microbial structures varies across the onshore–offshore transect. Nearshore sections at the Muddy Mountains Overton and Ute localities contain no microbial fabrics in carbonates, although small (<0.5 m) nearshore stromatolites have been reported from nearby localities (Blakely, 1972). Wrinkle structures have been described from bedding planes and talus at the Muddy Mountains Overton locality (Pruss et al., 2004). No microbialites have been documented from the Mountain Pass locality. At the Lost Cabin Springs locality, carbonate microbialites are noted from three horizons (Schubert and Bottjer, 1992; Pruss and Bottjer, 2004b) and wrinkle structures have been found in siltstone talus (Pruss et al., 2004).

### *6.1. Carbonate microbialites*

Three carbonate beds that display a variety of microbialite facies are present in the Virgin Limestone Member at the Lost Cabin Springs locality (Fig. 7A–D). Microbial reef mounds composed of stromatolites and thrombolites occur in the lowest bed and have been described in detail elsewhere (Pruss and Bottjer, 2004b). The microbial reef mounds range from 2 to 2.5 m in thickness. The reef mounds are laterally extensive at this locality. The larger mounds are composed of smaller domes. The second microbialite unit is located ~19 m above the reef mound bed and is composed of small microbial mounds 0.5–1 m in thickness that exhibit stromatolitic laminations (Fig. 7A). Approximately 15 m above this is a 1 m thick microbialite bed consisting of digitate stromatolites that nucleated on micritic intraclasts (Fig. 7B and C). This bed is not composed of individual mounds, but rather forms a laterally extensive but thin stromatolitic biostrome. The two beds containing microbialites above the reef mounds do not represent former reef mounds because they did not attain significant topographic relief above the seafloor.

The macrostructure of the microbialites shows both stromatolitic and thrombolitic characteristics (Pruss and Bottjer, 2004b). In outcrop, some clotted textures are visible as well as laminated columnar features (Fig. 7D). The stromatolitic fabrics are characterized by

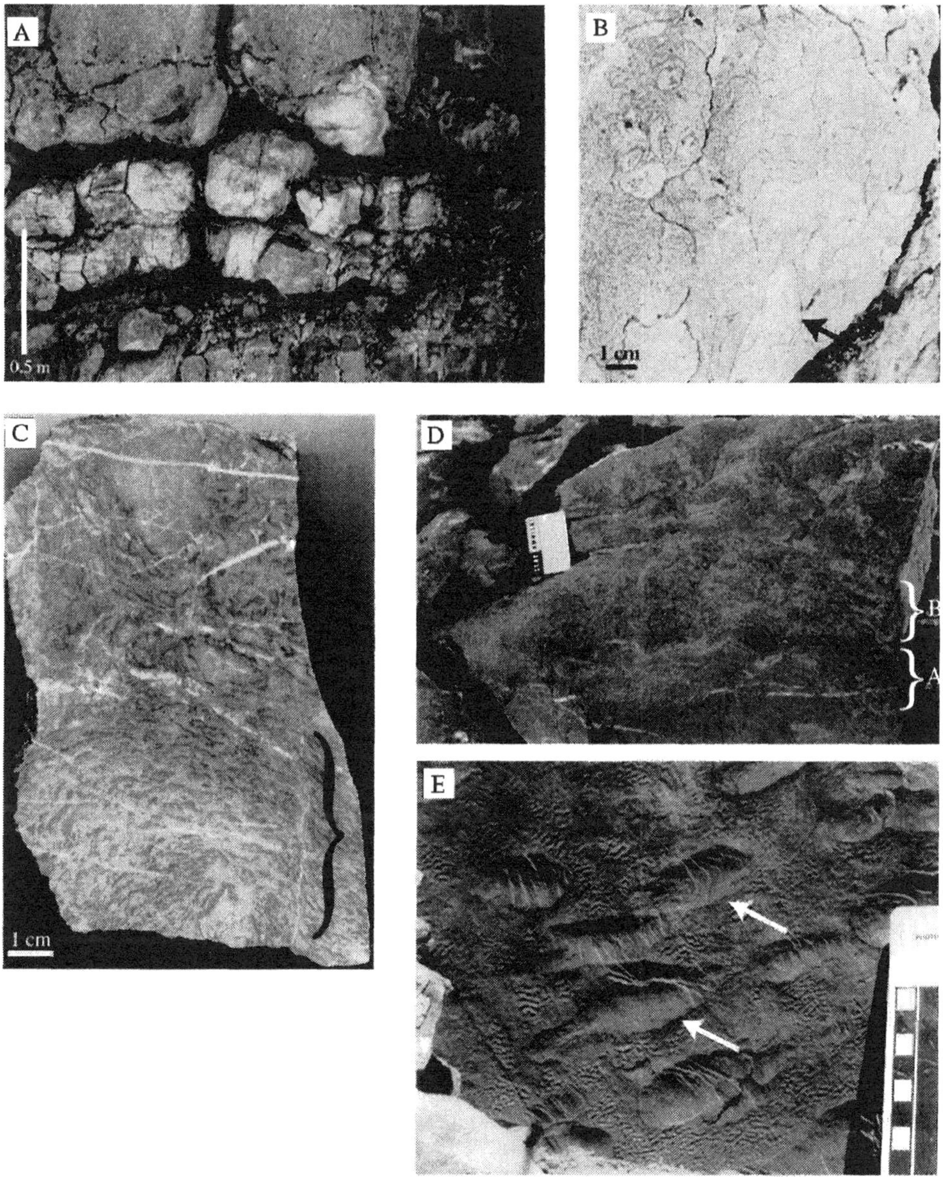


*Figure 7.* Photographs of microbialites from the Virgin Limestone Member. (A) Photograph of a bed consisting of small microbial mounds from the Lost Cabin Springs locality. This bed is stratigraphically higher than the microbial reef mounds of Pruss and Bottjer (2004). (B) Photograph of microbialite unit that consists of microbialites growing on micritic clasts from the Lost Cabin Springs locality. (C) Slab photo of a sample of the microbialites growing on clasts. Notice small microbial dome indicated by brace. (D) Photograph showing typical microbial fabrics seen in outcrop. Picture shows microbialites exposed in the reef mound bed of Pruss and Bottjer (2004) at the Lost Cabin Springs locality. Basal portion exhibits laminated columnar features (brace A) and upper portion shows clotted thrombolitic fabric (brace B). Card is ~7 cm for scale. (E) Photograph of wrinkle structures exposed on a bedding plane at the Muddy Mountains Overton locality. Note scour marks indicated by arrows. Modified from Pruss et al. (2004).

distinct laminations that can be seen in the field and are commonly enhanced by weathering. The thrombolitic fabrics display a clotted appearance that is commonly cryptic in outcrop but visible in cut slabs (Fig. 7C). The mounds are composed of dark gray micrite and are completely devoid of macrofossils in outcrop. Detailed analysis of the microstructure of the microbial reef mounds is described in Pruss and Bottjer (2004); these fabrics are also present in the other microbialite beds described here.

### *6.2. Wrinkle structures*

Wrinkle structures are preserved at two localities in the Virgin Limestone Member (Pruss et al., 2004). At the Lost Cabin Springs locality (Schubert and Bottjer, 1992), wrinkle structures have been found in dark red siltstone talus associated with hummocky cross stratification near the top of the unit, above the stromatolite-bearing interval (Pruss et al., 2004). Mica grains are concentrated in the troughs between wrinkle crests. The presence of wrinkle structures high in the section associated with hummocky cross stratification implies that the wrinkle structures were deposited below fair-weather wave base in a subtidal setting. At the Muddy Mountains Overton locality, wrinkle structures are found in more than one horizon; however, only one bedding plane is fully exposed (Pruss et al., 2004) (Fig. 7E). The exposed bedding plane containing wrinkle structures occurs atop a calcareous siltstone bed. These wrinkle structures are preserved in beds containing ripple marks, cross-bedding, and trace fossils that include *Rhizocorallium, Arenicolites, Planolites, Gyrochorte,* and *Asteriacites* (Pruss et al., 2004). The Muddy Mountains wrinkle structures were deposited in a more nearshore palaeoenvironment than the Lost Cabin Springs examples.

## 7. Discussion

A more complete picture of Early Triassic biofabrics can be gleaned from studying their occurrence across a nearshore intertidal-subtidal to offshore middle shelf transect. The highest diversity of trace fossils is found in nearshore siliciclastic environments (Fig. 8). This indicates that nearshore siliciclastic facies record a recovery of organisms that would not be detected by a study of coevally deposited carbonate strata. In offshore carbonate palaeoenvironments, ichnofabric indices were commonly as high as 5; however, in time-equivalent nearshore siliciclastic facies, ichnofabric indices rarely reach values of 5. This suggests that levels of bioturbation are facies-limited, and this is consistent with results from trace fossil studies of other time periods in the Phanerozoic (e.g. Droser and Bottjer, 1989b; Bottjer and Droser, 1991). Microbial fabrics such as wrinkle structures are noted in subtidal siliciclastic facies and microbial buildups, including microbial reefs (Pruss and Bottjer, 2004b), are found in middle shelf carbonate facies (Fig. 8). This trend shows that microbialite formation occurred in a variety of facies during the Early Triassic, which differs from the common post-Ordovician Phanerozoic distribution of microbialites.

### *7.1. Trends in trace fossils and ichnofabric indices*

At the Muddy Mountains Overton and Ute localities, the highest diversity of trace fossils is found with a total of seven ichnogenera preserved (Pruss and Bottjer, 2004a). At the

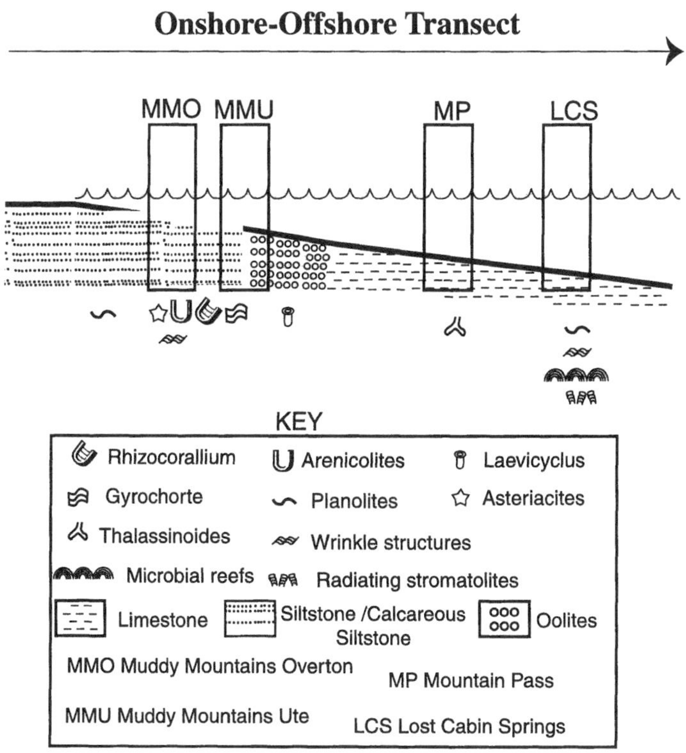


*Figure 8.* Diagram showing onshore–offshore transect of Virgin Limestone Member from Muddy Mountains Overton, Muddy Mountains Ute, Mountain Pass and Lost Cabin Springs localities. Trace fossils and microbialites are shown. Note high diversity of trace fossils in nearshore palaeoenvironments and concentration of microbial build-ups in more offshore palaeoenvironments. Modified from Pruss and Bottjer (2004a).

Mountain Pass and Lost Cabin Springs locality, only four ichnogenera are preserved (two of which occur in only one bed) (Pruss and Bottjer, 2004a). The abundance of trace fossils in the nearshore sections may be related to the exposure of facies that preserve these trace fossils. For instance, the lack of *Arenicolites* and *Rhizocorallium* at Mountain Pass and Lost Cabin Springs may be attributable to the dearth of exposed siliciclastic units at these localities. The preservation of *Laevicyclus* is likely linked to oolitic grainstones because this is the only sediment type in which these traces were found in the Virgin Limestone Member (Pruss and Bottjer, 2004a). It follows that these traces would be preserved at the Muddy Mountains Ute locality where there is an abundance of oolitic sediments relative to the other localities. In pioneering Early Triassic trace fossil studies, trace fossil diversity was assessed through time (e.g. Twitchett, 1999); however, it is clear that facies control plays a role in trace fossil assmeblage diversity and this phenomenon should be taken into account when interpreting such diversity data from other localities.

The trace fossil *Thalassinoides* is common in carbonates at the Mountain Pass and Lost Cabin Springs locality, and is less commonly preserved at the Muddy Mountains Overton and Ute localities (Pruss and Bottjer, 2004a). This suggests that *Thalassinoides* traces, like *Laevicyclus*, are facies-limited, and more commonly occur in shelfal carbonates. *Planolites* is the only trace fossil preserved at all localities and occurs in a variety of facies, suggesting that the presence of *Planolites* reveals less about palaeoenvironment than some of the other trace fossils. The facies-limitation of trace fossils has long been noted and played a role in the development of the ichnofacies concept (Seilacher, 1964).

Ichnofabric indices also vary with locality and lithologic association (Pruss and Bottjer, 2004a), and this is consistent with trends in ichnofabric noted for much of the Phanerozoic (e.g. Droser and Bottjer, 1986, 1989a; Droser, 1987; Bottjer and Droser, 1991). The nearshore localities exhibit few beds with high ichnofabric indices (ii of 4–5), indicating that the trace makers did not fully disrupt bedding in nearshore sediments. In offshore carbonates ichnofabric indices more frequently attained values of 5 (11 out of 35 measured beds, ~31%), which stands in contrast to the nearshore settings (3 out of 27 measured beds, ~11%). Interestingly, beds with the highest ichnofabric indices were bioturbated by horizontal traces such as *Planolites*. No vertical trace fossils were found to penetrate to depths beyond 5 cm (Pruss and Bottjer, 2004a), confirming earlier studies that concluded infaunal tiering had not fully recovered by the latest Early Triassic (e.g. Ausich and Bottjer, 2002).

### *7.2. Trends in microbial distribution*

Microbial structures were deposited in a variety of facies in the Lower Triassic Virgin Limestone. At the nearshore Muddy Mountains Overton locality, wrinkle structures were found preserved in siliciclastic sediments. At the more offshore Lost Cabin Springs locality, abundant microbialites were preserved in carbonates including microbial reefs (Pruss and Bottjer, 2004b), as well as smaller beds of microbialites. Wrinkle structures were also found at the Lost Cabin Springs locality in talus (Pruss et al., 2004). The occurrence of microbial structures in both carbonate and siliciclastic facies suggests that environmental stress that led to their formation was ubiquitous during this time (Pruss and Bottjer, 2004b; Pruss et al., 2004).

The presence of Lower Triassic wrinkle structures in a spectrum of palaeoenvironments indicates that unusual biological and environmental conditions must have existed to foster their preservation (Hagadorn and Bottjer, 1997, 1999; Noffke and Krumbein, 1999; Noffke et al., 2002). Models that describe the formation of wrinkle structures call upon the presence of microbial mats in siliciclastic settings (e.g. Hagadorn and Bottjer, 1997, 1999; Noffke et al., 2002). Wrinkle structures are thus considered good evidence for the former presence of seafloor microbial mats. Wrinkle structures were common in many palaeoenvironments in the Neoproterozoic and Cambrian, but are generally restricted to tidal flat and deep-water environments in the remaining Phanerozoic (Hagadorn and Bottjer, 1999). The occurrence of wrinkle structures in beds exhibiting hummocky cross-stratification in the Lost Cabin Springs section provides evidence for deposition in a storm-dominated subtidal palaeoenvironment. The storm events did not cause complete erosional disruption of the underlying mats (e.g. Noffke et al., 2002), but in some cases may have scoured portions of the cohesive sediment (see Fig. 7E). The wrinkle structures preserved in Lower Triassic strata exhibit a variety of geometries including crest heights that vary from 1 to 5 mm (Pruss et al., 2004), and these different sizes are also a common feature of Proterozoic–Cambrian wrinkle structures (Hagadorn and Bottjer, 1997).

The suite of trace fossils found in the Lower Triassic sections corroborates that wrinkle structures formed because of a lack of deep infaunalization. Vertical traces such as *Rhizocorallium* and *Arenicolites* only penetrated the sediment to shallow depths, and beds containing these traces never attained high ichnofabric indices (Pruss and Bottjer, 2004a). *Thalassinoides* traces were isolated to limestones and calcareous siltstones, and also never totally destroyed bedding (ii of 5 or 6) (Pruss and Bottjer, 2004a). The presence of only shallow burrowers in latest Early Triassic time suggests that infaunalization had not fully recovered by the close of the Early Triassic (Ausich and Bottjer, 2002). This means that substrates of the Early Triassic, which show a microbial influence in various facies as well as reduced infaunalization, differ from those of most of the Phanerozoic (Fig. 9).

Deposition of microbialites in more offshore carbonate palaeoenvironments of the Virgin Limestone indicates that conditions affecting carbonate environments were similarly influenced by deleterious conditions in outer-shelf to basinal settings (Woods et al., 1999; Pruss and Bottjer, 2004b). In particular, the only reefs known from the Virgin Limestone (or any Lower Triassic strata) are microbial buildups. Sponge-microbial buildups were common in these palaeoenvironments during the Permian and Middle Triassic (e.g. Leinfelder et al., 2002; Weidlich, 2002); however, the absence of these buildups from shelf environments for the entirety of the Early Triassic indicates environmental suppression of reef-building metazoans.

Microbialites are found in three limestone beds at the Lost Cabin Springs locality. The presence of microbialites at a variety of stratigraphic intervals suggests that the conditions favoring microbial development occurred periodically. One explanation would be that deep, euxinic ocean water that formed under anoxic conditions episodically flooded the shelf (Woods et al., 1999), suppressed metazoan activity and fostered the formation of microbialites. During other times, conditions may have returned to normal, as is suggested by the presence of *Thalassinoides* (Pruss and Bottjer, 2004a) and a variety of metazoan fossils found in close stratigraphic proximity to the microbial buildups. The occurrence of the microbial buildups in the latest Early Triassic indicates that suppression of metazoans lasted as long as 4–7 million years; however, the periodic nature of deleterious conditions exhibited by the Lower Triassic biofabrics of the Virgin Limestone indicates that oceanic environments may have begun to improve during this time, but that real biotic recovery

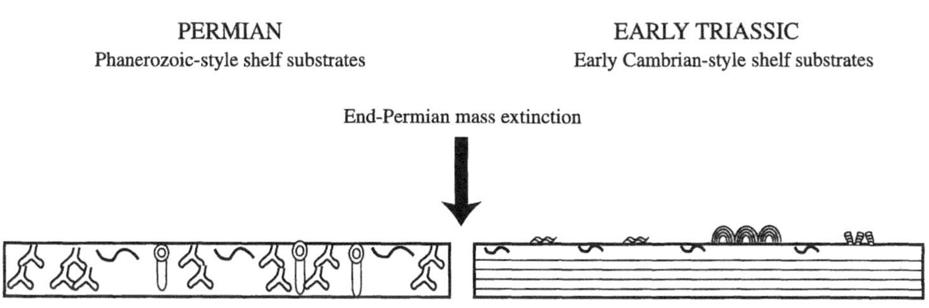


*Figure 9.* Diagram showing the effect of the end-Permian mass extinction on Early Triassic marine substrates. Permian substrates are shown as typical Phanerozoic-style substrates characterized by high levels of bioturbation and a pronounced mixed zone. Early Triassic substrates show a return to Cambrian-style substrates that are typified by low levels of vertical bioturbation and the development of microbial mats and build-ups.

was still delayed until the Middle Triassic (e.g. Hallam, 1991; Ausich and Bottjer, 2002). Recent work on the carbon isotope record of the Early Triassic shows large-scale oscillations interpreted to represent carbon cycle instability (Payne et al., 2004); this adds independent support to the hypothesis that the Early Triassic was a time of periodic environmental perturbations.

### *7.3. Palaeoenvironmental control on biofabrics*

The reduction in burrow depth and proliferation of microbialites during the Early Triassic suggest that substrate conditions were similar to those of the Cambrian. Early in the Palaeozoic, a substrate transition occurred from environments dominated by microbial mats exhibiting no bioturbation (Proterozoic-style soft substrates) to well-bioturbated substrates that noticeably lacked microbial mats (Phanerozoic-style soft substrates) (e.g. Droser, 1987; Hagadorn and Bottjer, 1999; Bottjer et al., 2000). This change also involved a shift in the dominant processes affecting substrates. For example, Proterozoic-style soft substrates consisted of mat-bound firmgrounds that were primarily affected by microbial and physical processes. Phanerozoic-style mixgrounds, however, are characterized by soupy substrates with a pronounced mixed layer and high water content, and these substrates are shaped primarily by metazoan activity and physical processes (Hagadorn and Bottjer, 1999; Bottjer et al., 2000; Dornbos et al., 2004). This substrate transition that began at the Precambrian–Cambrian boundary has been attributed to the advent of deep vertical bioturbation (Droser et al., 1999), and after the Cambrian, Proterozoic-style soft substrates become rare or absent from the geologic record (e.g. Hagadorn and Bottjer, 1999).

In the Lower Triassic Virgin Limestone, shallow burrow depths and evidence for the former presence of microbial mats have been documented (Pruss and Bottjer, 2004a, b; Pruss et al., 2004). This suggests that some Early Triassic substrates shared characteristics with typical Proterozoic-style firmgrounds (Fig. 9); however, vertical bioturbation did exist in the Early Triassic, and some reports suggest burrow depths may have penetrated >15 cm (e.g. Wignall et al., 1998; Ausich and Bottjer, 2002). For this reason, substrates of the Early Triassic do not represent a true return to those of the Proterozoic but rather more closely resemble substrates of the Early Cambrian. Subtidal substrates of the Early Cambrian show evidence for low levels of vertical bioturbation as well as the former presence of microbial mats and therefore represent the midpoint between substrates dominated exclusively by microbial and physical processes (Proterozoic substrates) and those affected by metazoan activity and physical processes (Phanerozoic substrates) (Fig. 10). Early Triassic substrates, as demonstrated by facies in the Virgin Limestone, mimic substrates of the Early Cambrian more so than those of the rest of the Phanerozoic because both microbial and metazoan processes affected Early Triassic benthic environments. Because evidence for Cambrian-style substrates was found in the uppermost Lower Triassic Virgin Limestone, it can now be concluded that the return to Cambrian-style substrates was not isolated to the Permian–Triassic boundary, but rather persisted for millions of years after the extinction event.

The variations in biofabrics exhibited by the Virgin Limestone across this onshore–offshore transect suggest that understanding the recovery interval from the end-Permian mass extinction requires examining a variety of facies. In this study, the carbonates

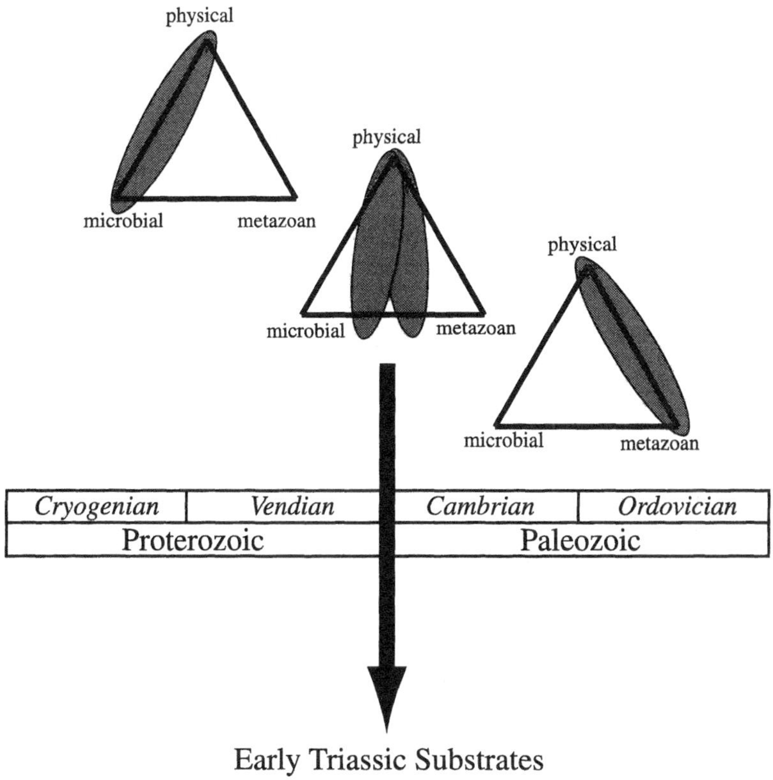


*Figure 10.* Diagram showing that Early Triassic substrates represent a confluence of factors that resulted in substrates similar to those of the Early Cambrian. Early Triassic substrates were therefore affected by microbial, metazoan, and physical processes unlike those of most of the Phanerozoic that were dominated primarily by metazoan and physical factors. Modified from Hagadorn and Bottjer (1999).

of the Mountain Pass and Lost Cabin Springs localities exhibited high ichnofabric indices and a low diversity of trace fossils. After examining the Virgin Limestone Member at the nearshore equivalents, it is apparent the high ichnofabric indices are not common at all of these localities, and a higher diversity of trace fossils is preserved than would be gleaned by study of the shelfal carbonates alone. It is only after investigation of all facies that one can paint a real picture of the biotic recovery during the latest Early Triassic.

## 8. Conclusion

Study of the Lower Triassic Virgin Limestone Member of the Moenkopi Formation, western United States has revealed that understanding the biotic recovery from the end-Permian mass extinction requires examination of a variety of facies in many palaeoenvironments. The Virgin Limestone Member provides a holistic picture of the biotic recovery in the western United States during the latest Early Triassic because it contains a variety of mixed siliciclastic and carbonate nearshore and shelfal palaeoenvironments in close proximity to one another. Analysis of biofabrics across an onshore–offshore transect has revealed that a very different picture of the recovery is observed in nearshore siliciclastic deposits versus more offshore carbonate units. In the Virgin Limestone, the nearshore siliciclastic biofabrics are dominated by a higher diversity of traces than the more offshore carbonates. In comparing the beds scored for ichnofabric across the onshore–offshore transect, the nearshore siliciclastic units contain more beds that scored with a low average ichnofabric index. The differences between these localities illustrate that, as in other times of the Phanerozoic (e.g. Droser and Bottjer, 1989a; Bottjer and Droser, 1991), only upon examination of all facies across an onshore–offshore transect can one determine the true diversity of trace fossils and the relative levels of bioturbation for any given time period. Microbial structures in the nearshore sections occur primarily as wrinkle structures. In the more offshore carbonate settings, biofabrics consist of a relatively low diversity of trace fossils and in comparing beds scored for ichnofabric indices, offshore carbonates contain more beds that scored with a higher index than beds scored at the nearshore localities. Microbial structures occur as well-developed microbial reefs, small stromatolites, and subtidal wrinkle structures. The evidence for microbial mats in subtidal environments in conjunction with low levels of vertical bioturbation suggest that some Early Triassic environments experienced a return to Cambrian-style substrates (e.g. Hagadorn and Bottjer, 1999) that had been absent from the rock record since the Early Palaeozoic. This fundamental change in substrates in the Early Triassic may have been fostered by environmental stress that prohibited the development of deep bioturbation and promoted the development of microbialites (e.g. Pruss and Bottjer, 2004a, b; Pruss et al., 2004). The variations in biofabrics from nearshore to offshore palaeoenvironments also suggest that a true picture of the biotic recovery can only be realized by examining many facies in a variety of palaeoenvironments.

## Acknowledgments

The authors would like to thank the Geological Society of America (grant to SBP), the USC Earth Science Graduate Student Research Fund (grant to SBP), and the USC Women in Science and Engineering (WISE) Program (grant to DJB) for research funds. SBP thanks C. Herbold, D. DeSwert and M. Pruss for their help in the field, and helpful reviews from A. Hallam and P. Wignall.

## References

Ausich, W.I., Bottjer, D.J., 2002. Sessile invertebrates. In: Briggs, D.E.G., Crowther, P.R. (Eds), Palaeobiology II. Blackwell Science, Oxford, pp. 384–386.

Belnap, D.W., 1971. Petrology and geochemistry of shoal water carbonates of the Virgin limestone member, Triassic Moenkopi Formation, Clark County, Nevada. Master's Thesis, Brigham Young University, Provo, UT, 184pp.

Blakely, R.C., 1972. Stratigraphy and depositional environments of Moenkopi formation in southeastern Utah. Am. Assoc. Petrol. Geol. Bull. 56(3), 604.

Bottjer, D.J., Droser, M.L., 1991. Ichnofabric and basin analysis. Palaios 6, 199–205.

Bottjer, D.J., Hagadorn, J.W., Dornbos, S.Q., 2000. The Cambrian substrate revolution. GSA Today 10(9), 1–7.

Dornbos, S.Q., Bottjer, D.J., Chen, J.Y., 2004. Evidence for seafloor microbial mats and associated metazoan lifestyles in Lower Cambrian phosphorites of Southwest China. Lethaia 37, 127–137.

Droser, M.L., 1987. Trends in extent and depth of bioturbation in Great Basin Precambrian-Ordovician strata, California, Nevada and Utah. Doctoral Thesis, University of Southern California, Los Angeles, CA.

Droser, M.L., Bottjer, D.J., 1986. A semiquantitative field classification of ichnofabric. J. Sediment. Petrol. 56(4), 558–559.

Droser, M.L., Bottjer, D.J., 1989a. Ichnofabric of sandstones deposited in high-energy nearshore environments; measurement and utilization. Palaios 4(6), 598–604.

Droser, M.L., Bottjer, D.J., 1989b. Ordovician increase in extent and depth of bioturbation; implications for understanding early Paleozoic ecospace utilization. Geology 17(9), 850–852.

Droser, M.L., Gehling, J.G., Jensen, S., 1999. When the worm turned; concordance of Early Cambrian ichnofabric and trace-fossil record in siliciclastic rocks of South Australia. Geology 27(7), 625–628.

Erwin, D.H., 1993. The Great Paleozoic Crisis; Life and Death in the Permian. Columbia University Press, New York, 327pp.

Fraiser, M.L., Bottjer, D.J., 2004. The non-actualistic Early Triassic gastropod fauna: a case study of the Lower Triassic Sinbad Limestone member. Palaios 19(3), 259–275.

Hagadorn, J.W., Bottjer, D.J., 1997. Wrinkle structures; microbially mediated sedimentary structures common in subtidal siliciclastic settings at the Proterozoic–Phanerozoic transition. Geology 25(11), 1047–1050.

Hagadorn, J.W., Bottjer, D.J., 1999. Restriction of a late Neoproterozoic biotope; suspect-microbial structures and trace fossils at the Vendian–Cambrian transition. Palaios 14(1), 73–85.

Hallam, A., 1991. Why was there a delayed radiation after the end-Palaeozoic extinctions? In: Brasier, M. (Ed.), Innovations and Revolution in the Biosphere. Harwood Academic Publishers, London, pp. 257–262.

Hallam, A., Wignall, P.B., 1997. Mass Extinctions and their Aftermath. Oxford University Press, New York, NY, 320pp.

Larson, A.R., 1966. Stratigraphy and paleontology of the Moenkopi Formation in southern Nevada. Ph.D. Thesis, University of California at Los Angeles, Los Angeles, CA, 256pp.

Leinfelder, R.R., Schmid, D.U., Nose, M., Werner, W., 2002. Jurassic reef patterns. In: Kiessling, W., Flügel, E., Golonka, J. (Eds), Phanerozoic Reef Patterns. SEPM Special Publication, Tulsa, pp. 465–520.

Martin, M.W., Lehrmann, D.J., Bowring, S.A., Enos, P. Ramezani, J., Wei, J., Zhang, J., 2001. Timing of Lower Triassic carbonate bank buildup and biotic recovery following the end-Permian extinction across the Nanpanjiang Basin, South China, GSA, Abstracts with Programs, vol. 33, p. 201.

Noffke, N., Knoll, A.H., Grotzinger, J.P., 2002. Sedimentary controls on the formation and preservation of microbial mats in siliciclastic deposits: a case study from the Upper Neoproterozoic Nama Group, Namibia. Palaios 17(6), 533–544.

Noffke, N., Krumbein, W.E., 1999. A quantitative approach to sedimentary surface structures contoured by the interplay of microbial colonization and physical dynamics. Sedimentology 46(3), 417–426.

Payne, J.L., Lehrmann, D.J., Wei, J., Orchard, M.J., Schrag, D.P., Knoll, A.H., 2004. Large perturbations of the carbon cycle during recovery from the end-Permian mass extinction. Science 305, 506–509.

Pruss, S.B., 2004. The unusual sedimentary rock record of the Early Triassic: anachronistic facies in the Western United States and Southern Turkey. Ph.D Thesis, University of Southern California, Los Angeles, CA, 286 pp.

Pruss, S.B., Bottjer, D.J., 2004a. Early Triassic trace fossils of the western United States and their implications for prolonged environmental stress from the end-Permian mass extinction. Palaios 19, 559–571.

Pruss, S.B., Bottjer, D.J., 2004b. Late Early Triassic microbial reefs of the Western United States: a description and model for their deposition in the aftermath of the end-Permian mass extinction. Palaeogeogr. Palaeoclimatol. Palaeoecol. 211(1–2), 127–137.

Pruss, S.B., Fraiser, M.L., Bottjer, D.J., 2004. The global proliferation of Early Triassic wrinkle structures: implications for environmental stress following the end-Permian mass extinction. Geology 32, 461–464.

Rodland, D.L., Bottjer, D.J., 2001. Biotic recovery from the end-Permian mass extinction; behavior of the inarticulate brachiopod Lingula as a disaster taxon. Palaios 16(1), 95–101.

Schubert, J.K., Bottjer, D.J., 1992. Early Triassic stromatolites as post-mass extinction disaster forms. Geology 20(10), 883–886.

Schubert, J.K., Bottjer, D.J., 1995. Aftermath of the Permian–Triassic mass extinction event; paleoecology of Lower Triassic carbonates in the Western USA. Palaeogeogr. Palaeoclimatol. Palaeoecol. 116(1–2), 1–39.

Seilacher, A., 1964. Biogenic sedimentary structures. In: Imbrie, J., Newell, N. (Eds), Approaches to Paleoecology. Wiley, New York, pp. 296–316.

Shorb, W.M., 1983. Stratigraphy, facies analysis and depositional environments of the Moenkopi Formation (Lower Triassic), Washington County, Utah, and Clark and Lincoln counties, Nevada. Master's Thesis, Duke University, Durham, 205pp.

Stanley, S.M., Yang, X., 1994. A double mass extinction at the end of the Paleozoic Era. Science 266, 1340–1344.
Twitchett, R.J., 1999. Palaeoenvironments and faunal recovery after the end-Permian mass extinction. Palaeogeogr. Palaeoclimatol. Palaeoecol. 154, 27–37.
Twitchett, R.J., Wignall, P.B., 1996. Trace fossils and the aftermath of the Permo-Triassic mass extinction: evidence from northern Italy. Palaeogeogr. Palaeoclimatol. Palaeoecol. 124, 137–151.
Weidlich, O., 2002. Middle-Late Permian reef patterns. In: Kiessling, W., Flügel, E., Golonka J. (Eds), Phanerozoic Reef Patterns. Society of Economic Paleontologists and Mineralogists Special Publication, Tulsa, pp. 339–390.
Wignall, P.B., Morante, R., Newton, R., 1998. The Permo-Triassic transition in Spitsbergen; $\delta^{13}C_{org}$ chemostratigraphy, Fe and S geochemistry, facies, fauna and trace fossils. Geol. Mag. 135(1), 47–62.
Wignall, P.B., Twitchett, R.J., 1999. Unusual intraclastic limestones in Lower Triassic carbonates and their bearing on the aftermath of the end-Permian mass extinction. Sedimentology 46(2), 303–316.
Woods, A.D., Bottjer, D.J., Mutti, M., Morrison, J., 1999. Lower Triassic large sea-floor carbonate cements; their origin and a mechanism for the prolonged biotic recovery from the end-Permian mass extinction. Geology 27(7), 645–648.

# Subject Index

FSC
www.fsc.org
MIX
Papier | Fördert
gute Waldnutzung
FSC® C083411

Kolibri 360 GmbH
Ferdinand-Jühlke-Straße 7
99095 Erfurt, Deutschland
produktsicherheit@kolibri360.de